T0313389

Thomas Banks

Quantum Mechanics: An Introduction

CRC Press
Taylor & Francis Group
Boca Raton London New York

CRC Press is an imprint of the
Taylor & Francis Group, an **informa** business

Cover Credit: Cover image "Dead or Alive?" (oil on wood panel, 18 × 24 inches) provided by Ricardo Mulero (www.ricardomulerodesign.com), used with permission.

CRC Press
Taylor & Francis Group
6000 Broken Sound Parkway NW, Suite 300
Boca Raton, FL 33487-2742

© 2019 by Taylor & Francis Group, LLC

CRC Press is an imprint of Taylor & Francis Group, an Informa business

No claim to original U.S. Government works

Printed on acid-free paper

International Standard Book Number-13: 978-1-4822-5506-5 (Hardback)

Library of Congress Cataloging-in-Publication Data
Names: Banks, Tom, 1949- author.
Title: Quantum mechanics: an introduction / Thomas Banks.
Description: Boca Raton: CRC Press, Taylor & Francis Group, 2018.
Identifiers: LCCN 2018025041 | ISBN 9781482255065
Subjects: LCSH: Quantum theory—Textbooks.
Classification: LCC QC174.12 .B35545 2018 | DDC 530.12—dc23
LC record available at https://lccn.loc.gov/2018025041

Visit the Taylor & Francis Web site at
http://www.taylorandfrancis.com

and the CRC Press Web site at
http://www.crcpress.com

Contents

Preface

Quantum mechanics is at the basis of most physical science and modern electronic technology, and has increasing relevance to the biological sciences. It's also the most confusing subject in the world, because it seems to deny the very foundations of logic. Logic is a precise distillation of our intuitive grasp of how things work. It seems to have nothing to do with particular physical situations where we've become used to the fact that our intuition is only an approximation. For example, Galileo appreciated that Aristotle's intuitive notion that rest is the natural state of bodies was wrong, and invented the relativity of frames moving at uniform velocity. Einstein realized that Galileo's laws relating the kinematics of two frames moving at uniform velocity were only approximately correct, valid when the velocity was much smaller than that of light. All of this is a bit confusing when you first encounter it, but it's not actually that mind boggling, and once you understand how the correct formulas reduce to the non-relativistic ones (which are straightforward and make intuitive sense) when the velocity is small compared to that of light, it actually can be pretty easy to come to terms with relativity, and even develop an intuition for it. The goal of this text is to help the reader develop a similar understanding of Quantum Mechanics (QM).

While there are many available textbooks on quantum mechanics, they almost uniformly present certain aspects of the subject in a manner that reflects the confusions encountered by inventors of quantum mechanics. These confusions include the meaning of wave–particle duality and the correct interpretation of measurements. *Quantum Mechanics: An Introduction* presents the subject from a modern perspective. It includes an elementary discussion of field quantization, the only proper way to understand wave particle duality, at a very early stage. The essentials of field quantization are not difficult, because fields are just collections of simple harmonic oscillators (the standard example used in elementary texts). The interpretation of particles as excitations of quantized fields is the *only* way to understand the identity of particles and the peculiar statistical properties of multi-particle states observed in the real world. All extant textbooks either give incorrect explanations of the origin of Bose and Fermi statistics, or introduce these laws as an additional postulate. On the other hand, all working quantum theorists know that the statistics of identical particles is a consequence of quantum field theory.

Field theory is introduced in Chapter 5, and used to simplify the discussion of quantum statistical mechanics in Chapter 12. It's invoked in Chapter 11 as well, in our brief discussion of density functional theory.

The book also explains the interpretation of measurements in terms of decoherence, including the correct explanation of how the classical world we experience emerges from the underlying quantum formalism (e.g., order of magnitude estimates of deviations from classical behavior) in a new way, to provide a more accurate and rounded picture for the reader. While detailed derivations of the principles of decoherence are difficult, the description of the results of those derivations is straightforward to understand at an elementary level.

The third major innovation in this book is the decision to include a brief discussion in Chapter 11 of the principal approximation methods used in many body physics. Students going on to careers in areas that use quantum mechanics will learn about these in more advanced courses, but all students need a glimpse of the way that quantum mechanics explains the world around us.

Throughout, I've attempted to emphasize the key principle that quantum mechanics is a probability theory, in which not all quantities that appear in the equations of motion of the fundamental variables can take definite values at the same time. As a consequence, histories cannot be predicted definitely, even if we have a precise account of a complete set of initial data. Moreover, the quantum formalism is mathematically inevitable, since given any list of data, we can introduce matrices, which change one data point into another. Once we do this, quantum probabilities are defined, even in systems where we're able to ignore them because the equations of time evolution only require the values of quantities that are simultaneously definite.

Those of a mathematically rigorous turn of mind will find my discussion of continuous spectra and unbounded operators lacking in precision. Von Neumann's famous book cleared up most of the issues, and there are many fine books on the mathematics of quantum mechanics to which one can turn for the details. Physics students, for the most part, are impatient about such things, and it would distract from their absorption of the already difficult conceptual and computational issues of quantum mechanics.

ORGANIZATION OF THE BOOK

I've chosen to begin with a careful explanation of the differences between classical and quantum probability. This takes place in terms of the simplest quantum system: a two state system or q-bit, where the algebra involved is elementary. I also describe why the two lowest energy states of the ammonia molecule form a good testing ground for comparing the two theories, and how quantum mechanics wins that test in a decisive manner.

The beginning of the text also introduces a key theme: the relation between symmetries and conservation laws (energy conservation is a consequence of time translation symmetry).

In classical mechanics, this is a sophisticated theorem, proven by Emmy Noether, but in quantum mechanics it follows from the very definition of symmetry. Finally, we'll understand at this very early stage, the relation between discrete energy levels and the spectrum of light emitted by matter.

My treatment of the free particle is based on symmetry principles: invariance under spatial translations and Galilean boosts. The harmonic oscillator is treated by algebraic methods, which enable us to obtain both the energy eigen-values and the eigen-functions with a minimum of computation. We do this through the introduction of coherent states, a simple topic that is usually reserved for more advanced courses. Apart from computational simplicity, the introduction of coherent states enables us to expose the real connection between particles and physical waves, and to dispel the false notion that the Schrödinger wave function is a physical wave, rather than a device for computing probabilities. Finally, this discussion enables us to introduce photon creation and annihilation operators at an early stage. This is conceptually useful in discussions of transitions among energy levels.

An important choice that must be made in any quantum mechanics text is where and when to go over the mathematics of Hilbert space. I do this by introducing it informally in Chapters 2, 3, and 4, providing a formal introduction in Chapter 6, and summarizing the important rules in a brief appendix. Similarly, group theory, which is not discussed in detail, is split between Chapter 6 and an appendix.

Chapter 7 covers the hydrogen atom. We first solve for the spherical harmonics using the algebraic techniques of angular momentum theory. Then we solve the radial problem two ways, first using the traditional power series solution, and secondly (in a problem set) with an algebraic method, which explains the accidental degeneracy of the hydrogen spectrum.

Chapters 8 and 16 are devoted to scattering theory, the first for the exactly soluble Coulomb potential, while the second is a more general discussion. The book introduces scattering theory in the spherically symmetric context, rather than using artificial one-dimensional examples. One-dimensional scattering is treated through an extensive worked problem set in Chapter 4, on the square well and barrier.

Chapter 9 is about Landau levels. This subject is often omitted from textbooks at this level, but it's the basis for an enormous amount of modern activity, so it's important to include.

Chapter 10 finally deals with the thorny problem of the proper interpretation of quantum mechanics, and with measurement theory. The point of view emphasized here is that, while QM always gives us a mathematical definition of probabilities for histories of any given complete commuting set of quantities, these probabilities do not satisfy the "sum over histories rule" for total probability, which leads to Bayes' notion of conditional probability. The interpretation of quantum predictions in terms of actual experiments depends on the existence of systems with large numbers of variables for which that history sum rule is satisfied with accuracy exponential in the number of atoms in the subsystem on which those variables operate. This is the phenomenon of decoherence, and we briefly review the order of magnitude estimates necessary to demonstrate its plausibility.

Chapters 13, 14, 17, and 18 sketch the main approximation methods that have been used to solve problems in QM. These chapters follow fairly standard lines. Chapter 15 is on the adiabatic approximation, Berry phases, and the Aharonov–Bohm effect. It also discusses anyon statistics in two spatial dimensions and the idea of changing statistics by flux attachment. We learn that fermions in any dimension can be thought of as bosons coupled to a Z_2 gauge field. In a one-semester course, I usually include the discussion of the Aharonov–Bohm effect along with Landau levels.

Chapter 19 discusses Feynman's path integral formulation of quantum mechanics in somewhat more detail than is found in most textbooks. Chapter 20 is a brief introduction to quantum information and quantum computer science. While far from complete, its aim is to enable the reader to get a head start on a fascinating, rapidly developing field. Lastly, the first appendix discusses a variety of attempts to interpret quantum mechanics in a "realist" fashion, while the others are devoted to technical and mathematical details.

FOR INSTRUCTORS

This textbook is intended for an advanced (junior or senior level) undergraduate quantum mechanics course, or a first year graduate course, for physics and math majors, depending on the level of preparation of the students. The whole book is intended for a full year course. A single semester course may be constructed using Chapters 1–7, 9 and 12, in addition to an optional lecture briefly presenting the main idea in Chapter 10. It may also be used in a physical chemistry or materials science course, as long as the students have had linear algebra, and will be useful to computer scientists who are interested in studying the subject from a physics standpoint. Readers should be comfortable with basic notions of linear algebra, and are encouraged to review the matrix representation of a linear operator, with particular emphasis on the fact that it depends on a choice of basis. The book uses operator algebra as its primary computational tool, rather than differential equations, because these methods involve much less mathematical manipulation, and are of much greater general utility. The differential equation form of the Schrödinger equation is actually only really useful for artificial problems involving a single particle.

INSTRUCTOR RESOURCES

Solutions to problems are available to course instructors upon request. Please visit the book's page at http://www.crcpress.com/9781482255065.

Author

Thomas Banks is a Distinguished Professor of Physics at Rutgers University and Emeritus Professor at University of California at Santa Cruz. He was born and brought up in New York City and got his undergraduate education at Reed College in Portland, Oregon, majoring in Physics and Mathematics, and his Ph.D. in physics at M.I.T. in 1973. He then joined the faculty at Tel Aviv University, rising to the rank of full Professor before he left in 1985. In 1986, he became a Professor at U.C., Santa Cruz, and, in 1989, moved to become a founding member of the New High Energy Theory Center at Rutgers University. From 2000 to 2015 Banks split his time between Santa Cruz and Rutgers, and returned to Rutgers full time in 2015. In 2014, Banks married Anne Elizabeth Barnes, a judge on the Georgia Court of Appeals. He has two children from a previous marriage. Banks has held visiting professorships at Stanford, the Stanford Linear Accelerator Center, the Weizmann Institute of Science, and the Institute for Advanced Study in Princeton, New Jersey. He has been a Guggenheim Fellow, a Fellow of the American Physical Society, and in 2011, was elected to the American Academy of Arts and Sciences. His research has mostly been in the fields of particle physics and cosmology.

CHAPTER 1

Introduction

1.1 WHAT YOU WILL LEARN IN THIS BOOK

The formalism of quantum mechanics (QM) was developed some 90 years ago. Since then, tens of textbooks on the subject have appeared. The reader deserves to know why she/he should choose to learn QM from this book, in preference to all the others. There is no better explanation than a list of the things you should be able to learn from reading it. Comparing it to one of the older texts, you will find some differences in emphasis and some differences in the actual explanations of the physics. The latter were inserted to correct what this author believes are errors, either conceptual or pedagogical, in traditional presentations of the subject. The following list includes both topics where our presentation differs from the traditional one and topics that parallel tradition.

- Any mathematical description of a physical system consists of a list of the possible states the system is in. Physical quantities characterizing the system can be thought of as functions on the space of states. A completely equivalent mathematical description is to view the set of states as the basis of a vector space and the physical quantities as matrices A, diagonal in that basis. This description is convenient because *operations* on the system then become a certain kind of nondiagonal matrix, U. If one does not know what state the system is in, one can introduce a probability distribution ρ whose diagonal matrix elements p_i give you the probability of being in the i-th state. The expected or average value of the quantity given by the matrix A is just given by the formula

$$\langle\langle A \rangle\rangle = \mathrm{tr}\,(\rho A). \tag{1.1}$$

The trace of a matrix means the sum of its diagonal matrix elements. The mathematical result that leads to QM is that we can extend this formula to

$$\langle\langle M \rangle\rangle = \mathrm{tr}\,(\rho M). \tag{1.2}$$

for *any* matrix M which is diagonal in a basis related to the original basis by a transformation that preserves the lengths of (complex) vectors. Such operators[1] are given the name *normal*. The resulting formula gives a probability for the quantity M to take on one of its *eigenvalues*. What is remarkable is that if we make this interpretation of the formula, then the quantity M has uncertain values *even when we know exactly which of the original states the system is in.* Conversely, if we accept the eigenvectors of M, for which M takes on a definite value, as allowed physical states of the system, then all of our original physical quantities are uncertain in that state. The essence of QM is that we accept all *normal* operators as possible physical quantities characterizing the system, so that the theory has an intrinsic uncertainty built into it, not related to our lack of knowledge or failure to measure details. This new kind of probability theory violates some of our intuitions about what a probability theory should do. We will explore this idea in more detail in the second part of this introduction and in Chapter 10.

Another, completely equivalent way of describing this new probability theory is illustrated in Figure 1.1: We view the answer to a Yes/No question as the vertical or horizontal position of a switch, along a pair of axes with positive orientation. If we draw another unit vector in the plane, it defines a pair of positive quantities, $p_{1,2}$, that sum to one: the squares of the dot products of that vector with the original axes. In QM, we consider each of these unit vectors as defining a possible state of the system, with p_i interpreted as *the probability that the system will be found to be in each of the Yes/No states.* p_1 is *also* the probability that, assuming the answer to our question is No, the system is in the state defined by the unit vector in the picture, and similarly for p_2 if the answer is Yes. Operators diagonal in different bases are just physical quantities that have definite values in the states described by those bases. Since the process

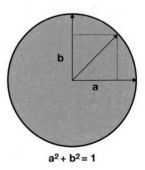

Figure 1.1 Pythagoras' theorem defines a new kind of probability.

---

[1] The same linear operation or *operator* has different matrices if we use different orthonormal bases of the vector space to describe vectors as n-tuples of complex numbers.

of diagonalizing matrices involves the solution of algebraic equations, we are forced to consider complex vector spaces, instead of the real space shown in the figure, in order to find the most general allowed state of our system.

- Equations of motion in physics relate the time rate of change of physical quantities to the present values of other physical quantities. Classical mechanics is the special case of QM in which all the physical quantities appearing in the equations of motion have definite values in the same state. Our current state of knowledge suggests that the equations that describe our world at the microscopic level do not have this property.

- QM's unification of physical quantities characterizing the state of a system and operations on that system leads to a transparent proof of Noether's theorem, which relates symmetry operations on a system to conserved physical quantities. In particular, time translation invariance leads to a conservation law called energy conservation. The corresponding matrix/operator is called the Hamiltonian H and leads to Heisenberg's form of the equations of motion of QM, for any physical quantity (normal operator) A, whose definition has no explicit time dependence.

$$i\hbar\partial_t A = HA - AH \equiv [H, A]. \tag{1.3}$$

- The eigenvalues of the Hamiltonian operator are the allowed energies of the system *in states that have definite energy*. In states that correspond to classical motions which extend only into finite regions of space, these eigen-energies are discrete, leading to the famous quantization laws that give the subject its name. For unbounded motions, the energy eigenvalues are continuous.

- A particular example, with only bounded motions, is the harmonic oscillator, a good first approximation to almost any system perturbed a little bit from its lowest energy state. Harmonic oscillators are particularly important because fields obeying linear field equations like Maxwell's equations are just collections of oscillators. We will see that oscillators have two natural kinds of excitations of their lowest energy state: small quantized oscillations with definite energy and large coherent excitations with indefinite energy. The coherent excitations are parameterized by classical solutions of the system, and if the parameters defining the coherent states are large, then the corresponding quantum states have small uncertainties in the classical values of the coordinates, at all times. When promoted to the field theory context, these two kinds of excitations correspond, respectively, to *particles, which are automatically identical and obey what is called Bose-Einstein statistics for multiparticle states* and *classical fields*. This duality between two kinds of states of a quantized field theory is the proper interpretation of the phrase *wave-particle* duality. Historically, and in many current textbooks, that phrase is applied to the description of particle states by Schrödinger "wave functions." You will learn why that use of the phrase is misleading by the fifth chapter of this book.

- Many of the particles in the real world, although identical, do not obey Bose–Einstein statistics, but rather another rule, called Fermi-Dirac (FD) statistics, which was motivated by Pauli's Exclusion Principle, one of the keys to understanding the Periodic Table of chemical elements. The explanation of FD statistics is quite a bit more complicated, and we will have to learn about the Aharonov–Bohm effect in Chapter 15 before we understand it completely. Leave this as a mysterious teaser for now.

- The application of these ideas to motion of a single particle in a spherically symmetric potential leads to the theory of angular momentum. By Noether's theorem, this is equivalent to studying how quantum states transform under rotations of coordinates. Since such motions are compact, they lead to quantization rules: in this case, the correct form of Bohr's rules for quantizing angular momentum. We will then move on to the study of radial motion, which will lead us to Bohr's formula for the energy levels of hydrogen.

- To study most interesting physical systems, we have to understand the dynamics of multiple particles, and we must abandon exact solutions of the equations. Before moving on to that, we pause to explain the relation between the mathematics of QM and our everyday experience of an apparently classical world. The key to understanding that is the notion of *a collective coordinate of a macroscopic system*. A macroscopic system is one containing of order 10^{20} atoms or more, which is in an energy range where a huge number of different, closely spaced, multiatomic energy levels are excited. A collective coordinate of such a system is an average over all of the atoms, like their center of mass. We will argue that the QM uncertainties in collective coordinates are of order 10^{-10} or smaller, and that the violations of the "intuitive" rules of probability theory by the quantum predictions for these quantities are smaller than $e^{-10^{20}}$. This means that the latter violations are unobservable, even in principle. The world we are used to corresponds to observations only of such collective coordinates. A proper understanding of this fact removes much of the mystery of QM, associated with phrases like "Schrödinger's Cat," "Collapse of the Wave Function," "Spooky Action at a Distance," etc.

Quantitatively detailed treatments of multiparticle QM require the use of large computers and a variety of approximations. The two most important approximations are the Born-Oppenheimer approximation, which exploits the fact that nuclei move much more slowly than electrons, and some form of Hartree's self-consistent field approximation. The modern form of the latter is called Density Functional Theory, and we will be able to give a quick sketch of it, but in very little detail. It will be enough to give us a rough explanation of the periodic table and the gross properties of solids.

- The quantum treatment of the statistical mechanics of multiparticle systems, which leads to the resolution of Gibbs paradox, the Planck black body spectrum, and the phenomena of Fermi surfaces and Bose–Einstein condensation, will be dealt with in Chapter 12.

- Much of the rest of the book is devoted to discussions of various analytic approximation schemes for quantum problems. Some of this material is presented from a fresh perspective, but for the most part, it follows roughly traditional lines.

- Chapter 15 on Berry phases and the Aharonov–Bohm effect is somewhat novel, in that it presents an explanation of Fermi statistics in terms of the A–B effect and a simple presentation of the properties of *anyons*, particles that can exist only in two space dimensions, and which obey statistical rules different from either bosons or fermions.

- Chapter 19 on Feynman's Path Integral formulation of QM presents a topic often omitted from textbooks at the undergraduate level. It also covers Schwinger's alternative functional differential equation derivation of the path integral as well as path integrals for fermionic variables and other systems with only a finite number of quantum states.

- Chapter 20 is a quick introduction to Quantum Computing. Its sole purpose is to enable you to pick up one of the good texts on the subject and get into it rapidly. For the most part, we have stuck to the notation and nomenclature of this book, rather than introducing a whole chapter written in the foreign language of quantum computer scientists.

1.2 WHAT YOU SHOULD KNOW ABOUT LINEAR ALGEBRA

The basic premise of the approach to QM used in this book is that the mathematics of linear algebra defines a new kind of intrinsic probability theory, in which not all quantities can take definite values at the same time. If the equations of motion relate quantities that cannot be definite at the same time, then the concept of *probabilities for histories*, which is central to the way that classical physicists and philosophers think about probability, can only be an approximate one, valid for certain systems containing many fundamental degrees of freedom, and only for certain average or collective properties of those systems. The key mathematical notion that defines QM probability theory is the notion of change of basis in a vector space over the complex numbers. It cannot be stressed often enough that the mathematical surprise that leads to QM is that the generalization of Pythagoras' theorem to N-dimensional complex vector space can be interpreted as saying that every unit vector defines a probability distribution over the set of all other unit vectors, the absolute square of the projection of one vector on another. This leads to the idea that any diagonalizable linear operator on the vector space is just as good a candidate for a quantity that can be measured, as any other. All of the basic properties of quantum systems, and their violation of classical logic, flow from this observation. It is therefore incumbent on any would-be cognoscenta of QM to have a thorough knowledge of linear algebra.

You should know the definition of a complex vector space and understand that the representation of a vector $|v\rangle^2$ in that space as a column of complex numbers:

$$|v\rangle = \begin{pmatrix} v_1 \\ \vdots \\ v_N \end{pmatrix},$$

depends on a choice of *orthonormal basis vectors* $|e_n\rangle$. The word orthonormal means that the scalar products (Dirac notation again) of these vectors with each other satisfy

$$\langle e_i | e_j \rangle = \delta_{ij}.$$

The components v_n are the coefficients in the expansion

$$|v\rangle = \sum_{n=1}^{N} v_n |e_n\rangle.$$

The notation $\langle v|$ for a given (column vector) $|v\rangle$ means the corresponding transposed row vector, but with its elements subjected to complex conjugation

$$\langle v| = (v_1^* \dots v_N^*).$$

This is the representation for $\langle v|$ in the transposed basis $\langle e_i|$, for row vectors, corresponding to the basis $|e_i\rangle$ for column vectors. The reader would do well to convince him/herself at this point that the expression $\langle v|w \rangle$ is equal to the complex number

$$\langle v|w \rangle = \sum_{i=1}^{N} v_i^* w_i.$$

A key mathematical fact that leads to QM is that if both vectors have unit length, the absolute square of this number obeys all the mathematical properties that one would need to call it "the probability that, if one were in the state of a system represented by the unit vector $|v\rangle$, then a measurement designed to detect whether one was in the state represented by $|w\rangle$ would give a positive answer." This violates only one rule of classical logic: *The Law of the Excluded Middle*. That law takes as the definition a state that one cannot be in

---

[2] We are here using Dirac's notation for vectors, which will be explained in Chapter 2, and more extensively in Chapter 6.

two states simultaneously. Ultimately, like any other law in a scientific theory, the Law of the Excluded Middle must be tested by experiment, and it fails decisively for experiments performed on microscopic systems. Mathematically, this law is equivalent to claiming that every state encountered in the world is an element of some particular orthonormal basis.

Orthonormal bases are not unique. In two- or three-dimensional space, we are familiar with the fact that any choice of orthogonal axes is related to any other by a rotation. The corresponding ambiguity in N-dimensional complex space is an N-dimensional *unitary transformation*. The most important thing to remember is that the column vector notation for $|v\rangle$ depends on the choice of basis, but the vector itself is independent of that choice. This is like saying that the position of your house on the earth is independent of the longitude latitude coordinates we generally designate it by. One can define another set of longitude latitude coordinates by choosing the "North Pole" to be any other point on earth and drawing the corresponding arcs. The ones we use are convenient because of their relation to the rotation axis and magnetic field of the earth, but the South Pole would be just as good. An observer looking at our solar system from the outside might have chosen the plane of the ecliptic to define Earth's equator, or the plane of galactic rotation, etc. These changes of the definition of the coordinate system change your house's designation by a pair of numbers, but they do not change where your house is.

Linear operations on a vector space are mappings that take vectors into vectors and satisfy

$$A(a|v\rangle) + b|w\rangle) = aA(|v\rangle) + bA(|w\rangle).$$

From now on, we will drop the round brackets around the argument of a linear operator and write $A(|v\rangle) \equiv A|v\rangle$. Once we have chosen a basis, A is represented by a *matrix*, an array of rows and columns of complex numbers. We will see that we can compute those numbers as

$$A_{ij} = \langle e_i|A|e_j\rangle.$$

In words, the number in the i-th row and j-th column of the matrix is given by the scalar product of the i-th orthonormal basis vector, with the action of A on the j-th orthonormal basis vector. It is the component of $A|e_j\rangle$ along the $|e_i\rangle$ axis. The matrix looks different for different choices of bases. In particular, for certain operators, called *normal*, there is a basis where the matrix is diagonal. This is the basis of *eigenvectors*, $|a_i\rangle$ of A satisfying

$$A|a_i\rangle = a_i|a_i\rangle.$$

The process of diagonalizing A, given its matrix in some random orthonormal basis, is simply the process of finding the orthonormal basis of eigenvectors.

To reiterate, the fundamental principle of QM is that Pythagoras' theorem in N-dimensional complex vector space can be thought of as a probability theory for unit vectors in that space. Each unit vector $|e\rangle$ is a state of a physical system, and it defines a probability distribution that tells you that if you are in the state $|e\rangle$, the probability to be in any other state $|f\rangle$ is the absolute square of the scalar product $\langle f|e\rangle$ (called a probability amplitude). The classical approach to the same system would only allow states that are elements of a particular orthonormal basis. In that case, the same rule tells us that the probability to be in some allowed state, given that one is definitely in another, is exactly zero. This is sometimes called *The Law of the Excluded Middle,* and it is not true if one allows *any* unit vector to be an allowed state. At this point, students should go back to our discussion of defining expectation values as traces of operators, and try to understand how that discussion of probability relates to the current one. The connection involves the *projection operator* $P(e)$ on a state $|e\rangle$, which is defined to equal 1 when applied to $|e\rangle$ and to give zero when acting on the subspace of vectors orthogonal to $|e\rangle$. It is better if you work out that connection for yourself, to learn it more thoroughly.

The intuitive reason that we think that knowing the state of a system should determine that there is no chance of being in any other state, is that we think of determining the state by "doing all possible measurements of the properties of the system." This is simply incorrect in QM. The quantum rule is that one determines a state by doing all possible *compatible* measurements. That is, the mathematics of the theory, and the way it assigns values to quantities, is such that not all possible quantities that *could be measured* can be known with precision at the same time. One can try to make this intuitively plausible by thought experiments first described by Heisenberg. One determines the position of a particle by "looking where it is." Maxwell's theory of light tells us that in order to do that with precision Δx you have to scatter light off it with wavelength less than or equal to Δx. Maxwell tells us that light carries momentum, but in classical electrodynamics one can transfer arbitrarily small momentum with any wavelength. However, the quantum theory of light (see Chapter 5) says that the minimum momentum carried by photons of wavelength Δx is $\frac{\hbar}{\Delta x}$, so position and momentum cannot be known simultaneously with arbitrary precision.

If there was anything you did not understand *about the mathematics* in the previous few paragraphs, you should probably learn it well before starting to try to learn QM from this book.

To summarize: in QM, every vector $|s\rangle$ of length one in the vector space representing the system, is considered a possible state of the system. If that vector is an eigenstate of the operator A, with eigenvalue a, then the theory predicts that a measurement of the quantity represented by A will find the value a with probability one. If $|s\rangle$ is not an eigenstate of A, then the probability of finding the eigenvalue a_i is $|\langle s|a_i\rangle|^2$.[3] In Chapter 10, and elsewhere

---

[3] This assumes all the eigenvalues are different. If some of them are the same, the formula is summed over all eigenvectors with the same eigenvalue.

throughout the book, we will discuss at some length the meaning of the words probability and measurement that we used in the preceding paragraphs.

1.3   THE ESSENTIALS OF PROBABILITY THEORY

Before moving on, let us review the essentials of the classical theory of probability. We will do this for a system that has only a finite number of states, labeled by an integer $1 \leq i \leq N$. For a simple coin flipping experiment, we have $N = 2$. The mathematical definition of infinity and of continuous variables always involves a limit from finite systems, so that measurements of any finite precision will never be able to distinguish a system described by an infinite number of states from such a finite system.

A mathematical probability distribution for a finite system is simply a function $p(i)$ defined on the finite set of integers, such that $p(i) \geq 0$ and $\sum_i p(i) = 1$. There are, roughly speaking, two approaches to the physical interpretation of $p(i)$, which we will call the *Bayesian* and *frequentist* interpretations of probability. The Bayesian approach is tied to psychology. $p(i)$ represents the *expectation* that if one tries to determine the state of the system, one will find the state i. In the frequentist approach, one defines the probability in terms of repeated trials. One flips a coin K times, determines the fraction of times it comes up heads $f_K(1)$ and defines $f_K(2) = 1 - f_K(1)$, and then takes the limit $K \to \infty$ of $f_K(i)$ to be $p(i)$. One can think of the Bayesian interpretation as the theoretical model of the probability distribution and the frequentist definition as the experimental method for testing the theory.

The problem is that we can never really take K to infinity. So no actual experiment can carry out the rigorous frequentist definition of probability. If one has a theory of $p(i)$, like "the coin is not weighted, so $p(1) = p(2)$," and one finds 20 million heads in a row, one cannot say that the theory is wrong, because the theory predicts that there is a probability $2^{-20,000,000}$ that the first 20 million tosses will come up heads. All you can say is that "the probability that the unweighted theory is wrong is very close to one." From a strictly logical point of view, this means that the frequentist definition of probability is circular for any finite number of experiments. On the other hand, there is a clear sense in which, for K this large, one is close to the required limit, and one should simply say that the "equally weighted" theory is wrong.

Indeed, in most real systems, one tries to determine the state one is in by measuring variables that nominally take on all possible real values. This means that the possible values one can find by doing the measurement are distributed more densely than the precision of the measuring apparatus. In such systems, there is an unavoidable measurement error in determining what the state of the system is. Thus, experimental results are quoted with "error bars." This means that the results of any experiment are themselves given by a probability distribution. Experimental physicists work hard to eliminate or estimate "systematic errors," which are caused by defects in the apparatus or biased changes in the environment which skew the results in a particular direction. They then assume that the environmental factors over which they do not have systematic control are given by a Gaussian probability distribution $P(x) \propto e^{-a(x-x_0)^2}$, where a is estimated from the results of multiple trials of the experiment. In the exercises below, you will see one example of how Gaussian probabilities arise in a very general way when one is dealing with a large number of random events.

If one has only a probabilistic theory of what the results should be, this probability distribution has to be combined with the experimental probability distribution of what the results actually were. One displays the comparison of theory to experiment as a graph with various bands surrounding a line of theoretical predictions for the functional relationship between two measurable quantities. The bands represent *confidence intervals*, which take into account both the theoretical and experimental uncertainties in the problem.

As an example, the Large Hadron Collider at the CERN laboratory in Geneva, Switzerland, announced in 2012 that "the Higgs boson had been discovered at 5 standard deviation level with a mass (times the square of the velocity of light) of about 125×10^9 eV," the level particle physicists consider a significant discovery of a new particle. Five standard deviations correspond to probability of about 1 in 3.5 million. This is the probability that if the Higgs particle does not exist at a mass close to that, the data that CERN scientists collected in Geneva, Switzerland, would be at least as far from the predictions of a theory where the Higgs particle had a significantly different mass as this data is, because of a random statistical fluctuation. The theoretical predictions here depend on QM, which is intrinsically probabilistic and so that random statistical fluctuation could be either a reflection of that theoretical probability or of some uncontrolled random event in the experimental apparatus.

Readers should note the extreme care and almost legalistic precision with which one must phrase scientific conclusions, if one wants to be completely accurate. Usually, scientists use shorthand phrases like "5 sigma" to refer to such a discovery. The assumption is that anyone professional will understand the small print and ultimately probabilistic guesses that go into any statement about a discovery.

In classical probability theory, the distribution $p(i)$ represents uncertainty about some of the factors that determine the state of the system at a given time. One is given a box into which 6 red, 5 blue, and 17 black balls have been inserted, and blindly reaches in and pulls out a ball. It could be any color, but the probabilities of blue, red, and black are $\frac{5}{28}$, $\frac{6}{28}$, and $\frac{17}{28}$, respectively. Here, the uncertainty is encoded in the word blindly. If the person reaching in were able to feel colors with their hand, then this would not be a problem in probability theory.

A more amusing example is recounted in the book *The Eudaemonic Pie* [4]. Some of the inventors of chaos theory had discovered experimentally that roulette wheels obeyed a form of "low-dimensional chaos." That is simply the statement that the apparently random fall of the ball into different numbered slots was actually fit to high accuracy, by an equation with a small number of parameters, which one could determine by observing a reasonably small number of spins of the wheel. The odds in the casinos in Las Vegas and Reno were set by assuming that the probability distribution for the fall of the ball was completely random. The principle of odds making is to set them so that the customer wins sometime, but assuming the distribution is random, the net flow of money goes to the casino. One does not want to skew them so badly that no one will play the game. The chaos theorists were able to beat the odds because they could come up with a better ansatz for the distribution, which overcame the slight edge the casino owners had built in to the odds.

So, in classical probability theory, the initial distributions themselves must be determined experimentally, or guessed on the basis of some theory of completely random events. There is a variety of such general theoretical models, appropriate to different assumptions about the randomness.

Given a probability distribution for the initial state of some system, one makes predictions in physics by writing down equations of motion. These take the form

$$i(t + 1) = g(i(t)). \tag{1.4}$$

Here, the 1 refers to some unit of time. Our insistence that there are only a finite number of states means that we can only contemplate discrete time evolution, but the time unit could be as small as we like. One usually studies evolution equations that are invertible. This means that one can follow the evolution backwards or forwards, starting from any time t. Note that this is a much weaker requirement than time inversion symmetry, which says that the evolution in the backward direction is the same operation as that in the forward direction. Invertibility is the same as saying that $g(i) = S(i)$, where S is some permutation which exchanges each state label i with exactly one other label.

Classical physicists assumed that in such a situation, the only detectable properties that the system has were simply functions $f(i)$ which take values on each state. The evolution law for such a function is just

$$f(i, t + 1) = f(i(t + 1)) = f(g(i)). \tag{1.5}$$

The probability distribution $p(i)$ looks just like another function, satisfying some constraint equations, but its evolution law is different.

$$p(i, t - 1) \equiv p(g(i)). \tag{1.6}$$

That is, the time evolution for the probability distribution goes backwards, relative to the evolution of the detectable quantities. To see why, note that the average or expectation value of any function at time t is calculated in terms of the value of the probability distribution at the initial time, $t = 0$ via the equation

$$\langle f(i, t + 1) \rangle = \sum_i p(i) f(g(i)) = \sum_j p(g^{-1}(j)) f(j) = \sum_j p(j, t + 1) f(j), \tag{1.7}$$

where we have used the fact that the evolution is invertible to redefine the summation variable by $j = g(i)$. This says that we can view the time evolution of the system either in terms of the time dependence of physical quantities with a fixed probability for initial conditions, or as a time-dependent probability distribution to be in any given state with the value of a given quantity $f(j)$ completely fixed at the initial time. In QM, as we will see, these two different ways of thinking about time evolution are called the Schrödinger and Heisenberg pictures.

The second point of view allows us to think about a more general situation in which some of the parameters that determine the function $g(j)$ are themselves uncertain. Most generally, we can make these transformations depend on time, which we denote by $g_t(j)$ and let the time-dependent transformation be random, with some prescribed probability distribution. One can show that if the probability distributions for those variables at different times are independent of each other, then one gets a similar evolution equation for the time dependence of the probability distribution. Similar means that the equation still relates the probability distribution at time $t + 1$ to that at time t, and that it is *linear* as a function of $p(j, t)$. The linearity in p is the crucial property that allows us to define probabilities for histories and formulate Bayes' rule for conditional probabilities.

A history $i(t)$ is simply some particular sequence of states. The probability of such a history, stretching from $t = 0$ to $t = k$ is simply

$$P[i(t)](k) = \prod_{t=0}^{k} p(i(t), t).$$ (1.8)

The square brackets in $P[i(t)]$ indicate that this probability depends on all the intermediate points. We will now show that one can write $p(i, t)$ as a sum over all possible histories, weighted by the probabilities for individual histories.

The space of all complex valued functions $f(i)$ is a vector space.[4] A *basis* in this space is the set of functions

$$e_j(i) = \delta_{ij}.$$

We can write the probability distribution as

$$p(i, t) = \sum_{j=1}^{N} p_j(t) e_j(i).$$ (1.9)

The mapping $i \to g(i)$ acts on the basis by

$$e_j(g(i)) = \sum_l S_j^l e_l(i),$$ (1.10)

---

[4] For the present section, we need to invoke only real-valued functions, but in QM, we will need the more general space.

where S_j^l is a permutation matrix. Its nonzero elements are a 1 in the j-th row and l-th column, where $j = g(l)$. There is exactly one nonzero element in each row, and they are all in different columns.

Now let us rewrite our time evolution equation

$$p(i, t+1) = p(g^{-1}(i), t) = \sum_{j,l} p_j(t)(S^{-1})_j^l e_l(i). \tag{1.11}$$

In terms of the coefficients in the expansion of $p(i, t)$ in terms of basis functions

$$p_j(t+1) = \sum_l (S)_j^l p_l(t). \tag{1.12}$$

We can now iterate this equation to get the full solution in terms of initial conditions

$$p_j(t) = \sum_{l(1)...l(t)} S_j^{l(t)} S_{l(t)}^{l(t-1)} \ldots S_{l(2)}^{l(1)} p_{l(1)}(0). \tag{1.13}$$

You will verify in Exercise 1.8 that the multiple sum over indices in this formula is precisely a sum over probabilities of histories, assuming that the histories are related to the initial condition by the equation of motion $i(t+1) = i(g(t))$. In Exercise 1.9, you will show that if the matrices S at different times are not necessarily the same, but are picked independently from a random ensemble of permutations whose probability distribution is chosen independently at each time, then the same sum over histories formulation is correct as long as we introduce the same randomness into the evolution law of the histories.

The sum over histories solution to the evolution law for probability leads directly to Bayes' law of conditional probability. Suppose we consider some intermediate time $0 < t_i < t$. Then we can divide all histories into those whose state at t_i lies in some particular subset Σ of the set of all states, and those which do not. We can then say that if we make an observation at t_i, which verifies that the state at that time lies in Σ, then we can throw out all the histories that do not satisfy that condition, and define a conditional probability distribution over the subset Σ. That distribution is the sum over restricted histories, *divided by a factor that accounts for the fact that the total probability of the restricted histories is less than one in the original distribution*. Bayes' conditional probability rule is the instruction to construct such a distribution, based on observation.

Readers should be alert to the fact that the discussion above contains the seeds of QM. It shows that we can reformulate all of conventional classical physics in terms of linear transformations, the matrices S, on a vector space. In Exercise 1.10, you will show that as a consequence of the special properties of permutation matrices, all functions of the probability distribution satisfy the same equations we have derived above. QM replaces the permutation matrices by more general unitary matrices acting on a complex vector space. Functions of the original distribution no longer satisfy the sum over histories rule and the functions that do satisfy the rule are complex functions called *probability amplitudes* and cannot be thought of directly as probability distributions. There is a unique sensible definition of a probability distribution that one can construct from these complex functions. This definition is called Born's Probability Rule, and it says that a probability is the absolute square of a probability amplitude. Since the absolute square of a sum of complex numbers is not, in general, the sum of their absolute squares, probabilities in QM will not satisfy the sum over histories rule, which allows us to define conditional probabilities. The fact that Born's rule has the form of an absolute square is just a consequence of the fact that Pythagoras' theorem in complex inner product spaces defines, for every unit vector, $|e\rangle$, an infinite number of functions, one for each orthonormal basis in the space. These functions are just the absolute squares of the projections $\langle e_i|e\rangle$ of the vector $|e\rangle$ on the basis vectors $|e_i\rangle$. They are all nonnegative, and for each basis they sum up to one. That is, they have the mathematical properties that we would assign to the phrase "$|\langle e_i|e\rangle|^2$ is the probability to find that the system is in the state represented by $|e_i\rangle$, assuming that we have determined it to be in the state represented by $|e\rangle$". QM follows from assuming that this sentence applies to systems in the real world. It violates the assumption of traditional logic, that we *define* different states of the system by saying that we have determined that if we are in one state, then we cannot be in another. This assumption is based on an incorrect extrapolation of macroscopic experience, namely that we can always measure all of the properties that might determine the state of a system, simultaneously. The task of "understanding QM" really amounts to demonstrating that certain systems obeying the rules of QM can behave like the idealized systems of classical logic, to a sufficient degree of accuracy to account for our missing the correct rules. We will see that the key to demonstrating this is that typical macroscopic objects are composed of $> 10^{20}$ atoms. Such a system has $c^{10^{20}}$ quantum states with $c \geq 2$, and under normal conditions (temperatures far removed from absolute zero), the system explores a double exponentially large number of those states. Collective variables, averages over all the atoms like the center of mass position, then obey the sum over histories rule of classical probability theory, with double exponential accuracy.

1.4 PHILOSOPHICAL INTRODUCTION

The rest of this introduction is important, but it is not important to read it before you start the meat of the book. If the philosophizing makes you impatient, skip to the next chapter and come back to this at your leisure. Its main message can be summarized in a couple of sentences. QM is an intrinsically probabilistic theory. The randomness of the world at the microscopic level cannot be attributed to our inability to measure everything with sufficient accuracy, but stems from the mathematical definition of the theory. The theory of QM identifies certain properties of macroscopic objects, made of large numbers of constituents, which obey the laws of classical probability theory (where all probability is attributed to ignorance/measurement error) with incredible accuracy. That accounts for the apparent classical nature of the macroworld we live in, which shapes all of our intuitions. Now, go on or skip, at your discretion (but please first read the two Feynman chapters, which are mentioned at the end of this chapter).

1.4.1 The Essentials of Quantum Mechanics

QM is the most confusing subject in the world, because it seems to deny the very foundations of logic. Logic is a precise distillation of our intuitive grasp of how things work. It seems to have nothing to do with particular physical situations where we have become used to the fact that our intuition is only an approximation. For example, Galileo appreciated that Aristotle's intuitive notion that rest is the natural state of bodies was wrong, and invented the relativity of frames moving at uniform velocity. Einstein realized that Galileo's laws relating the kinematics of two frames moving at uniform velocity were only approximately correct, valid when the velocity was much smaller than that of light. All of this is a bit confusing when you first encounter it, but it is not mind boggling, and once you understand how the correct formulae reduce to the nonrelativistic ones, which make "intuitive sense," when the velocity is small compared to that of light, it is pretty easy to come to terms with relativity, and even develop an intuition for it.

But how could a similar situation hold for LOGIC? How could logic be "a little bit wrong"? The key to answering these questions involves the notions of probability and uncertainty. We are used to assuming that at any one time, measurable quantities have definite values, and that given values of enough quantities, one can predict what the values of anything else will be in the future. The essence of QM is that this is not true. The fact that a quantity, which has been measured, had a definite value at the time it was measured, is tautological. However, our experience tells us that in order to predict the future value of something, we need to know not only its current value, but also something else, its "rate of change." For example, Newton's equations predict the future motion of a particle, given the present values of both its position and velocity. It turns out that in QM you cannot determine the precise value of both of these quantities at the same time. Measuring one with precision introduces an UNAVOIDABLE uncertainty into the other. This uncertainty has nothing to do with the clumsiness of our measuring apparatus. It is built into the fundamental mathematical structure of QM and the definition of what velocity is in that structure.

At the time QM was invented, probability was a concept familiar to classical physicists, gamblers, and insurance brokers. The world seems to be full of events that seem unpredictable, and we are naturally interested in assessing the risk that some disaster or piece of good luck will occur. Classical physicists always assumed that the reason for this unpredictability was our inability *in practice* to measure all of the variables necessary in order to make predictions.

This was particularly plausible given the picture of ordinary objects as large collections of microscopic constituents, whose miniscule size and vast numbers made the measurement of their properties problematic. So, by the 19th century, there was a well-developed theory of probability, based on the assumption that everything was precisely measurable and predictable, *in principle*, but rarely in practice. One of the problems of this theory is that the probability distributions cannot usually be known in advance. One must make guesses about what they are, and refine those guesses based on data collected about the system under observation. However, it turns out that fairly general mathematical assumptions about probability distributions enable one to make successful statistical predictions about uncertain events, and to estimate the likelihood of those predictions being right. In the limit of a large number of independent trials, most distributions are well approximated by a normal or Gaussian distribution. A great popular account of this theory can be found in [1] and a classic treatise on the mathematics is the two volume work of Feller [2].

One of the fundamental laws of this classical probability theory is Bayes' law of conditional probability. Without going into the details, Bayes' law says, essentially, that once one has measured a definite value of some quantity, one can define new probability distributions in which that quantity has no uncertainty and proceed to calculate the probabilities of other things. Philosophically, this fits perfectly with the assumption that everything could have been known in principle, and that it is only our laziness and incompetence, which prevent us from making completely accurate predictions. A classic example of Bayesian conditional probabilities is the probability that a given hurricane will hit a particular city. Early on in the evolution of hurricane Katrina in 2005, there was some probability, according to the equations used by the National Weather Service of the United States, that it would hit Galveston, Texas instead of New Orleans, Louisiana. These are two cities about 400 miles apart on the Gulf Coast of the United States. Once the hurricane hit New Orleans, one could define a new probability distribution in which the probability of hitting Galveston was zero, and that of hitting New Orleans was one, and make future predictions for the evolution of the storm given that new piece of data.

There is a mathematical fact, associated with the use of Bayes' law, which will be useful in understanding QM. If we write the equations of probability theory as differential equations, then they have the kind of locality we are used to from the equations for fields in classical physics. The time derivative of the probability distribution at a point depends only on the behavior of the distribution at very nearby points, so that the influence of a change at a point takes time to propagate to distant points. However, when we use Bayes' law, an event at one point (observation of a hurricane hitting New Orleans) *immediately* changes the value of the probability distribution at a distant point (Galveston), changing the value there to zero in a way that is not prescribed by the equation. There is of course nothing funny going on here. The probability distribution is not a physical field, it is an expression of the uncertainty in the position of the storm. Changing it to zero at Galveston is just adding another piece of data, to our algorithm for predicting what will happen in the future in this particular sequence of events. The theory does not predict what will happen in that particular sequence, but only the frequency of occurrence of different outcomes in the limit of an infinite sequence of reruns of the same initial data. The philosophy behind the use of Bayes' rule in this instance is that in principle we could have predicted that the hurricane would hit New Orleans. The reason that we did not is that we did not know all the initial conditions and so our probabilistic equations included initial conditions that would have led the hurricane to hit Galveston.

Mathematically, the reason that we are able to use Bayes' law on solutions of the National Weather Service equations is that these are *linear equations* for the probability that the hurricane will be in a certain place at a certain time. If we think about all possible *histories* of the system, the linearity of the equations leads to a rule

- The probability $P(x, t)$ for the hurricane to be at a particular place at a particular time is equal to the sum of the probabilities of all histories $x(s)$ (for $0 \leq s \leq t$) that have initial conditions whose probability is nonzero in the original distribution $P(x, 0)$, and that have $x(t) = x$.

Probabilities in QM do not obey the laws of classical probability theory, and in particular, they do not in general obey Bayes' rule or the history sum rule above. Given any system, there is a complete set of compatible detectable quantities,[5] which could, in principle, be measured with absolute precision at the same time. There are actually a continuous infinity of different compatible sets, but when one set is in a state where its members have definite values, the others all have uncertain probability distributions. There are two striking things about these probability distributions, when compared to those of classical probability theory. The first is that they are all quite definite mathematical functions, completely determined by the values of the measured detectables. In classical probability theory, the distributions must be discovered from experiment and only take on *a priori* functional forms in certain limits. More striking and much more peculiar is the fact that QM probability distributions do not, in general, satisfy the Bayesian conditional probability rule, which allows us to replace uncertainty by certainty when a measurement has been made.

We will see that the last feature has profound philosophical consequences, which are best summarized in a phrase invented by Rutgers professor, Scott Thomas, "Objective Reality is an Emergent Phenomenon." Emergent concepts are basic objects in an approximate theory of nature, which have no counterpart in a more exact theory which underlies it. The notion of water as a continuous fluid is a prime example. Real water is made up of discrete molecules. The nature of the emergence of reality in QM is still a matter of some debate, and there are many researchers who hope to find at least an interpretation of QM which allows for

---

[5] The traditional word to use here is *observable*, rather than *detectable*. Detectable is preferable because it does not hint that the "observation" must be performed by a conscious observer.

some kind of underlying reality. We will defer most of our discussion of these abstruse issues to Appendix A. However, it is important to stress that all popular discussions of nonlocal action of things on each other in QM has to do with such interpretations, rather than with the use of QM as a probability theory to make predictions about the results of experiments.

Let us be a little more precise here about the meaning of the phrase Objective Reality in Thomas' aphorism. In classical probability theory, there are probabilities for *histories* of any system. If z_i represent all possible detectable quantities characterizing the system, then a history is a time-dependent set $z_i(t)$. Even if we divide the system's characteristics up into visible quantities $v_a(t)$ and hidden ones $h_A(t)$, there will still be probabilities for histories of the $v_a(t)$. Those probabilities will satisfy the "obvious" rule that if one makes an observation at time t_0, one can divide up the histories according to whether they agree with that observation or not and base future predictions on the result of that observation, throwing out those histories which did not agree with that observation. This statement is true in QM as well, *as long as we actually make the observation*. The italicized phrase is the source of much of the confusion about the meaning of QM, because it seems to imply some connection between human intervention and the basic laws of physics. We will learn later that a more proper form of the phrase is *as long as the "observation" is a quantum entanglement between the microscopic property being observed and the collective coordinate of some macroscopic object*. The point of this long winded phrase is that the microscopic rules of QM probability theory do not have the property that probabilities of histories can be manipulated in this way. The probability of finding the values $v_a(0)$ and $v_a(T)$ is not the sum of the probabilities of all histories $v_a(t)$ with those initial and final values. However, there are special systems, called macroscopic objects, and certain variables, like the center of mass position of such an object, for which the violations of this probability sum rule are incredibly small ($< 10^{-10^{20}}$). "Realist" interpreters of QM believe that a notion of probabilities for histories, obeying the usual rules, is essential to a notion of Objective Reality. This is clearly untrue if we take the commonplace meaning of objective reality, which is merely a record of the things that actually happen in the universe, with no predictive power. It is only if we want to make a predictive theory of what will happen given the maximum information about what has happened, that we might want to use the notion of probabilities for histories obeying the classical sum rule. Experiment shows that those predictions are wrong. QM is a mathematically beautiful (and somewhat inevitable) alternative theory of prediction, which does agree with experiment.

To those who agree with Prof. Thomas' epigram, objective reality in the sense described above, is not an exact property of any quantum variable. However, objects that we call macroscopic are made of huge numbers of atoms $N > 10^{20}$. Such macroscopic objects have *collective coordinates*, like the center of mass, which are defined in terms of weighted averages over all the atoms in the object. One can show that the uncertainties in the values of the collective coordinates are of order $N^{-1/2}$ or smaller. Furthermore, the violations of the rules of classical probability theory for the distributions of these collective variables, are, *under normal circumstances*[6], of order e^{-N}. So, collective variables behave a lot like classical objects were supposed to behave in classical physics. Their values are not very uncertain, and the uncertainty that there is can, with extraordinary accuracy, be mistaken for uncertainty due to measurement error. According to this view, our brains and our bodies live in this fictitious macroworld of certainty and it was not until we became sophisticated enough to probe the atomic constituents of the matter around us, that we were forced to recognize the correct, quantum mechanical, rules, which govern the world.

1.4.2 Unhappening

One of the most disturbing features of the emergent nature of the concept of "happening" in QM is what can only be called "unhappening." In classical probability theory, when we use Bayes' rule to throw away part of the probability distribution, and renormalize the part we keep, we are doing the correct thing physically. That is, if we are only using probability because of our ignorance, then every new piece of data about the world, which tells us "what really happened," should be used to reduce the uncertainty in our probability distribution. In QM, this is not true. In QM, when we use Bayes' rule, upon a single observation of a particular value for some collective coordinate, to throw out the part of the probability distribution that predicted another value, this only gives the correct answer for the probabilities of later observations, as long as the macroscopic object (or some other macroscopic object whose collective coordinates were determined by it) continues to exist. If the object disintegrates into elementary particles, we must go back to the initial probability distribution, before its truncation by the use of Bayes' rule, to get correct predictions for future observations. We will give a particularly poignant example of this in Chapter 10.

It is important to note that classical probabilities are just a special case of the probabilities defined by QM, and classical mechanics is a special case of QM, defined by the requirement that all quantities appearing in the equations of motion can be definite at the same time. This means that in classical mechanics, we can define probabilities for *histories* of the system, which allows us to say that some things definitely occurred in the past, by using Bayes' law to consider only probabilities conditioned on the results of past experiments. No real system exactly obeys these classical rules, but the collective coordinates of macroscopic systems, under "normal" conditions, obey them with such fantastic accuracy that experiments to detect the deviations would take far longer than the current age of the universe (see Chapter 10) and require impossible amounts of isolation of the system and accuracy of the measuring device. As noted in the previous paragraph, these statements only remain valid as long as the macrosystems in question do not disintegrate into elementary particles.

---

[6] We will be more precise about what we mean by normal circumstances in Chapter 10.

1.4.3   Quantum Mechanics of a Single Bit

We will begin our discussion of QM with the simplest possible system, one which has only two states, corresponding classically to a single Yes/No question. As we will see, the molecule of ammonia, NH_3 can be approximated by such a system, in a certain energy regime. We will see that all the machinery of QM, the mathematics of linear algebra, can be introduced in a purely classical discussion of this system, as if we were computer scientists, discussing a single bit. Quantities which a classical physicist/logician would think of as measurable can be modeled as diagonal matrices, while classical operations which change the state of the system are off diagonal matrices. A general classical probability distribution is a diagonal matrix, and the usual formula for the expectation value of a quantity is written as a trace of the product of the matrix for the quantity with that for the distribution. This trace formula immediately generalizes to all matrices, even if they are not diagonal. For those matrices which are diagonal in *some* orthonormal basis, the trace can be interpreted as a probability distribution for that matrix to take on one of its eigenvalues.

This discussion will show that QM is, in a certain sense, inevitable. That is, for any system, even one that we think of classically, we can introduce quantum variables, which have uncertainty even when we have completely fixed the values of the classical variables. The choice of which variables are definite corresponds, for a two state system, to a choice of basis in a two-dimensional vector space. From the point of view of linear algebra, this choice seems arbitrary. The difference between classical and quantum mechanics is in the nature of their equations of motion. Classical mechanics has initial conditions which can all be definite at the same time, while in more general quantum mechanical systems, only some of the initial variables are compatible with each other. The collective coordinates of macroscopic quantum systems do not obey classical mechanics, but they do, up to fantastically small corrections of size $< e^{-10^{20}}$, obey a classical stochastic theory, in which the uncertainties are very small (of order 10^{-10} or smaller). These mathematical facts about the quantum theory are sufficient to explain why our intuitions about the logic of the world are incorrect.

The present author is among those who believe that we will never find a consistent interpretation of the facts of the quantum world, which admits the concept of an underlying reality with exact probabilities for histories. This is not a settled question, and we will try to avoid injecting our prejudices into most of this text.

1.4.4   What is Probability?

QM adds confusion to what one may worry is already a complicated issue in classical probability theory, namely how we should think about probabilities. The inventors of the theory, particularly those whose primary interest was in gambling or other financial transactions, clearly thought of it as a sophisticated way of guessing an unpredictable future. This interpretation is clearly tied up with human psychology. A discussion of probability from this point of view, which attempted to make very precise rules about how to guess and how to use additional data to assess and improve the quality of one's guesses, was given by Bayes [3] in the 18th century. In modern times, this view of probability is given the label *Bayesian interpretation*. If you are a gambler or a financier, this is certainly the way you think of probability. Experimental physicists and theorists who follow their work closely also use Bayesian reasoning quite frequently. Looking at experimental lectures you will often see plots, which include lines indicating the predictions of a theory, and colored stripes following the lines, which indicate things like "the 95% confidence interval." Translated into full English sentences, this means the region of the graph where, with 95% probability the data actually lie, given all the possible random and systematic errors. In graphs referring to the behavior of microscopic systems, these errors include the fact that the quantum theory does not make definite predictions for the number of events of a certain type, but only predicts (see below) the ratio of the number of events to the number of runs of the experiment in the limit that the number of runs goes to infinity. There may also be different bands telling you that the theoretical calculation has "uncertainty," but this is a completely different sort of error and stems from the fact that we can usually solve the equations of QM only approximately. The use of the word confidence interval in this context is the reflection of the Bayesian outlook on the meaning of probability.

As probability theory became more and more important in science, scientists searched for a more "objective" way of thinking about it, which removed the human psyche and words like confidence from the unbiased description of nature by the combination of mathematics and observation. This led to what is called the *frequentist* interpretation of probability. According to this paradigm, you test a prediction which gives probabilistic answers by repeating your experiment N times, and recording the fraction of times the experiment produces each possible result. As $N \to \infty$, these fractions converge to the predicted probabilities *if* the theory in question is correct. This is indeed an objective definition of probabilities, but it is problematic, because it is impossible to take N to infinity, even if the universe lasts for an infinite amount of time. To illustrate the problem, flip a coin 2000 times and observe that it always comes up heads. The probability for that, assuming the coin is unbiased is $2^{-2000} \sim 10^{-500}$, a pretty small number, but this does not *prove* that someone has weighted the coin. If you think it does, would you bet your life on it? Would you bet the lives of all your loved ones? Would you bet the lives of the entire human race? Obviously, these questions all have subjective answers, which depend on who you are and what your mood is. This is the reason that experimental physicists, who test probabilistic predictions (or even definite predictions that are tested with imprecise machinery) by applying the frequentist rule with a finite number of trials, cite their results in terms of a Bayesian confidence interval. We can try very hard to be completely objective about the data, but no finite amount of effort can completely eliminate the need for "leaps of faith."

These interpretational problems have nothing to do with QM. They would be there for a completely deterministic theory about which we were ignorant of some of the initial data, and since some of the initial data have to do with the performance of the measuring apparatus itself, or external influences interfering with the machinery (cosmic rays, sound waves, the electromagnetic field generated by a radio 4 km away, etc.), we are always ignorant of some of the data. We continue to do experimental and theoretical physics despite these obstacles, because we believe that we can control these sources of error well enough that we are happy with the small size of the required leap of faith.

The interpretation of QM as a new kind of probability theory is certainly correct, and is the only interpretation that has been tested by experiment. If it is the final word on how to interpret the mathematics, then we will just have to live with the intrinsically indefinite nature of probabilistic predictions. We will explore alternative explanations in Appendix A.

1.5   A NOTE ON MATHEMATICS

This subsection is addressed to two different groups of people. First to teachers of a more conventional course in QM: When the Schrödinger equation for a nonrelativistic QM is written in position representation, it is a partial differential equation. If we are discussing the QM of a single particle, the Schrödinger equation is similar to Maxwell's equations: it is a differential equation in time and space. It is tempting, and some would argue pedagogically preferable, to utilize the students' familiarity with electrodynamics as a crutch. Indeed, some students have definitely expressed a preference for this approach. In the long run, it is a mistake. The methods of partial differential equations (PDEs) are practically useless for understanding complicated QM problems involving many particles or an indefinite number of particles, and these are the vast majority of systems of interest in particle physics, nuclear physics, and condensed matter physics. The solution of problems via operator algebra, which is the approach taken in most of this book, is simpler (but more abstract and less familiar) and introduces methods, which are more useful in real applications of QM.

Perhaps equally important is the fact that concentration on the single particle Schrödinger equation and its mathematical analogy to Maxwell's wave equation, misleads students completely about the nature of what is called wave–particle duality. *Let me say it clearly once and for all. The Schrödinger wave function is NOT a classical wave, but instead defines a probability distribution.*[7] *For N particle systems, the Schrödinger equation is a differential equation in $6N+1$ variables, and is not a wave in space. Wave particle duality has to do with the fact that multiparticle systems can be described by quantum fields: operators which obey*

---

[7] More precisely, the density matrix, which contains the entire physical content of the wave function, defines a probability distribution for all normal operators in the Hilbert space of wave functions.

wave equations in space and time. These quantum fields have states that behave approximately like classical fields, and other states that behave approximately like particles.

Secondly, for those with a mathematically rigorous turn of mind: I have known a lot of brilliant mathematicians who had a hard time reading QM texts because of the nonrigorous treatment of operators and Hilbert spaces. This book will be no exception. Physicists find that excessive attention to mathematical rigor slows us down, and is difficult for most of our students, who are more interested in the use of physics to understand the world. In my opinion, the correct approach is that of von Neumann: if you are bothered by a statement in a physics book, work out the correct explanation yourself. Much of the necessary rigor is supplied in von Neumann's famous book [5] of which a beautiful new translation has appeared recently [6].

Finally, it should be emphasized that linear algebra is a *prerequisite* for this course. We will review it, using the Dirac notation beloved by most physicists and rather less popular among mathematicians, in Chapter 6, but you will be expected to know enough to follow the first five chapters. You can of course skip to Chapter 6 to brush up on things, or *in extremis*, to learn linear algebra from scratch, but that is not the best way to profit from this book. Another part of mathematical physics that would be helpful to know is the subject called *analytical mechanics*, which is to say, everything about Lagrangians, Hamiltonians, Poisson brackets, and the Hamilton-Jacobi equation. We will not use much of this in the book, but there is a short summary of it in Chapter 4, and the nomenclature will be helpful in our discussion of particles in magnetic fields and in the chapters on Path Integrals and the JWKB approximation.

1.6 FEYNMAN'S LECTURES

Before you go on any further, you should read the first two chapters of Volume 3 of the Feynman lectures on physics. You can find them here: www.feynmanlectures.caltech.edu. They illustrate the puzzling nature of QM in a beautifully simple way.

1.7 ACKNOWLEDGMENTS

In writing this book, I have benefited from a lot of help from colleagues and students. I would like to thank Eliezer Rabinovici, Patrick Hayden, Daniel Harlow, Subir Sachdev, Sriram Shastry, Michael Dine, and Jacob Barandes for important comments and advice on portions of the manuscript. I would especially like to thank Satish Ramakrishna for providing very careful solutions to many of the problems. Pouya Asadi and Conan Huang are responsible for the figures. Ricardo Mulero provided me with the beautiful cover art. Finally, I would like to thank my beautiful wife Anne for the patience she has shown as I worked on this book and all the love and support she gives me every single day.

1.8 EXERCISES

1.1 *The Let us Make a Deal Problem*: On this famous quiz show a contestant is shown three doors and told that behind one of them there is a fabulous prize. After the contestant has chosen a door, one of the other doors is opened, and shown to have no prize (or a booby prize) behind it. The contestant is then given the option to change the door he/she has chosen. What is the best strategy for getting the prize?

1.2 If a couple has three children, what is the probability, given no further information, that two of them are girls? Suppose you know that one of the children is a girl named Florida. What is the probability that two of the children are girls? Does this depend on the choice of name? FYI, Florida is a girl's name that used to be popular, but has gone out of fashion.

1.3 Drop a needle on a plane surface. Take the origin to be one end of the needle and draw radial lines at angles $\frac{2\pi k}{N}$, starting in a random direction. What is the probability that the direction of the needle is between the k and $k+p$th radial line? How does the answer change if one takes the origin to be at another point?

1.4 Here is a problem that is easy to state, but hard to analyze. Given a fairly weighted coin, what is the probability that in 100 flips there will be a run of at least K heads in a row? What if we ask for exactly K heads in a row? The way to set this up is to define a quantity $P(i)$ which is 1 if the ith throw is a head and zero if it is a tail. Each run is characterized by a set of values for these 100 variables, and there are 2^{100} possible values. The question we are trying to answer is, out of all of these possible "states of the system," how many of them have $P(i)P(i+1)\ldots P(i+K-1) = 1$ for some value of i. The trick is that we have to worry about double counting. E.M. Purcell, the Nobel Prize winning Harvard physicist and author of a classic text on electrodynamics, solved the much harder problem of determining the probability given, "dumb luck" of a player having an n game hitting streak in baseball. At the time he did the calculation in the 1980s, the only streak in the history of baseball, which was not, "what we could expect from random probabilities for hits and outs" was Joe DiMaggio's streak in 1941.[8]

1.5 The probability of getting K heads in N throws of a fair coin is given by

$$\frac{N!}{2^N K!(N-K)!}$$

---

[8] This calculation is reported in a lovely essay by S.J. Gould [7]. You will also be touched by Gould's essay [7] on estimating his own chance of beating the odds after his cancer diagnosis. Note that the Wikipedia article on Dimaggio cites some work disagreeing with Purcell's calculations.

This is maximized at $K = N/2$ if N is even. Show that as $N \to \infty$ the probability of getting $\frac{N}{2}(1 - x)$ heads, with x kept constant as N goes to infinity, is a Gaussian distribution of the form

$$P(x) = Ae^{-bx^2}.$$

Calculate the value of b. What happens if N is odd?

1.6 Suppose the probability of finding some quantity Q to have the value x is

$$P(x) = e^{-f(x)},$$

where $f(x) > 0$. Given N independent copies of the system, show that as $N \to \infty$, the fluctuation of Q away from its expectation value goes to zero. For the system with multiple copies, Q_{AV} is defined to be the average of the Q of the individual copies. That is

$$Q_{AV} = \sum \frac{x_i}{N}.$$

Show that fluctuations of Q away from that expectation value have, *generically*, a Gaussian distribution with a width that scales to zero like $N^{-1/2}$. What characterizes the *nongeneric* exceptions? This result is called the *Central Limit Theorem*.

1.7 Political polls often quote a "margin of error." This is defined in the following way. Suppose, in a very large population N, a fraction p of the population prefers candidate Thomas Jefferson and the rest prefer his opponent. Imagine you take a random sample of $n \ll N$ voters and find a fraction p_n prefer Jefferson. There are $\frac{N!}{n!(N-n)!}$ different random samples, which give a distribution of p_n. A sample of n people voting for Jefferson's opponent would give $p_n = 0$, whereas a sample containing only Jefferson voters would give $p_n = 1$. Show that for $N \gg n \gg 1$, the distribution of p_n is Gaussian and find its center and width. The margin of error usually quoted is one half of the 95% confidence interval for the true percentage p. The margins of error for different confidence intervals are simply related by the value of the inverse error function at those confidence levels.

1.8 Use the fact that the only nonzero matrix element of a permutation matrix is a 1 in each row, with a different column for each row, to show that the formula for the time evolved probability distribution can be thought of as a sum over histories. The contribution from a fixed value of each of the summed matrix indices $l(k)$ is the contribution of a single history.

1.9 Show that the sum over histories formulation still works, if at each time the permutation matrix S is replaced by a time-dependent matrix $S(t)$, chosen from some probability distribution of with no correlation between different times. The dynamics of the histories is simply replaced by $i(t) = g(i(t-1), t)$, where the permutation g is chosen from the same random distribution.

1.10 Use the fact that the only nonzero matrix element of a permutation matrix is a 1 in each row, with a different column for each row, to show that $p^2(j, t)$, or indeed any function $f(p(j, t))$ satisfies the same sum over histories formula as $p(j, t)$ itself.

Two State Systems: The Ammonia Molecule

2.1 INTRODUCTION AND MOTIVATION

In this chapter, we will use a simple system with two states to motivate the claim that the mathematical structure of Hilbert space (a vector space equipped with a positive definite scalar product), the basis of quantum mechanics (QM), is implicit in even the classical view of such a system. We will see that both the classical notion of a detectable quantity, whose values differentiate among the states of a physical system, and the classical notion of an operation, which changes the physical state of the system, are special cases of *linear operators on the Hilbert space of states of the system*. Both detectable quantities and operations come equipped with a natural notion of product and the algebra one gets by taking all possible linear combinations of products of detectables and operations is the algebra of all linear operators. Among those, the so-called *normal operators* are those whose matrix in *some* orthonormal basis is diagonal. The original detectable quantities are simply those which are diagonal in a particular basis. Any normal operator can be said to take on values if we allow for the possibility that any normalized vector in the Hilbert space is a possible state of the system.

Even more remarkably, any such normalized vector defines a probability distribution for *every* normal operator to take on each of its allowed values. This separates the notion of a probability theory from the classical context in which it was invented, i.e., as a way of codifying our *lack of knowledge* about the state of all relevant variables, which determine the time evolution of the system in question. In QM, probability and uncertainty are intrinsic to the mathematical form of the theory, rather than concepts we introduce to parameterize our ignorance.

Figure 2.1 Model of the ammonia atom.

Classical physics is then seen as the special case of QM where the evolution of the state of the system with time simply permutes the elements of a particular basis, so that the subalgebra of the algebra of all linear operators, consisting of operators diagonal in that basis, always takes on definite values.

2.2   THE AMMONIA MOLECULE AS A TWO STATE SYSTEM

In any chemistry department, you can find little models of molecules made out of colored balls and sticks. The balls represent atoms, the sticks the chemical bonds between them. These models, as we will learn, are only crude visualizations of the real properties of molecules, but the model of ammonia, a pyramid with a nitrogen atom at its apex and three hydrogen atoms arrayed in an equilateral triangle beneath it (Figure 2.1), will allow us to set up the QM of ammonia in a certain energy regime. We imagine a situation where the molecule is isolated from the rest of the world. Excitations of the molecule around its lowest energy state could be classified according to its states of motion. First of all, we could imagine excitations of the individual electrons in its constituent atoms.[1] It turns out that such excitations have a characteristic energy scale ranging from 1 to 10^3 electron-volts (eV). Next come vibrational excitations, in which individual atoms oscillate around their equilibrium positions in the molecule (which are the positions assigned to the atoms in the Chem Lab Model), and overall rotations of the molecule, as if it were a rigid body. We can also move the center of mass of the molecule, but in the absence of external forces, its motion will have constant velocity, and we can always work in an inertial reference frame where the center of mass is at rest. The excitation energies of rotational and vibrational motions have characteristic energy scales which are smaller than atomic excitations by powers of the electron mass divided by the masses of the nuclei.

In classical physics, one could get arbitrarily low energy of rotation or vibration by letting the atoms oscillate slowly, or continuously lowering the angular momentum. We will see later that QM quantizes these energy levels so that any transitions between rotational, vibrational, or electronic excitation energies are quantized multiples of the characteristic scales. However,

---

[1] At a deeper level, we could imagine excitations of the nuclei and of the individual protons and neutrons in the nuclei. These may be neglected for reasons similar to those we will discuss in Chapter 11 for electronic motions.

if we assume that the physics of an isolated ammonia molecule is invariant under space reflection $\mathbf{x} \to -\mathbf{x}$, then classical reasoning leads us to the conclusion that there are, for each state of atomic, rotational, or vibrational motion, actually two states of the molecule, degenerate in energy, corresponding to configurations related by reflecting the nitrogen atom through the plane of the hydrogens (Figures 2.2 and 2.3).

Let us denote these two states by the symbols $|\pm\rangle$, where we suppress the information about the particular state of vibration, rotation, and atomic excitation. $|+\rangle$ is the state where the nitrogen atom is in the positive 3 direction. This direction is defined to be perpendicular to the plane of the hydrogens, and having an orientation defined by the right-hand rule. $|-\rangle$ is the mathematical representation of the state with the opposite orientation. This funny notation for a state of a system was invented by Dirac. It is an incredibly useful way of encoding the notion of what a state is, in QM, but it is an equally valid tool in classical mechanics. The reason it is so useful is that, mathematically, states are really lines in a huge vector space, and Dirac's notation is a very quick way to mechanize the operations you can do in vector spaces. But we do not yet know any of that, so let us just accept the notation as the professor's quirk, for the moment. Alternatively, you can consult the last appendix of the book, where you can find a quick summary of Dirac Notation and linear algebra, or jump to Chapter 6, where you will find an extensive exposition of these subjects.

According to classical reasoning, these two states would have definite energies, and the two energies would be equal, by reflection invariance. A detectable quantity which distinguishes between the two states is the electric dipole moment, D. Since it is odd under reflection, D has the values $\pm d$ in the two states.

$$|+\rangle_3 =$$

Figure 2.2 The up state of ammonia.

$$|-\rangle_3 =$$

Figure 2.3 The down state of ammonia.

2.3 PHYSICAL QUANTITIES AS MATRICES

A crucial step on the way to QM is the observation that we can think of the two states as vectors in a two-dimensional vector space

$$|+\rangle = \begin{pmatrix} 1 \\ 0 \end{pmatrix}, \tag{2.1}$$

$$|-\rangle = \begin{pmatrix} 0 \\ 1 \end{pmatrix}, \tag{2.2}$$

and the measurable quantities as matrices

$$E = e\mathbf{1} \qquad D = d\sigma_3,$$

where

$$\mathbf{1} = \begin{pmatrix} 1 & 0 \\ 0 & 1 \end{pmatrix}, \tag{2.3}$$

$$\sigma_3 = \begin{pmatrix} 1 & 0 \\ 0 & -1 \end{pmatrix}. \tag{2.4}$$

The rule is that the value of a given quantity in a given state is gotten by acting with the matrix representing that quantity on the vector representing the state, obtaining the value of the quantity, times the original vector. That is, the states are eigenstates of the energy and dipole moment matrices, with eigenvalues equal to the values of the corresponding quantities.

Our notation $|\pm\rangle$ for states is known as *Dirac Notation*. Dirac invented it as a symbol for column vectors, and called such vectors *kets*. We will eventually be discussing systems with more than two states, so we might as well introduce the notation in that more general context. For any n-dimensional column vector

$$|v\rangle \equiv \begin{pmatrix} v_1 \\ \vdots \\ v_n \end{pmatrix},$$

there is a corresponding row vector

$$\langle v| \equiv \begin{pmatrix} v_1^* & \cdots & v_n^* \end{pmatrix},$$

and the scalar product between two vectors is denoted

$$\langle v|w\rangle \equiv \sum_{k=1}^{n} v_k^* w_k = \langle w|v\rangle^*.$$

Dirac called row vectors *bras*, so that the scalar product is a *bra-ket* or bracket. It is one of those jokes that only sound funny to physicists. The real virtue of Dirac notation is that it allows us to construct a self-explanatory notation for the linear operator D_{vw} called the outer product or dyadic formed from two vectors. Dirac's notation is

$$D_{vw} = |v\rangle\langle w|.$$

In words, what this operator does to a vector $|u\rangle$ is to take its scalar product with $|w\rangle$ and multiply that number times the vector $|v\rangle$. In particular, if $|e\rangle$ is a vector with norm 1 ($\langle e|e\rangle = 1$) then

$$|e\rangle\langle e|$$

is the projection operator, which, for any other vector $|v\rangle$ gives the projection of $|v\rangle$ along $|e\rangle$.

Let us take a moment to recall the definition of the term *linear operator*, which was used in the previous paragraph. In any vector space over the complex numbers, we can construct linear combinations of a pair of vectors $|v\rangle$ and $|w\rangle$ according to the formula

$$a|v\rangle + b|w\rangle.$$

Thinking about vectors as n-tuples of complex numbers, you know exactly what this means. A linear operator is a mapping that takes any vector $|v\rangle$ into another vector $T|v\rangle$. Linearity means that the map satisfies the rule

$$T[a|v\rangle + b|w\rangle] = aT|v\rangle + bT|w\rangle. \tag{2.5}$$

The representation of vectors as n-tuples of complex numbers depends on a choice of what is called a *basis* in the space of all vectors. We can consider the vectors $|e_k\rangle$ whose representation as a column of numbers has a 1 in the k-th row and zeroes everywhere else. Then the complex column representing $|v\rangle$ can be thought of as the numbers that appear in the expansion

$$|v\rangle = \sum_k v_k |e_k\rangle.$$

With our definition of the scalar product, the vectors $|e_n\rangle$ satisfy $\langle e_m|e_n\rangle = \delta_{mn}$, which is the definition of a basis that is *orthonormal*, shorthand for "orthogonal and normalized to 1." The concept of basis should be familiar to you from two- and three-dimensional geometry. Unit vectors in the x, y, and z directions form an orthonormal basis of three-dimensional space. From these examples, you are familiar with the fact that orthonormal bases are not unique. We can rotate them in any direction in three-dimensional space. The same is true of the n-dimensional complex vectors that are of interest in QM. We call a linear operator U, a *unitary transformation or unitary operator*, if the vectors $U|e_n\rangle \equiv |f_n\rangle$ also form an orthonormal basis. We will explore the conditions for a transformation to be unitary in the following chapters.

Given an orthonormal basis and a linear transformation T, we can construct the square array of numbers

$$T_{mn} = \langle e_m | T | e_n \rangle. \tag{2.6}$$

This array is called *the matrix of the transformation (or the operator) T in the $|e_n\rangle$ basis.* You should convince yourself that you can think of T_{mn} as the coefficient of $|e_m\rangle$, in the expansion of the vector $T|e_n\rangle$ in the $|e_n\rangle$ basis.

Returning to our two state ammonia molecule, we say that we have identified the energy and electric dipole moment as linear operators in a space with orthonormal basis $|+\rangle$ and $|-\rangle$, and that the matrices of these operators are diagonal in that basis. We can also think of a probability distribution for the different states as a diagonal matrix ρ, with eigenvalues $p_\pm \geq 0$, $p_+ + p_- = 1$. For any probability distribution, the expectation value of a quantity, which takes the values A_i in the i-th state of a system, is just

$$\langle A \rangle \equiv \sum_i A_i p_i.$$

Another term for this is the average value of the quantity A, but in QM, we always use the term expectation value. Exercise (2.5) asks you to show that the expectation value, $\langle P \rangle$, of any polynomial in the quantities E and D is just

$$\langle P(E, D) \rangle = \text{Tr } P(E, D)\rho.$$

Recall that the trace of a matrix is the sum of its diagonal elements. It is important to note that, despite the appearance of matrices, all of these formulae are just a compact mathematical way of discussing the properties of a bit in classical logic. A bit in classical computer science (Turing) is the same as an irreducible Yes/No question. This is a question, which completely characterizes the state of a system. By definition, if you can characterize a system by the answer to a single Yes/No question, then that system has two states. So, our low-energy ammonia molecule provides a mathematical model applicable to *any* two state system, in physics, logic, or computer science.

$P(E, D)$ is just a general function of the two physical variables, energy and dipole moment, one of which takes the same value in both states of the system and the other of which takes opposite values. Any quantity, which takes definite values in the two states is a function of these two (prove this in Exercise 2.15). The eigenvalues of ρ are just the probabilities that the system is in each of its possible states, and the trace formula is just the usual statement that the expectation value is the sum of the possible values of a quantity, weighted by the probability that it takes on each of these values. At this point, you should do Exercises 2.1 and 2.2, which allow you to refresh your memory of matrix multiplication.

Exercise 2.1 The action of a matrix on a vector is denoted by the algebraic formula $v^i \rightarrow \sum_j M^i_j v^j$. Use this formula to evaluate the action of the matrix

$$\begin{pmatrix} 0 & a \\ b & 0 \end{pmatrix} \tag{2.7}$$

on the vector

$$\begin{pmatrix} c \\ d \end{pmatrix}. \tag{2.8}$$

Exercise 2.2 The product of two matrices M and N is denoted MN and is defined by letting M act on each of the column vectors making up N, according to the rule in Exercise 2.1. In symbols,

$$(MN)^i_j = \sum_k M^i_k N^k_j. \tag{2.9}$$

Evaluate the product of two general 2×2 matrices and show that $MN \neq NM$. The difference $MN - NM \equiv [M, N]$ is called the commutator of the two matrices. The inverse of a matrix is defined by the equation $MM^{-1} = 1$, where the matrix 1 is the *unit matrix*, which has ones on its diagonal and zero everyplace else. Find the inverse for a general 2×2 matrix, if it exists, and describe the criterion for the inverse to exist. Show that the commutator of a matrix and its inverse vanishes.

Answer to Exercise 2.2: The equation for the inverse is

$$MM^{-1} = \begin{pmatrix} a & b \\ c & d \end{pmatrix} \begin{pmatrix} e & f \\ g & h \end{pmatrix} = \begin{pmatrix} 1 & 0 \\ 0 & 1 \end{pmatrix}. \tag{2.10}$$

The product of the two matrices is

$$\begin{pmatrix} ae + bg & af + bh \\ ce + dg & cf + dh \end{pmatrix}. \tag{2.11}$$

The equations for the inverse are thus

$$ae + bg = 1 = cf + dh, \quad af + bh = 0 = ce + dg. \tag{2.12}$$

So, $g = -(c/d)e$, $h = -(a/b)f$, and $1 = (a - \frac{bc}{d})e = (c - \frac{ad}{b})f$. The last two equations have solutions if and only if $ac - bd \neq 0$. This combination of matrix elements is called the *determinant* of the matrix M and denoted det M. It measures whether the rows and columns of the matrix are linearly independent vectors. The inverse matrix is

$$M^{-1} = (ad - bc)^{-1} \begin{pmatrix} d & -b \\ -c & a \end{pmatrix},$$ (2.13)

and it is easy to verify that $M^{-1}M = 1$.

In classical probability theory, a state is called *pure* if it corresponds to probability one for one of the elementary states, and zero for all of the others. If a state is not pure, it is called *mixed*. In matrix language, pure states are characterized by the matrix equation

$$\rho^2 = \rho.$$

Although we have introduced it in the context of a two state system, this description in terms of matrices is completely general, for any finite number of elementary pure states. N states corresponds to a theory of $N \times N$ matrices. In classical logic, all matrices representing data about the system are simultaneously diagonal, as are all probability distributions.

2.4   OPERATIONS AS MATRICES

Things become a little more interesting when we recognize that the classical operation of reflecting the nitrogen through the plane of the hydrogens is also represented by a matrix. The matrix is

$$\sigma_1 = \begin{pmatrix} 0 & 1 \\ 1 & 0 \end{pmatrix},$$ (2.14)

and Exercise 2.3 asks you to verify that it indeed does what it was claimed to do.

Exercise 2.3 Use the rules for acting with matrices on vectors to show that

$$\sigma_1 |\pm\rangle = |\mp\rangle.$$

From this equation, it follows that $\sigma_1^2 = 1$. Verify that this is true using matrix multiplication. This algebraic equation for σ_1 expresses the fact that if we do two successive reflections, we return the system to its original state.

The operation of evaluating the dipole moment does not commute with this reflection

$$\sigma_1 \sigma_3 = -\sigma_3 \sigma_1 \equiv -i\sigma_2,$$

from which it also follows that $\sigma_2^2 = 1$. Calculation (Exercise 2.4) shows that

$$\sigma_2 = \begin{pmatrix} 0 & -i \\ i & 0 \end{pmatrix}.$$ (2.15)

We can also write the general product relation

$$\sigma_a \sigma_b = \delta_{ab} + i\epsilon_{abc}\sigma_c,$$

where the sum on c from 1 to 3 is left implicit (this is called the Einstein summation convention). The symbol ϵ_{abc} is totally antisymmetric in its three indices and has the value $\epsilon_{123} = 1$. It is the object we use to define the cross product of two vectors,

$$(\mathbf{A} \times \mathbf{B})_a = \epsilon_{abc} A_b B_c.$$

Again we are summing over repeated indices, and from now on we will use the Einstein summation convention without comment. Do Exercise 2.6 if you want some practice with the epsilon symbol.

The fact that these matrices do not commute is obvious, from the point of view of classical logic, if we think of them both as operations on the system. One operation evaluates the dipole moment in a particular state, while the other changes which state the molecule is in. *Of course* you get a different result if you evaluate the dipole moment before or after you change the state!

It is because we can think of both the diagonal matrices, which act on a state by multiplying it by a number, and matrices like σ_1, as *operations* on the system, that we use the words *matrix* and *operator* almost interchangeably. A matrix is just a way of characterizing what a linear operator does to a particular orthonormal basis.[2]

Now let us consider something peculiar. Evaluate

$$\text{Tr } \rho \sigma_1 = 0. \tag{2.16}$$

This is true no matter what the probabilities are to be in $|\pm\rangle_3$, and in particular it is true when we know that we are certainly in one of those states. On the other hand, since $\sigma_1^2 = 1$ it is natural to think that it might "take on the values ± 1," just like σ_3. Indeed, we can find vectors of length 1

$$|\pm\rangle_1 = \frac{1}{\sqrt{2}}(|+\rangle_3 \pm |-\rangle_3), \tag{2.17}$$

such that

$$\sigma_1 |\pm\rangle_1 = \pm |\pm\rangle_1. \tag{2.18}$$

For diagonal matrices, the trace with the density matrix gave the expectation value for the operator σ_3, in a probability distribution for it to take on one of its possible values. We can interpret the formula in the same way for σ_1, *and come to the conclusion that σ_1 is uncertain even when we know the precise state we are in.* That is, we could accept the proposition that the normalized vectors $|\pm\rangle_1$ defined possible states that the system could be in, in which

---

[2] One can also associate matrices with the action of an operator on a basis which is not orthonormal. We will have little or no use for these in the finite dimensional case, although we will discuss matrix elements of operators in continuous, *overcomplete* bases in Chapters 5 and 19.

σ_1 had definite values. The density matrix saying we are definitely in the state $|+\rangle_1$ would be the projection operator on this vector.

$$P_1^+ = \frac{(1 + \sigma_1)}{2}. \tag{2.19}$$

It is easy to see that

$$\sigma_3 |\pm\rangle_1 = |\mp\rangle_1, \tag{2.20}$$

so that

$$\text{Tr } (P_1^+ \sigma_3) = 0. \tag{2.21}$$

That is, in the state $|+\rangle_1$, σ_1 has a definite value, while σ_3 is maximally uncertain.

This observation and conjecture lead one to ask which normalized vectors can be thought of as states of the system, and which operators correspond to quantities that can take on definite values. A quick review of the theory of diagonalization of matrices, will lead us to the conclusion that *any* normalized vector is acceptable, and any operator whose eigenvectors form a complete orthonormal basis is a potentially observable quantity.

2.5   THE EIGENVALUES OF A MATRIX

The fact that matrices satisfy algebraic equations like $\sigma_1^2 = 1$, is not a surprise. In fact, there is a general theorem, for any square matrix in any number of dimensions. Recall that the determinant of a square $n \times n$ matrix $\det(M)$ is the sum of all possible products of n distinct elements, with a sign that you probably learned in terms of an algorithm where you go down the left-hand column of the matrix, changing sign at every step, and evaluate the determinant in terms of products of the matrix element in the k-th row of the left-hand column times determinants of cofactor matrices obtained by omitting the k-th row and first column. For a diagonal matrix, this algorithm just gives the product of diagonal matrix elements. Given any matrix M, its *characteristic polynomial* is

$$P_M(x) = \det(xI - M),$$

where I is the $n \times n$ unit matrix. If the matrix is diagonal, with diagonal matrix elements m_i, this just gives

$$P_M(x) = \prod_i (x - m_i).$$

We can easily see from this that if we make the same polynomial of the *matrix*, then

$$P_M(M) = 0.$$

This is (the simplest case of) the Cayley-Hamilton theorem. Now recall a simpler theorem, namely that for any two square matrices

$$\det(AB) = \det(BA),$$

so that

$$\det(SBS^{-1}) = \det(B),$$

for any invertible matrix S. This generalizes Cayley–Hamilton to a large class of nondiagonal matrices, namely all those that are diagonalizable by a similarity transformation S. If $M = SDS^{-1}$, where D is diagonal, then

$$\det (xI - M) = \det (S[xI - D]S^{-1}) = \det (xI - D), \tag{2.22}$$

so M and D have the same characteristic polynomial. It is also true for any polynomial $q(x)$ that $q(M) = Sq(D)S^{-1}$, so that $P_M(M) = SP_D(D)S^{-1} = 0$. In fact, the Cayley–Hamilton theorem is true (but we will not use it) for any matrix.

At this point let us recall that linear algebra is a *prerequisite* for this course, stop treating you like a babe in the matrix woods, and freely use concepts from linear algebra. If you are uncomfortable with your understanding of linear algebra, you should skip to Chapter 6 at this point. This will have the added advantage that you will review linear algebra using the Dirac notation that we will employ throughout this book. A shorter summary of Dirac notation and linear algebra can be found in Appendix F.

The matrices σ_a are all examples of matrices that can be diagonalized by conjugation by a unitary transformation. That is, there is a unitary transformation, U_i (i.e., $U_i^\dagger U_i = 1$) such that $U_i^\dagger \sigma_i U_i$ is diagonal. Equivalently there is an orthonormal basis of eigenvectors such that

$$\sigma_i|\pm\rangle_i = \pm|\pm\rangle_i.$$

For example,

$$|\pm\rangle_1 = \frac{1}{\sqrt{2}}(|+\rangle_3 \pm |-\rangle_3).$$

In the basis $|\pm\rangle_i$, σ_i is diagonal, while the other two have off diagonal matrices. In fact, any of the matrices $\sigma(\mathbf{n}) \equiv \sum n_a\sigma_a$ where n_a is a real three-dimensional vector, can be diagonalized. In this new notation, the states we called $|\pm\rangle$ are now $|\pm\rangle_3$. You can easily verify that for the vectors $|\pm\rangle_1$,

$$\sigma_1|\pm\rangle_1 = \pm|\pm\rangle_1.$$

The factor of $\frac{1}{\sqrt{2}}$ is inserted so that these vectors are orthonormal. These words refer to the scalar product on our vector space. We will discuss this important concept in more detail below, but you should start trying to remember things about scalar products from your linear algebra class.

The reason that it is interesting to look at orthonormal bases is that they are the quantum version of a complete set of independent alternatives in classical probability theory. The projection operators $P_n \equiv |e_n\rangle\langle e_n|$ have the property of density matrices. Their eigenvalues are positive and sum to 1. Furthermore,

$$\mathrm{Tr}\,(P_m P_n) = \delta_{mn}. \tag{2.23}$$

This equation is interpreted as meaning that if one is in the state specified by P_m, then the probability of being in the state P_n is zero unless $m = n$. The normalization condition on states,

$$1 = \langle s|s\rangle = \mathrm{Tr}(|s\rangle\langle s|), \tag{2.24}$$

is the statement that probabilities sum to one in any state. What distinguishes QM from classical mechanics is that we allow ourselves to contemplate the quantities

$$\mathrm{Tr}\,(|e_n\rangle\langle e_n||f_m\rangle\langle f_m|) \tag{2.25}$$

for two different orthonormal bases and interpret this number as the probability that the system is in one of the f states, given that we know that it is in one of the e states (or vice versa: the formula is symmetric under interchange of the two bases). If the time evolution law for states does not preserve a particular orthonormal basis, we are forced to work with these more general probabilities.

To say this all in another way, in addition to the original matrices, which we identified with quantities detectable or measurable for the up/down bit of the ammonia molecule, there are other matrices, either operations on the up/down states, or algebraic combinations of operations and measurable quantities, which satisfy algebraic equations in terms of matrix multiplication and addition. For some of them we can, using the rules of linear algebra, find linear combinations of the vectors we originally identified with states of the system, in which the different solutions of those algebraic equations are "realized." Linear algebra treats all of these realizable operators (which are called *normal operators*) in a democratic fashion. QM results from exploring the consequences of this democracy for physics, that is, of viewing every normal operator as something that can take on a definite value in some state of the world.

2.6   THE BORN RULE AND UNCERTAINTY

To reiterate: the essence of QM lies, in the fact that we can generalize the trace formula for expectation values, to a large class of nondiagonal matrices A. The formula

$$\langle A \rangle = \mathrm{Tr}\,(A\rho)$$

may be evaluated as

$$\langle A \rangle = \sum_{r,k} a_k p_r |\langle a_k | r \rangle|^2,$$

for any matrices A and ρ, which are diagonalizable by a unitary transformation (Exercise 2.7). The vectors labeled a_k and r are the orthonormal eigenvectors of A and ρ, respectively. The a_k are the two eigenvalues of A. If the eigenvalues p_r satisfies the probability (in)equalities above, then so does the probability distribution

$$P(a_k) \equiv \sum_r p_r |\langle k | r \rangle|^2,$$

as a consequence of completeness and orthonormality of the two bases (Exercise 2.8). *Note in particular that even if $p_r = 1$ for some r, which would, according to classical logic, imply that all quantities were known with absolute certainty, $P(a_k) < 1$ unless the two bases k and r are identical (up to permutation).* Note also that, while we have introduced all of this notation for a simple bit, with only two states and a single nontrivial question, our remarks about matrices are completely general, and apply to a system with any number of states. Finally, note that if $p_r = 1$ (and thus all other probabilities zero) then the probability of finding A with eigenvalue a_k is

$$|\langle a_k | r \rangle|^2, \tag{2.26}$$

the square of a complex number called the *probability amplitude that A takes on the value a_k.*

This equation is known as *Born's Rule* for QM probabilities. It is usually introduced as an *ad hoc* postulate, or following a discussion like Feynman's lecture on the double slit experiment. Here, we have derived it by generalizing classical probability formulae, written in matrix form, to matrices that are diagonalizable, but not diagonal in the same basis as the (probability) density matrix ρ. Born's formula for probability is one of the two most important equations in QM, the other one being

$$1 = \sum_n |e_n\rangle\langle e_n|, \tag{2.27}$$

which is called the *resolution of the identity*, and is true for *any* choice of the orthonormal basis $|e_n\rangle$. The individual terms, $|e_n\rangle\langle e_n|$, in the sum are projection operators onto the one-dimensional spaces spanned by a single member of the orthonormal basis (we have said this before, but prove this to yourself, if it is not obvious to you). What this equation says, in words, is that the unit operator is the sum of the projection operators onto the elements of *any* orthonormal basis.

2.6.1 Born's Rule in Pictures

Let us consider Born's Rule when the state of the system is pure, and the operator A is the projection operator onto some pure state $|f\rangle$. Then

$$A = |f\rangle\langle f|, \tag{2.28}$$

$$\rho = |e\rangle\langle e|, \tag{2.29}$$

$$A\rho = (\langle f|e\rangle)|f\rangle\langle e|, \tag{2.30}$$

$$\rho A = (\langle e|f\rangle)|e\rangle\langle f|. \tag{2.31}$$

In the example where

$$|e\rangle = \begin{pmatrix} 1 \\ 0 \end{pmatrix} \quad \text{and}$$

$$|f\rangle = \frac{1}{\sqrt{2}} \begin{pmatrix} e^{i\alpha} \\ 1 \end{pmatrix},$$

you should verify that these formulae give

$$\rho A = \frac{e^{i\alpha}}{\sqrt{2}} \begin{pmatrix} \frac{e^{-i\alpha}}{\sqrt{2}} & 0 \\ 1 & 0 \end{pmatrix}, \tag{2.32}$$

$$A\rho = \frac{e^{-i\alpha}}{\sqrt{2}} \begin{pmatrix} \frac{e^{i\alpha}}{\sqrt{2}} & 1 \\ 0 & 0 \end{pmatrix}. \tag{2.33}$$

Note that the trace of these matrices is the same, a consequence of the general theorem that $\mathrm{Tr}\ AB = \mathrm{Tr}\ BA$, and is equal to $|\langle f|e\rangle|^2$, which is equal to $1/2$ for our example.

This leads to a geometrical interpretation of Born's rule, illustrated in Figure 2.4.

If $|e\rangle$ and $|f\rangle$ are two unit vectors, then the quantity $\langle e|f\rangle$ is the *projection of the $|e\rangle$ vector on the $|f\rangle$ vector*, which is the name for the component of the $|e\rangle$ vector along $|f\rangle$, in a basis, one of whose elements is $|f\rangle$. If the vectors were real, this would be $\cos(\theta)$, the cosine of the angle between them. Since they are complex, the projection is a complex number. Its complex conjugate $\langle f|e\rangle$ is the projection of $|f\rangle$ along $|e\rangle$.

The projections f_i of any unit vector along the elements of any orthonormal basis $|e_i\rangle$ satisfy Pythagoras' theorem

$$\sum |f_i|^2 = 1.$$

Born's rule tells us that we should interpret this geometrical formula as defining a probability distribution. Each unit vector is a possible physical state of the quantum mechanical system. $|f_i|^2$ is interpreted as *the probability that, if the system is in the physical state represented by*

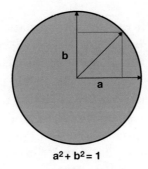

$$a^2 + b^2 = 1$$

Figure 2.4 Geometric interpretation of Born's rule.

$|f\rangle$, *that a measurement which attempts to find out whether it is in one of the physical states* $|e_j\rangle$, *will find that* $j = i$. In order to truly understand the implications of this sentence, we will have to make a quantum model of what we mean by the word measurement. We will postpone this difficult question until Chapter 10.

The picture also helps us to understand the meaning of the name *projection operator*. Given the unit vector $|f\rangle$, the projection operator $|f\rangle\langle f|$ operates on any vector and gives us the projection of that vector along $|f\rangle$. This is illustrated in Figure 2.5.

To summarize: one can formulate all of the classical information about the two apparently degenerate states of the ammonia molecule, and operations which change from one state to the other, in terms of the algebra of two by two matrices, acting in a vector space whose basis vectors represent the two states. The formulae for expectation values of classical measurable quantities generalize to arbitrary matrices diagonalizable by a unitary transformation and give us a probability theory in which, in any state of the system, not every quantity can be known definitely.

Another way to think about quantum probability theory is that it uses the mathematical similarity between probability formulae and Pythagoras' theorem for the projections of one unit vector on an orthonormal basis of others, to define a notion of probability in which there is uncertainty for *something*, even in states of the system where *a complete set of things is completely determined.*

Figure 2.5 A projection operator.

Probability theory was invented to deal with situations in which things were definite, and evolved in a definite manner, but where we lacked knowledge, whether through sloth or inability to measure, about all the relevant initial conditions. We think of it as a way of guessing intelligently, about the things we have not measured. The initial probability distributions in a classical probabilistic theory are guessed, and then revised on the basis of experiment, to fit the data. In contrast, the intrinsic quantum probability theory has nothing to do with guessing. Once we have specified the state completely, by measuring the maximal set of compatible quantities, *the probability distributions for all other quantities are exactly computable*. One might say that QM takes the guesswork out of probability. At this point, the reader should do Exercise 2.9, and find the probabilities, according to Born's rule, for the operator $n_a \sigma_a$, where n_a is a real three vector, to take on its two possible values $\pm|\mathbf{n}|$, assuming that the system is in one of the states $|\pm\rangle_3$.

The essence of QM is that we accept Born's rule, which defines, for every normalized vector in our space (our two-dimensional space in the present example) and every matrix A which is diagonalizable by a unitary transformation, a mathematical function, $P_\psi(a_k) \equiv |\langle a_k|\psi\rangle|^2$, which has the properties of a probability distribution for the quantity A to take on one of its eigenvalues, a_k when the system is in the state ψ. QM is the statement that this mathematical function should be considered a real probability distribution for physical quantities.

2.7 WHICH OPERATORS ARE DIAGONALIZABLE?

Now let us take up the question of which linear operators A, are candidates to be measurable properties of our system. Our fundamental probability assumption is that the quantity

$$\langle\langle(A^p)\rangle\rangle = \text{Tr}\left(A^p|\psi\rangle\langle\psi|\right) = \langle\psi|A^p|\psi\rangle = \sum_k a_k^p P_\psi(a_k), \tag{2.34}$$

is the expectation value of the value of A^p when we make repeated measurements of it, after having prepared the system in the state represented by the unit norm vector $|\psi\rangle$. We will have to explain what we mean by measurement and state preparation, in order to make contact between this mathematical probability theory, and things that happen in the real world, but we will put off that discussion until Chapter 10, when we have a lot more experience with the mathematics of QM.

In order for this formula to have the properties of a probability distribution, there must be an orthonormal basis $|a_k\rangle$ such that

$$A|a_k\rangle = a_k|a_k\rangle. \tag{2.35}$$

Now consider the matrix of A in any other orthonormal basis

$$A_{mn} \equiv \langle e_m|A|e_n\rangle = \sum_{kl} \langle e_m|a_k\rangle\langle a_k|A|a_l\rangle\langle a_l|e_n\rangle, \tag{2.36}$$

where we have used the resolution of the identity for the basis $|a_k\rangle$ twice.

We now use the eigenvalue equation and the orthonormality relation $\langle a_k|a_l\rangle = \delta_{kl}$, to write this as

$$A_{mn} = U_{mk}a_k U_{nk}^*, \tag{2.37}$$

where $U_{mk} = \langle e_m|a_k\rangle$. As a matrix equation, this is $A = UDU^\dagger$, where the Hermitian conjugate $U^\dagger$ is the matrix obtained by both complex conjugating and transposing (flipping row and column indices) the matrix U. D is the diagonal matrix of A in the $|a_k\rangle$ basis. The resolutions of the identity and orthonormality relations for the two bases $|a_k\rangle$ and $|e_m\rangle$ are (prove it!) equivalent to the matrix equation

$$UU^\dagger = U^\dagger U = 1. \tag{2.38}$$

As we have said, such matrices are called unitary matrices. Their rows and columns are the two sets of orthonormal basis vectors. They are the analog of ordinary rotations in the vector space representing three-dimensional Euclidean space.

Now note that

$$AA^\dagger = U|a_k|^2 U^\dagger = A^\dagger A. \tag{2.39}$$

Any operator that commutes with its Hermitian conjugate is called *normal*. The fact that $[A, A^\dagger] = 0$ is obvious in the original basis where the matrix of A is diagonal. On the other hand, if we have a polynomial involving some collection of matrices M_i, then $P(U^\dagger M_i U) = U^\dagger P(M_i)U$, for any unitary matrix. One can prove this identity by noting that all the "intermediate" factors of $U^\dagger U$ can be replaced by 1 due to unitarity. (Informal Exercise: Prove this statement.)

The converse statement, that any normal operator can be written as $A = UDU^\dagger$, is called the *spectral theorem*, and the set of a_k is called the spectrum of the operator. To prove it, note that $\det(A - a)$ is an N-th order polynomial in the complex parameter a and so has N roots a_k. This means that $A - a_k$ is not an invertible operator, and so must satisfy

$$(B_k)^p|a_k\rangle = 0, \tag{2.40}$$

where $B_k = A - a_k$, for some value of p and some state $|a_k\rangle$. If A is normal, this means that

$$\langle a_k|[B_k^\dagger B_k]^p|a_k\rangle = 0. \tag{2.41}$$

If p is even, this is the norm of the state $[B_k^\dagger B_k]^{p/2}|a_k\rangle$, while if p is odd, it is the norm of $B_k[B_k^\dagger B_k]^{(p-1)/2}|a_k\rangle$. The only state with vanishing norm is the zero state, so we find that $B_k^q|a_k\rangle = 0$, where q is $p/2$ or $(p-1)/2$. We can continue reducing the value of p in this way, until we conclude that $B_k|a_k\rangle = 0$, which is the statement that $|a_k\rangle$ is an eigenvector of A with eigenvalue a_k. The eigenvectors with different eigenvalues are orthogonal because

$$\langle a_k|A|a_m\rangle = \langle a_k|a_m\rangle a_m = \langle a_k|a_m\rangle a_k, \tag{2.42}$$

if we act with either A to the right or $A^\dagger$ to the left. Note that $\langle a_k|A$ is the complex conjugate row vector corresponding to $A^\dagger|a_k\rangle = a_k^*|a_k\rangle$. For a degenerate subspace with the same eigenvalue, we can always choose an orthonormal basis. So, a general normal operator is diagonalizable by a unitary matrix.

In most books on QM, one is told that operators corresponding to measureable quantities must be *Hermitian*: $H = H^\dagger$. Hermitian operators are obviously normal. Furthermore, for any operator

$$A = (A + A^\dagger)/2 + i\frac{1}{i}(A - A^\dagger)/2 \equiv H_1 + iH_2, \tag{2.43}$$

where $H_i = H_i^\dagger$. This is analogous to breaking a complex number into its real and imaginary parts. If A is normal, then $[H_1, H_2] = 0$. So, a normal operator is equivalent to a pair of commuting Hermitian operators. There is no reason not to allow any normal operator as a "physically measurable" quantity in QM, and it is convenient because it allows us to solve any algebraic equation involving physical quantities, and get another physical quantity.

2.7.1 Vectors, Functions, and Diagonal Matrices

Later on in this book, we are going to encounter vector spaces, in which the vectors are functions of continuous variables. A lot of students get confused transferring what they know about linear algebra into the context of function spaces, so it is worth explaining that we can always think of linear algebra in terms of functions. If v_n are the components of a vector $|v\rangle$ in some basis, then we can associate with them a complex valued function, defined on the finite set of integers from 1 to N, by the rule

$$f_v(n) = v_n. \tag{2.44}$$

The space of all such functions is a vector space, since we can add them and multiply them by complex numbers, and the scalar product takes the form

$$\langle w|v\rangle = \sum_{n=1}^{N} f_w^*(n) f_v(n). \tag{2.45}$$

The basis vectors are functions which vanish on all integers between 1 and N except for one. Linear operations become operations on functions. A particularly interesting one is the finite difference operator

$$[\Delta f](n) = f(n) - f(n - 1). \tag{2.46}$$

If we take a limit where we think of N going to infinity and define $x = \frac{n}{N}$, then x becomes a continuous variable in the limit and $N\Delta$ approaches the derivative operator, when it acts on differentiable functions of x. You should write out the matrix corresponding to the difference operator Δ to make sure you understand this. Note that this correspondence between vectors, and functions defined on the discrete set $1 \ldots N$, depended on a choice of basis. Different choices of basis will give different functions for the same vector $|v\rangle$. When we get to continuous variables, we will see that the famous Fourier Transform is just a relation between the functions that represent a vector in two different bases.

Another confusion that arises when thinking about functions as elements of the space is that functions also act as *multiplication operators on the space of functions*. This also has a finite dimensional analog. Given a vector v_n, or the associated function $f_v(n)$ one can construct a diagonal matrix, whose n-th diagonal element is v_n. When one acts with that matrix on a vector whose components are w_n, then one gets the vector whose components are $v_n w_n = f_v(n) w_n$. The function corresponding to this new vector is the product $f_v(n) f_w(n)$. These remarks may seem sort of silly in the finite dimensional context, but they are the bridge that allows you to make the transition to thinking about function spaces.

2.8 QUANTUM DYNAMICS OF THE AMMONIA MOLECULE

The reader is entitled to be confused at this point, since we started out with a seemingly classical description of the two low-energy states of the ammonia molecule and have already introduced quantum mechanical variables, which cannot take all definite values at the same time. How is it possible then, to have a classical description of ammonia, even in principle?

Physics is all about dynamics. It is a set of *equations of motion*, which tell us how to predict the state of a system at a later time, given its state now. In Newtonian mechanics, this prescription consists of a set of differential equations[3]

$$\dot{z}^i = f^i(z).$$

Time flows continuously. The variables z^i are differentiable functions of t. The values of $z(t_0)$ enable us to predict/retrodict all past or future values of $z(t)$. The key point in these equations is that, we *assume* that all of the different z^i can take on definite values simultaneously. QM emerges upon realizing that this assumption is not necessary, if we are willing to live with an *intrinsically* probabilistic description of the world, as opposed to one in which probability is introduced merely as a device to compensate for our lack of ability or desire to measure all of the $z^i(t_0)$ precisely. The only way to see whether the special case of C(lassical) or C(ertainty) mechanics describes the real world, is to do experiments.

For our simple low-energy ammonia molecule, a classical theory must describe the evolution of the system in discrete jumps. We started out by assuming that there is no continuous way to go between two states that differ by a reflection of the nitrogen atom through the plane

---

[3] We can always reduce ordinary differential equations with higher derivatives to first order equations by simply calling the derivatives names and adding them to the list of variables. An N-th order equation for a single function is equivalent to N first-order equations for that function and its derivatives up to the $N-1$ order.

of the hydrogens, without having much more energy than we have available. For simplicity, let us imagine that the jumps happen every fixed time interval τ. Introduce the frequency $f = 1/\tau$. Then the state of the system at time t is

$$|\psi(t)\rangle = (\sigma_1)^{[ft]}|\psi(0)\rangle.$$

The square bracket notation in the exponent stands for "greatest integer in." In this classical version of the theory, the energy of both states is the same. It is definite and conserved during the motion. σ_3, which determines the value of the dipole moment, switches sign every time an interval τ passes. A classical physicist would not want us to talk about measuring the variables $\sigma_{1,2}$ (though the classicist *would* let us talk about σ_1 as an operation on the system), but we can use the trace formula to evaluate their expectation values. The initial density matrix is the projection matrix onto whichever of the two states $|\pm\rangle_3$ is the initial state, and it switches to the projector on the other state every time interval τ. The expectation values of $\sigma_{1,2}$ are zero in either of these states, so we conclude that these two variables are always maximally uncertain. The most general diagonalizable 2×2 matrix has the form

$$A + Bn_a\sigma_a,$$

where n_a is a real unit three vector (prove this in Exercise 2.10), and A, B are complex numbers. With the classical evolution law we have given, the components of this matrix proportional to 1 and σ_3 remain definite for all time, while $\sigma_{1,2}$ are maximally uncertain.

That is, if we interpret the formula

$$\mathrm{Tr}\rho A$$

in the probabilistic manner suggested above, and choose ρ to be the projection operator on either of the vectors $|\pm\rangle_3$, then we find the expectation values of $\sigma_{1,2}$ are both zero. Given the possible values, ± 1, of those two operators, and the fact that the system only has two states, we learn that the probability to be in each of the eigenstates of $t_a\sigma_a$ is $\frac{1}{2}$, if t_a is perpendicular to the 3 axis.

One thing that is a little annoying about this classical system is that the time evolution has jumps, whenever t passes through an integer times τ. There is nothing wrong with this. In particular, if our experimental time resolution is much larger than τ, we would never see the discrete jumps. However, it is amusing to note the purely mathematical result that if $\omega \equiv \frac{\pi}{2}f$, then

$$\sigma_1^N = e^{-i\omega t(\sigma_1 - 1)},$$

when $t = N\tau$. So, we can view the discrete classical evolution as a sequence of snapshots of a continuous evolution of the state

$$|\psi(t)\rangle = e^{-i\omega t(\sigma_1 - 1)}|\psi(0)\rangle \equiv U(t)|\psi(0)\rangle,$$

if we contemplate expanding our notion of what states the system can be in, to include all the two-dimensional complex vectors that are swept out by this evolution. Mathematically this extension of the concept of state is simple, and in a certain sense obvious. Conceptually it is a profound deviation from the logic of classical physics. The idea of allowing the system to be in any normalized state is responsible for all of the uncertainty principles of QM.

The evolution operator is unitary (Exercise 2.11), and so the two vectors $|\pm\rangle_3$ are transformed into two other orthonormal vectors at any given time. That means that, at each time, there is some matrix, which is diagonal in that instantaneous orthonormal basis, but it is not the original matrix σ_3, which we identified with the detectable electric dipole moment of the molecule.

Let us step back and review the formula $e^{-i\omega t(\sigma_1)}$ that we just used. What does it mean? Matrix multiplication allows us to define polynomials of matrices, so in principle, up to questions of convergence, it also allows us to define any function of a matrix that is given by a power series. So,

$$e^{-i\omega t(\sigma_1)} = \sum_{k=0}^{\infty} \frac{(-i\omega t \sigma_1)^n}{n!}.$$

Using $\sigma_1^2 = 1$, the matrix appearing in the even terms in this sum is always the unit matrix, while the odd terms are all proportional to the matrix σ_1. The even and odd terms in an exponential of a complex number just sum up to the cosine and i times the sine, so

$$e^{-i\omega t(\sigma_1)} = \cos(i\omega t) - i\sin(i\omega t)\sigma_1.$$

Another way to think about functions of operators, if they are diagonalizable, is to think in the basis where the operator A is diagonal, and just define a function $f(A)$ to be the diagonal matrix whose matrix elements are $f(a_i)$. We can then express this function in any other orthonormal basis by using the unitary operator U_A, which rotates the basis where A is diagonal, into that other basis. In other words, we define functions of operators so that $f(U^\dagger A U) = U^\dagger f(A)U$. This equation is obviously correct for any function defined as a power series, and we simply take it to be the definition of $f(A)$ for any diagonalizable A.

The density matrix of the system evolves according to

$$\rho(t) = U(t)\rho(0)U^\dagger(t). \tag{2.47}$$

This is "obvious" for pure states and follows for general mixed states by linearity of the time development operation:

$$U(t)\sum b_i A_i U^\dagger(t) = \sum b_i U(t) A_i U^\dagger(t).$$

Note that the overall phase $e^{i\omega t}$ disappears from the formula for $\rho(t)$. We can study the time development of expectation values in the system by noting that (see Exercise 2.7), by the cyclicity of the trace ($\operatorname{Tr} AB = \operatorname{Tr} BA$)

$$\text{Tr } A(t)\rho = \text{Tr } A\rho(t), \tag{2.48}$$

where the time evolution of the *Heisenberg picture operator* $A(t)$ is defined by

$$A(t) \equiv U^\dagger(t)A(0)U(t). \tag{2.49}$$

That is, we can either evaluate expectation values of fixed operators in a time-dependent state (the Schrödinger picture) or time-dependent operators in a fixed state (the Heisenberg picture). Whichever way we do it, we find that, even if we start in a state where σ_3 has a definite value, it will not have a definite value at a generic time. Indeed

$$\sigma_3(t) = (\cos(\omega t) + i\sin(\omega t)\sigma_1)\sigma_3(\cos(\omega t) - i\sin(\omega t)\sigma_1)$$

$$= \cos(2\omega t)\sigma_3 - \sin(2\omega t)\sigma_2.$$

Note, for future (Chapter 7) reference, that this looks like a rotation by an angle $-2\omega t$ in the $2-3$ plane, if we think of the three Pauli matrices as a three-dimensional vector. Since the expectation value of σ_2 in the initial state vanishes, we have

$$\langle \sigma_3(t) \rangle = \cos(2\omega t) = \cos^2(\omega t) - \sin^2(\omega t).$$

That is, *it appears that the system has a time-dependent probability distribution*, with probability $p(t) = \cos^2(\omega t)$ to have a positive value of the dipole moment, and $1 - p(t)$ to have a negative value.

We can now try to compare the predictions of the classical and quantum models for the evolution of the system with data. In principle, in order to do this, we have to understand what it means to measure a microscopic property of a molecule. In classical physics, this is a conceptually straightforward, if practically difficult operation. In QM, we will have to discuss the concept of measurement more carefully, but we will wait till we understand the subject a bit better before doing that (Chapter 10). For the moment, let us just imagine that we know how to do the measurement.

The classical prediction is that the molecule jumps from positive to negative electric dipole moment (EDM), or vice versa, every $\frac{\pi}{\omega}$ seconds. It will always be found to have the same magnitude and only its sign will flip. At intermediate times, it will remain in whatever state it got to after the last flip. The QM prediction is the same[4] at precisely these discrete times, but the prediction for intermediate times is different. The quantum prediction is that at an intermediate time, the EDM will be either $\pm$, the same classical magnitude, but that the choice of sign is random. There is no way to predict precisely what will happen at intermediate times. However, if we repeat the experiment over and over again, with exactly

---

[4] Actually there is a subtle difference. See below.

the same conditions, we will find that the frequency of plus versus minus signs follows the $\cos^2(\omega t)$ law. Needless to say, we would not be talking about this if it were not for the fact that experiments show the QM prediction to be the correct one for ammonia molecules.

An important subtlety appears when we think carefully about the quantum versus classical predictions for what will happen at the discrete times. Since the quantum prediction of *anything* is always probabilistic, we really have to do the experiment many times if we want to test the quantum prediction, *even when QM predicts certainty.* After all, a single measurement at an intermediate time does not prove the quantum law is correct, since it is only a probabilistic law. The prediction of probability one for the discrete times can only be checked by showing that the dipole *never* has the wrong orientation at those times. In the classical theory, by contrast, we take the stance that, since we have completely measured the initial conditions, the theory makes a definite prediction, whose truth can be determined by a single experiment. It is in this subtlety that we see the real distinction between a theory which considers probability to be a mere consequence of failure to measure things precisely, and one in which probability is intrinsic.

QM can explain an even more striking experimental fact if we make a hypothesis about how to identify *energy* in the QM formalism. In order to do this, we have to remember the connection between energy conservation and time translation invariance. There is a very general theorem about this connection, called Noether's theorem, which is proved in the Lagrangian formulation of classical mechanics (see Appendix C). For simplicity, we can just think about Newton's law for a classical particle moving in one dimension, under the influence of a potential:

$$m\frac{d^2x}{dt^2} = -\frac{dV}{dx}.$$

If either the mass or the potential depend explicitly on time, then energy is not conserved. Time translation invariance is necessary for conservation of energy and Noether showed that it is also sufficient.

Noether's theorem, the connection between symmetries and conservation laws, is one of the most profound results in all of theoretical physics. One of the attractive features of QM is that, as a consequence of its unification of detectable quantities for a system and operations on that system, Noether's theorem is almost a definition of a symmetry. With the notable exception of time reflection symmetry, which we will discuss in Chapter 6, both detectable quantities and symmetry transformations are normal operators. Symmetries, however, obey two restrictions, which are not required of general detectable operators. Given any pair of states, the quantity $|\langle s_1|s_2\rangle|^2$ is the probability, assuming the system is in one of those states, that one finds the projection operator on the second, equal to 1. This should still be true after subjecting each state to a symmetry operation $|s_i\rangle \rightarrow (|W(s_i)\rangle)$. That is,

$$|\langle s_1|s_2\rangle|^2 = |\langle W(s_1)|W(s_2)\rangle|^2. \qquad (2.50)$$

One general way to satisfy this is for W to be implemented by a linear unitary transformation

$$|W(s)\rangle = W|s\rangle; \quad WW^\dagger = 1. \tag{2.51}$$

The other possibility exploits the fact that the probability formula is invariant under complex conjugation of $\langle s_1|s_2\rangle$. We will explore it in Chapter 6.

The other requirement for the quantum mechanical version of a symmetry transformation is analogous to the requirement in classical mechanics. If we transform the initial state of some quantum system, it should lead to an equivalent time evolution. This means that the symmetry operator U has to commute with the time translation operator. In equations, the Heisenberg operator

$$W(t) \equiv U^\dagger(t, t_0)W(t_0)U(t, t_0) = W(t_0). \tag{2.52}$$

is a conserved quantity! This is the quantum mechanical version of Noether's theorem. If you know the classical derivation of the theorem, or read the appendix, you will conclude that the quantum version is simpler.

Our classical and quantum evolution laws for the ammonia molecule are both time translation invariant (in the classical case, by discrete multiples of τ). A more general set of laws would allow the variation of the state at each time to be arbitrary. In the classical case, this would just mean that at each time step we could choose to flip the EDM, or not. Quantum mechanically, we could have

$$|\psi(t)\rangle = U(t, t_0)|\psi(t_0)\rangle,$$

with an arbitrary time-dependent unitary matrix. At any time, we could diagonalize

$$U(t, t_0) = \text{diag}(e^{-i\omega_1(t)t}, e^{-i\omega_2(t)}).$$

The two frequencies $\omega_i(t)$, would be time dependent. So, it is tempting to identify the time-independent parameter ω of the time translation invariant system with the conserved energy. Frequency and energy have different units, so we must introduce a constant $\hbar$, with dimensions of energy times time, or *action* and define

$$E = \hbar\omega.$$

It turns out that this conversion factor is universal for all systems.[5] The constant $\hbar$ is called Planck's constant (it is actually the constant Planck originally defined, divided by 2π because of the conversion between cycles per second and radians per second). In centimeter-gram-second units, it has the value $1.05457266(63) \times 10^{-27}$ erg-s. Theoretical physicists often use *natural units* in which $\hbar = 1$ and energy and frequency are simply identified.

---

[5] This is a consequence of energy conservation. It turns out that the laws of QM say that *frequency* is always conserved, so if different systems had different conversion factors, energy would not be conserved.

Energy conservation is more than just the fact that ω is constant in time. In the time-independent case, $U(t)$ is diagonal in the basis where σ_1 is diagonal, for all time. Thus, the *Hamiltonian operator*

$$H \equiv \hbar\omega(\sigma_1 - 1),$$

is time independent, $H(t) = H(0)$. This is the full statement of energy conservation. The Hamiltonian, H, as it is called, is the quantum mechanical version of energy, and like everything else in QM, it is a matrix/operator. What is more interesting is that this matrix has two different eigenvalues 0 and $-2\hbar\omega$. The corresponding eigenstates are

$$|\pm\rangle_1 = \frac{1}{\sqrt{2}}(|+\rangle_3 \pm |-\rangle_3).$$

Notice that under the space reflection operation, which changes the sign of the dipole moment, both these states go into themselves, up to a phase ± 1. Note also that we are perfectly free to define the reflection operation on the states by

$$|\pm\rangle_3 \to -|\mp\rangle_3,$$

since the extra minus sign does not affect the transformation property of the expectation values of any operator.

In the classical theory, where the only allowed states were $|\pm\rangle_3$, we were forced to conclude that reflection invariance forced the energies of the two states to be exactly equal to each other. Here, we see that these two eigenstates of the Hamiltonian can have different energies, in a way that does not violate reflection symmetry. So, we have a new prediction: there should be a small splitting, $2\hbar\omega$ between the classically degenerate energy levels of the ammonia molecule, and this splitting should be related to the frequency of oscillation of the probability distribution for the electric dipole moment.

We can test this prediction by exposing the molecule to light. One of the most characteristic and puzzling features of atomic systems from a classical point of view was the fact that they absorb and emit light at discrete characteristic frequencies. The historical origin of quantum theory was Planck's work on the blackbody spectrum. His resolution of the classical paradox was that light of a given frequency carries discrete amounts of energy, given by the formula $E = \hbar\omega$. This relation was used by Einstein to understand the photoelectric effect, and by Compton to understand the scattering of light by electrons. And we indeed find that ammonia molecules absorb and emit light of a fixed (microwave) frequency, much lower than most other characteristic frequencies of atoms and molecules. Later, we will understand the small size of this energy difference in terms of the concept of quantum mechanical tunneling.

We said earlier that it was *tempting* to identify $(\hbar\times)$ the frequencies with which the eigenstates of the time-independent evolution operator oscillate, as energies. In fact, even if we would never heard of the concept of energy, the general discussion of symmetries in QM tell us that these frequencies are indeed constants of the motion and we would have been led to define a conserved quantity, the frequency operator $H/\hbar$ for any time-independent quantum system.

2.9 SUMMARY

Let us summarize all this by saying it in a different way. The quantifiable data about *any* system that we think of classically, can be organized by making a list of all possible states of n of the system and then listing all functions $f(n)$ on that space of states, which could conceivably correspond to measurable properties of the system. This is mathematically identical to inventing a vector space, whose basis is labeled by the different states, with the functions realized as diagonal matrices. A probability distribution is also a diagonal matrix ρ, with nonnegative matrix elements summing up to one. Operations that change the state of the system are nondiagonal matrices. One can then make the mathematical observation that the formula

$$\text{Tr } f\rho,$$

for computing the expectation value of a function in a given probability distribution, can be generalized by replacing f by any diagonalizable matrix, even if it is not diagonal in the original basis. If the matrix is diagonalizable by a unitary transformation (which is equivalent to the requirement that $[A, A^\dagger] = 0$), then the above formula defines a probability distribution $p(a)$ for A to take on one of its eigenvalues. Even if the original probability density ρ specified a definite value for each of the functions f, the probabilities for diagonalizable matrices which do not commute with the original functions do not give definite values for these variables.

The question of whether this intrinsically uncertain probability theory is relevant to a particular physical system depends on its dynamics. If time development always takes elements of the original basis into other elements of the basis (i.e., if it just permutes the states), then we can just declare that the only measurable operators are those diagonal in that basis and ignore the fact that many other quantities are uncertain. This is the analog of classical mechanics for a system with a finite number of independent projection operators. However, the most general probability conserving operation on states is given by a unitary operator $U(t, t_0)$, which can continually change the basis. Future predictions of the history of a general quantity will be probabilistic, because the time derivative of some variable which is definitely known cannot be known with certainty. This is QM. Experiment shows that it is the relevant form of dynamics for systems of atomic scale and smaller. We will learn later that our illusion that macroscopic objects do not suffer from such uncertainties is a consequence of the fact that they are made of a large number $> 10^{20}$ of atomic constituents, and that we only observe coarse grained average properties of the states of these atoms.

2.10   SKIP THIS UNLESS YOU ARE INTERESTED IN PHILOSOPHY

The probability theory defined by QM differs in two important ways from the probability familiar to classical physicists and gamblers. In the latter, we are forced to use probability because we cannot (or at least have not) measured all of the variables needed to make predictions about the future of our system. As a consequence, the probability distributions are not known *a priori*, but must be established by experiments, or by making assumptions like "the coin is not biased, the dice are not loaded," or that the distributions follow certain laws that have been established by observation of large classes of systems. In contrast, *once one has established a definite initial state in QM, the probability distributions for all quantities are calculable with no recourse to observation.* So, in a certain sense, QM takes the uncertainty out of probability theory, but introduces intrinsic uncertainty into our predictions about the future, even for initial conditions measured with infinite precision.

The second difference is the violation of Bayes' rule of conditional probability. Consider two mutually exclusive alternative states A and B of the system at some time, t_0. This means that if A is true, then B is false and vice versa. Then in either a classical or quantum theory of probability, we have

$$P(x = A \text{ or } x = B) = P(x = A) + P(x = B),$$

at time t_0. Bayes' rule assumes that this linearity will persist in time. In a classical theory, this is reasonable, since $\dot{x}$ is a function of x, which is simply to say that it can be definite when x is. So,

$$\dot{P} = \dot{x}\frac{\partial P}{\partial x} + P(x)\dot{V}(x),$$

and similarly for higher time derivatives. Here, V is the volume element in the space of x variables. The second term comes from the fact that the time evolution might not preserve the volume element in the space of x variables, and so the probability *density* must change in order to keep the total probability of being in a fixed region constant. These equations are linear in P. Even if the x variables are in constant interaction with a random medium, we always make the approximation that the medium's effect comes through modifications of the laws of motion of x by friction, random forces, etc., and we obtain a linear evolution equation for P. As a consequence, we can say that if we measure $x = A$ at t_0, then we can, without making a mistake, ignore the fact that the theory predicted that x could have been B, and model the future evolution of the system by saying that x was definitely equal to A. In QM, this rule is simply invalid.[6] The evolution equation *is* linear, but it is not an equation

---

[6] In the Bohmian approach to QM, described in the appendix, one rewrites quantum mechanics in a manner that looks like a probability theory for particle trajectories. However, the evolution equation for the trajectories depends on the probability distribution itself, in a bizarre manner. Bayes' rule is *not* satisfied.

for the probability but for the "Schrödinger wave function," or probability *amplitude*. As we have seen, the rules for probabilities that we sketched above imply that probability is the absolute square of the complex probability amplitude (this is Born's rule). So, it does not evolve linearly and does not in general satisfy the Bayes rule of conditional probabilities. We will see later that the essence of the measurement process in QM is to correlate the microscopic state of an atomic system with a macroscopic variable (a pointer), for which the quantum rules of probability obey Bayes' rule with exquisite accuracy. All of the "paradoxes" associated with "collapse of the wave function" are really just a consequence of the application of Bayes' rule to probabilities for macroscopic objects. The quantum violations of Bayes' rule for macrovariables are exponentially small and cannot be measured, even given a time much longer than the history of the universe.

Note that because the Schrödinger equation

$$i\hbar\frac{\partial}{\partial_t}|\psi(t)\rangle = H|\psi(t)\rangle$$

is linear, we can apply a sort of quantum Bayes' rule to the Schrödinger equation directly. This is what is called *collapse of the wave function upon measurement*. We will see that it only makes sense when the wave function is a linear combination of states with different histories of macroscopic collective coordinates, in which case it implements Bayes' rule for probabilities, and defines what it means for *something to "definitely happen" to a macroscopic variable*. The mathematically precise meaning of that phrase is that QM in general predicts only probabilities,[7] but it can define conditional probabilities for macrovariables (a shorthand phrase we will use from now on to denote "the collective coordinates of macroscopic systems, which are defined so that interference terms between different histories are too small to be measured, even in principle") for which one can define probabilities for conventional histories by using Bayes' rule. The approximate histories in this emergent classical probability theory are called *decoherent histories*, because the phase coherence between parts of the wave function describing different histories is wiped out. However, this use of Bayes' rule is both approximate and evanescent, in the sense that once the last collective variable involved in a decoherent history has gone out of causal contact with a part of the system that has disintegrated into elementary particles, then the correct predictions for future measurements done on those particles must use the original wave function derived from the Schrödinger equation, rather than the one gotten by applying Bayes' rule after observing individual histories of the collective coordinates. That is, we use the approximate Bayes' rule obeyed by collective coordinates to split the wave function into parts that correspond to fixed histories of the collective coordinates. This procedure only makes sense if we can monitor the collective

---

[7] These probabilities can be used in two ways: the Bayesian approach, which uses them to inform human judgment about the validity of theories, and the frequentist approach, which requires us to do the same experiment over and over again, to eliminate as much as possible the reliance on subjective criteria.

coordinates in question. If they are inaccessible, then we must go back to the probabilistic predictions of the original wave function. This is all perfectly reasonable if we think of the wave function only as a device for computing probabilities. It is troublesome if we think of it as a physical object. In Chapter 10, we will describe a thought experiment called Schrödinger's Bomb, which illustrates the question in a quite dramatic fashion.

The foregoing was a lot to swallow, and you should not be surprised if you do not understand it at this point. The examples in Chapter 10 will clarify these difficult ideas. We conclude this chapter by reiterating the important lessons we have learned:

- When we think about vectors in ordinary space, we know that we can refer them to different orthogonal coordinate systems, and that the same vector has different representations as a collection of numbers in different coordinate systems. The same is true of vectors $|v\rangle$ in a general N-dimensional complex vector space. If $|e_n\rangle$ is an orthonormal basis, then the coefficients in $|v\rangle = \sum_n v_n |e_n\rangle$ are the column vector representation of the vector in that basis. Similarly, operators have different matrix forms in different bases. The set of operators diagonal in a given basis $|a_n\rangle$ have the form

$$A = \sum_n f(a_n)|a_n\rangle\langle a_n|,$$

  since $|a_n\rangle\langle a_n|$ is the projection operator on $|a_n\rangle$.

- Given a vector $|s\rangle$ of length 1, the positive numbers $P(a_n) = |\langle a_n|s\rangle|^2$ have the mathematical properties associated with a probability distribution for the operator A to take on the eigenvalue a_n. We call vectors of length 1, *states* of the system, and describe these numbers as "the probability of being in the state $|a_n\rangle$ when the system is in the state $|s\rangle$." QM is essentially the statement that we take these probability distributions as making predictions about actual physical systems. The key feature of these distributions is that not all variables can be certain at the same time, because not all operators are diagonal in the same basis. When the equations of motion relate operators with mutual uncertainty, then we are doing QM.

- We can study QM with either time-dependent states and time-independent operators (Schrödinger picture) or time-dependent operators and time-independent states (Heisenberg picture). The physical predictions are the time dependence of expectation values of operators.

The way this math applies to physics can be understood from the following table, which contrasts the classical and quantum treatments of a system with N states

	Classical	**Quantum**
Observables	Operators Diagonal in Basis $|c_i\rangle$	Operators Diagonal in ANY Orthonormal Basis
Pure States	Projection Operators on $|c_i\rangle$	Projection Operators on ANY Unit Vector
Mixed States	Diagonal Matrices $\mathrm{Tr}\,(\rho) = 1$ w/ nonnegative eigenvalues	Matrices $\mathrm{Tr}\,(\rho) = 1$ w/ nonnegative eigenvalues
Expectation Values	$\mathrm{Tr}\,(\rho A)$	$\mathrm{Tr}\,(\rho A)$
Time Evolution	Sequence of Permutations of Basis $|c_i\rangle$	Discrete or Continuous Set of Unitary Operators $U(t)$

In both frameworks, we can allow the time dependence to be carried by the states *or* the observables. It is often easiest to think in the Schrödinger picture, where the vectors corresponding to pure states are simply evolved into $|c_i\rangle \to S(t_n)|c_i\rangle$ or $|v(0)\rangle \to U(t)|v(0)\rangle$. Both frameworks contain observables whose value cannot be certain at all times. These are diagonalizable operators whose eigenstates are not (at any given time) the unit vectors $|c_i(t_n)\rangle$ or $|v(t)\rangle$. However, in the classical case, we can simply ignore these operators and declare they are not observable.

2.11   FOUR STATE SYSTEMS

In order to understand the formalism, we will now study the case of a system with four states. This is quite complicated, but simple enough to actually allow for an exact analytic solution. Finding the eigenvalues of a matrix requires us to solve the equation

$$\det\,(M - e) = 0,$$

which is a polynomial equation of order N, where N is the dimension of the Hilbert space. It is only possible to give analytic formulae for the solutions of a polynomial when $N \leq 4$. The four-dimensional case is also the first case in which dimension is not a prime number. In all such composite dimensions one can construct a description of the space as what is called a *tensor product* of two smaller spaces whose dimensions involve fewer prime factors. We will discuss this concept in more detail in Chapter 6, and use the four state system to understand the nonintuitive nature of quantum probability theory in Chapter 10.

To have a physical picture of a four state system, consider two different[8] molecules, each of which has, like ammonia, two low lying energy states. The detectable quantities of the two individual molecules are two commuting copies of the Pauli matrix algebra.

$$\sigma_a \sigma_b = \delta_{ab} + i\epsilon_{abc}\sigma_c. \tag{2.53}$$

$$\tau_a \tau_b = \delta_{ab} + i\epsilon_{abc}\tau_c. \tag{2.54}$$

$$\sigma_a \tau_b = \tau_b \sigma_a \tag{2.55}$$

Again, we are using the summation convention over the three valued indices we denote by $a, b, c, d \ldots$

Now consider the following 4×4 matrices, which act in the combined Hilbert space of the two molecule system. If the interaction energies between the two molecules are of the same size as the splittings between their two lowest lying states, and thus much smaller than the energy gap to higher molecular states, we can approximate the two molecule system by this four state system. The matrices we consider are $1, \sigma_a, \tau_a, \sigma_a \tau_b$, and there are $1 + 3 + 3 + 9 = 16$ of them. 1 is the four by four unit matrix. We think of it as a *tensor product* of two 2×2 unit matrices, $1_4 = 1_2 \otimes 1_2$ defined in the following way: for each matrix element of the matrix on the right of the tensor product symbol, write the two by two matrix on the left of the tensor product symbol, multiplied by that matrix element. In symbols,

$$\begin{pmatrix} a & b \\ c & d \end{pmatrix} \otimes \begin{pmatrix} e & f \\ g & h \end{pmatrix} = \begin{pmatrix} ae & af & be & bf \\ ag & ah & bg & bh \\ ce & cf & de & df \\ cg & ch & dg & dh \end{pmatrix}. \tag{2.56}$$

Exercise 2.15 Show that

$$A \otimes B = B \otimes A$$

for any pair of 2×2 matrices. Note that if you write tensor product matrices in block form, so that

$$A \otimes B = \begin{pmatrix} A_{11}B & A_{12}B \\ A_{21}B & A_{22}B \end{pmatrix},$$

then the two different orders use a different labeling for the basis of the tensor product space, so the two orders are only conjugate by a permutation matrix. Avoid this by considering the action on the basis $|i, I\rangle$.

All 16 of the matrices we wrote above should be thought of in this way. That is $\sigma_a^{4 \times 4} = \sigma_a \otimes 1_2$, etc. Using this fact you should be able to do

---

[8] We make them different in order to avoid the constraints of *identical particle statistics*, which we will discuss in Chapter 5.

Exercise 2.16 Show that the 16 matrices formed by tensor products of 2×2 matrices are linearly independent. Hint: Consider all of the matrices in the basis where τ_3 is diagonal. Then the matrices $(1_2 \mathrm{or} \sigma_a) \otimes (\tau_\pm, \mathrm{or}(1 \pm \tau_3)$ each have a different 2×2 block of their 4×4 matrix not equal to zero. The result then reduces to the 2×2 case.

This result means that we can write any 4×4 matrix as a linear combination of these 16 matrices. Note that all of the matrices are Hermitian, so any Hermitian matrix, which could be the Hamiltonian for the interacting molecules, is a real linear combination of these 16. If the system is time translation invariant, the coefficients will be independent of time. The unit matrix term simply shifts the overall zero of energy, so it has no physical consequences. We will save the case of the most general Hamiltonian for the last exercise, and concentrate on easily soluble examples of the form

$$H = m_a \sigma_a \otimes 1_2 + 1_2 \otimes n_a \tau_a + m_a n_b / E \sigma_a \otimes \tau_b. \tag{2.57}$$

Exercise 2.17 Write the Hamiltonian above explicitly, as a 4×4 matrix.

To solve for the eigenvalues of this Hamiltonian, we note that the operators $Q_1 \equiv m_a \sigma_a \otimes 1_2$ and $Q_2 \equiv 1_2 \otimes n_a \tau_a$, both commute with H, as well as with each other. If several diagonalizable operators commute with each other, then they are all diagonal in the same basis. Equivalently, the eigenvectors of H can all be chosen to be eigenvectors of both Q_i. In fact, the eigenvalue of H is completely determined by the simultaneous eigenvalues of $Q_{1,2}$. That is

$$h = q_1 + q_2 + q_1 q_2 / E. \tag{2.58}$$

The eigenvalues q_i have dimensions of energy, and the energy scale E is put in so that the same is true for the eigenvalues of the Hamiltonian. We can label the states by the eigenvalues of $Q_{1,2}$. Those eigenvalues are determined by the equations

$$Q_1^2 = m_a m_a 1_4, \tag{2.59}$$

$$Q_2^2 = n_a n_a 1_4. \tag{2.60}$$

This tells us that each of these operators has two degenerate states with eigenvalues $q_1 \pm \sqrt{m^2}$ for Q_1 and $q_2 \pm \sqrt{n^2}$ for Q_2. The eigenstates can be computed using your knowledge of the two state problem. You can call them

$$|q_1, q_2\rangle \equiv |q_i\rangle.$$

Notice the labeling of the states by a pair of two valued indices, rather than a single four valued index. This is a very general property of tensor product Hilbert spaces. A choice of basis in each of the two spaces involved in the product determines a basis in the full space.

Exercise 2.18 Find the eigenvectors $|\pm, \pm\rangle$ as four-dimensional column vectors, in the basis where $1_2 \otimes \tau_3$ and $\sigma_3 \otimes 1_2$ are diagonal.

The physical significance of the Hamiltonian eigenvalues is two fold. First of all, they give us the four possible values of the energy for the two molecule system. In the real world, the molecules can make transitions between these states by emitting light quanta (photons) and the energies (and thus frequencies/colors) of the emitted photons will be differences between these values. The second role of the eigenenergies is to determine the time evolution of a general initial state

$$|s(0)\rangle = \sum_{q_i} v(q_i)|q_i\rangle. \tag{2.61}$$

It evolves to

$$|s(t)\rangle = \sum_{q_i} v(q_i) e^{-i \frac{q_1+q_2+q_1 q_2/E}{\hbar} t} |q_i\rangle. \tag{2.62}$$

The time-dependent density matrix is

$$\rho(t) \equiv |s(t)\rangle\langle s(t)| = \sum_{q_i, p_j} e^{-i \frac{q_1+q_2+q_1 q_2/E - p_1 - p_2 - p_1 p_2/E}{\hbar} t} |q_1\, q_2\rangle\langle p_1, p_2|. \tag{2.63}$$

In this equation, q_1 and p_1 are independently summed over the values $\pm\sqrt{n_a n_a}$, while q_2 and p_2 are independently summed over $\pm\sqrt{m_a m_a}$.

Exercise 2.19 Write out the density operator $\rho(t)$ explicitly as a 4×4 matrix in the basis where $Q_{1,2}$ are diagonal. Now write the same density operator in the basis where τ_3 and σ_3 are diagonal, using your knowledge of the 2×2 case to find the form of the eigenvectors $|q_i\rangle$ in this basis.

Given the time-dependent density operator $\rho(t)$, we can compute the time-dependent expectation value of any normal operator A as $\text{Tr}(A\rho(t))$. This gives the answer to any prediction possible for this time-dependent system in QM. We could, for example, start in an eigenstate of A, a state where A has a definite value a_1. Then the expectation value formula tells us that the probability to find one of the eigenvalues a_i when we measure A at time t is $\text{Tr}\left(|a_i\rangle\langle a_i|\rho(t)\right)$.

The traditional classical view of time dependence is that a detectable quantity has some value $A(t)$ at all times. The analog of this in QM is to write $\rho(t) = U(t)\rho(0)U^\dagger(t)$, where $U(t) = e^{-i\frac{H}{\hbar}t}$ and note that (by cyclicity of the trace: $\text{Tr}(AB) = \text{Tr}(BA)$)

$$\text{Tr}(A\rho(t)) = \text{Tr}(A(t)\rho(0)), \tag{2.64}$$

where the Heisenberg operator

$$A(t) \equiv U^\dagger(t)AU(t). \tag{2.65}$$

Thus, the time dependence of the detectable's value can equally well be computed as the expectation value of a time-dependent operator in an initial probability distribution. Note that the time dependence of a product of operators, according to the Heisenberg equation, is

$$AB(t) = A(t)B(t), \tag{2.66}$$

which means that we only have to solve the Heisenberg equations for a set of operators, which generate the full operator algebra by matrix addition and multiplication.

Exercise 2.20 For our four state system, define operators $\psi_{1,2}$ by insisting that

$$\psi_1| - \sqrt{n^2}, q_2\rangle = 0,$$

$$\psi_1|\sqrt{n^2}, q_2\rangle = |-\sqrt{n^2}, q_2\rangle.$$

ψ_2 performs the analogous lowering operation on the q_2 eigenvalue. Prove that $\psi_{1,2}$ together with their Hermitian conjugates, generate the operator algebra. Find the explicit matrix forms of these operators in the basis where Q_i are diagonal.

Exercise 2.21 Write and solve the Heisenberg equations of motion, which express time derivatives of the ψ_1 in terms of functions of themselves and their conjugates.

2.12 FURTHER EXERCISES

2.4 Evaluate the matrix σ_2 defined by the matrix equation $\sigma_1\sigma_3 = -i\sigma_2$ using explicit matrix multiplication. Show that $\sigma_2^2 = 1$ follows both from the algebraic definition and the explicit matrix form.

2.5 Using standard probability theory, evaluate the expectation value of a general polynomial $P(E, D)$ of the energy and dipole moment of the ammonia molecule. Show that this is equivalent to the formula $\text{Tr}P(E, D)\rho$.

2.6 The three-dimensional Levi-Civita symbol, or ϵ symbol is denoted ϵ_{abc}. $\epsilon_{123} \equiv 1$ and for any other permutation of the three indices it is ± 1 according to the evenness or oddness of the permutation of 123. Show that

$$\epsilon_{abc}\epsilon_{cde} = \delta_{ad}\delta_{be} - \delta_{ae}\delta_{bd}.$$

Use this equation to evaluate $\nabla \times (\nabla \times \mathbf{V})$, for a vector function $\mathbf{V}(\mathbf{x})$.

2.7 Show that the trace of a matrix is independent of the basis used to compute the trace. First show that the diagonal matrix element of A in some orthonormal basis is equal to $\langle e_i|A|e_i\rangle$, and then use the fact that $1 = \sum_i |e_i\rangle\langle e_i|$ for any choice of basis. Also, prove that $\text{Tr } AB = \text{Tr } BA$ for any two matrices.

2.8 Consider a matrix A in an n-dimensional complex vector space and a density matrix $\rho = \sum_i p_i |e_i\rangle\langle e_i|$, where $\langle e_i|e_j\rangle = \delta_{ij}$, $p_i \geq 0$ and $\sum p_i = 1$. Assume that A has an orthonormal basis of eigenstates $A|a_k\rangle = a_k|a_k\rangle$, $\langle a_l|a_k\rangle = \delta_{kl}$. Evaluate $\mathrm{Tr}\,(A\rho) = \mathrm{Tr}\,(\rho A) = \sum_k \langle a_k|\rho A|a_k\rangle \equiv \sum_k a_k P(a_k)$. Show that the numbers $P(a_k)$ are nonnegative and sum to one. That is, they satisfy the rules for a probability distribution for A to take on one of its eigenvalues.

2.9 According to Born's rule, if the ammonia molecule is in one of the states $|\pm\rangle_3$, the probabilities for the operator $n_a\sigma_a$ to take on the values $\pm\sqrt{n_a n_a}$ are given by the absolute squares of the scalar products of these states with the corresponding normalized eigenvectors of the operator. Find these normalized eigenvectors as linear combinations of $|\pm\rangle_3$ and compute the probabilities.

2.10 Prove that the most general 2×2 matrix diagonalizable by a unitary transformation has the form $A + B n_a\sigma_a$, where A and B are complex numbers and n_a is a real three vector. First prove that any matrix M that is diagonalizable by a unitary must satisfy $[M, M^\dagger] = 0$. In proving this you will have to argue that *every* matrix can be written as a linear combination of the unit matrix and the three Pauli matrices. Then use the multiplication law of Pauli matrices to show that only the indicated form satisfies this commutator equation. To show that any matrix of this form can indeed be diagonalized, find the eigenstates of $n_a\sigma_a$ and show that they are orthonormal. The basis formed by these eigenstates is thus related to the basis where σ_3 is diagonal by a unitary transformation (we will see this quite generally in Chapter 6, but you can try to prove it in the two-dimensional case if you would like). Finally show that $A + B n_a\sigma_a$ is diagonal in this same basis.

2.11 Prove that every operator of the form $e^{i\alpha}e^{in_a\sigma_a}$, where n_a is a real 3-vector, and α is real, is unitary. One way to do this is to evaluate the exponential matrix explicitly using the algebra of the Pauli matrices. You should find a diagonalizable matrix of the form studied in Exercise 2.10. Show that the eigenvalues are pure phases and argue that this implies that matrix is unitary.

2.12 Generalize 2.10 to n dimensions, in the following way. Argue that the commutator of two Hermitian matrices is antihermitian, which means that it is i times a Hermitian matrix. If λ_a is a basis for the space of all hermitian $n \times n$ matrices (show that there are n^2 independent elements in this basis), the previous sentence means that

$$[\lambda_a, \lambda_b] = if_{abc}\lambda_c.$$

It is obvious that f_{abc} is antisymmetric in its first two indices. Multiply the equation by λ_d and take the trace to show that it is in fact totally antisymmetric. Use this to prove a statement analogous to the first step in Exercise 2.9, which involves two complex constants and a real vector of dimension $n^2 - 1$.

2.13 Show in general dimension that e^{iH}, where H is a hermitian matrix, is unitary. Assume that H is diagonalizable by a unitary transformation (sketch of the proof in Chapter 6).

2.14 Show that for any matrix M diagonalizable by a unitary transformation

$$\det M = e^{\mathrm{tr}\ln M}.$$

2.22 Consider a general 4×4 Hermitian matrix $H_{ij} = H_{ji}^*$, in the basis where τ_3 and σ_3 are diagonal. Use Mathematica, Maple or Sage to find the eigenvalues and eigenvectors of the matrix. Write an explicit form for the time-dependent density matrix corresponding to this Hamiltonian, starting from a density matrix diagonal in the original basis. Evaluate the time-dependent expectation values of the projection operators on each of the common τ_3 and σ_3 eigenstates.

2.23 Apply the general solution of Exercise 2.22 to the special Hamiltonian studied in the text. At what point in the analysis does one see evidence for the conservation laws of $Q_{1,2}$?

CHAPTER 3

Quantum Mechanics of a Single Particle in One-Dimensional Space I

3.1 INTRODUCTION

In this chapter, we will derive the quantum mechanics (QM) of a free particle moving on an infinite line, from invariance properties. This will be another example of the quantum version of Noether's theorem. We will show that translation invariance implies the existence of an operator P, which commutes with the Hamiltonian, and makes infinitesimal changes in the value of the coordinate x of the particle on the line. We will derive the spectrum of P and show that the Hamiltonian must simply be a function of it. Since the system is also invariant under space reflection ($x \to -x$) the expansion of $H(P)$ in powers of P/m must begin as $\frac{P^2}{2m}$. That formula suggests that we identify the conserved quantity P with the classical quantity called momentum. It leads to a system invariant under Galilean boosts to frames moving with constant velocity. The simplest description of a particle moving under the influence of external forces is to add a term $V(X)$ to the Hamiltonian, which breaks translation invariance.

We will also show that these considerations lead to the fundamental Heisenberg commutation relation

$$[X, P] = i\hbar, \tag{3.1}$$

which tells us that position and momentum cannot simultaneously have definite values. This implies that there are no probabilities for histories of particle motion. We will briefly discuss an alternative Hilbert space treatment of free particle motion, which does allow for the notion of particle histories. In that "quantum theory of classical mechanics," momentum and position commute and can take on definite values simultaneously. Only experiment can decide which of these two theories to use for a given system. Experiment confirms that the quantum, rather than the classical description of particles, is the one relevant to real atoms and molecules, as well as more elementary particles.

3.2 TRANSLATION INVARIANCE AND FREE PARTICLE MOTION

3.2.1 No Experimental Definition of Infinity

The real line is infinitely long and contains infinitesimally small intervals. Both of these introduce mathematical complications. However, as physicists, we know that we will never be able to know for sure if space is infinite, or infinitesimally divisible. Mathematicians know that the definition of the infinite real line can be thought of by taking two kinds of limits of a clock with a finite number of "minutes." We can first take the limit where the minutes become seconds, microseconds, picoseconds . . . and then the limit where the circumference of the clock goes from 1 to 10 m to 100 million km . . . If you now ask an experimental physicist to test the mathematician's notion of infinity, she will tell you that she cannot. The mathematician defines an infinite continuous line by studying the behavior of a finite discrete set of points, and taking an imaginary limit in which the number of points gets larger and larger and the distances between them smaller and smaller. The physicist, with finite resources and a finite amount of time to do measurements can only verify the mathematical properties of the infinite line with limited precision. In other words, she cannot tell the difference between an infinite continuous line and a finite and/or discrete one, by discovering that the line is not really infinite or continuous. For any given experiment, there will be a finite discrete model of the line, which fits the data as well as the infinite continuous one.

It is useful to think of the infinite continuous line as the limit of a clock, because it illustrates some very general basic features of QM. Look at Figure 3.1, which illustrates a sequence of clocks of radius R, divided up into N equal "minutes," for a range of N. We are discussing motion on the rim of the clock. One of the detectable quantities associated with motion on the rim of the clock is its position. We can describe this as a complex number $z_n = Re^{\frac{2\pi i n}{N}}$ where n ranges from zero to $N - 1$. A more abstract way of thinking of the different values of position is that they correspond to functions of n defined by $e_m(n) = \delta_{mn}$.

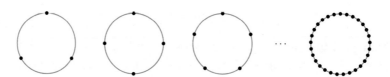

Figure 3.1 Clocks with N minutes.

You should think of this equation as defining the n-th component of a vector $|e_m\rangle$. The set of all complex valued functions of n is a complex vector space, and the functions $e_m(n)$ are a basis for that space. That is,

$$f(n) = \sum_m f(m)e_m(n). \tag{3.2}$$

There is a position operator, U, in this vector space defined by

$$RUf(n) = z_n f(n), \tag{3.3}$$

and you should convince yourself that it is a linear operator. You will study all the properties of this operator, and the translation operator we are about to introduce, in the Exercises. For the moment, we are just using them to get some intuitions. We are thinking about the position operator as defining a point on a circle of radius R, centered at the origin of the complex plane. The functions $e_m(n)$ are eigenstates of this operator, with eigenvalue z_m. As N gets large, the dimension of the space increases, and the numbers z_n start to fill up the rim of the clock densely.

We now want to think about an operator V, which moves you from one point on the clock to the one with next larger n. That is

$$Ve_m(n) = e_{m+1}(n) = \delta_{m+1,n}. \tag{3.4}$$

On a general function $f(n)$, we have

$$Vf(n) = \sum f(m)Ve_m(n) = f(n-1). \tag{3.5}$$

Obviously, if we apply V N times, we get back to where we started, so $V^N = 1$. This means that the eigenvalues of V are $e^{\frac{2\pi i k}{N}}$, $k = 0 \ldots N-1$, and they form a "dual clock." The eigenvectors corresponding to these eigenvalues satisfy

$$f_k(n-1) = e^{\frac{2\pi i k}{N}} f(n). \tag{3.6}$$

The solution to this is

$$f_k(n) = \frac{1}{\sqrt{N}} e^{-\frac{2\pi i k}{N}}. \tag{3.7}$$

We have chosen the normalization factor so that

$$\sum_n |f_k(n)|^2 = 1. \tag{3.8}$$

The eigenstates of V are discrete analogs of standing waves on the circle and k is what we would call the wave number of the wave. Note that as N gets large, the spectrum of values of k gets infinitely large, just like the spectrum of values of n. However, if we want to think of $2\pi\frac{n}{N}$ as a continuous variable, the values of k remain discrete. Thus, a continuous circle corresponds to a discrete infinite lattice of wave numbers. For finite N, both the clock rim and the space of wavenumbers are discrete finite sets. As N goes to infinity, they both become infinite, but in different ways. The circle remains compact but becomes continuous, while wave number space becomes an infinite lattice.

We can also contemplate the opposite limit, in which space becomes an infinite regular lattice, while wave number space is a circle, which in this context is called the Brillouin zone. This limit has applications to the study of solids which have a periodic crystal structure. We will discuss this briefly in Chapter 11. See Figure 3.2 for an illustration of how the two complementary limits of continuous circle and infinite lattice are related in position and wave number space.

For finite N, it is easy to see that the set of all wave number eigenfunctions is a basis for the space. The easiest way to do this is to compute the scalar product of two different wave numbers. You should do this for yourself before reading the next few lines. The scalar product is

$$\sum_n f_k^*(n) f_l(n) = \frac{1}{N} \sum_n e^{2\pi i \frac{(k-l)n}{N}}. \tag{3.9}$$

For $k = l$ this is 1. For $k \neq l$ it is the sum of all of the Nth roots of unity, which equals zero. So we have N orthonormal vectors in an N-dimensional space, and these vectors form a basis. As we go through the next two chapters you will realize that this little computation is the basis for the proof of Fourier and Plancherel's theorems about writing any square integrable

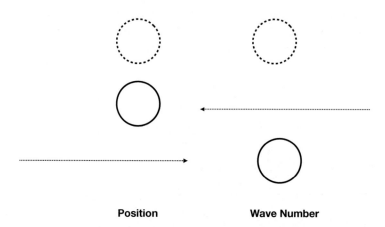

Position   **Wave Number**

Figure 3.2 The relation between continuous/discrete circles in position and wave number. Arrowheads indicate the line extends indefinitely.

function as a Fourier series or Fourier integral. It will also be the key to understanding what momentum is in QM and to Heisenberg's Uncertainty Relation for position and momentum.

Now we return to the continuous circle, which has circumference $2\pi R$. We can parameterize it by a variable x which goes from $-\pi R$ to πR. The geometry of the circle is invariant under translations of x if we make the identification $x \equiv x + 2\pi Rm$, where m is an integer. The Hilbert space of the theory becomes the space of periodic functions of x satisfying

$$\int_{-\pi R}^{\pi R} |f|^2 < \infty. \tag{3.10}$$

The variable x has dimensions of length, so a function e^{ikx} will be periodic if $k = n/R$, where n is an integer. These functions just get multiplied by a phase e^{ika} when we translate $x \to x + a$, so they are eigenfunctions of the translation operator $V(a)$, which is defined by

$$V(a)f(x) \equiv f(x + a). \tag{3.11}$$

This one parameter set of operators is the limit of the discrete set V^n in the $N \to \infty$ limit.

So far we have not said anything about dynamics. We can make some very interesting general conclusions about free particle motion on discrete or continuous circles by making two *assumptions*:

- We do not need any more variables to specify the states of our particle than we have already described.

- The dynamics of a *free* particle is invariant under translations around the circle.

The second assumption has some immediate mathematical consequences. It says that the Hamiltonian matrix

$$\langle e_n|H|e_m \rangle \tag{3.12}$$

is the same as the matrix in the translated basis $V|e_n\rangle$. That is

$$\langle e_n|V^\dagger HV|e_m \rangle = \langle e_n|H|e_m \rangle. \tag{3.13}$$

Since this is true for every matrix element, it implies the operator identity

$$V^\dagger HV = H, \tag{3.14}$$

or

$$HV = VH. \tag{3.15}$$

Since H and V commute, they are diagonalizable in the same basis. On the continuous circle, the same arguments go through with the operator V replaced by $V(a)$.

What this means is that in the basis where $V(a)$ is an infinite diagonal matrix with matrix elements $e^{i\frac{an}{R}}$, with n running over all integers, H is a diagonal matrix with matrix elements $H(n)$. Remarkably, this means that in any quantum theory of free particle motion, based on these two assumptions, *the allowed energies of a free particle on a continuous circle are discrete*. This is the famous quantization of continuous classical variables, which gives QM its name. We will see later that it occurs any time the particle is confined in a fixed region of space, or even when we insist that the probability of finding the particle at infinity goes to zero rapidly enough for the function representing its state to actually be normalizable.

If we take the model of an infinite discrete line instead of a continuous circle, we can see that quantization of energy no longer occurs. Notice that this system is exactly the same Hilbert space as that of a particle on a circle. We simply think of the circle as wave number eigenstates and the discrete line as positions. It is clear then that the translation operator from one point to the other on the line has a continuous spectrum, so the argument above will say the same for the energy operator.

We learn a general lesson: quantization of energy has to do with motions that are bounded in space. In QM, where we can only talk about probabilities, this is equivalent to saying that the probability of finding the particle at any finite point in space, goes to zero sufficiently fast that the total probability is one. Systems in infinite space can also have discrete energy eigenstates if they have *bound states*. The states with continuous energy eigenvalues, generally look like free particle states at infinity and one can investigate what happens when one throws a free particle into finite regions, where it interacts. This is called a *scattering problem* and the corresponding eigenstates are called *scattering states*. We will learn more about them in Chapters 4, 8, and 16.

In the previous Chapter, we argued that a "classical" theory of the two states of the ammonia molecule was a special case of the quantum theory, in which time was taken discrete and the time translation operator was always the matrix σ_1. One might have expected the same would be true for a particle moving on a circle, but it is not. The classical physics of a particle moving on a circle has a continuous energy spectrum. The reason is that it does not satisfy the first assumption above. In classical physics, the velocity or momentum is an independent variable at fixed time, which determined the future history of the system. So a state is characterized by two parameters x, p, rather than only x. We will see below that there is a particular form of QM on the Hilbert space of functions of these two variables, which reduces to the standard formulation of classical mechanics. We are now ready to start a more formal treatment of the QM of particles in one dimension.

3.2.2   Free Particles on an Infinite Continuous Line

Despite the issues of principle that we have pointed out in connection with the operational definition of the notion of infinity, the use of continuous and infinite sets is useful to physicists because it simplifies mathematical calculations and throws away nonuniversal behaviors of discrete finite systems, which might not show up in a given set of experiments. In this section, we will treat some infinities cavalierly, referring to Chapter 4 for more careful definitions.

We start by thinking about a particle moving *freely* on the real line. The salient fact about this system is that it is invariant under translations of the position x of the particle. Among the detectable quantities in this system are functions, $A(x)$ of its position. For any state $|\psi\rangle$, the probability amplitude to find all of these functions evaluated at a particular value $x = y$ is a complex number $\psi(y)$ called *the wave function* of the particle. The continuous analog of probabilities summing to one is

$$\int_{-\infty}^{\infty} dx |\psi(x)|^2 = 1.$$

We can think of the meaning of $\psi(x)$ in Dirac notation as $\psi(x) = \langle x|\psi\rangle$, where $|x\rangle$ is "the state where the particle is definitely at the point x." If the real line consisted of discrete points, this would be a perfectly sensible normalized state, but because we have taken the imaginary limit of a continuous line, it turns out that its squared norm $\langle x|x\rangle$ is infinite. Nonetheless, the formal expression

$$P(a, \epsilon) = \int_{a-\epsilon}^{a+\epsilon} dx |x\rangle\langle x|, \tag{3.16}$$

defines a perfectly sensible operator in the Hilbert space of a free particle on a line. It is an operator whose value is one if the particle is between the two points $a \pm \epsilon$ and zero otherwise. Its expectation value in a general state is

$$\langle \psi|P(a, \epsilon)|\psi\rangle = \int_{a-\epsilon}^{a+\epsilon} dx \, |\psi(x)|^2. \tag{3.17}$$

To see this, consider the discrete line for a moment. A particle hopping from point to point has a space of states which is infinite dimensional. A particular orthonormal basis consists of precisely those states $|n\rangle$ $(-\infty \leq n \leq \infty)$ where the particle is definitely at the n-th point. A general normalized state will be characterized by its projections $\langle n|\psi\rangle$ on this basis. The projections satisfy

$$\sum_n |\langle n|\psi\rangle|^2 = 1. \tag{3.18}$$

The operator

$$P_M(b) = \sum_{n=b-M}^{n=b+M} |n\rangle\langle n|, \tag{3.19}$$

is a projection operator which gives zero on a state $|m\rangle$ unless $b - M \leq m \leq b + M$. Now imagine taking the limit as the spacing s between the discrete points goes to zero, and that in this limit we replace ns by a continuous variable x. Then, recalling your first year calculus course, you will realize that sums get replaced by integrals. The projector $P_M(b)$ converges to $P(a = bs, \epsilon = Ms)$, and the normalization condition is just $\int_{-\infty}^{\infty} dx |\psi(x)|^2 = 1$. The "states" $|y\rangle$ are not normalizable vectors in our Hilbert space, because their wave function is the Dirac delta function $\delta(x - y)$. We will explain this in more detail in Chapters 4 and 6. A quick introduction to these delta function normalizable states can be found in Appendices B and F.

For a *free* particle in infinite space, translation invariance should be a property of the energy operator H. That is, a translated state, with wave function $\psi(x + a)$, should have the same expectation values of any function of the energy operator, as the state with wave function $\psi(x)$. Furthermore, as we discussed in the previous chapter translation should be represented by an operator $V(a)$ on $|\psi\rangle$, which satisfies

$$V(a)V(b) = V(b)V(a) = V(a + b),$$

as well as $V(0) = 1$. V should also be unitary, in order to conserve the probability interpretation of the translated wave function (see Chapter 6 for an extensive discussion of the necessity for choosing symmetry operations to be unitary, as well as the single exception to this rule).

So we must have $V(a) = e^{iKa}$ (prove this in Exercise 3.1), where K is an operator with real spectrum (the spectrum of an operator is the set of its eigenvalues). K is called *the wave number operator*. It is Hermitian and the translation operators are unitary.

Exercise 3.1 Prove that a one parameter family of unitary operators satisfying $U(a)U(b) = U(b)U(a) = U(a + b)$, as well as $U(0) = 1$, has the form e^{iKa}, where K is an operator with real spectrum. Hint: the operators $U(a)$ are unitary and commute with each other, so can all be diagonalized in the same basis.

The probability distribution, which says that the state of the system is definitely $|\psi\rangle$, is the projection operator $P = |\psi\rangle\langle\psi|$. The translated projection operator is $P(a) = V(a)PV^\dagger(a)$, so the condition that the Hamiltonian have the same expectation value after translation is

$$\text{Tr}\,(HV(a)PV^\dagger(a)) = \text{Tr}\,(V^\dagger(a)HV(a)P). \tag{3.20}$$

This must be true for every a and every projection operator P, and that is only true if

$$[H, K] = 0. \tag{3.21}$$

To see this, note that, for a projection operator on the state $|\psi\rangle$ we have

$$\text{Tr}\,(V^\dagger(a)HV(a)P) = \langle\psi|V^\dagger(a)HV(a)|\psi\rangle. \tag{3.22}$$

So the expectation value of the difference between H and the translated Hamiltonian vanishes in *every* state. If we now apply this rule to a state which is a superposition $|\psi\rangle = |\alpha_1\psi_1 + \alpha_2\rangle|\psi_2\rangle$ and insist that the expectation value be independent of a for every pair of states and arbitrary coefficients, we conclude that

$$\langle\psi_1|V^\dagger(a)HV(a)|\psi_2\rangle = \langle\psi_1|H|\psi_2\rangle, \qquad (3.23)$$

which is the same as the operator identity

$$V^\dagger(a)HV(a) = H.$$

This can only be true for all a, if K commutes with H.

In the previous section, we established that the time evolution of operators is

$$A(t) = e^{\frac{i}{\hbar}Ht}A(0)e^{-\frac{i}{\hbar}Ht}, \qquad (3.24)$$

from which it follows that

$$\dot{A}(t) = \frac{i}{\hbar}[H, A]. \qquad (3.25)$$

This is called the Heisenberg equation of motion and is obeyed by every operator in the Heiseberg picture, whose corresponding Schrödinger picture operator is independent of time.

The Heisenberg equation of motion then implies

$$\frac{dK}{dt} = \frac{i}{\hbar}[H, K] = 0. \qquad (3.26)$$

In other words, K is a *conserved quantity*. This connection between symmetries and conservation laws is one of the most important general principles in theoretical physics. In QM, it follows naturally from the idea that operations on the system are also measurable quantities. A symmetry is, by definition, an operation on the system, which does not change the Hamiltonian. In QM, operations are implemented by unitary transformations, so this is equivalent to saying that there is a group of unitary transformations, which has the same[1] multiplication laws as the group of symmetries. Every element of the group commutes with the Hamiltonian, which, by Heisenberg's equations, tells us that it is a conservation law. Groups that depend on continuous parameters have an infinite number of different elements. However, as we will see, they are all gotten by concatenating infinitesimal symmetry transformations, so if there are a finite number of independent parameters, we only get a finite number of conservation laws.

---

[1] Actually, there is freedom for what is known as a central extension of the group, stemming from the fact that quantum states are only defined up to an overall phase. We will not talk about this in this book, except when discussing particles of half-integral spin.

72 ■ Quantum Mechanics

Let us stop and reiterate that we have just proven the quantum mechanical version of Noether's famous theorem[8] from classical Lagrangian mechanics (see Appendix C). Almost nothing we have said so far distinguishes the group[2] of one-dimensional spatial translations, from any other group of symmetries that depends on a continuous parameter a. Every such group is represented in QM by a one parameter group of unitary operators $U(a) = e^{iaG}$. G is a Hermitian operator called the infinitesimal generator of the symmetry group. Invariance of the expectation value of the Hamiltonian in an arbitrary state, under the symmetry transformations then implies that the Hermitian operator G commutes with the Hamiltonian.

Returning to the example of the translation group, we note that since H and K commute, they are diagonal in the same basis. Since the spectrum of K is nondegenerate (all the different real numbers), this means that H is a function of K. That is, H just has some number in each of the diagonal places where K takes on the real value, k, so its values are labeled by $H(k)$. We are being somewhat cavalier about the idea of matrices with continuous indices at this point, and we will be much more careful about this in Chapter 6.

We will use one more fact about the classical mechanics of free particles, namely that the velocity

$$\dot{X} = \frac{P}{m},$$

is a translation invariant quantity. This follows simply from the fact that it is a limit of a difference of positions, which does not depend on the choice of origin. We conclude that the momentum P commutes with K and so is also a function of K, which commutes with H. The last fact deserves to be written down, because it implies

$$\frac{dP}{dt} = \frac{i}{\hbar}[H, P] = 0, \tag{3.27}$$

which is Newton's equation for free particle motion. This is pretty amazing. We have derived Newton's equation for free particles by just using symmetries. We are not quite done yet.

Now let us examine the action of K on wave functions. The state $e^{iKa}|\psi\rangle$ is supposed to be translated by an amount a. That means that its amplitude to be at the point x is the same as the amplitude of the state $|\psi\rangle$ to be at $x + a$. If we take a to be infinitesimal, this implies that K acts on $\psi(x)$ like $\frac{1}{i}\partial_x$ (we use a partial derivative in this one-dimensional space because we will also want to talk about the time derivative of wave functions). If we think of $\psi(x)$ as "the column vector corresponding to $|\psi\rangle$ in the $|x\rangle$ basis", then this partial derivative representation of K is like a matrix representation of an operator in a finite dimensional Hilbert space.

The operators $V(a)$ have eigen-states which satisfy

$$V(a)|k\rangle = e^{ika}|k\rangle, \tag{3.28}$$

---

[2] If you know nothing about group theory at this point, please take a moment to read the appendix on that subject before going on. We will not use much of the math, but you should get used to the words.

and are also eigenstates of $K, H(K)$ and $P(K)$, with eigenvalues given by $k, H(k)$ and $P(k)$.

Exercise 3.2 Prove that the wave functions of K eigenstates are

$$\psi_k(x) = \frac{1}{\sqrt{2\pi}} e^{ikx}. \tag{3.29}$$

Solution of Exercise 3.2: Using the partial derivative representation of K we have

$$K\langle x|k\rangle = \frac{1}{i}\partial_x\langle x|k\rangle = \langle x|K|\rangle = k\langle x|k\rangle. \tag{3.30}$$

The exponential function is the unique solution to this homogeneous differential equation and we have chosen a conventional normalization for the exponential. You might think that we should have determined the normalization by insisting that the total probability be 1. However, in a wave number eigenstate, the probability density for finding the particle at position x is uniform over the real line. This implies that these states *cannot* be normalized. We can understand this a little better, and relate it to the analogous inability to normalize the states $|x\rangle$ by putting the system on a circle, rather than the real line. Then, the allowed values of k are $\frac{n}{R}$ where n is an integer and R the radius of the circle. The states $|k = \frac{n}{R}\rangle$ *are normalizable* because the range of integration in the formula for the squared norm is finite. The analogy becomes more appealing when we recognize that on the discrete infinite line, where the group of translations consists of discrete shifts by an integer number of lattice spacings, the wave numbers, e^{iks}, appearing in the eigenstates of the discrete translation operator are continuous, but periodic under $k \to k + \frac{2\pi}{s}$. If wave number states live on a circle, position eigenstates are discrete and normalizable. If position states live on a circle, wave numbers are discrete and wave number eigenstates are normalizable.

The relationship between the amplitudes to be in position or wave number eigenstates is now given by

$$\langle k|\psi\rangle = \int_{-\infty}^{\infty} dx\,\langle k|x\rangle\langle x|\psi\rangle = \int_{-\infty}^{\infty} \frac{dx}{\sqrt{2\pi}}\,e^{-ikx}\psi(x). \tag{3.31}$$

This means that the *wave function in wave number space is the Fourier transform of the wave function in position space.* The next few paragraphs will assume you know all about Fourier transforms. The next subsection contains a primer for those readers to whom that assumption does not apply.

Note that mathematics textbooks give a variety of choices of the factors of 2π in the definition of Fourier transform. We have chosen a particular one based on our desire to have the Fourier transform be symmetric between wave number space k and position space x. Indeed, this discussion shows that the Fourier transform is the continuous analog of a change of basis in a finite dimensional vector space, so it should be implemented by a unitary transformation. The inverse of the Fourier transform expresses position space wave functions in terms of wave functions in wave number space

$$\langle x|\psi\rangle = \int_{-\infty}^{\infty} dk \, \langle x|k\rangle\langle k|\psi\rangle = \int_{-\infty}^{\infty} \frac{dk}{\sqrt{2\pi}} e^{ikx}\psi(k). \tag{3.32}$$

We can make wave functions more localized than the wave number eigenstates by taking superpositions

$$\psi(x) = \int_{-\infty}^{\infty} \frac{dk}{\sqrt{2\pi}} f(k) e^{ikx}, \tag{3.33}$$

where

$$\int \frac{dk}{2\pi} |f(k)|^2 = 1.$$

The last equality is called Plancherel's theorem[9], which says that the squared norm of a function and its Fourier transform are the same. We want $\int dx|\psi(x)|^2 = 1$, because that is the total probability for finding the particle anywhere on the line.

In calling these integrals the norm of a wave function, we are using the relation described above between quantum states and their wave functions, and anticipating a definition of the Hilbert space in terms of square integrable functions, which we will give in the next chapter. At this point, the reader may benefit from reviewing the discussion in Chapter 2 of finite dimensional vector spaces, as function spaces on the set of integers from 1 to N. And while we are on the subject of mathematical theorems, Fourier's theorem[10] tells us that we can represent any function on the real line as a Fourier transform, including those corresponding to localized lumps.

Using the general rules of QM, we can now evaluate the time evolution of this general wave function, assuming that the equation above gives its form at $t = 0$. It is

$$\psi(x,t) = \sum_{k} e^{-iH(k)t} f(k) e^{ikx}. \tag{3.34}$$

For a localized wave function, localized near x_0, $f(k)$ can be concentrated near some k_0 but must have lots of other nonzero values. The general rule of thumb is that the more localized the Fourier coefficients $f(k)$ are in k space, the less localized the position space wave function is.

For example, we could try a Gaussian falloff

$$f(k) = Ne^{-(k-k_0)^2}.$$

The normalization factors are determined by $\int_{-\infty}^{\infty} dx \, |\psi(x)|^2 = 1$, and can be easily calculated. For any such function, concentrated around k_0, but containing enough weight for other values that $\psi(x)$ is localized at x_0, then, assuming that $H(k)$ is a smooth function of k near $k = k_0$, we can approximate the time evolution by

$$\psi(x,t) \sim \int dk f(k) e^{-i\frac{H(k_0)}{\hbar}t} e^{-i\frac{H'(k_0)}{\hbar}(k-k_0)t+ikx} = e^{-i(\frac{H(k_0)}{\hbar}-k_0\frac{H'(k_0)}{\hbar})t}\psi(x - \frac{H'(k_0)}{\hbar}t). \quad (3.35)$$

That is, the center of the wave "packet" moves through space according to the equation

$$x(t) = x_0 - \frac{H'(k_0)}{\hbar}t, \quad (3.36)$$

which is the equation of motion for a free particle if

$$\frac{H'(k_0)}{\hbar} = \frac{P(k_0)}{m}. \quad (3.37)$$

If we compare this to the Heisenberg equation of motion

$$\frac{P(K)}{m} = \dot{X} = \frac{i}{\hbar}[H(K), X], \quad (3.38)$$

we learn that

$$i[H(K), X] = \hbar H'(K) = \hbar \frac{P(K)}{m}. \quad (3.39)$$

This is solved by

$$X = i\hbar \frac{\partial}{\partial k}, \quad (3.40)$$

in the basis where K is diagonal.

To fix the function $H(K)$, we recall from our discussion of the ammonia molecule, the principle that frequency is the same as energy divided by $\hbar$. Using the classical formula for energy,

$$E = \frac{P^2}{2m},$$

and the formulae above, we derive the relations

$$P = \hbar K, \quad (3.41)$$

$$H = \frac{\hbar^2 K^2}{2m}. \quad (3.42)$$

Alternatively, we could have derived the same formulae by insisting that the theory of a free particle be invariant under Galilean transformations (see Exercise 3.2). Yet a third way of deriving this formula is to just assume that we are interested in k much smaller than some specified inverse length L_c^{-1}. Then $H(k)$ will have a power series expansion. The term linear in

k vanishes by invariance under reflections, while the term quadratic in k must have dimension of inverse mass. The factor of two may be thought of as a conventional normalization of the definition of mass. Higher order terms involve (by dimensional analysis) a new parameter with dimensions of velocity, which Einstein discovered had to be the velocity of light in vacuum.

We also record the fundamental commutation relation we have derived:

$$[X, P] = i\hbar. \tag{3.43}$$

This says that position and momentum are not compatible observables. Measuring one of them introduces uncertainty in the other. Since both are needed in order to predict the history of particle motion, the histories will be uncertain. Note however that when m is very large, the uncertainty in the *velocity* $\dot{X} = \frac{P}{m}$ is very small, so the history will not be that uncertain. This is *part* of the explanation of why macroscopic objects behave classically. We will explore this in more detail in Chapter 10.

When we discussed the ammonia molecule, we said that there was a choice to be made between having a classical theory of its motion and a quantum theory. Only experiment could tell us which was right. The same is true for free particle motion. In the classical theory, X and P are simultaneously measurable, and you will show in Exercise 3.3 that the Hamiltonian operator, in the basis where both X and P are diagonal, is

$$\mathcal{H} = i\hbar \left(\partial_x E \partial_p - \partial_p E \partial_x \right), \tag{3.44}$$

where $E = \frac{p^2}{2m} + V(x)$ is the classical energy. That is, classical mechanics is QM in a Hilbert space whose vectors are functions of x and p, with Hamiltonian operator $\mathcal{H}$. Note that in this case the Hamiltonian operator is *not* the energy $E(p, q)$, although it does commute with it, so that both quantities are conserved. As you will show in the exercises, we can choose initial states in which both x and p are known with precision, and these evolve into other eigenstates of x and p. More general operators, involving the derivatives w.r.t. x or p are uncertain, but if we agree not to measure them, then the theory makes only precise predictions.

For microscopic particles, experiment shows that our quantum form of the equations for a free particle is the correct one. We will see later that the classical theory is approximately valid for collective coordinates of macroscopic objects.

To summarize: translation invariance of free particle dynamics in one dimension leads, via general principles of QM, to the conclusion that the system has a conserved quantity K which takes on all possible real values. The momentum and energy must be functions of K. Using the classical formula for energy as a function of momentum, or imposing Galilean symmetry, we conclude that $P = \hbar K$, and that $[X, P] = i\hbar$.

3.2.3 Primer on the Fourier Transform and Fourier Series

Given an infinite one-dimensional lattice, we can define a Hilbert space with orthonormal basis vectors $|n\rangle$. A general vector is given by

$$|\psi\rangle = \sum_{n=-\infty}^{\infty} \psi(na)|n\rangle. \tag{3.45}$$

$$\sum_{-\infty}^{\infty} |\psi(na)|^2 < \infty. \tag{3.46}$$

You can think of $\psi(na)$ in two ways: it is both an infinite column vector and a function of the lattice points. These are two different names to describe the same mathematical object.

We can define a linear operator on this space by translation through a single lattice point:

$$V|\psi\rangle = \sum_{n=-\infty}^{\infty} \psi((n+1)a)|n\rangle. \tag{3.47}$$

It has a continuous set of eigenvalues/eigenvectors

$$V|\alpha\rangle = e^{i\alpha}|\alpha\rangle. \tag{3.48}$$

They are given explicitly by

$$|\alpha\rangle = \sqrt{N} \sum_{n=-\infty}^{\infty} e^{in\alpha}|n\rangle, \tag{3.49}$$

where N is a normalization constant. Note that α lives on a circle

$$\alpha \equiv \alpha + 2\pi m. \tag{3.50}$$

Note that the function of na corresponding to this "vector" is

$$\psi_\alpha(na) = \sqrt{N} e^{i\frac{\alpha}{a}na}. \tag{3.51}$$

We put air quotes around the word vector because it is not normalizable

$$\langle\alpha|\alpha\rangle = \sum_n N = \infty. \tag{3.52}$$

More generally:

$$\langle\alpha|\beta\rangle = N \sum_n e^{in(\beta-\alpha)}. \tag{3.53}$$

Given a vector $|\psi\rangle$ we can define a function of α by

$$\psi(\alpha) = \langle\alpha|\psi\rangle = \sum_{n=-\infty}^{\infty} e^{-in\alpha}\langle n|\psi\rangle = \sum_{n=-\infty}^{\infty} e^{-in\alpha}|\psi(na)\rangle. \tag{3.54}$$

Now if we evaluate

$$\delta_{mn} = \langle m|n\rangle = \int_0^{2\pi} d\alpha \, \langle m|\alpha\rangle\langle\alpha|n\rangle = N \int_0^{2\pi} d\alpha \, e^{i(n-m)\alpha} = 2\pi N\delta_{mn}, \tag{3.55}$$

we see that we must choose $N = \frac{1}{\sqrt{2\pi}}$. We can also derive from this (do it!) that

$$\delta(\alpha - \beta) = \sum_{-\infty}^{\infty} e^{in(\alpha-\beta)}. \tag{3.56}$$

You should now think about all of these formulae and realize that you have proved Fourier's theorem: It is all about a unitary transformation mapping the space of square integrable periodic functions $\psi(\alpha)$ into the space of square summable infinite sequences, which can be thought of as the space of functions $\psi(na)$ on an infinite lattice. We use the same letter to denote the function on the lattice and the function on the circle, even though they have different functional forms. The different nature of the argument tells you which one you are talking about.

The Fourier transform is the limiting form of this Fourier series story, when we take the limit of zero lattice spacing, so that $x = na$ becomes a continuous variable.

$$\text{Now } a \to 0 \quad x = na. \tag{3.57}$$

In this limit, the periodicity of α goes away.

$$e^{-in\alpha} \to e^{-ikx}, \quad k = \frac{\alpha}{a}. \tag{3.58}$$

The range of k now becomes infinite, $-\infty < k < \infty$ and the periodicity is gone. Sums over the discrete variable n are replaced by integrals over x in the usual way, but we now have the freedom to rescale x and k keeping kx fixed. In QM we do this to make the Fourier transform look completely symmetrical between x and k.

$$\psi(k) = \int \frac{dx}{\sqrt{2\pi}} e^{-ikx}\psi(x). \tag{3.59}$$

The normalization is determined by Fourier's theorem, which is equivalent to

$$\delta(x - y) = \int \frac{dk}{2\pi} e^{ik(x-y)}. \tag{3.60}$$

You should understand how to prove Fourier's theorem, as well as Plancherel's theorem about the equality of the norms of a function and its Fourier transform, using this equation. In the exercises, you will learn how to do the Fourier transform of a Gaussian. You should also practice by figuring out the Fourier transform of the function $e^{-b^2|x|}$ and of a function which is equal to 1 on the interval $[b, c]$ and zero everywhere else. These last two depend only on knowing how to integrate an exponential function between finite limits.

3.3 LAGRANGIAN AND HAMILTONIAN MECHANICS: A REMINDER OR A PRIMER

In the best of all possible worlds, students would take a course in Lagrangian and Hamiltonian mechanics before taking an advanced undergraduate course in QM. In this book, we will use the actual formalism of analytical mechanics only sparingly, but like notions from group theory, we will use the language of that subject more extensively. What follows is a bare bones introduction to the subject.

The solution to Newton's equations for a free particle is a straight line

$$\mathbf{x}(t) = \mathbf{x_0} + \frac{\mathbf{P}}{m}t. \tag{3.61}$$

The fact that straight lines minimize distance, and Fermat's principle that light travels in such a way as to minimize the time, led 18th century physicists to formulate this equation as a minimum principle. Namely, the path minimizes the integrated kinetic energy

$$0 = \delta \int_{t_0}^{t} ds \; \frac{m}{2}(\frac{d\mathbf{x}}{ds})^2. \tag{3.62}$$

This can be generalized to get Newton's equation in the presence of a force

$$m\frac{d^2\mathbf{x}}{ds^2} = \mathbf{F}, \tag{3.63}$$

if the force is conservative

$$\mathbf{F} = -\nabla V,$$

by writing

$$0 = \delta S \equiv \delta \int_{t_0}^{t} ds \; [m(\frac{d\mathbf{x}}{ds})^2 - V]. \tag{3.64}$$

Both of these results can be derived from Lagrange's more general formula

$$\delta \int_{t_0}^{t} ds \; L(q^i(s), \dot{q^i}(s)) = \int_{t_0}^{t} ds \; [\delta q^i(s)\frac{\partial L}{\partial q^i} + \delta \dot{q^i}(s)\frac{\partial L}{\partial \dot{q^i}}]. \tag{3.65}$$

If the infinitesimal variation of $q^i(s)$ vanishes at the endpoints of the integral, we can integrate by parts to write

$$\delta \int_{t_0}^{t} ds \, L(q^i(s), \dot{q}^i(s)) = \delta \int_{t_0}^{t} ds \, \delta q^i(s) [\frac{\partial L}{\partial q^i} - \frac{d}{ds} \frac{\partial L}{\partial \dot{q}^i}]. \tag{3.66}$$

The variational principle $\delta S = 0$ says that the correct path is determined by insisting that this formula vanish for every differentiable function $\delta q^i(s)$, which vanishes at the endpoints. This leads to *Lagrange's equations*

$$0 = [\frac{\partial L}{\partial q^i} - \frac{d}{dt} \frac{\partial L}{\partial \dot{q}^i}], \tag{3.67}$$

because we can choose $\delta q^i(s)$ to vanish outside of a tiny interval centered on any point in the interval.

Defining the *canonical momentum* $p_i \equiv \frac{\partial L}{\partial \dot{q}^i}$, these equations take the form

$$\frac{\partial L}{\partial q^i} = \frac{dp_i}{dt}. \tag{3.68}$$

The definition of p_i can be viewed as an equation that determines $\dot{q}^i$ in terms of q^i and p_j. This is made more explicit by defining the Legendre transform of L, which replaces it with a function $H(p, q)$, via

$$H(p, q) \equiv \sum \dot{q}^i p_i - L. \tag{3.69}$$

The statement that H depends on $\dot{q}^i$ only through its dependence on p_i is what we have taken as the definition of p_i. On the other hand, the statement that L depends on p_i only via its dependence on $\dot{q}^i$ reads

$$\dot{q}^i = \frac{\partial H}{\partial p_i}. \tag{3.70}$$

The q dependence of L and H is not Legendre transformed, so $\frac{\partial H}{\partial q^i} = -\frac{\partial L}{\partial q^i}$, and we have

$$\frac{dp_i}{dt} = -\frac{\partial H}{\partial q^i}. \tag{3.71}$$

These are Hamilton's form of the equations of classical mechanics. The function L is called the *Lagrangian* and its Legendre transform H, the (classical) Hamiltonian. The integral of L is called the *action*. For a particle moving in a potential in one dimension, the Lagrangian is

$$L = \frac{1}{2} m \dot{x}^2 - V(x), \tag{3.72}$$

the canonical momentum is

$$p = m\dot{x}, \tag{3.73}$$

and the classical Hamiltonian is

$$H = \frac{p^2}{2m} + V(x). \tag{3.74}$$

The Hamiltonian equations of motion have two very interesting structures associated with them. The first is the Poisson bracket, defined for two functions of p and q by

$$[F, G]_{PB} = \frac{\partial F}{\partial q^i} \frac{\partial G}{\partial p_i} - \frac{\partial G}{\partial q^i} \frac{\partial F}{\partial p_i}. \tag{3.75}$$

For the variables p_i and q^j themselves, we find that the Poisson brackets with themselves vanish, while

$$[q^i, p_j]_{PB} = \delta^i_j. \tag{3.76}$$

Furthermore, the Poisson bracket satisfies (Exercise 3.11):

$$[F, G]_{PB} = -[G, F]_{PB},$$

$$[AB, C]_{PB} = A[B, C]_{PB} + [A, C]_{PB},$$

$$[A, [B, C]_{PB}]_{PB} + [C, [A, B]_{PB}]_{PB} + [B, [C, A]_{PB}]_{PB} = 0.$$

These are identical to the algebraic identities satisfied by commutators in QM, and this coincidence led Dirac to propose a general method for "quantizing" a classical system written in Hamiltonian form, by simply replacing the Poisson bracket by the commutator of the canonical variables, by multiplying the Poisson bracket formula by $i\hbar$. For nonlinear functions of the canonical variables, there are ambiguities in how one orders the products of p and q to form quantum detectables from classical functions.[3] This is, in particular, true for the Hamiltonian function, so the Dirac procedure assigns many different quantum theories to the same classical system. This is, as it should be. We will see below that classical behavior arises in certain limits of the rules of QM, and it makes sense that details are lost in taking a limit.

The second general property of these equations is that they can be formulated as a single nonlinear partial differential equation. Consider the classical action, for a classical path that ends at some point $q^i(t)$ at time t. It has the form $S(q(t), t)$. From the definition $S = \int_{t_0}^t L$, we have

$$\frac{dS}{dt} = \frac{\partial S}{\partial q^i} \dot{q}^i + \frac{\partial S}{\partial t} = L = p_i \dot{q}^i - H(p_i, q^j). \tag{3.77}$$

---

[3] The ambiguity is ameliorated, but not removed, by insisting that real functions be mapped into Hermitian operators.

Thus,

$$\frac{\partial S}{\partial t} = -H(\frac{\partial S}{\partial q^i}, q^j). \tag{3.78}$$

This is called the Hamilton–Jacobi equation. A solution of the H–J equation is determined by giving the value of $p_i = \frac{\partial S}{\partial q^i}$ for all q^i, at time t_0. This is the same information as the initial conditions for Hamilton's equations.

For a one-dimensional system, with Hamiltonian $\frac{p^2}{2m} + V(x)$ the H–J equation reads

$$\partial_t S = -[\frac{S'^{\ 2}}{2m} + V(x)]. \tag{3.79}$$

We solve this by making the ansatz $S = -Et + S_0$, which reduces it to

$$S_0'^{\ 2} = 2m(E - V), \tag{3.80}$$

whose solution is

$$S_0 = \int_{x_0}^x dy \ \sqrt{2m(E - V(y))}. \tag{3.81}$$

In this solution, we view x_0 as an initial condition, while x is the position to which the classical trajectory with energy E has gotten in time t. That time is determined by the equation

$$t - t_0 = \int_{x_0}^x \frac{dy}{\dot{y}}. \tag{3.82}$$

The equation

$$\dot{y} = \frac{\partial H}{\partial p} = \frac{p}{m} = \frac{\partial S}{m \partial y} = \sqrt{\frac{2(E - V(y))}{m}}, \tag{3.83}$$

completes the determination of the classical trajectory in terms of the solution of the H–J equation. By a similar, but somewhat more abstract argument, one shows that a complete solution of the H–J equation determines all the classical trajectories for a general problem in analytical mechanics.

The significance of the H–J equation for QM is that it is a partial differential equation in the variables t and q^i, just like the Schrödinger equation. In Chapter 17, we will explore the JWKB approximation, in which the H–J equation is derived as a first order approximation to the Schrödinger equation, analogous to the description of light waves by ray tracing.

There is a philosophical comment about classical mechanics that is relevant to some of the material in Appendix A about interpretations of QM. The H–J equation contains all of classical mechanics in the sense that once we have found its general solution, a particular motion of the system is just a solution of the equations

$$\dot{q}_i = \partial_i S, \tag{3.84}$$

which goes through some initial point $q_0 = q(0)$. So we have a "field on the multidimensional configuration space, which guides the motion of particles in that space." Of course, we never think of classical physics this way, because for any given initial q_0, the values of the field $S(q,t)$ at points away from the unique trajectory are irrelevant. However, this is a consistent interpretation of classical physics, because nothing singular happens if we insist that only one particular trajectory occurs in the real world.

Schrödinger invented his equation by thinking of the H–J equation as the geometrical optics approximation to a wave equation, much as the eikonal approximation to Maxwell's Equations is used to derive the Newton-Fermat picture of light in terms of corpuscles following fixed trajectories. The equation for the phase of the complex Schrödinger wave function looks like the H–J equation, with corrections proportional to powers of $\hbar$. However, in QM, the correction terms become singular in regions where the magnitude of the wave function vanishes, and of course the phase is ill-defined there. Since the situation of a particle traveling only along a fixed trajectory corresponds to a wave function that vanishes almost everywhere at fixed time, the QM extension of the H–J equation does not allow for such a situation.

3.4   FURTHER EXERCISES

3.3   Consider a system with energy $E = \frac{P^2}{2m} + V(X)$. We want to define a quantum theory in which P and X are compatible variables (commute with each other), and satisfy Newton's equations. Show that this is achieved with the Hamiltonian

$$H_{cl} = \pm i\hbar \left[\frac{\partial E}{\partial p} \frac{\partial}{\partial x} - \frac{\partial E}{\partial x} \frac{\partial}{\partial p} \right], \tag{3.85}$$

for one choice of the overall sign. This Hamiltonian is written in the basis where X and P are both diagonal. Show that if we agree to measure only expectation values of functions $f(X, P)$, then these are computed in terms of the square of the wave function

$$\rho(X, P) = \psi^*(X, P)\psi(X, P),$$

which satisfies

$$\frac{\partial \rho}{\partial t} = \pm \left[\frac{\partial E}{\partial X} \dot{X} + \frac{\partial E}{\partial P} \dot{P} \right],$$

where $X(t)$ and $P(t)$ satisfy the usual classical equations of motion. It is *your* job to find the right choice of sign in *both* of these ambiguous equations. You will also find that the choice of sign in the equation for ρ is *opposite* to what you would expect for a function $f(X(t), P(t))$ and you should also explain why this is the *right* choice of sign for a probability distribution. This shows that, as we saw for the ammonia molecule, there is a classical theory of particle motion, which fits into the framework of QM, and one must distinguish between the "classical quantum theory" and the real quantum theory *experimentally.*

a. In the classical quantum theory of Exercise 3.3, we had to agree not to measure the operators $\Pi_{P,X} = \frac{\hbar}{i}\frac{\partial}{\partial X,P}$. Show that the Schrödinger equation for H_{cl} is invariant under the "gauge transformation"

$$\psi(X, P) \to e^{i\theta(X,P)}\psi(X, P),$$

and that the only "gauge invariant operators" in the system are functions $f(X, P)$. This observation is the mathematical excuse for a person who wanted to believe in classical mechanics to disregard the quantum version of classical mechanics. If we introduce the idea that the phase of the wave function is redundant (this is the meaning of the phrase *gauge invariance*), then the quantum formulation of Classical Mechanics is just adding redundant information. Below we will see that there is a completely different way to think about gauge invariance in regular QM, which leads to Maxwell's electrodynamics. What is the analog of this gauge invariance for a two state system?

3.4 Compute the commutator

$$[X, P^n],$$

in two ways. First, use the representation where X is diagonal and $P = \frac{\hbar}{i}\partial_x$ and act with this commutator on a function $\psi(x)$ and use the definition of the commutator as the difference between the action of operators in two different orders. You will need to evaluate $\partial_x^n[x\psi(x)]$ to do this. Now evaluate it by using Leibniz' rule

$$[A, BC] = [A, B]C + B[A, C],$$

repeatedly.

3.5 Find the expectation values of P, P^2, and P^3 in the states described by the (unnormalized) wave functions $e^{ik_0x}(x^2 + a^2)^{-p}$. You can use Mathematica, Maple, or Sage to do the integrals. Remember to divide by the norm of the wave function ($\frac{\langle\psi|A|\psi\rangle}{\langle\psi|\psi\rangle}$ is the formula for the expectation value if you do not normalize your states). How large does p have to be in order for the wave functions to be normalizable?

3.6 Show that if a wave function is real, then the expectation value of P^{2n+1} vanishes for all nonnegative n. Hint: Use integration by parts to write the expectation value as the total derivative of a function that vanishes at infinity.

3.7 Show that the operator defined in the representation where X is diagonal by

$$C(\psi(x)) = \psi^*(x),$$

is *antilinear*

$$C(a\psi + b\phi) = a^*C(\psi) + b^*C(\phi),$$

and satisfies $C^2 = 1$. We will see in Chapter 6 that C is an essential part of the time reversal operation in QM.

3.8 a. Show that the scalar product of $C(\phi)$ with $C(\psi)$ is the complex conjugate of the scalar product of ϕ with ψ.

b. Show that

$$\langle\phi|C(|\psi\rangle)\rangle = \langle\psi|C(|\phi\rangle)\rangle,$$

for any two vectors.

c. Show that the operator $A^C = CAC$, has the same matrix elements between $C(\psi)$ and $C(\phi)$ as $A^\dagger$ has between ϕ and ψ. In order to do this, you will have to show that A^C is an ordinary linear operator, so that you know how to define its Hermitian conjugate. Hint: Insert $1 = C^2$ twice into the expression for the matrix element of A.

d. Show that for the momentum operator $P^C = -P$ and use this to explain the result of Exercise 3.6.

3.9 Use Fourier series to solve the Schrödinger equation for a free particle moving on a circle. Assume that the initial wave function has Fourier coefficients $f(n)$.

3.10 Galilean transformations transform momentum and particle position according to $P \to P + mv$ and $X \to X + vt$, where m is the mass and v the velocity of the Galilean boost transformation. An operator which generates these changes in the sense of

$$\delta O = -i[N, O]\delta v,$$

for infinitesimal $v = \delta v$, is

$$N = \frac{m}{\hbar}[X - \frac{P}{m}t].$$

The commutator of N with the Hamiltonian gives the Heisenberg equations of motion for the Heisenberg operator $N(t)$, which coincides with the Schrödinger picture operator N at $t = 0$. In this case, we must take into account the fact that the Schrödinger picture operator N depends explicitly on time. The correct Heisenberg equation is then

$$\frac{dN(t)}{dt} = \frac{i}{\hbar}[H, N] + \frac{\partial N}{\partial t}, \tag{3.86}$$

where the partial derivative symbol means derivative w.r.t. the explicit time dependence in the Schrödinger picture. Show that the requirement that the equations of motion of X and $P = m\dot{X}$ take the same form in different Galilean reference frames[4] implies that $\frac{dN(t)}{dt} = 0$. Show that, assuming translation invariance, this gives the usual form of the free particle equations and determines the Hamiltonian as a function of P.

3.11 Find the normalization constant N for a Gaussian wave function

$$\psi(x) = N e^{-\frac{(x-x_0)^2}{4\Delta^2}},$$

such that

$$\int dx\ \psi^*(x)\psi(x) = 1.$$

In solving this problem you should encounter the integral

$$\int du\ u^{-1/2} e^{-u},$$

which is Euler's Gamma function $\Gamma(z)$ evaluated at $z = 1/2$. $\Gamma(z)$ satisfies

$$\Gamma(z)\Gamma(1-z) = \frac{\pi}{\sin(\pi z)},$$

which will allow you to evaluate this.

a. Do the Gaussian integral

$$\int dx dy\ e^{-a(x^2+y^2)},$$

in two different ways. First, it is the product of two one-dimensional Gaussians of the type you encountered in Exercise 3.11. However, we can also do it by going to radial coordinates in the x, y plane. Show that the second route leads to an easy derivation of the fact that $\Gamma(1/2) = \sqrt{\pi}$.

3.12 Find the expectation value of the operator X^n in the normalizable wave function of Exercise 3.11. The expectation value can be written as a ratio of two integrals. Evaluate the numerator integral by differentiating the denominator.

3.13 Repeat exercise 3.12 for the operators P^n.

---

[4] You will have to think about what this phrase means in QM.

3.14 In an n-dimensional Hilbert space, define two operators U_n and V_n by the equations

$$U_n^n = V_n^n = 1,$$

$$U_n V_n = V_n U_n e^{\frac{2\pi i}{n}}.$$

In the basis where U_n is diagonal, its eigenvalues are $e^{\frac{2\pi i k}{n}}$. In this basis, U_n is called the *clock operator*. Show that in this basis, V_n is the matrix that "shifts the clock by one tick." Write it explicitly and verify that it satisfies the remaining two equations.

3.15 Show that every $n \times n$ matrix M can be written as

$$M = \sum_{k,l=1}^{n} a_{kl} U_n^k V_n^l.$$

3.16 In the limit $n \to \infty$, the spectrum of U_∞ becomes dense on the circle $\theta \in [0, 2\pi]$. Write the eigenvalues of U_∞ as $e^{i\theta}$ and consider the set of functions $e^{ik\theta}$, the eigenvalues of U^k. The fact that this is a complete set of complex square integrable functions on the circle (see the next chapter and Chapter 6 for definitions and more discussion) is the content of a famous theorem by Fourier. The operator $P_\theta = \frac{\hbar}{i} \frac{\partial}{\partial \theta}$ has an integer valued spectrum (in units of $\hbar$) and the exponentials are its normalizable eigenvalues. Define $V_\alpha = e^{i\alpha \frac{P_\theta}{\hbar}}$ and evaluate $U^k V_\alpha U^{-k} V_{-\alpha}$.

3.17 Now write $\theta = \frac{x}{R} 2\pi$, where x and R have dimensions of length. Find the spectrum of the operator $\frac{P_\theta}{R}$ and show that in the limit $R \to \infty$ at fixed x, this gives us the Hilbert space of a single particle on an infinite line.

3.18 Prove that the Poisson bracket satisfies the identities:

$$[F, G]_{PB} = -[G, F]_{PB},$$

$$[AB, C]_{PB} = A[B, C]_{PB} + [A, C]_{PB},$$

$$[A, [B, C]_{PB}]_{PB} + [C, [A, B]_{PB}]_{PB} + [B, [C, A]_{PB}]_{PB} = 0.$$

3.19 Solve for the eigenstates of the free particle Hamiltonian on the interval $[-a.a]$, with the boundary condition $\psi(\pm a) = 0$. These are called Dirichlet boundary conditions.

3.20 Repeat the previous exercise, but with the boundary condition $\partial_x \psi(\pm a) = 0$. These are called Neumann boundary conditions.

3.21 Solve the free particle on a finite interval with one Neumann and one Dirichlet boundary condition.

CHAPTER 4

Quantum Mechanics of a Single Particle in One-Dimensional Space II

4.1 INTRODUCTION

This chapter adds more mathematical detail to the description of one-dimensional particle motion. It also discusses the double slit experiment in some detail. The chapter ends with an important worked exercise, in which the reader is guided through the solution of both the bound state and scattering state equations for a particle in a square well or barrier potential. It is crucially important for every reader to work through this exercise in exhaustive detail. One should understand why the bound state spectrum is discrete, and what the scattering matrix is, as well as the relation between bound states and the analytic continuation of scattering matrix elements.

4.2 A MORE MATHEMATICAL DESCRIPTION OF ONE-DIMENSIONAL MOTION

Now let us add a little more mathematical detail to our description. As we will see in Chapter 6, the proper description of the space of states of a particle moving on an infinite one-dimensional line is the set of all complex valued functions $\psi(x)$ satisfying the condition

$$\int dx \, |\psi(x)|^2 < \infty. \tag{4.1}$$

This is a vector space, in an obvious way: we can make linear combinations of functions with complex coefficients. The resulting functions are square integrable, because of the inequality

$$\left| \int dx\ f^*(x)g(x) \right| \le \left(\int |f|^2 \int |g|^2 \right)^{1/2}. \qquad (4.2)$$

The quantity on the left-hand side

$$\int dx\ f^*(x)g(x) \equiv \langle f|g \rangle, \qquad (4.3)$$

defines the scalar product on this space. The inequality above is then precisely the Cauchy–Schwarz inequality for this scalar product (see Chapter 6). The definition of distance between two vectors in a finite dimensional Hilbert space is $D^2 \equiv \sum |v_n - w_n|^2$, which is the length of the vector $|v\rangle - |w\rangle$. The generalization of this to the distance between two functions is

$$D^2(f,g) = \int dx\ |f(x) - g(x)|^2. \qquad (4.4)$$

This space of functions is infinite dimensional. For example, the functions $P(x)e^{-x^2}$, where $P(x)$ is an arbitrary polynomial, all belong to it. Saying that a finite linear combination of these vanishes (i.e., that there is a finite basis for the space) is equivalent to saying that some polynomial vanishes everywhere on the real line. Of course, polynomials of order N have at most N distinct real zeroes, so this means the polynomial is zero. The monomials x^k for different k are linearly independent (using the same argument) so we have an infinite number of linearly independent square integrable functions.

On the other hand, there is a discrete or countable basis for this space. Again, the argument goes through polynomials. Consider a continuous complex valued function $f(x)$ on an interval $[a, b]$. Divide up the interval into N, not necessarily equal segments, of size about $\frac{b-a}{N}$, and think about the equations

$$P_{N-1}(x_m) \equiv \sum_{k=0}^{N-1} a_k^{(N-1)} x_m^k = f(x_m),$$

for x_m the middle of each of these segments. This is a set of N linear equations for the N coefficients of P_N. It will generally have a solution as long as the determinant of the matrix $M_m^k \equiv x_m^k$ is nonzero. If it is not, we can shift the x_m slightly off the midpoints to make the determinant nonvanishing. Since both the polynomial and f are continuous, they cannot deviate too much from each other at other points on the interval. In the limit $N \to \infty$, they cannot differ at all. This is a nonrigorous proof that every continuous function on an interval can be uniformly approximated by a polynomial.

Now consider a continuous square integrable function $g(x)$ and look at

$$D^2 = \int_{-\infty}^{\infty} |g(x) - P_N(x)e^{-x^2}|^2.$$

Thinking of the space of square integrable functions as an infinite dimensional Hilbert space, D^2 is the squared length of the vector difference between the vector $g(x)$ and its approximation by polynomials times a Gaussian. Note that here we are approximating $g(x)e^{x^2}$ uniformly by polynomials.

Now choose a large number Y. The contributions to D^2 outside the interval $[-Y, Y]$ obviously go to zero as $Y \to \infty$. Inside the interval, we can find a polynomial of order $N(Y)$, with large enough $N(Y)$ that the contribution to D^2 from the interval is $< \frac{1}{N(Y)}$, because we can uniformly approximate the continuous function $g(x)e^{x^2}$ by polynomials. It follows that we can find a polynomial P_N, with large enough N that makes $D^2 < 1/N$. As we have said, D is the infinite dimensional analog of the distance between two points in Euclidean space, according to the Pythagorean formula. When the distance D goes to zero we say that the sequence of functions $P_N e^{-x^2}$ "converges to g in the L^2 norm." Now not all square integrable functions are continuous, but they are all limits of piece-wise continuous functions, and you can convince yourself by drawing pictures that every piece-wise continuous function can be approached arbitrarily closely (in the sense of D going to zero) by continuous functions. So, we have shown that polynomials times Gaussians are a countable basis of the space of square integral functions. There are many other discrete, complete bases. Mathematically, what we have shown is that the Hilbert space of square integrable functions is a *separable* Hilbert space, which is the mathematical name for a space with a countable basis.

If we have two different bases of the same Hilbert space, there is a mathematical isomorphism that maps the space into itself, by taking $|e_k\rangle$ into $|f_k\rangle$. The *best thing about Dirac notation* is that it provides a simple formula for this mapping (and for many other transformations on Hilbert space). Recall that a dyadic operator determined by two vectors $|v\rangle$ and $|w\rangle$ is

$$D_{vw}|f\rangle = |v\rangle\langle w|f\rangle. \tag{4.5}$$

In words, we take the scalar product of $|w\rangle$ with $|f\rangle$ and then multiply this number by the vector $|v\rangle$. We also define the operation of Hermitian conjugation of operators by $D_{vw}^{\dagger} = D_{wv}$. Given this definition, it is easy to write down the transformation that takes one basis into another:

$$|f_k\rangle = U(f, e)|e_k\rangle \equiv \sum_l |f_l\rangle\langle e_l|e_k\rangle. \tag{4.6}$$

In other words,

$$U(f, e) = \sum_l |f_l\rangle\langle e_l|. \tag{4.7}$$

We just have to use the orthonormality of the $|e_k\rangle$ vectors to verify this. The Hermitian conjugate of the operator $U(f, e)$ is also its inverse, $U(f, e)^{\dagger} = U(e, f)$, as you will verify in Exercise 4.1. This is usually written $U(e, f)^{\dagger}U(e, f) = 1$, and we have agreed to call operators satisfying this property *unitary operators*. We will go into all of this in more detail in Chapter 6, which will give you a chance to brush up on your linear algebra, while learning Dirac notation more thoroughly. Eventually, if you want to become a quantum physicist, you would like to be able to do manipulations of Dirac notation in your sleep.

Exercise 4.1 Consider a unitary *matrix* U_{ij}. Show that its rows and columns, *both* form orthonormal bases. Interpret this fact in terms of the dyadic construction of Exercise 4.2.

Exercise 4.2 Prove that the operators $\sum_k |e_k\rangle\langle f_k|$, are unitary, for any choice of the two orthonormal bases $|e_k\rangle$ and $|f_k\rangle$. Prove that *any* unitary transformation[1] has this form. (Hint: Consider the action of U on the elements of an orthonormal basis. In the course of this proof, you will realize that the dyadic representation of the unitary operator is not unique.)

We now want to introduce a particular unitary transformation, which is extremely useful for any quantum mechanics (QM) problem involving particles. It is actually a formula you probably know from your math-physics class, and which we used at the end of the last chapter, the Fourier transform. This is defined as a mapping from functions to functions:

$$f(x) = \int_{-\infty}^{\infty} \frac{dp}{\sqrt{2\pi\hbar}} e^{ikx} \tilde{f}(p) \equiv F[\tilde{f}] = \int \frac{dp\,dy}{2\pi\hbar} e^{ik(x-y)} f(y). \tag{4.8}$$

Here, $\hbar k \equiv p$, and from now on all integrals without limits go from $-\infty$ to ∞. Actually, we have written both the definition of the Fourier transform *and* Fourier's theorem, which says that if you apply the Fourier transform twice, you get back to the function you started with, but evaluated at a negative value of the argument. In other words, the Fourier operator F satisfies $F^4 = 1$ and $F^2 = R$, where $Rf(x) \equiv f(-x)$.

Physicists like to summarize the Fourier theorem in a pair of equations

$$\int \frac{dk}{2\pi} e^{ik(x-y)} = \delta(x - y). \tag{4.9}$$

and

$$\int \frac{dx}{2\pi} e^{ix(k-k')} = \delta(k - k'). \tag{4.10}$$

In these equations, k is the wave number, related to the momentum by $p = \hbar k$. These serve to introduce the Dirac delta function, which is defined by the property.

$$\int dx\, \delta(y - x) f(x) = f(y) \tag{4.11}$$

---

[1] The phrases *unitary transformation* and *unitary operator* mean the same thing.

for any continuous function $f(x)$. We will reserve the discussion of the mathematical properties of this object, which is not really a function but a *distribution*, to Appendix B. Its defining property is that

$$\int dy\, f(y)\delta(x-y) = f(x). \tag{4.12}$$

From these equations, it follows that

$$\int dx\, f^*(x)g(x) = \int \frac{dx\, dp\, dq}{2\pi\hbar} e^{i(k_p - k_q)x}\tilde{f}^*(p)\tilde{g}(q) = \int dp\tilde{f}^*(p)g(p). \tag{4.13}$$

This shows that the Fourier transform is a unitary operator. It preserves the scalar product on the space of square integrable functions. In the previous chapter, we called this Plancherel's theorem. We will be using the Fourier transform extensively in this book. Start getting used to it by doing Exercise 4.3.

Exercise 4.3 Compute the Fourier transform of a Gaussian $\psi(x) = e^{-ax^2+bx}$, as well as an exponential function $\psi(x) = e^{-a|x|}$. Compute the Fourier transform of a step function $S(x) = 0$ for $x > a$ or $x < b < a$, and $S(x) = s$ on the interval $[b, a]$.

4.3 CONTINUUM OR DELTA FUNCTION NORMALIZATION

We have emphasized that we can expand every state in the Hilbert space of square integrable functions in terms of a countable basis. In equations, this says

$$f(x) = \sum \psi_n(x) \int dy\, \psi_n^*(y)f(y), \tag{4.14}$$

which we can now write formally as

$$\delta(x-y) = \sum \psi_n(x)\psi_n^*(y). \tag{4.15}$$

This is called the completeness relation for a complete orthonormal set of functions. It is a continuous version of the relation

$$\delta_{kl} = \sum \langle f_k|e_n\rangle\langle e_n|f_l\rangle, \tag{4.16}$$

where $|f_n\rangle$ and $|e_n\rangle$ are any two orthonormal bases.

In a finite dimensional Hilbert space, any normal operator, satisfying $[A, A^\dagger] = 0$ has a complete basis of orthonormal eigenfunctions. This is not true in infinite dimensions. If we take a multiplication operator

$$F_g\psi(x) \equiv g(x)\psi(x), \tag{4.17}$$

then this could only have a fixed numerical value if $\psi(x)$ vanishes everywhere except at some fixed $x = x_0$. Such a function could not be in the Hilbert space of square integrable functions, since if its value at x_0 is finite, its norm is zero and so it should be identified with the zero vector. On the other hand, if we take $\psi(x) = \delta(x - x_0)$, then this does satisfy the eigenvalue equation for F_g, with eigenvalue $g(x_0)$. The delta function is not a square integrable function but it does satisfy

$$\int dy \, \delta(x - y)\delta(y - z) = \delta(x - z), \tag{4.18}$$

which means that "eigenfunctions" corresponding to different eigenvalues of the multiplication operator are orthogonal. This also says that those "eigenfunctions" have infinite norm, and are not actually in the Hilbert space. We say that such states are *delta function normalized*, and introduce the notation $|x\rangle$ for the abstract state corresponding to the delta function located at x. Then $(p \equiv \hbar k)$,

$$\langle x|y \rangle = \delta(x - y) = \int \frac{dp}{2\pi\hbar} e^{ik(x-y)}. \tag{4.19}$$

For any other state $|\psi\rangle$, we can interpret the function $\psi(x)$ that represents this state as the overlap $\psi(x) = \langle x|\psi \rangle$, with these delta function normalized eigenstates of the position operator x. The proper mathematical treatment of these delta function normalized states is complicated[11]. We will use the physicist's approach to mathematical rigor. We ignore it and treat the position eigenstates as if they were discrete, replacing sums by integrals in equations like the decomposition of the operator F_g into projection operators on a complete orthonormal basis:

$$F_g = \int dx \, g(x)|x\rangle\langle x|. \tag{4.20}$$

The equation for a momentum eigenstate

$$\frac{\hbar}{i} \frac{\partial}{\partial x} \langle x|p \rangle = p\langle x|p \rangle \tag{4.21}$$

is solved by

$$\langle x|p \rangle = \frac{1}{\sqrt{2\pi\hbar}} e^{i\frac{px}{\hbar}}, \tag{4.22}$$

where we have chosen the normalization so that

$$< p|q > = \delta(p - q) = \int \frac{dx}{2\pi\hbar} e^{ix(k_p - k_q)}. \tag{4.23}$$

The factors multiplying x in the exponent are the wave numbers corresponding to the indicated momentum. So, momentum eigenstates also have delta function normalization on the real line. In Exercise 4.4, you will see how this gets modified on the circle (where we have seen the eigenvalues are discrete) and in Exercises 3.7–3.9, you saw how all of this mysterious infinite stuff can be realized in terms of limits of finite matrices.

Exercise 4.4 The Hilbert space of functions on a circle consists of periodic square integrable functions $f(\theta) = f(\theta + 2\pi)$ on the interval $[0, 2\pi]$. The eigenfunctions of the wave number operator $K = \frac{1}{i}\frac{\partial}{\partial\theta}$ are

$$\psi_k = \frac{1}{\sqrt{2\pi}}e^{ik\theta}.$$

What are the allowed values of k? The fact that these are a complete orthonormal set of functions is the content of the famous theorem by Fourier:

$$f(\theta) = \sum_k \frac{f_k}{\sqrt{2\pi}}e^{ik\theta}.$$

$$f_k = \frac{1}{\sqrt{2\pi}}\int_0^{2\pi} f(\theta)e^{-ik\theta}.$$

For a circle of radius R, simply make the replacement $2\pi \to 2\pi R$. The momentum operator is $\hbar K = P$. Write an equation for the Dirac delta function on the circle, using these formulae.

Answer to Exercise 4.4: $e^{ik\theta}$ is periodic, if and only if k is an integer. The delta function is defined by

$$f(\theta) = \int_0^{2\pi} f(\alpha)\delta(\theta - \alpha).$$

We use Fourier's theorem in the form

$$f(\theta) = \sum_k \frac{1}{2\pi}e^{ik\theta}\int_0^{2\pi} f(\alpha)e^{-ik\alpha}$$

to conclude that

$$\delta(\theta - \alpha) = \sum_k \frac{1}{2\pi}e^{ik(\theta-\alpha)}.$$

While we are on the subject of "generalized functions" like the Dirac delta function, let us introduce the Heaviside step function $\theta(x)$, which is defined to be zero for $x < 0$ and 1 for $x > 0$. Its real definition (see Appendix B) is in terms of its action on functions. Namely,

$$\int_{-\infty}^{\infty} \theta(x)f(x) = \int_0^{\infty} f(x). \tag{4.24}$$

Sometimes it is useful to define $\theta(0) = 1/2$. Note that

$$\int dx \frac{df}{dx}\theta(x - a) = \int_a^{\infty} \frac{df}{dx} = -f(a), \tag{4.25}$$

where f is any differentiable square integrable function. Square integrability implies that $f(\pm\infty) = 0$. Doing a formal integration by parts on the integral from $-\infty$ to ∞ with the Heaviside function, and recalling the definition of the delta function, we conclude that

$$\frac{d\theta(x - a)}{dx} = \delta(x - a). \tag{4.26}$$

4.4   TIME EVOLUTION OF A FREE PARTICLE WAVE FUNCTION

We are now ready to use all of this math to study the propagation of free particles through space in QM. At time $t = t_0$, the position space wave function $\langle x|\psi\rangle$ is the Fourier transform of the momentum space wave function, $\langle p|\psi\rangle$. We also learn that

$$\langle p|x\rangle = \frac{e^{-ikx}}{\sqrt{2\pi\hbar}}, \tag{4.27}$$

from the identity $\langle p|\psi\rangle = \int dx \langle p|x\rangle\langle x|\psi\rangle$.

We will take

$$< p|\psi(t_0) >= Ne^{-(\frac{p-p_0}{2\Delta p})^2}.$$

The normalization constant N is determined by insisting that the integral over all momenta is 1:

$$N^{-2} = \int dp e^{-2(\frac{p-p_0}{2\Delta p})^2}.$$

Introducing $z \equiv \frac{p-p_0}{\Delta p}$, we write this as

$$N^{-2} = \Delta p \int dz\ e^{-\frac{1}{2}z^2} = 2\Delta p \int_0^\infty dz e^{-\frac{1}{2}z^2}.$$

If we define $\frac{1}{2}z^2 = u$ and note that $dz = (\sqrt{2u})^{-\frac{1}{2}}du$, this is

$$N^{-2} = \Delta p \sqrt{2}^{-\frac{1}{2}} \int_0^\infty du u^{-\frac{1}{2}}e^{-u}.$$

This integral is the Euler Gamma function $\Gamma(s)$, at $s = \frac{1}{2}$. If you do not know the Euler Gamma function, look it up on the Web. If you know, or look up, the identity

$$\Gamma(s)\Gamma(1-s) = \frac{\pi}{\sin(\pi s)},$$

we see that this integral is just $\sqrt{\pi}$.

We introduce the Gamma function here because we will run into it in a few other places, but the result $\Gamma(1/2) = \sqrt{\pi}$ can be proven simply by the following argument.

$$[\int dx\ e^{-\frac{1}{2}x^2}]^2 = \int dx dy\ e^{-\frac{1}{2}(x^2+y^2)} = 2\pi \int_0^\infty dr\ re^{-\frac{1}{2}r^2} = \pi \int_0^\infty du\ e^{-u} = \pi. \tag{4.28}$$

So,

$$N = \left(\frac{\pi}{2(\Delta p)^2}\right)^{\frac{1}{4}} = \left(\frac{\pi}{2(\hbar\Delta k)^2}\right)^{\frac{1}{4}}.$$

To see the time development of the wave function in position space, we remember that for the free particle, energy eigenstates are also momentum eigenstates, so we can write

$$\langle x|\psi(t)\rangle = N \int dp \frac{e^{ikx}}{\sqrt{2\pi\hbar}} e^{-i\frac{p^2}{2m\hbar}(t-t_0)} e^{-(\frac{p-p_0}{2\Delta p})^2}. \tag{4.29}$$

We convert this into an integral over k, defining $\hbar k_0 = p_0$ and $\hbar \Delta k = \Delta p$,

$$< x|\psi(t) >= \hbar^{\frac{1}{2}} N \int dk \frac{e^{ikx}}{\sqrt{2\pi}} e^{-i\frac{\hbar k^2}{2m}(t-t_0)} e^{-(\frac{k-k_0}{2\Delta k})^2}. \tag{4.30}$$

This is the same as

$$e^{-(\frac{k_0}{2\Delta k})^2} \hbar^{\frac{1}{2}} N \int dk \frac{1}{\sqrt{2\pi}} e^{-\frac{1}{2}\alpha^2 k^2 + ak}, \tag{4.31}$$

where

$$\alpha^2 = \frac{1}{2(\Delta k)^2} + i\frac{\hbar \Delta t}{m}.$$

$$a = \frac{k_0}{2(\Delta k)^2} + ix.$$

By completing the square, shifting, and rescaling the integration variable, we find

$$< x|\psi(t) >= e^{-(\frac{k_0}{2\Delta k})^2} e^{\frac{a^2}{2\alpha^2}} \alpha^{-1} \sqrt{2\pi} \hbar^{\frac{1}{2}} N. \tag{4.32}$$

This is a Gaussian in position space. To find its center and width, we need to compute the real part of $\frac{a^2}{2\alpha^2}$, which is

$$\frac{-(x^2 - 2x\frac{\hbar k_0}{2m})}{2(\Delta x)^2},$$

with

$$(\Delta x)^2 = \frac{1}{2(\Delta k)^2}(1 + \frac{4(\Delta k)^4 \hbar^2 (\Delta t)^2}{m^2}).$$

The wave packet is initially narrow and concentrated near the classical trajectory, if Δk is big, and wide if Δk is small. It spreads with time, and the spread becomes important more quickly if Δk is large, and less quickly if m is large.

This spread can be understood intuitively in terms of the probabilities of momentum and position and the fact that they cannot both be sharp at the same time. We tried to localize the wave function in position space at the initial time. However, the more we try to do so the larger we must take Δk, the uncertainty in momentum. But the time derivative of the position is the momentum divided by the mass, so large Δk means a big uncertainty in velocity, which of course leads to a larger uncertainty in position at a later time. The velocity uncertainty is inversely proportional to the mass, for fixed Δk, which explains the slower spread of the wave packet as the mass increases. Roughly speaking, the same thing happens for any localized initial condition. In Exercise 4.5, you will relate the rate of falloff of a function in position space to the rate of falloff of its Fourier transform in momentum space. These results are all a consequence of the Heisenberg uncertainty relations, which we will study in Chapter 6.

Indeed, we can understand the spread of the wave packet that we have found in a much more general way, by thinking in the Heisenberg picture, where the wave function $\psi(x)$ is independent of time, but the operator $X(t)$ satisfies

$$X(t) = X + \frac{P}{m}t. \tag{4.33}$$

We can choose $\psi(x)$ such that X has very little uncertainty. Such a tightly focused function has to have large derivatives, since it must fall rapidly to zero outside of a small interval in x. The uncertainty in P is the average of the absolute square of $d\psi/dx$, so this will be large. This means that at large t, the uncertainty in $X(t)$ is large $\sim \Delta Pt/m$, which is equivalent to the spread of the wave packet, in the Schrödinger picture.

In Exercise 4.11, we will estimate the consequences of these results for a variety of macroscopic bodies, from baseballs to the moon, and see that the predicted uncertainties are very small. This is a part, but not the most important part, of the explanation of how classical physics emerges as an approximation to the more fundamental quantum rules, which are obeyed by baseballs as much as they are by electrons.

There is one point that one always has to keep in mind when thinking about quantum uncertainties for macroscopic objects. Einstein once said that "I cannot believe that the Moon exists only because a mouse looks at it." In saying this, Einstein was objecting to an interpretation of QM in which we think of the quantum state of a system as describing what really happens to a physical system in any one history. As we will emphasize over and over again, no such interpretation is correct.[2] QM only tells us about probabilities, and the only way to test a probabilistic theory, even when it says that the probability for something is very close to one, is by doing the experiment over and over again, and comparing the frequencies of occurrence to the probabilities predicted by the theory. For systems where we cannot repeat the experiment, we can only treat the probabilities in a psychological, Bayesian sense. They tell us, sometimes, what we expect to see in a single run of an experiment, but only if the probability for a certain history is very close to one. So, the uncertainty predicted in the motion of the moon just tells us that after a long enough time we cannot predict what will happen to the moon, given a measurement of its current position and velocity. Einstein never

---

[2] Modern attempts to find a set of "hidden variables," such that quantum probabilities follow from definite histories for the hidden variables, give up the possibility of predicting histories for the quantities we actually measure. We will discuss this in Appendix A on interpretations of QM.

objected to this statistical interpretation of QM. However, he believed, because of his deep understanding of classical mechanics, that all uncertainty had to be explained in terms of probabilities for definite histories of systems. The statistical interpretation of QM denies the existence of such probabilities, unless the systems in question are the collective coordinates of macroscopic objects.

The other thing we have to emphasize about this computation of the spread of the wave packet for a free particle is that real particles are never exactly free. In particular, whenever we try to measure the position of a particle with a detector, we are introducing complicated interactions between the particle and the detector, which change the predictions for the spread of the wave packet. We will see that when particles move under the influence of forces, the spread of their wave packets can be dramatically slower than the computation we have just done would indicate. For example, for simple harmonic motion, certain wave packets show no spread whatsoever.

4.5 THE DOUBLE SLIT EXPERIMENT

We are now in a position to study the double slit experiment that you were asked to read about in Feynman's Lectures, before starting this book. The basic idea is illustrated in Figure 4.1, but we will use Schrödinger wave function language to describe what is happening.

We have a particle moving in a two-dimensional plane. It is described by some incoming normalizable wave packet from the left. It encounters a barrier in the vicinity of $x = \pm a$.

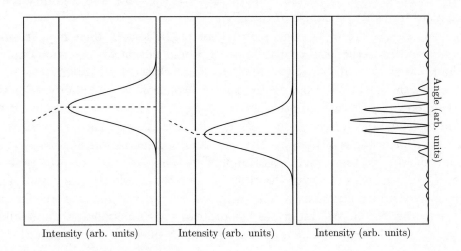

Intensity (arb. units)   Intensity (arb. units)   Intensity (arb. units)

Angle (arb. units)

Figure 4.1 The Double Slit Experiment.

The barrier is a potential V that is infinitely high between $x = \pm a$, except for two slits in y, of width D, where the potential is zero. The slits are centered at $y = \pm b$. For simplicity, we are going to consider a Hamiltonian for the particle of the form

$$\frac{p_x^2}{2m_x} + \frac{p_y^2}{2m_y} + V,$$

with $m_x \gg m_y$. This means that we can consider a very narrow wave packet in the x direction, localized in a distance $\ll a$, without worrying about rapid spreading in the x direction. An adequate approximation to the problem is found by simply starting the system off at $t = 0$ at a time when the wave function is "just emerging from the barrier." This phrase makes sense because of the control we have over the width of the wave packet in the x direction.

If we want to be considering states for which the expectation value of the energy is finite, then the wave function has to vanish identically everywhere that the potential is infinite. This means that at $t = 0$, we have

$$\psi(x, y, 0) = \psi_-(x, y, 0) + \psi_+(x, y, 0),$$

where $\psi_\pm$ vanishes everywhere but in the slit centered at $y = \pm b$. At this time, we can say for this quantum particle, as we could for a bullet, that the particle either went through the plus slit or the minus slit. The total probability that the particle is somewhere is the sum of a probability P_+ that it went through the plus slit, and a probability P_- that it went through the minus slit, and these two probabilities add up to one. The probabilities $P_\pm$ are calculated by integrating $|\psi_\pm|^2$ over the slits where they are nonzero. The total probability is calculated by integrating the absolute square of ψ over all space, and is normalized to one. The interference terms $\psi_\pm^* \psi_\mp$ vanish identically.

For bullets, we use "logic" to conclude that for predictions of what happens for $t > 0$ we continue to add together the probabilities we would obtain for the conditional events that the bullet went through one slit or another. For a classical probability theory, we can do this because the equations that determine the evolution of probability with time are linear in the probability density. Another way of saying this is that we are using Bayes' Law. Linearity of the equations for the probability density in a classical theory means that there are probabilities for histories of the variables. We can divide all histories up into those which pass through one slit and those which pass through another, at $t = 0$. The total probability is the sum of the probabilities of one or the other of these classes of histories, and the linearity of the equation for probability density tells us that this additivity remains true for all times.

This is not true in QM. We can define "conditional wave functions" $\psi_\pm$, and follow their evolution separately, because the Schrödinger equation is linear. It is even true that

$$\int dx dy \, (|\psi_+|^2 + |\psi_-|^2) = 1,$$

at all times, if it is chosen so initially. Nonetheless, given an initial wave function $\psi = \psi_+ + \psi_-$, when we ask what the probability is, at some $t > 0$ to find the particle at some point (x, y), it is not the sum of $P_+(x, y, t) + P_-(x, y, t)$. The interference term $\psi_+^* \psi_- + \psi_-^* \psi_+$, integrates to zero, but is nonzero point by point because of spreading of the wave packet.

This would follow from calculations we have already done, if the wave functions were Gaussians, but these wave functions actually vanish outside of a finite interval. In fact, one can show that this makes the spreading even worse. You will do that in Exercise 4.5, but let us just try an example. Suppose we have a wave function

$$\psi_c = e^{-(\frac{y}{b})^{2c}}.$$

For very large c, this approaches a function that vanishes for $|y| > b$. Its Fourier transform is proportional to

$$\int dy \, e^{iky} e^{-(\frac{y}{b})^{2c}}.$$

The result is an even function of k, so we take k positive and large. For large k we can do the integral by the method of steepest descents: we find the stationary point of the exponent. This is

$$y* = [\frac{ikb^{2c}}{2c}]^{\frac{1}{2c-1}}].$$

At the stationary point, the real part of the exponent behaves like $-dk^{\frac{2c+1}{2c-1}}$, where d is a positive constant of order 1 in the large c limit. Thus, the larger the value of c, the more slowly the momentum space wave function falls off at large momentum. The uncertainty in the momentum grows larger, and at large times, this leads to an uncertainty in position worse than that of the Gaussian.

Many readers find that the most mystifying part of Feynman's discussion of the double slit experiment is the disappearance of the interference pattern when one tries to determine experimentally the slit through which the particle goes. There are two reasons for this mystery. The first is that, despite Feynman's care in saying that he is talking about probability, and that in any given run of the experiment the detector goes "clunk" at some particular point on the screen, his pictures give the impression that the Schrödinger wave function is a physical wave. In fact, the picturesque wave pattern is something we put together to summarize the results of many experiments. It is *not* a physical wave in space, measured at a particular time. If it were, it would indeed be mysterious that the value of this physical thing near the plus slit could affect the detector near the minus slit. This is a lot less mysterious when we say, correctly, that what we are doing is measuring the frequency of occurrence of hits on different parts of the screen in two different situations, one with a detector sitting at the minus slit and one without it.

Conditional probabilities often behave in a "nonlocal" manner. If the equations for the weather predict on a certain day that a particular storm has a probability to hit one of two cities a few days later, then once a few days have passed we know that one of those two predictions was wrong. If we thought of the probability distribution as a physical thing, it would seem insane to suddenly change it in Galveston, because we see a storm hitting New Orleans. Bayes' rule of conditional probability tells us this is exactly the right thing to do.

Bayes' rule is not in general applicable in QM, because probabilities only add when interference terms are negligible. If we want to understand *why* the interference terms suddenly become negligible when we put a detector near one of the slits, then we have to understand more about the properties of the detector than Feynman's discussion supplies. We will see in Chapter 10 that the crucial property is that the detector is a complicated system, made of many atoms, with a huge number of microscopic atomic states corresponding to each position of the "needle on the detector's dial." As a consequence, the interference terms in the probability distribution for seeing the detector's dial either register a hit or not, are unimaginably small, and Bayes' rule is approximately valid. This is a general lesson: *the quantum rules for the probabilities that refer to macroscopic objects differ from the rules of a classical statistical theory, which satisfies Bayes' rule, by amounts which are of order* e^{-cN} *where* N *is the number of atoms out of which the object is composed.* These differences are so small that they are impossible to measure, even in principle.

The fact that the rule for computing probabilities of future events depends on whether the detector is there or not is not mysterious. Indeed, we are describing very different experiments in the two cases. In the first, we ask for a prediction of the probability of getting a hit on the screen. In the second, we are asking for probabilities of hits, *conditioned on the detector's behaving a certain way at an intermediate time.* As we will see, the fact that the violation of Bayes' rule for observations done on the macroscopic detector is vanishingly small, completely accounts for the difference in the predictions.

4.6 A WORKED EXERCISE

Exercise 4.6 As we will see in the next chapter, the generalization of the Schrödinger equation to particles moving under the influence of a force is simply to add the potential energy $V(x)$ to the Hamiltonian operator. In the following exercises, we will consider potentials, which are locally constant, so that we can use free particle solutions to solve the equation. To begin, solve the Schrödinger equation for a step function potential $V(x) = V_0\theta(x+a)\theta(a-x)$. The real line divides into three intervals, in each of which the equation reduces to that of a free particle. By taking the integral of the Schrödinger equation in small intervals around the points of discontinuity, you can figure out how to join the solutions in the three regions and impose the condition of normalizability. For one sign of V_0, you should find only delta function normalizable eigenstates, whereas for the other, you will find both delta function normalizable and truly normalizable eigenfunctions. Show that the spectrum of the latter is discrete and describe how it depends on V_0 and a. The normalizable wave functions are called bound states. Show that the bound state spectrum depends only on the parameter $s_0 \equiv \frac{a}{\hbar}\sqrt{2mV_0}$. By examining the equation for bound states graphically, show that the number of such states is finite, but goes to infinity as $s_0 \to \infty$. For small s_0, show that there is only a single bound state. Now find the eigenfunctions for $E > 0$. There are two linearly independent solutions, which can be characterized by saying that the associated time-dependent solution moves toward or away from the center, near $+\infty$. Alternatively, we can characterize the solutions as incoming or outgoing at $-\infty$. There must be two linear relations between these four solutions, since only two of them can be linearly independent. Compute the 2×2 *Scattering matrix* or *S*-matrix, which expresses the two outgoing solutions in terms of the two incoming solutions.

Exercise 4.7 Consider the square well potential of Exercise 4.6. For the case of the normalizable wave functions, show that they decrease exponentially for large $|x|$. Think about classical mechanics with the same potential, and show that motion in the region where the quantum wave function is exponentially decreasing is simply not allowed. The fact that there is nonzero probability for the particle to be found in this region is called *quantum tunneling through a barrier*. We will study it more extensively in Chapter 17 on the JWKB approximation. Show that the equation for bound states depends only on the parameter s_0.

Exercise 4.8 For the continuum (i.e., scattering) wave functions in Exercise 4.6, we have two possible behaviors $e^{\pm ikx}$ for positive k in each of the asymptotic regions $x \to \pm\infty$. By considering the time-dependent Schrödinger equation show that the choices of the sign of k correspond to left and right moving waves in the two regions. Argue that the two different signs in a given asymptotic region, correspond to two linearly independent solutions of the second order ordinary differential equation for fixed k^2. Argue that there are exactly two linearly independent solutions. Thus, there must be linear relations between the four possible asymptotic behaviors. Using your exact solutions, find these linear relations. Your tool for finding these relations is continuity of the wave function and its derivative at the points where the potential, and thus the *second* derivative of ψ is discontinuous.

Answer to Exercises 4.6–4.8: The solution to the Schrödinger equation with energy E is

$$\psi_E = e^{\pm\sqrt{\frac{2m}{\hbar^2}(V(x)-E)}\,x}, \tag{4.34}$$

where $V(x) = V_0\theta(a-x)\theta(x+a)$. Integrating the Schrödinger equation near the discontinuities in $V(x)$, we find that the wave function and its first derivative must be continuous there Figure 4.2. If $E > 0$, we have delta function normalizable solutions, while if $E < 0$, we can get normalizable bound state solutions by picking only the falling exponential as $x \to \pm\infty$. The problem is invariant under reflections, so for the bound states, we can choose to look at even and odd solutions. We can impose the continuity conditions only at $x = a$ and they

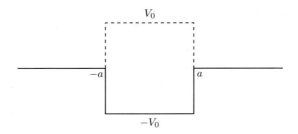

Figure 4.2 The potential for Exercises 4.6–4.8.

will automatically be satisfied for negative x. In the region near the origin, the wave function depends on the crucial quantity

$$ik_0 = \frac{\sqrt{2m(V_0 - E)}}{\hbar}, \tag{4.35}$$

which gives k_0 imaginary. The imaginary part of ak_0 is equal to the classical action in units of $\hbar$ of a particle traveling from $-a$ to a, if $E - V_0 > 0$. When the opposite inequality holds, k_0 is imaginary. In either case, we define k_0 with the positive (imaginary) square root. We define

$$ik = \frac{\sqrt{-2mE}}{\hbar}. \tag{4.36}$$

In addition, define $r \equiv \frac{k_0}{k}$. In the region $|x| > a$, the wave function grows or falls exponentially if $E < 0$. We must choose the falling solution at both positive and negative values of x, to obtain a normalizable wave function. The continuity conditions for even and odd wave functions at $x = a$ are:

$$A\cosh(|k_0|a) = Be^{-|k|a}, \quad Ar\sinh(|k_0|a) = -Be^{-|k|a}, \tag{4.37}$$

$$A\sinh(|k_0|a) = Be^{-|k|a}, \quad Ar\cosh(|k_0|a) = -Be^{-|k|a}, \tag{4.38}$$

when k_0 is imaginary. The hyperbolic sine and cosine are both positive, and these equations have no solution. Thus, we must have the intuitively obvious condition $E - V_0 > 0$, which implies a potential well rather than a barrier and asymptotic energy less than the depth of the well, in order to have a bound state. For k_0 real, the matching conditions become

$$A\cos(|k_0|a) = Be^{-|k|a}, \quad Ar\sin(|k_0|a) = Be^{-|k|a}, \tag{4.39}$$

$$A\sin(|k_0|a) = Be^{-|k|a}, \quad Ar\cos(|k_0|a) = -Be^{-|k|a}. \tag{4.40}$$

Motivated by the bound on E, we define $E = yV_0$, with $0 \le y \le 1$. The conditions become

$$\sqrt{\frac{y}{1-y}} = -\tan(s_0\sqrt{1-y}). \tag{4.41}$$

$$\sqrt{\frac{y}{1-y}} = -\cot(s_0\sqrt{1-y}). \tag{4.42}$$

These depend only on a single parameter $s_0 \equiv a\sqrt{\frac{-2mV_0}{\hbar^2}}$. When s_0 is small, we can expand the trigonometric functions. The equation for odd solutions becomes $y = 1/s_0^2$, which is inconsistent with $0 \le y \le 1$. The even solution is $y = s_0^2$ and is consistent. There is a single bound state. When s_0 is large, we can satisfy the equation with $s_0\sqrt{1-y} \sim \frac{n\pi}{2}$, where n is odd for even solutions and vice versa, as long as $\frac{n\pi}{2s_0} < 1$. The point is that the tangent and cotangent take on any large value in the vicinity of their poles, so that as long as y is sufficiently close to 1 we can match. These explicit formulae are only valid for large n, but there are bound states for every value of n satisfying the inequality. As s_0 goes to infinity, we get an infinite number of states and the explicit formula becomes more exact. In this limit, after adding a constant to make all the energies positive, the spectrum approaches that of the infinite square well. Note, however, that s_0 can be large for a shallow, but very broad, well also. In that situation, we would still have continuum eigenstates.

Turning now to $E > 0$, we define

$$\psi^+ = A_{out}^+ e^{ikx} + A_{in}^+ e^{-ikx}, \quad x > a, \tag{4.43}$$

$$\psi^+ = A_{out}^- e^{-ikx} + A_{in}^- e^{ikx}, \quad x < -a. \tag{4.44}$$

The subscripts *in* and *out* refer to the fact that, when multiplied by $e^{-i\frac{\hbar^2k^2}{2m}t}$, the relevant part of the solution becomes a traveling incoming or outgoing wave. Note that the sign of the exponential changes when we compare an incoming (outgoing) wave on the left and right. To write the continuity conditions at $x = \pm a$ compactly, it is convenient to define $z_0 = e^{ik_0a}$, $z = e^{ika}$, and $r^\pm = 1 \pm r$. Then we have

$$A_{in}^- z^{-1} + A_{out}^- z = Az_0^{-1} + Bz_0, \tag{4.45}$$

$$A_{in}^- z^{-1} - A_{out}^- z = r(Az_0^{-1} - Bz_0), \tag{4.46}$$

$$A_{out}^+ z + A_{in}^+ z^{-1} = Az_0 + Bz_0^{-1}, \tag{4.47}$$

$$A_{out}^+ z - A_{in}^+ z^{-1} = r(Az_0^{-1} - Bz_0). \tag{4.48}$$

These are equivalent to a pair of 2×2 matrix equations expressing the vectors

$$A_{in/out} \equiv \begin{pmatrix} A_{in/out}^- \\ A_{in/out}^+ \end{pmatrix}$$

in terms of the vector

$$A_m \equiv \begin{pmatrix} A \\ B \end{pmatrix}$$

$$A_{in} = \frac{z}{2}[\frac{r_+}{z_0} + r_- z_0 \sigma_1] A_m, \tag{4.49}$$

$$A_{out} = \frac{1}{2z}[\frac{r_-}{z_0} + r_+ z_0 \sigma_1] A_m. \tag{4.50}$$

The *Scattering matrix* or S-matrix is the transformation that takes A_{in} into A_{out}, so we have

$$S = \frac{1}{2z}[\frac{r_-}{z_0} + r_+ z_0 \sigma_1]\frac{2}{z}[\frac{r_+}{z_0} + r_- z_0 \sigma_1]^{-1}, \tag{4.51}$$

These equations reflect the fact that there are only two linearly independent solutions of the Schrödinger equation. The *in* and *out* states represent two independent bases for the Hilbert space. These equations tell us how to transform between them.

Note that the two factors in the formula for the S-matrix commute with each other. Thus, since $S = S_+ S_-^{-1}$, we have $S^\dagger S = S_+^\dagger S_+ [S_-^\dagger S_-]^{-1}$. Now compute

$$S_+^\dagger S_+ = \frac{1}{4}[r_- z_0 + r_+ z_0^{-1}\sigma_1][\frac{r_-}{z_0} + r_+ z_0 \sigma_1] = \frac{1}{4}[r_+^2 + r_-^2 + r_+ r_- (z_0^2 + z_0^{-2})], \tag{4.52}$$

$$S_- S_-^\dagger = \frac{1}{4}[\frac{r_+}{z_0} + r_- z_0 \sigma_1][\frac{r_+}{z_0} + r_- z_0^{-1}\sigma_1] = \frac{1}{4}[r_+^2 + r_-^2 + r_+ r_- (z_0^2 + z_0^{-2})]. \tag{4.53}$$

Therefore, $S^\dagger S = 1$. The S-matrix is a unitary transformation between two bases of the Hilbert space. Of course, the states of fixed k are not normalizable states in the Hilbert space, but this calculation guarantees that the mapping of normalizable incoming wave packet states into normalizable outgoing wave packet states, defined as superpositions of the fixed k states, is a genuine unitary operator in the Hilbert space.

We can write the eigenvalues of the unitary S-matrix as $e^{2i\delta_\pm(k)}$ where $\delta_\pm$ are called the phase shifts.

$$e^{2i\delta_\pm} = z^2 \frac{(r_- z_0^{-1} \pm r_+ z_0)}{(r_- z_0^{-1} \pm r_- z_0)}. \tag{4.54}$$

They are analytic functions of k, which will have poles whenever the denominator has a linear zero. The equation for the poles is

$$\frac{r_+}{r_-} = -z_0^2. \tag{4.55}$$

For real k, there are no solutions of this equation because the left-hand side is a real number greater than 1 and the right-hand side is a phase. For imaginary k, the left-hand side is also a phase. If $k = il$ the equation for a pole is

$$r_- = \pm z_0^2 r_+. \tag{4.56}$$

Some simple algebra shows that this is identical to the equation for bound states, and has solutions only when $V_0 < 0$. The fact that the S-matrix is an analytic function of energy, with poles at the bound state energies, is a very general result of immense importance.

4.7 FURTHER EXERCISES

4.4 Consider the Fourier transform of a smooth function $f(x)$ (see Appendix B for a definition of smooth).

$$\tilde{f}(k) = \int dx \ f(x)e^{ikx},$$

for large values of k. Show that smoothness implies that the Fourier transform falls off faster than any power of k. Compute the Fourier transform of $(x^2 + a^2)^{-b}$ for those values of b for which the function is square integrable. How does the falloff in wave number space depend on b?

4.5 Compute the Fourier transform of a step function, $\theta(a - x)\theta(a + x)$, where $\theta(x)$ is 1 for $x > 0$ and 0 for negative x, and demonstrate the relation between the width of the interval over which the function is nonzero and the falloff in wave number space. Use this to calculate the time evolution of a wave function that is initially confined in a fixed interval.

4.9 Consider the plane wave normalizable solutions for the potential of Exercises 4.6–4.8, in the case where there is no left moving or incoming wave in the region of large positive x. Thus, $\psi = A_{out}^+ e^{ikx}$ in this region. For $x < -a$ we have

$$\psi = A_{in}^- e^{ikx} + A_{out}^- e^{-ikx}.$$

A_{in}^- is called the incident amplitude. It is a wave moving toward the barrier from the left. A_{out}^- is called the reflected amplitude. A_{out}^+ is the transmitted amplitude. The *transmission coefficient*,

$$T = |\frac{A_{out}^+}{A_{in}^-}|^2,$$

gives us the probability of transmission through the barrier. Calculate it, for all values of V_0, E, and a. Also *define* and calculate the *reflection coefficient R* and show that $R + T = 1$. Show that T oscillates as a function of E and that there are values of E for which it is equal to 1. That is, there are discrete energies for which the barrier becomes completely transparent! This is called the Ramsauer–Townsend effect, and has been observed in the laboratory.

4.10 Consider the potential $V = -a^2\theta(x)$. Consider a particle of mass m and energy E traveling to the right from large negative x. Assume that for $x > 0$ there is no left moving wave. Show that there is a reflected wave and calculate the reflection coefficient. A classical particle would not of course experience such a reflection. Is it possible that $R = 1$, i.e., for some values of energy, a quantum particle "cannot fall off a cliff" ?

4.11 Consider an initial state $\psi(x, 0)$ which is normalizable and even so that $\langle X \rangle = 0$ and $\langle X^2 \rangle = (\Delta X)^2$ is finite. Using the Heisenberg picture, find the minimal value of the uncertainty in the position of the particle after time t. For a freely moving baseball of mass 0.5 kg and an initial uncertainty $\Delta X = .001$ m, how long will it be before the uncertainty is of order a meter? Assume the initial velocity is 100 km/h. Treat the baseball as a point particle. Do the same calculation for the moon, whose mass is about 7.35×10^{22} kg and whose orbital velocity is about $3,683$ km/h. Note that for motion in a Newtonian potential, the spreading of the wave packet is even slower[12].

4.12 Show that Gaussian wave functions saturate the Heisenberg uncertainty bound on $\Delta X \Delta P$. Prove that these are the only wave functions that do this.

4.13 Let $r(x) \geq 0$ be a function such that $\int_{-\infty}^{\infty} x^n r(x) = m_n$ is finite for every $n \geq 0$. This function defines a positive definite scalar product on the space of polynomials $P(x)$. By the Gram–Schmidt process, we can find an orthonormal basis of polynomials

$$\int r(x) P_n(x) P_m(x) = \delta_{mn}.$$

Show that

$$P_n(x) = c_n \det \begin{pmatrix} m_0 & m_1 & m_2 & \dots & m_n \\ m_1 & m_2 & m_3 & \dots & m_{n+1} \\ & & \dots & & \\ m_{n-1} & m_n & m_{n+1} & \dots & m_{2n-1} \\ 1 & x & x^2 & \dots & x^n \end{pmatrix}.$$

Find the value of c_n.

4.14 Find conditions on the weight function r and the function g such that the Rodrigues formula

$$P_n(x) = \frac{1}{r} \frac{d^n}{dx^n} [rg^n],$$

defines a set of orthogonal polynomials with respect to $r(x)$.

4.15 Consider a differential operator of the form

$$H = a(x) \frac{d^2}{dx^2} + b(x) \frac{d}{dx} + c(x).$$

Find the condition that the operator H is Hermitian in the scalar product space defined in the previous exercise. Assume that $r(x)$ is differentiable. This should give you two conditions on the functions a, b, c, for a fixed choice of r.

4.16 Consider a non-Hermitian operator a, which satisfies $[a, a^\dagger]_+ \equiv aa^\dagger + a^\dagger a = 1$ as well as $a^2 = 0$ and define $Q = (P - iW(X))a$, where X, P are the coordinate and momentum of a one-dimensional particle. Show that

$$[Q, Q^\dagger]_+ = 2H = P^2 + W^2 + W'(aa^\dagger - a^\dagger a).$$

Show that the operator in parentheses has eigenvalues ± 1, and that a can be represented by a two-dimensional matrix. The full Hilbert space consists of two component wave functions $\Psi = \psi_i(x)$. Show that if a state exists, satisfying

$$Q\Psi = 0,$$

then it is the ground state of the system. Find the explicit solution to this equation, and the criterion that this candidate ground state wave function is normalizable. Find the lowest eigenvalue. This system is called *supersymmetric quantum mechanics* [13].

4.17 Show that $\frac{d}{dx}\theta(x) = \delta(x)$ by integrating the derivative against a smooth integrable function and using integration by parts.

4.18 Show that

$$\int_0^\infty ds \, \frac{e^{ixs}}{2\pi i(s + i\epsilon)} = \theta(x). \tag{4.57}$$

Here ϵ is a small positive number, which is taken to zero after doing the integral. Use Cauchy's theorem from complex analysis.

4.19 Consider the infinite square well, Exercise 4.6 with $V_0 \to -\infty$. Shift the energy by an infinite amount so that the energy inside the well is defined to be zero, and that outside to be positive infinity. Show that there are only bound state solutions, and that they correspond to solving the free Schrödinger equation with boundary conditions at $x = \pm a$.

4.20 Show, for *every* solution of the time-dependent Schrödinger equation of the infinite square well, that there is a time T such that $\psi(x, 0) = \psi(x, T)$. You can use the expansion of a general solution in energy eigenstates. Now consider the classical motion in this potential. A particle will return to the same state at some time T_{cl}. Show that T_{cl} depends on the energy. See [14] for an explanation for this discrepancy between quantum and classical recurrences.

4.21 Solve for the bound states in the potential

$$V(x) = -\frac{\hbar^2 k_1^2}{2m}[\theta(x+a+b)\theta(a-b-x) + \theta(a+b-x)\theta(x-b+a)],$$

where $0 < a < b$. This has the form of a double square well, symmetric around the origin. We can view it as a very primitive model of an electron attracted to two different nuclei, considered as infinitely heavy.[3] Set up the equations that solve for the ground state energy, and argue that the ground state energy is lowered as b is made smaller with a fixed (corresponding to moving the nuclei, without altering the potential each nucleus exerts on the electron. The argument here should go back to the original Hamiltonian). If we now restore the finite mass of the nuclei, the variation of the ground state energy will correspond to a force on the nuclei. Is it attractive or repulsive? We will do more realistic problems of this type when we study the Born–Oppenheimer approximation in Chapter 11.

4.22 In the previous problem consider $b \gg 1$, with a fixed and of order 1. You should be able to solve the bound state equations by hand in this limit. There is another, very interesting, way to approach the solution. Argue that a solution, where the wave function is approximately the bound state wave function for a single well, is a good approximate solution of the one well problem in this limit. There are two such solutions, which are degenerate in energy, because of the reflection symmetry of the potential. Consider arbitrary linear combinations of these solutions, and determine which linear combination minimizes the expectation value of the Hamiltonian. What is the orthogonal linear combination? How do these two linear combinations transform under the reflection symmetry? This discussion should remind you of the ammonia molecule. Explain why.

4.23 Consider a general potential $V(x)$ which has the property that $V(x) \to 0$ rapidly as $x \to \pm\infty$. We will discuss the question of how rapid the falloff has to be when we talk about scattering in a Coulomb potential. At $\pm\infty$ we can use solutions

$$\psi_{\pm} = A_{\pm}e^{ikx} + B_{\pm}e^{-ikx}.$$

Since the Schrödinger equation only has two linearly independent solutions, there have to be two linear relations. The time dependence of all of these solutions is $e^{-i\frac{\hbar^2 k^2 t}{2m}}$, so the coefficients B_+ and A_- represent incoming waves, while the other two coefficients are outgoing waves. We write the two linear relations as expressions for the outgoing waves in terms of the incoming waves.

$$\begin{pmatrix} B_- \\ A_+ \end{pmatrix} = \begin{pmatrix} S_{11} & S_{12} \\ S_{21} & S_{22} \end{pmatrix} \begin{pmatrix} A_- \\ B_+ \end{pmatrix}.$$

[3] The lightest nucleus weighs $2,000$ times as much as an electron.

The 2×2 S-matrix or *scattering matrix* relates incoming waves from either side to outgoing waves on either side. Write formulae relating reflection and transmission coefficients to S-matrix elements. Argue that the S-matrix is unitary $S^\dagger S = 1$, as a consequence of conservation of probability.

4.24 Consider the Hamiltonian

$$H = \frac{P^2}{2} + \frac{c}{X^2},$$

where we have used units such that $\hbar^2 = m$, which you can call *unnatural units*. Show that the operator $XP + PX \equiv D$ satisfies

$$[H, D] = -4iH.$$

Use this fact to solve the Heisenberg equation of motion for D. You will want to recall that the operator H is time independent. Also show that if H has an eigenstate $\psi_1(x)$ with eigenvalue E_1, which is normalizable, or delta function normalizable, then it has another one for every other real value of E with the same sign as E_1. What is the corresponding eigenstate?

4.25 Argue that for $E = 0$, the eigenfunctions H are also eigenfunctions of the rescaling operator D, and are therefore power laws. Show that they are not even delta function normalizable, so $E = 0$ is not in the spectrum of H.

4.26 Argue that for $c > 0$ the expectation value of H in any normalizable state $\psi(x)$ is positive, and that this means that all eigenvalues are positive.

4.27 Solve the Schrödinger equation for the Hamiltonian in Exercise 4.24 exactly in terms of Bessel functions. Use Mathematica, Maple, SAGE, or any handbook of functions to determine the behavior of the solutions at the origin and at infinity, as well as the conditions for normalizability. Show that for $c > 0$, there are a continuum of delta function normalizable solutions, but no normalizable ones.

CHAPTER 5

The Harmonic Oscillator

5.1 INTRODUCTION

This is probably the most important chapter in this book. It introduces the harmonic oscillator, the single most important soluble problem in physics. It also contains a very brief introduction to the theory of quantized fields, since quantized fields satisfying linear field equations are nothing but collections of quantum harmonic oscillators. The real importance of quantized fields is that they are the correct description of multiparticle states of all the particles we have observed in the world, and explain why those states are either totally symmetric (Bose–Einstein particle statistics) or totally antisymmetric (Fermi–Dirac statistics) under interchange of particle labels. The theory of quantized fields also explains the correct meaning of the phrase "wave–particle" duality, which pervades much of the old quantum mechanics (QM) literature and many current textbooks. Bosonic quantized fields have two different kinds of states: their energy eigenstates are interpreted as collections of noninteracting particles, while *large amplitude coherent states* behave like classical waves. We will explain the concept of coherent state first in the context of a single harmonic oscillator. Coherent states turn out to be the most efficient tool for solving the dynamical equations of an oscillator.

5.2 QUANTIZING THE SIMPLE HARMONIC OSCILLATOR

Simple harmonic motion is the most useful textbook problem in the history of physics textbooks. Its utility comes from the fact that *almost every* problem in physics has a lowest energy state or *ground state*, and small deviations in energy from the ground state can be decomposed into simple harmonic motions, with some spectrum of frequencies. The harmonic oscillator is not just an artificial exactly soluble problem, but forms the basis for investigation of a very wide range of systems.

Force is defined in classical mechanics as the rate of change of momentum of a particle. In QM,

$$\dot{P} = \frac{i}{\hbar}[H, P],\tag{5.1}$$

so in order to have a force, we must add a term to the Hamiltonian that does not commute with P. The simplest possibility is to add a function of X, which we call *the potential* and denote by $V(X)$. V has dimensions of energy. It leads to the Heisenberg equation of motion

$$\dot{P} = \frac{i}{\hbar}[V(X), P].\tag{5.2}$$

We can understand how to compute this by taking $V = X^n$, and observing that, for any three operators,

$$[AB, C] = ABC - CAB = (AC - CA)B + A(BC - CB) = [A, C]B + A[B, C].\tag{5.3}$$

This is called Leibniz' rule for commutators. You might note the similarity to the action of a derivative on a product, but you have to be careful of the ordering of the terms. Applying Leibniz' rule with $A = X^{n-1}, B = X$, and $C = P$, we get

$$[X^n, P] = [X^{n-1}, P]X + X^{n-1}i\hbar = in\hbar X^{n-1},\tag{5.4}$$

where the last step follows by induction. It follows that for any potential that has a power series expansion around some point x_0,

$$[V(X), P] = i\hbar \frac{dV}{dX},\tag{5.5}$$

so that

$$\dot{P} = -\frac{dV}{dX}.\tag{5.6}$$

This is the quantum mechanical version of Newton's second law.

If V has a minimum at some point (which without loss of generality we can take to be $x = 0$), then near that point we can expand $V \sim V_0 + \frac{1}{2}kX^2$. The harmonic oscillator problem approximates a more general problem by just taking V to be given exactly by this formula. Also, we usually set $V_0 = 0$ since this just shifts the zero of energy, which is not observable. The Heisenberg equations of the harmonic oscillator are

$$\dot{X} = \frac{P}{m},\tag{5.7}$$

$$\dot{P} = -kX,\tag{5.8}$$

from which we conclude that

$$ci\dot{X} + \dot{P} = ci\frac{P}{m} - kX = \frac{ci}{m}(P + i\frac{mk}{c}X), \tag{5.9}$$

for any c. If we choose $c = \sqrt{km}$, then the time derivative of this complex linear combination is proportional to itself, so that

$$P(t) \pm i\sqrt{km}X(t) = e^{\pm i\omega t}(P(0) \pm i\sqrt{km}X(0)), \tag{5.10}$$

where

$$\omega = \sqrt{\frac{k}{m}}. \tag{5.11}$$

Notice that since we never had to multiply operators together, these manipulations are equally valid in classical and quantum mechanics.

We define the creation and annihilation[1] operators by

$$a^\dagger \equiv \frac{1}{\sqrt{2\hbar m\omega}}(m\omega X - iP), \tag{5.12}$$

$$a \equiv \frac{1}{\sqrt{2\hbar m\omega}}(m\omega X + iP). \tag{5.13}$$

The operators X and P are Hermitian, because they have real eigenvalues, so a and $a^\dagger$ are Hermitian conjugates of each other. Their commutation relations with each other are *really interesting*. We have

$$a^\dagger a = \frac{1}{2\hbar m\omega}(m\omega X - iP)(m\omega X + iP), \tag{5.14}$$

$$= \frac{1}{2\hbar m\omega}(m^2\omega^2 X^2 + P^2 - im\omega[P, X]). \tag{5.15}$$

Also,

$$aa^\dagger = \frac{1}{2\hbar m\omega}(m\omega X + iP)(m\omega X - iP), \tag{5.16}$$

$$= \frac{1}{2\hbar m\omega}(m^2\omega^2 X^2 + P^2 + im\omega[P, X]). \tag{5.17}$$

If we use $[X, P] = -[P, X] = i\hbar$, then these two equations can be written

$$H = \frac{1}{2}(\frac{P^2}{m} + kX^2) = \hbar\omega(a^\dagger a + \frac{1}{2}), \tag{5.18}$$

---

[1] These are also called creation and destruction, raising and lowering, and ladder operators.

and

$$[a, a^\dagger] = 1. \tag{5.19}$$

Using Leibniz' rule again, we can derive from these that

$$[H, a] = -\hbar\omega a, \tag{5.20}$$

and

$$[H, a^\dagger] = \hbar\omega a^\dagger, \tag{5.21}$$

which allow us to solve Heisenberg's equations as

$$a(t) = e^{-i\omega t}a(0). \tag{5.22}$$

We should have expected this because the creation and annihilation operators are just multiples of the complex linear combinations of X and P that we discussed above. Finally, note that the commutator between a and $a^\dagger$ is the same as that between the multiplication operator w by a complex variable, and the partial derivative $\frac{\partial}{\partial w}$, with respect to that variable.

The commutators of the Hamiltonian with the creation and annihilation operators, when they are applied to an eigenstate of the Hamiltonian, give

$$[H, a]|E\rangle = -\hbar\omega a|E\rangle. \tag{5.23}$$

This can be rewritten

$$Ha|E\rangle = (E - \hbar\omega)a|E\rangle, \tag{5.24}$$

which says that $a|E\rangle$ is proportional to an eigenstate of H with eigenvalue lowered by $\hbar\omega$. We say proportional to, rather than equal, because eigenstates are defined to have norm 1. A similar calculation shows that $a^\dagger|E\rangle$ is proportional to an eigenstate of the Hamiltonian with energy $E + \hbar\omega$.

The norm of $a|E\rangle$ is

$$||a|E\rangle||^2 = \langle E|a^\dagger a|E\rangle = \langle E|\frac{1}{\hbar\omega}(H - \frac{1}{2})|E\rangle = \frac{(E - \frac{1}{2})}{\hbar\omega}. \tag{5.25}$$

This fact is responsible for the alternate name of raising and lowering operators for $a^\dagger$ and a. In a couple of paragraphs, we will see that the allowed values of E are $E_n = (n + \frac{1}{2})\hbar\omega$, where n is a nonnegative integer. Let us denote the eigenstate with eigenvalue E_n by $|n\rangle$. Then we can write

$$a|n\rangle = \sqrt{n}|n - 1\rangle, \tag{5.26}$$

$$a^\dagger|n\rangle = \sqrt{n + 1}|n + 1\rangle, \tag{5.27}$$

and by induction,

$$|n\rangle = \frac{(a^\dagger)^n}{\sqrt{n!}}|0\rangle. \tag{5.28}$$

On the other hand, the Hamiltonian is a sum of squares of Hermitian operators, which means (Exercise 5.1) that all of its eigenvalues are positive. We cannot keep lowering the energy. That is, there must be a lowest energy, or *ground* state, with positive energy, such that

$$a|0\rangle = 0. \tag{5.29}$$

The formula for the Hamiltonian in terms of creation and annihilation operators tells us that this lowest energy state has energy $\frac{1}{2}\hbar\omega$. We can obtain other states, with energies $(n + \frac{1}{2})\hbar\omega$, with n a positive integer, by acting n times on the ground state with $a^\dagger$. Since X is proportional to $a + a^\dagger$, the states that we generate by acting on the ground state with powers of $a^\dagger$ generate all products of a polynomial times the ground state wave function. In a moment, we will show that the ground state wave function is a Gaussian, so that all of the states $|n\rangle$ are normalizable, because of the sort of calculation that we did in the previous chapter. Furthermore, those calculations show that they form a complete basis for the Hilbert space. This means that we have found the entire spectrum of the Hamiltonian, from these simple algebraic considerations. The energy spectrum is $\hbar\omega(n + \frac{1}{2})$, where n is a nonnegative integer.

In fact, we can do a lot more than that, by introducing what are known as *coherent states*. It is worth doing this because these states have many uses, and illuminate the relation between classical waves and quantum particles, which we will call *wave–particle duality*.[2] Coherent states are defined by the formula:

$$|z\rangle = e^{za^\dagger}|0\rangle, \tag{5.30}$$

where $|0\rangle$ is the ground state of the oscillator, satisfying[3]

$$a|0\rangle = 0.$$

---

[2] In much of the literature on QM, wave–particle duality is used for a different notion: the supposed duality between the Schrödinger wave function and the particles for whose properties that wave function computes probabilities. This use of the phrase is misleading. In the rest of physics and mathematics, the word duality refers to two equivalent descriptions of the same system, in terms of different variables. The Schrödinger wave function is not another description of particles. Particles are entities whose properties we measure in single experiments in the real world. The wave function is a machine for computing probabilities that those measurements will give certain results. It cannot be determined by any single experiment, but only by tabulating the frequencies of repeated trials. Our use of the phrase wave–particle duality refers to an actual equivalence between a particle and a wave description of the space of states of quantum fields. The intent is to supplant the older, misleading, usage.

[3] The 0 on the right-hand side of this equation means the zero vector, while $|0\rangle$ is a nonzero state whose corresponding wave function is a normalized Gaussian.

Thus, the ground state is the coherent state with $z = 0$. We will see in a moment that the vectors $|z\rangle$ all have finite norm, but that norm is not equal to one. Although we should really reserve the term *state* for vectors of norm 1, we will follow convention and refer to $|z\rangle$ as a coherent state, and the unit vector $\frac{|z\rangle}{\sqrt{\langle z|z\rangle}}$, as a *normalized coherent state*. The coherent states are of obvious utility as a generating function for the energy eigenstates; the n-th term in the power series expansion in z of the coherent state is $\frac{1}{\sqrt{n!}}|n\rangle$. This is only the first of their remarkable properties.

Let us solve the Schrödinger equation with $|z\rangle$ as an initial condition.

$$|z, t\rangle = e^{-\frac{i}{\hbar}Ht}|z\rangle = \sum_{n=0}^{\infty} \frac{z^n}{n!} e^{-i\omega t(n+\frac{1}{2})}(a^\dagger)^n|0\rangle = e^{-i\omega t/2}|z(t)\rangle, \qquad (5.31)$$

where $z(t) = e^{-i\omega t}z$. That is, *the quantum evolution of a coherent state is, up to a phase which just measures the ground state energy, exactly give by the classical evolution of the coherent state parameter z.* Indeed, we will see in a moment that both the position and momentum space wave functions of the coherent state are Gaussians of time-independent width, centered around a classically evolving position and momentum. Although this result is special to the harmonic oscillator, it shows that the spreading of wave packets in position space, which we found for free particles, can be drastically modified by forces acting on those particles. This observation that coherent states satisfy classical equations of motion is at the basis of the connection between particles and classical fields in QM.

The scalar product of two coherent states $\langle w|z\rangle$ is easily computed as

$$\langle w|z\rangle = \langle 0|e^{w^* a}e^{za^\dagger}|0\rangle = \sum_{m,n} \frac{(w^*)^m z^n}{\sqrt{m!n!}}\langle m|n\rangle = e^{w^* z}, \qquad (5.32)$$

where we have used $\langle m|n\rangle = \delta_{mn}$. Thus, the norm of a coherent state is $e^{z^* z}$. We also learn that different coherent states are not orthogonal to each other.

The final important property of coherent states is that

$$a|z\rangle \equiv z|z\rangle. \qquad (5.33)$$

The annihilation operator is not a normal operator, since it does not commute with its Hermitian conjugate, but this does not mean it cannot have eigenstates. What it does mean is that the eigenstates are not orthonormal, as we have just seen. To prove this identity, we use the operator equation

$$[a, (a^\dagger)^n] = n(a^\dagger)^{n-1}, \qquad (5.34)$$

which we prove by applying the Leibniz rule for commutators

$$[A, BC] = B[A, C] + [A, B]C, \qquad (5.35)$$

repeatedly. Then

$$a|z\rangle = \sum a\frac{(za^\dagger)^n}{n!}|0\rangle = \sum [a, \frac{(za^\dagger)^n}{n!}]|0\rangle = z|z\rangle . \qquad (5.36)$$

We can find the position ($\psi(x,z)$) or momentum ($\psi(p,z)$) space wave functions for coherent states by solving the equation $a|z\rangle = z|z\rangle$ in the appropriate basis:

$$\frac{1}{\sqrt{2\hbar m\omega}}(m\omega x + \hbar\frac{\partial}{\partial x})\psi(x,z) = z\psi(x,z).$$

$$\frac{1}{\sqrt{2\hbar m\omega}}(i\hbar m\omega\frac{\partial}{\partial p} + ip)\psi(p,z) = z\psi(p,z).$$

It is easy to solve these first-order linear Ordinary Differential Equations (ODEs), and the answers are both Gaussians. You will work out the details, as well as the proper normalizations of these functions, in Exercise 5.2. The correctly normalized position space wave function is

$$\psi(x,z) = (\frac{m\omega}{\pi\hbar})^{1/4}e^{-\frac{m\omega}{2\hbar}(x-x_0)^2}e^{ik_0 x}, \qquad (5.37)$$

where

$$x_0 = \text{Re } z\sqrt{\frac{2\hbar}{m\omega}},$$

and

$$k_0 = \text{Im } z\sqrt{\frac{2m\omega}{\hbar}}.$$

This is the wave function corresponding to the normalized coherent state

$$\psi(x,z) = \frac{\langle x|z\rangle}{\sqrt{\langle z|z\rangle}}. \qquad (5.38)$$

We have seen that the coherent states are not orthogonal. The statement that they are *over complete* is:

$$1 = \int \frac{d^2z}{\pi} e^{-zz^*}|z\rangle\langle z|. \qquad (5.39)$$

To prove this, we use the power series expansions of e^{az^*} and its complex conjugate and find that the right-hand side of this equation is

$$\sum_{m,n}\frac{1}{n!m!}(a^\dagger)^n|0\rangle\langle 0|a^m \int \frac{d^2z}{\pi}(z^*)^m z^n e^{-z^*z}. \qquad (5.40)$$

If we write $z = re^{i\theta}$, then the angular integral of the terms with $m \neq n$ vanish. Recalling the normalization

$$|n\rangle = \frac{(a^\dagger)^n}{\sqrt{n!}}|0\rangle,$$

we find that the coefficient of the projection operator $|n\rangle\langle n|$ in this sum is

$$c_n = \frac{2}{n!} \int_0^\infty r dr e^{-r^2} r^{2n}. \tag{5.41}$$

Introducing $u \equiv r^2$, we rewrite

$$c_n = \frac{1}{n!} \int_0^\infty du e^{-u} u^n = 1. \tag{5.42}$$

Thus, the right-hand side is equal to $\sum_n |n\rangle\langle n| = 1$.

As a final bonus, we can compute the position space eigenfunctions of the harmonic oscillator Hamiltonian. The power series expansion in z of the coherent state is

$$\sum_{n=0}^\infty \frac{z^n}{n!}(a^\dagger)^n|0\rangle. \tag{5.43}$$

From our discussion above, we recognize the term proportional to z^n as being proportional to the n-th eigenstate. We have

$$|z\rangle = \sum_{n=0}^\infty \frac{z^n}{\sqrt{n!}}|n\rangle. \tag{5.44}$$

Thus,

$$\psi(x, z) \equiv \langle x|z\rangle = \sum_{n=0}^\infty \frac{z^n}{\sqrt{n!}}\psi_n(x). \tag{5.45}$$

Since $\psi(x, z)$ is a Gaussian, with z appearing only in the term linear in x, each of these functions has the form

$$\psi_n(x) = H_n(x\sqrt{\frac{m\omega}{2\hbar}})e^{-\frac{m\omega}{2\hbar}x^2},$$

where $H_n(x)$ is a polynomial. It is called the n-th Hermite polynomial. These eigenfunctions were originally found by solving the second-order ODE

$$\langle x|H|\psi\rangle = E\langle x|\psi\rangle,$$

subject to boundary conditions of normalizability.

Many older textbooks emphasize that the Schrödinger equation is a differential equation, and exploit the fact that similar methods are used to solve Maxwell's equations, with which students already have some familiarity. You can follow this route in Exercise 5.3. The truth is, it has very little utility. The operator methods we have exploited are far simpler and, more importantly, they generalize to a large class of many body systems. There is no calculation for the harmonic oscillator, which is more transparent in the language of differential equations than it is in that of coherent states. For the most part, the utility of differential equations in QM is restricted to problems that can be reduced, exactly or approximately, to *ordinary* differential equations. Although such problems abound in textbooks, they are not easily found in applications of QM to important problems in modern physics. On the other hand, the use of operator algebra, particularly the algebra of creation and annihilation operators, is a standard tool in most real world applications of QM.

5.3 QUANTIZATION OF FIELDS AND WAVE–PARTICLE DUALITY

Wave–particle duality is a phrase, which was thrown around a lot in the early days of QM and still appears in many contemporary textbooks. It was almost universally interpreted as referring to the Schrödinger wave function in coordinate space, as if it were a wave in physical space. One cannot emphasize too strongly that this interpretation is misleading. The Schrödinger wave function for anything more than a single particle, obeys a "wave equation" in the $3N + 1$ dimensional configuration space of particle coordinates (plus time), which is not physical space.

The desire to redefine the phrase wave–particle duality is based on many years of teaching QM to juniors, seniors, and graduate students. The vast majority of students come into class thinking of the Schrödinger wave function as a wave in three-dimensional space and are shocked to learn the obvious fact that it is not, for multiparticle systems. The historical concept of wave–particle duality was based on a conflation of false thinking about the Schrödinger wave function, with the obvious fact that the electromagnetic field manifested both as a particle and a wave.

Indeed, as we have seen, the Schrödinger wave function is just a device for keeping track of the *probabilities* of a whole collection of variables. Its entire physical content consists of the formula

$$\langle A \rangle = \text{Tr} \, (A P_\psi),$$

for all normal operators on Hilbert space. In this formula, the left-hand side refers to the statistical expectation values of the quantities associated with A, some of which are subject to measurement. P_ψ is the projection operator in Hilbert space on the physical state whose coordinate space representation is the Schrödinger wave function $\psi(x_i)$. In coordinate space, P_ψ is an integral operator, with kernel

$$K(x, y) = \psi(x)\psi^*(y).$$

We test the predictions of the theory by preparing the system in the same state over and over again, and measuring each of the variables. Then we compare the frequencies of occurrence of each value of each variable, with the above formula for the expected value. That is the only role of the wave function in the theory. We do not measure the wave function directly.[4] We can extract knowledge of the wave function only by doing repeated experiments with the same initial state.

By contrast, a classical field, like Maxwell's electromagnetic field, is a variable that we measure in single experiments. In this sense, it is like a particle coordinate or momentum and it is certainly not a probability distribution. In order to understand the relation of such fields to particles, we have to study the quantum mechanical treatment of fields. This is usually considered a very advanced subject and is often discussed only in graduate courses. In fact, the rudiments of the theory of quantized fields are just a simple extension of our discussion of harmonic oscillators. Discussing them here will enable us to understand a concept that really deserves the name of wave–particle duality, the Planck–Einstein–Compton (PEC) notion of photons, and the peculiar notion of *identical particle statistics*, with a very small incremental effort. Physicists should cease to use the term wave–particle duality, when referring to the relation between the Schrödinger wave function and the observables of particles.

The electromagnetic field is a complicated object, with six components, and a host of subtle properties related to gauge invariance. We will restrict our attention instead to a field with one component, satisfying the same wave equation as each component of the electric and magnetic fields of a light wave

$$\partial_t^2 \phi(\mathbf{x}, t) - c^2 \nabla^2 \phi(\mathbf{x}, t) = 0. \tag{5.46}$$

Most of the physics discussed in this book is invariant under Galilean transformations, or can be thought of as a set of equations one obtains by going to the center of momentum reference frame by a Galilean boost. The above wave equation is not. It involves a particular velocity c of propagation of waves, whereas Galilean transformations add a constant to the velocity. There are two reasons for studying this equation here. The most important one is that the physics of atoms, molecules, and condensed matter systems is crucially dependent on the properties of photons, and in particular on the PEC relation between photon energy-momentum and frequency-wave number. The Planck distribution (Chapter 12) also depends on the properties of photons/electromagnetic waves and the duality between them. Finally, we should note that exactly the same wave equation describes the properties of sound waves and the quasiparticles called phonons, in condensed matter systems. In that context, the "contradiction" between Lorentz invariant wave equations and Galilean invariant physics is resolved because the phonon wave equation is valid only in the rest frame of the material. The

---

[4] See Chapter 10 on Measurement Theory for a more precise and detailed discussion of the meaning of this sentence.

mathematical treatment of this wave equation at this point in the book is natural because, as we are about to see, it follows simply by copying our treatment of the harmonic oscillator.

The Fourier transform

$$\phi(\mathbf{x}, t) = \int \frac{d^3k}{(2\pi)^3} e^{i\mathbf{k}\cdot\mathbf{x}} \phi(\mathbf{k}, t), \tag{5.47}$$

turns Equation 5.46 into a continuous infinity of harmonic oscillator equations

$$\int \frac{d^3k}{(2\pi)^3} e^{i\mathbf{k}\cdot\mathbf{x}} [\ddot{\phi}(\mathbf{k}, t) + c^2\mathbf{k}^2\phi(\mathbf{k}, t)] = 0. \tag{5.48}$$

Remember that if the Fourier transform of a function vanishes, then the function itself vanishes.

If a continuous infinity of oscillators makes you nervous, we can put the system in a box, with boundary conditions on ϕ. For example, if we impose periodic boundary conditions on a box of size L, the wave numbers will be restricted to $\mathbf{k}_n = \frac{2\pi\mathbf{n}}{L}$, where $\mathbf{n}$ is a vector of integers. Any sort of boundary condition will reduce the continuous infinity to a discrete set of $\mathbf{k}_n$. If you are still nervous about an infinite number of oscillators, we can agree that our experiments are never going to probe wavelengths shorter than some minimum size r_0. This puts an upper cutoff on the wave $|\mathbf{k_n}| < \frac{1}{r_0}$. None of this detail distorts the main message, which is that a field satisfying a linear, translation invariant, field equation is nothing but a collection of decoupled harmonic oscillators, one for each wave number. For our particular wave equation, which is called the D'Alembert equation, or the massless Klein–Gordon equation, the relation between the oscillator frequency and the wave number is $\omega = c|\mathbf{k}|$, which is the same as that for light waves.

We can now quantize all these oscillators, using the techniques of the previous section. The energy eigenstates consist of the ground state, and states obtained by acting on it with some number of creation operators. These eigenstates satisfy

$$(H - E_0)a^\dagger(\mathbf{k}_1)\ldots a^\dagger(\mathbf{k}_n)|0\rangle = [\sum_{j=1}^{n} \hbar\omega(\mathbf{k_j})]a^\dagger(\mathbf{k}_1)\ldots a^\dagger(\mathbf{k}_n)|0\rangle. \tag{5.49}$$

That is, *the eigenstates of the quantized field look like the states of a collection of noninteracting "photons" obeying the PEC relation between energy, momentum and wave number.* We say this because the states are labelled by a finite set of wave numbers, the energy is a sum of contributions from independent wave numbers, and the energy and wave number are related by the PEC formula. Note however that these "particles" are automatically indistinguishable, and that only states symmetric under particle interchange are allowed, because the different creation operators all commute with each other.

On the other hand, time-dependent coherent states of these oscillators are parameterized by functions $z(\mathbf{k}, t)$ *satisfying the Fourier transformed classical field equations.* This is just our result from the previous section that the coherent state parameters satisfy the classical oscillator equations, combined with the Fourier transform. The function

$$z(\mathbf{x}, t) = \int \frac{d^3k}{(2\pi)^3} e^{i\mathbf{k}\cdot\mathbf{x}} z(\mathbf{k}, t), \qquad (5.50)$$

satisfies the classical field equations and completely characterizes a combined coherent state of the oscillators with fixed wave number.

This then is the concept that deserves the name wave–particle duality. A single system, the quantized field, has two different kinds of states. Its energy eigenstates are multiphoton states, and have a particle interpretation. Its coherent states are associated with classical fields. The two kinds of states are incompatible with each other, in the sense that eigenstates of the matrices σ_1 and σ_3 were incompatible with each other for a two state system. In an energy (and particle number) eigenstate, the probability for finding a coherent state characterized by a fixed classical field history is

$$|\langle z(\mathbf{k}, t)|a^\dagger(\mathbf{k_1}) \dots a^\dagger(\mathbf{k_n})|0\rangle|^2. \qquad (5.51)$$

A fixed energy eigenstate does not correspond to a definite classical field. Conversely, if we say we have a fixed classical field, then energy and particle number have statistical fluctuations. You will explore these fluctuations in Exercises 5.8 and 5.9. It does not really make sense to treat a coherent state as a classical field unless the fluctuations in its energy are small compared to the expectation value of the energy. In the exercises, you will see that this is equivalent to saying that the field is large. More precisely, it says that the number of particles in each classical mode of the field is much larger than one. Coherent states that do not satisfy this condition are better thought of as superpositions of states of a small number of particles.

It should be emphasized that the classical nature of the coherent state parameters is not approximate. For fields satisfying linear field equations, the exact quantum equations of motion lead to classical field equations for the coherent state parameters. For strongly nonlinear field theories, both the existence of approximate classical fields and the particle interpretation of low-energy excitations about the ground state have to be rethought, but for linear field theories, the properties of the simple harmonic oscillator cement the connection between waves and particles.

As we noted above, the particle interpretation of the multiphoton states leads to another surprise. In classical particle mechanics, particles are distinguishable, even if their Hamiltonian has an exact permutation symmetry. Speaking colloquially, we can imagine that we watch the particles move around, starting from their original positions, and keep track of which was which. When this kind of reasoning is applied to classical statistical mechanics, it leads to the "Gibbs Paradox" in the entropy of mixing of two identical volumes filled with identical classical particles.

If one takes a box filled with equal volumes of two different types of gas, separated by a partition, and removes the partition, the entropy goes up when the box has returned to equilibrium.

According to Boltzmann's statistical mechanics, this is explained by the fact that each individual gas molecule has twice as much volume to move around in. However, this argument makes no reference to whether the gases are the same or different. Experiment shows that there is no corresponding decrease in temperature when the two sides of the partition are filled with identical particles.

Gibbs solved this problem by postulating that identical particles were indistinguishable. The particle interpretation of the excitations of quantum fields automatically produces particles that are indistinguishable. If we think of the field as the basic entity, then obviously exchanging two localized excitations of the field does not change the field configuration, and leads to the same physical state. Permutations of identical particles are, in this case, not merely symmetries of the system, but redundancies of a particle description of the underlying field degrees of freedom.[5] The experimental fact that there is no increase in entropy when we remove a partition between two volumes filled with the same gas is evidence that this identical particle picture is correct. However, it is not by itself evidence for QM.[6]

With small modifications, we can perform a similar quantization of the Schrödinger equation, *viewed as a classical field theory in space time.* If we write the Fourier transform

$$\Psi(\mathbf{x}, t) = \int \frac{d^3 k}{(2\pi)^3} e^{i\mathbf{k}\cdot\mathbf{x}} b(\mathbf{k}, t), \qquad (5.52)$$

then the Schrödinger equation for Ψ, combined with its complex conjugate, is equivalent to

$$\partial_t b(\mathbf{k}, t) = -i\frac{\hbar k^2}{2m} b(\mathbf{k}, t), \qquad (5.53)$$

$$\partial_t b^\dagger(\mathbf{k}, t) = i\frac{\hbar k^2}{2m} b^\dagger(\mathbf{k}, t). \qquad (5.54)$$

Again, these look like the equations for the creation and annihilation operators of an infinite set of oscillators (where we have used the † notation for Hermitian conjugation of operators,

---

[5] In modern high-energy physics, such redundancies are called gauge symmetries, and are discrete analogs of the gauge invariance of the description of the electromagnetic field in terms of potentials.

[6] For many years, Einstein advocated models in which particles would appear as localized solutions of nonlinear classical field equations. In such models, particle exchange would also be a redundancy. Modern quantum field theory [15] unifies these two points of view. Localized solutions to nonlinear field theories indeed exist. When these models are quantized, one finds that both the localized solutions and small fluctuations around a translation invariant background are particles. The analysis is valid in an approximation analogous to the JWKB approximation we will study in Chapter 17, which is dependent on the existence of a small dimensionless expansion parameter g. In the small g limit, particles that arise as small fluctuations, have much lower energy than particles that are localized classical solutions.

rather than the ordinary complex conjugate, in anticipation of the operator interpretation of the Fourier coefficients of the field). We quantize the *classical Schrödinger field* by imposing

$$[b(\mathbf{k}), b^\dagger(\mathbf{k}')] = \delta(\mathbf{k}, \mathbf{k}'), \tag{5.55}$$

where the right-hand side is a Kronecker delta. We have again put the field in a box with boundary conditions that make the allowed set of wave numbers discrete. The solution of the time-dependent Schrödinger equation,

$$\Psi(\mathbf{x}, t) = \int \frac{d^3 k}{(2\pi)^3} e^{i\mathbf{k} \cdot \mathbf{x}} e^{-i\frac{\hbar k^2}{2m}} b(\mathbf{k}), \tag{5.56}$$

is now interpreted, as was the field $\phi(\mathbf{x}, t)$ above, as a quantized field operator, in the Heisenberg picture, acting in the Hilbert space of multiparticle states, which are the eigenstates of the energy operator

$$H = \sum_{\mathbf{k}} \frac{\hbar^2 k^2}{2m} b^\dagger(\mathbf{k}) b(\mathbf{k}). \tag{5.57}$$

The multiparticle states will all be symmetric under any permutation of the particles.

Let us end this section by explaining why the misleading notion that wave–particle duality had to do with the Schrödinger wave function arose. To this end, consider states of *single particles*. These have the form

$$e^{-i\frac{\hbar \mathbf{k}^2 t}{2m}} a^\dagger(\mathbf{k}) |0\rangle, \tag{5.58}$$

and represent a single particle with momentum $\hbar \mathbf{k}$. The time-dependent factor comes from acting with $e^{-i\frac{H}{\hbar} t}$ on $a^\dagger(\mathbf{k}) |0\rangle$. We can consider superpositions of such states, defined by a complex function of wave number $f(\mathbf{k})$:

$$|f\rangle \equiv \int \frac{d^3 k}{(2\pi)^3} e^{-i\frac{\hbar \mathbf{k}^2 t}{2m}} f(\mathbf{k}) a^\dagger(\mathbf{k}) |0\rangle. \tag{5.59}$$

These are states with a single particle in some superposition of momentum eigenstates, which can be localized in $\mathbf{x}$. The traditional single particle Schrödinger wave function is just the Fourier transform

$$\psi(\mathbf{x}, t) = \int \frac{d^3 k}{(2\pi)^3} e^{-i\frac{\hbar \mathbf{k}^2 t}{2m}} f(\mathbf{k}) e^{-i\mathbf{k} \cdot \mathbf{x}}. \tag{5.60}$$

It then follows that the Schrödinger wave function satisfies the same wave equation as the field $\Psi(\mathbf{x}, t)$ (indeed, one can think of it as the matrix element of the Heisenberg picture field between the ground state and the state with a single particle). Richard Feynman was a great advocate of using the principle that "the same equations have the same solutions," to relate the solutions of one physics problem to those of an already solved problem. However, the fact that two different quantities satisfy the same equation does not imply that they have the same meaning. In this case, the quantized Heisenberg field operator, the classical coherent state parameters, and the single particle Schrödinger wave function all satisfy the same field equation, but they all have different meanings. The operator is an observable in QM and can, in principle, be measured at each time. If the system is in a particular coherent state then, as you will show in Exercise 5.10, the expectation value of the field operator will be equal to the coherent state parameter, at all times. The statistical fluctuations around this expectation value may be either large or small compared to the expectation value itself. You will find that when the classical field value is large, the fluctuations are small.

The Schrödinger wave function is something else entirely. It tells us what the probabilities are, in the single particle state $|f\rangle$ to find the particle with either a given position or a given momentum. It cannot be probed by a single measurement, but only by repeated measurements of identically prepared states. The same equations have the same solutions, but the different uses of those solutions can have meanings that are conceptually different.

Let us summarize the important things we have learned in this section. First and foremost, we have derived the PEC relation between the energy of the "smallest quantum of light of a given frequency" and that frequency. We used this in our discussion of the ammonia molecule, but now we really understand it. We have also learned that photons are identical particles, because they are excitations of a quantized field. Gibbs' classical discussion of the entropy of mixing suggests that all particles should arise in this way, and this turns out to be correct. There is, however, one subtlety in the discussion. There turn out to be two different kinds of identical particles in QM (and in the world), called *bosons* and *fermions*. Our discussion so far applies only to bosons, we will introduce fermions in a moment.

Finally, we learned that classical electromagnetic fields are coherent states of bosons. We first introduced coherent states for a single harmonic oscillator as a sort of technical trick for doing calculations. Their real importance is in showing how classical particle physics and classical field theory both emerge from the single concept of a quantized field, the real meaning of wave–particle duality. Not bad for a little algebraic trick for solving a simple harmonic oscillator!

5.4 FERMI–DIRAC STATISTICS

The formula for an n-particle state with each particle in a distinct single particle state:

$$|k_1 \ldots k_n\rangle = b^\dagger(k_1) \ldots b^\dagger(k_n)|0\rangle \tag{5.61}$$

guarantees complete symmetry of the multiparticle states under interchange of the particles, because the $b^\dagger(k)$ operators commute with each other. It turns out that many familiar particles do not obey this symmetry rule, but rather have states that are totally antisymmetric under interchange. Remember that any permutation is a product of transpositions, in which only two particles are interchanged. The new rule is that if the number of transpositions is odd, the state gets multiplied by a minus sign. Particles obeying the symmetric rule are called Bose particles or bosons, and are said to obey Bose–Einstein statistics. Particles obeying the rule with the minus sign are called *fermions* and obey Fermi–Dirac statistics. One can implement the rule easily by insisting that any pair of creation operators $a^\dagger(k_{1,2})$ *anticommute* with one another

$$a^\dagger(k_1)a^\dagger(k_2) + a^\dagger(k_2)a^\dagger(k_1) \equiv [a^\dagger(k_1), a^\dagger(k_2)]_+ = 0. \tag{5.62}$$

Extending this, as in the Bose case, to $k_1 = k_2$ we get

$$a^{\dagger\,2}(k) = 0. \tag{5.63}$$

Those of you who have had a class including elementary atomic physics will have heard of the *Pauli Exclusion Principle*. The previous equation is the mathematical expression of that principle. For bosons, we can make a state with an arbitrary number of particles, in the same single particle state. Fermions "don't like neighbors": you cannot put more than one in any single particle state. This property is the key to the stability of atoms and nuclei, as well as the solidity of solids and many other facts about the real world.

In Exercise 5.11, you will show that in order to be consistent with the anticommutation relations for creation operators, their adjoints have to satisfy anticommutation relations as well:

$$[a(k_1), a^\dagger(k_2)]_+ = \delta(k_1, k_2), \tag{5.64}$$

where the right-hand side is a Kronecker delta, because we have put the system in a box to make the wave numbers discrete. In fact, for a fixed value of $k_1 = k_2$, we have already run across operators with these properties when we studied two state systems, as you will recall in Exercise 5.12. In Exercise 5.13, you will learn a trick, called the *Jordan Wigner* transformation, for constructing the full algebra of Fermion creation and annihilation operators out of a collection of Pauli spin operators.

Given a classical wave equation, like

$$(\partial_t^2 - \nabla^2)\phi = 0, \tag{5.65}$$

or

$$(i\hbar\partial_t - \nabla^2)\Psi = 0, \tag{5.66}$$

you can expand it in Fourier components and the corresponding "creation and annihilation operators" which evolve with negative and positive frequencies. One can then quantize the system as one would a harmonic oscillator, obtaining a theory of Bose particles, *or* one can choose the alternative Fermi–Dirac quantization, obtaining Fermi particles. This raises three questions:

- How do we choose which statistics to use for each type of particle?

- Is there anything weird about the minus signs for Fermions?

- Are there other possibilities besides the Bose–Fermi alternative?

It turns out that the combination of relativity and QM resolves the first question. We will learn later that every elementary particle carries an internal angular momentum called spin. Spin describes the way the states of a particle at rest respond to rotations. It turns out that the spin state of a particle can change by a minus sign under a 2π rotation, because that will not affect the density matrix, which computes all physical expectation values in that state. Particles exhibiting that peculiar minus sign are said to have half integral spin, while those whose states are invariant under 2π rotations have integral spin. It turns out that when we combine QM with relativity, and with the principle that signals cannot be sent faster than the speed of light, half integral spin particles must be a Fermions, and integral spin particles Bosons. This is called the Spin Statistics Theorem [16].

The odd thing about the minus sign is the following: The creation operators for two fermions in different states do not commute with each other. Suppose that the states are localized millions of miles apart. If we can create single fermions locally, then an experiment right here and now, could be affected by an experiment done simultaneously millions of miles away. The resolution of this potential paradox is that the rules of the game, and in particular the Spin Statistics theorem, only allow even numbers of fermions to be created locally. As we will see, given the Spin-Statistics connection, this follows from conservation of angular momentum. Thus, if we insist on angular momentum conservation, then it is impossible to change the number of fermions from even to odd, or vice versa.

Finally to the question of other types of statistics, it turns out that if the world only had two space dimensions, then other kinds of particles, called *anyons*, are possible, but in higher dimension they are not allowed. Nonetheless, there are certain condensed matter systems, confined to a plane, which exhibit anyonic excitations. This is all connected to a wonderful purely quantum mechanical phenomenon, called the Aharonov–Bohm effect, and we will explore it in Chapter 15.

5.5   EXERCISES

5.1   Prove that the eigenvalues of the square of a Hermitian operator are all nonnegative, by proving the stronger theorem that H^2 has nonnegative expectation value in *any* state. The latter theorem is true for any sum of squares of Hermitian operators, and proves the statement about eigenvalues for such sums as well.

5.2 Solve the equations for a coherent state in position and momentum representations:

$$\frac{1}{\sqrt{2\hbar m\omega}}(m\omega x + \hbar\frac{\partial}{\partial x})\psi(x, z) = z\psi(x, z).$$

$$\frac{1}{\sqrt{2\hbar m\omega}}(i\hbar m\omega\frac{\partial}{\partial p} + ip)\psi(p, z) = z\psi(p, z).$$

Write down the solutions that have the normalization $\langle w|z\rangle = e^{w*z}$, used in the text.

5.3 Solve for the harmonic oscillator eigenstates by direct solution of the second-order equations.

a. The equation is

$$[\frac{-\hbar^2}{2m}\frac{d^2}{dx^2} + \frac{kx^2}{2}]\psi = E\psi,$$

with the boundary condition that $\int|\psi|^2 < \infty$. Define $x \equiv (\frac{\hbar^2}{k^2m})^{1/4}$ and $E \equiv \hbar\sqrt{\frac{k}{m}}\epsilon$ and show that

$$[-\frac{d^2}{dy^2} + y^2]\psi = 2\epsilon\psi.$$

b. Show that at large $|y|$, the two linearly independent solutions behave like $e^{\pm\frac{y^2}{2}}$. A solution is normalizable only if it is a falling Gaussian at both $\pm y \to \infty$.

c. Write the solution as

$$\psi = \sum_{k=0}^{\infty} a_k y^k e^{-\frac{y^2}{2}},$$

and convert the differential equation into a recursion relation for the coefficients a_k. Show that unless the recursion terminates, and the prefactor is a polynomial, the solution will behave like $e^{\frac{y^2}{2}}$. Do this by approximating the recursion relation for large k. This polynomial condition is only achieved for quantized values of ϵ.

5.4 Calculate the expectation value and the fluctuations of the number operator $a^\dagger a$ in the coherent state $|z\rangle$. Remember that the coherent state is not normalized to 1 so you have to divide by the norm to calculate expectation values.

5.5 Calculate the overlaps of the ground states of two harmonic oscillators with different frequencies.

5.6 The Klein–Gordon equation in d spatial dimensions is

$$\left(\frac{\partial^2}{c^2 \partial t^2} - \sum_i \frac{\partial^2}{\partial x_i^2}\right)\phi + \frac{1}{\ell^2}\phi,$$

where l is the Compton wavelength $l = \frac{\hbar}{mc}$. Calculate the frequencies of modes of wavenumber $\mathbf{k}$ (a d-dimensional vector). Use the result of the previous problem to calculate the ground state overlap of the Klein–Gordon field with two different values of l. You will have to put the system in finite volume in order to get a nonzero answer. In high enough dimension (how high?), you will find the overlap is zero even in finite volume if there is no ultraviolet cutoff on large wave numbers.

5.7 Use the relations

$$a^\dagger|n\rangle = (n+1)^{1/2}|n+1\rangle$$

and

$$a|n\rangle = n^{1/2}|n-1\rangle$$

to write recurrence relations for the Hermite *polynomials* and their derivatives.

5.8 Compute the uncertainty in the energy operator $H = \hbar\omega(a^\dagger a + \frac{1}{2})$ in a coherent state $|z\rangle$. Remember that you have to divide by the norm of the state. Note that the constant ground state energy cancels out of this computation.

5.9 Compute the uncertainties in particle number $N = \sum_{\mathbf{k}} a^\dagger(\mathbf{k})a(\mathbf{k})$ and energy $H = \sum_{\mathbf{k}} \hbar\omega(k)a^\dagger(\mathbf{k})a(\mathbf{k})$ in a coherent state of the Klein–Gordon field $\phi(x,t)$.

5.10 Show that the expectation value of the operators P and X in a coherent state are proportional to the imaginary and real parts of the coherent state parameter. Generalize this computation to the Klein–Gordon field.

5.11 Show that the anticommutation relation

$$[a(k_1), a^\dagger(k_2)]_+ = \delta_{k_1\,k_2}$$

is required for anticommuting creation and annihilation operators.

5.12 Show that the anticommutation relations

$$a^2 = 0, \quad [a, a^\dagger]_+ = 1$$

are realized in a two-dimensional Hilbert space.

5.13 Given an infinite collection of Pauli matrices $\sigma_a(i)$, $-\infty < i < \infty$ show that the operators

$$\psi(i) = \sigma_+(i) \prod_{j>i} \sigma_3(j)$$

$$\sigma_+(i) = \frac{1}{\sqrt{2}}(\sigma_1(i) + i\sigma_2(i))$$

satisfy

$$[\psi(i), \psi(j)]_+ = 0,$$
$$[\psi(i), \psi^\dagger(j)]_+ = \delta_{ij}.$$

This is called the Jordan–Wigner transformation.

5.14 A different way to solve for the time evolution of a coherent state is the following: the time evolution of the system, starting at $t = 0$ in a coherent state is given by

$$|z\rangle \to e^{-i\frac{H}{\hbar}t}|z\rangle. \tag{5.67}$$

We act on the time evolved state with the Schrödinger picture operator a (this is the Heisenberg picture operator, at $t = 0$).

$$ae^{-i\frac{H}{\hbar}t}|z\rangle = e^{-i\frac{H}{\hbar}t}(e^{i\frac{H}{\hbar}t}ae^{-i\frac{H}{\hbar}t})|z\rangle. \tag{5.68}$$

The quantity in parentheses is just the Heisenberg picture operator $a(t)$. Use the Heisenberg equations of motion to evaluate $a(t)$ and use this to reproduce the formula in the text for the time evolution of $|z\rangle$.

CHAPTER 6

Review of Linear Algebra and Dirac Notation

6.1 INTRODUCTION

This chapter provides a more formal introduction to the mathematics of Hilbert space, using Dirac notation. It also contains a general discussion of symmetries in quantum mechanics (QM) and a derivation of the generalized uncertainty relations for any pair of operators that do not commute. Remember that the key idea of QM is to view the complex N-dimensional generalization of Pythagoras' theorem for a vector of length 1:

$$1 = \sum_n |\langle v|e_n\rangle|^2, \tag{6.1}$$

as the definition of a probability distribution for each such unit vector. The individual terms in the sum are interpreted as *the probability that, given that the system is in the state represented by $|v\rangle$, a measurement to determine whether the system is in the state $|e_n\rangle$ will find that it is in fact in that state.* This probability distribution over all orthonormal bases is equivalent to a probability distribution over all normal operators, defined as operators that are diagonal in *some* basis. The equivalent formula for operators is that the expectation value of A^k in the state $|v\rangle$ is given by

$$\langle A^k \rangle = \mathrm{Tr}\, A^k P_v, \tag{6.2}$$

where $P_v \equiv |v\rangle\langle v|$ is the projection operator on $|v\rangle$. The theory of Hilbert space generalizes all of this to the limiting case of $N \to \infty$. There are lots of subtleties, the most significant of which is that operators can have continuous spectra and there are no normalizable eigenvectors for those eigenvalues.

6.2 SEPARABLE HILBERT SPACES

It is about time to review linear algebra a bit more formally than we have up till now. We will discuss linear algebra in separable Hilbert spaces. A *separable* Hilbert space is an infinite dimensional vector space, with a countably infinite basis and a scalar product. We will include the finite dimensional spaces you know and love from linear algebra in our definition of a Hilbert space. We will use Dirac's notation $|e_n\rangle$ for some particular basis, and replace e by other lower case Latin letters to denote other bases. We will also mostly follow the convention that the basis of eigenstates (eigenvectors normalized to one) of a normal operator labelled by a capital Latin letter like A, will be denoted $|a_k\rangle$. The major exception to this is the Hamiltonian operator H, whose eigenvalues are energies, and whose eigenstates are labeled $|E_k\rangle$.

Given a basis $|e_n\rangle$, we can define a scalar product such that the basis vectors are orthonormal

$$\langle e_n | e_m \rangle = \delta_{mn}. \tag{6.3}$$

A general vector is given by

$$|v\rangle = \sum v_n |e_n\rangle, \tag{6.4}$$

where

$$\|v\|^2 \equiv \sum |v_n|^2 < \infty.$$

The scalar product of two such vectors is

$$\langle v | w \rangle = \sum v_n^* w_n = \langle w | v \rangle^*, \tag{6.5}$$

which identifies the components of $|w\rangle$ as a column vector and the components of $\langle v|$ as a Hermitian conjugate (transpose followed by complex conjugate) row vector. Dirac's terminology for column and row vectors is *kets* and *bras*, because they combine together to make a *bra(c)ket*.[1] Mathematicians call column vectors simply *vectors*, and row vectors are called *covectors*, and thought of as complex valued linear functions defined on the vector space. The scalar product is an identification of the vector space with the vector space of linear functions defined on it. Given a basis of vectors $|e_n\rangle$, there is a canonical basis of linear functions defined by $F_n(|e_m\rangle) = \delta_{mn}$. If we define the identification of vectors with covectors/linear functions by associating each vector with the function that is equal to one on it, and zero on all other basis vectors, then we have defined the scalar product so that this particular basis is *orthonormal*, a portmanteau word that says that each vector in the basis is normalized to one and orthogonal to all the other vectors. We will always work with orthonormal bases, except when using coherent states, which form an overcomplete basis.

---

[1] One supposes that the c in bra(c)ket stands for the vertical line in the notation for scalar product.

It is easy to prove (Exercise 6.1) the Schwarz inequality, which states that

$$|\langle v|w\rangle| \leq \sqrt{\|v\|\|w\|}. \tag{6.6}$$

This inequality shows that the scalar product is finite, for all vectors with finite norm. It is the generalization of the intuitive statement from two-dimensional space that the projection of one vector on the direction of another is no longer than the vector itself. It also follows easily that

$$\|v\|^2 = \langle v|v\rangle.$$

The final axiom defining a separable Hilbert space is that it is complete. That is, the scalar product defines a norm or distance, by Pythagoras' formula[2]

$$\|v - w\| = \sqrt{\langle v - w|v - w\rangle} = \sqrt{\sum |v_n - w_n|^2}. \tag{6.7}$$

A sequence of vectors whose norm converges to zero is said to converge to zero, and completeness means that the limit $|v\rangle$, in this sense, of any sequence of vectors, $|v_i\rangle \to |v\rangle$, is in the Hilbert space.

If we have two different separable Hilbert spaces, with bases $|e_n\rangle$ and $|f_n\rangle$, respectively, then the map $|e_n\rangle \to |f_n\rangle$ maps one into the other in a way that preserves the vector space structure and the scalar product, so there is really only one infinite dimensional separable Hilbert space, just as there is only one finite dimensional Hilbert space for each dimension. This sounds pretty remarkable when you realize it means that the whole standard model of particle physics is somehow contained in the space of states of a single nonrelativistic free particle.

What this result is really telling us is that physics is really about the choice of particular operators, which are measurable in a particular system. We often talk about "measuring" *any* operator in our Hilbert space, but this is an exaggeration. Each system has certain simple operators, which we have a prescription for measuring, and those are really the only ones of interest. We are generally going to ignore this subtlety, but you should remember it.

6.3 UNITARY TRANSFORMATIONS, UNITARY MATRICES, AND CHANGES OF BASIS

The column and row vector representations of vectors $|v\rangle$ and covectors $\langle w|$ depend on the choice of orthonormal basis. If we have another orthonormal basis $|f_n\rangle$, we can expand the $|e_n\rangle$ basis in terms of it

---

[2] Notice that in this equation, we have made a convention we will follow from now on, which is that sums without limits on them are taken from 1 to ∞. Similarly, integrals without limits will be taken from $-\infty$ to ∞.

$$|e_n\rangle = \sum_k |f_k\rangle\langle f_k|e_n\rangle \tag{6.8}$$

Then,

$$|v\rangle = \sum_{k,n} v_n\langle f_k|e_n\rangle|f_k\rangle. \tag{6.9}$$

Thus, the column vector that represents $|v\rangle$ in the $|f_k\rangle$ basis is

$$v_k^{(f)} = \langle f_k|e_n\rangle v_n^{(e)}, \tag{6.10}$$

where we have left the sum over n implicit, using the Einstein summation convention.

The matrix

$$U_{kn} \equiv \langle f_k|e_n\rangle \tag{6.11}$$

is thus the matrix that transforms the coefficients of a vector in the e basis into its components in the f basis. It satisfies the defining equation for a *unitary* matrix

$$UU^\dagger = U^\dagger U = 1, \tag{6.12}$$

where the 1 in this matrix equation is the unit matrix. To prove this identity, we use the most important formula in Dirac notation, called resolution of the identity

$$1 = \sum_n |e_n\rangle\langle e_n|. \tag{6.13}$$

It is true for *any* orthonormal basis. The resolution of the identity is a special case of Dirac's notation for dyadic operators $D(v, w) = |v\rangle\langle w|$. The dyadics $|e_k\rangle\langle e_l|$ are operators whose matrix in the $|e_k\rangle$ basis has a 1 in the k-th row and l-th column and zeros everywhere else. The operator $|e_k\rangle\langle e_k|$ (no sum) is the projection operator on the vector $|e_k\rangle$, and the resolution of the identity follows from this.

Now recall that the Hermitian conjugate operation † is the same as complex conjugation followed by transposition (or the other way around). So,

$$(U^\dagger)_{kn} = U_{nk}^* = \langle f_n|e_k\rangle^* = \langle e_k|f_n\rangle. \tag{6.14}$$

Thus,

$$\sum_n U_{kn}(U^\dagger)_{np} = \sum_n \langle f_k|e_n\rangle\langle e_n|f_p\rangle = \delta_{np}. \tag{6.15}$$

The other direction of this multiplication is equally easy to prove (Exercise 6.2).

The defining equation for unitarity is the same as the statement that both the rows and the columns of the unitary matrix form an orthonormal basis. We can think of these as the form of the e basis vectors in the f basis, and vice versa. This shows that the most general unitary matrix represents the transformation between two different orthonormal bases. Conversely, we can get any orthonormal basis by acting with all possible unitary transformations on some fixed orthonormal basis. The relationship between a pair of bases and a unitary operator is best understood with an example. Consider the 4×4 unitary matrix

$$
\begin{pmatrix}
U_{11} & U_{12} & U_{13} & U_{14} \\
U_{21} & U_{22} & U_{23} & U_{24} \\
U_{31} & U_{32} & U_{33} & U_{34} \\
U_{41} & U_{42} & U_{43} & U_{44.}
\end{pmatrix}
\tag{6.16}
$$

The matrix elements should be thought of as the form of an operator U in a particular basis: $U_{mn} = \langle e_m | U | e_n \rangle$. The basis vectors in that basis have the usual form: a one in a single row and zeroes elsewhere. Consequently, the columns of the unitary matrix are the form of the new basis vectors $|f_n\rangle$, expressed in the $|e_n\rangle$ basis. Correspondingly, the columns of $U^\dagger$, which are the complex conjugates of the rows of U, are the forms of the $|e_n\rangle$ basis vectors in the $|f_n\rangle$ basis. You should make sure that you understand how this paragraph relates to the formulae

$$
U = \sum_k |f_k\rangle\langle e_k|,
\tag{6.17}
$$

$$
U^\dagger = \sum_k |e_k\rangle\langle f_k|.
\tag{6.18}
$$

It is worth rewriting all of this using the notion of *linear operators*. A linear operator on a vector space is simply a function $A(|v\rangle)$, from the vector space to itself, which satisfies

$$
A(a|v\rangle + b|w\rangle) = aA(|v\rangle) + bA(|w\rangle).
\tag{6.19}
$$

We denote the action of a linear operator by

$$
A(|v\rangle) \equiv A|v\rangle,
\tag{6.20}
$$

in order to distinguish it from any old function from the vector space to itself.

A linear operator is completely defined by its action on a basis $|e_n\rangle$, because linearity then fixes its action on any linear combination of basis vectors. Let us define such an operator by

$$
U^\dagger |e_n\rangle = |f_n\rangle.
\tag{6.21}
$$

Then,

$$
\langle e_k | U^\dagger | e_n \rangle = \langle e_k | f_n \rangle = U^*_{nk}.
\tag{6.22}
$$

That is, the *matrix* $U^\dagger$ is gotten by sandwiching the *operator* $U^\dagger$ between a pair of basis vectors of the e basis. We will call such a sandwich a *matrix element* of the operator $U^\dagger$. Now let us compute the matrix elements of the same operator in the $|f\rangle$ basis.

$$\langle f_k|U^\dagger|f_n\rangle = \sum_{p,r}\langle f_k|e_p\rangle\langle e_p|U^\dagger|e_r\rangle\langle e_r|f_k\rangle \tag{6.23}$$

$$\langle f_k|U^\dagger|f_n\rangle = U_{nk}U_{kl}^\dagger U_{lm}^\dagger. \tag{6.24}$$

Note that in writing this equation, we could have replaced $U^\dagger$ by any other operator. Thus,

$$A_{mn}^{(f)} = U_{nk}A_{kl}^{(e)}U_{lm}^\dagger. \tag{6.25}$$

In words, this equation says that we get the matrix of an operator in the $|f_k\rangle$ basis, by *conjugating* its matrix in the $|e_k\rangle$ basis, by the unitary matrix relating the components of vectors in the two bases.

6.4   NORMAL OPERATORS ARE DIAGONALIZABLE OPERATORS

If an operator A has a matrix that is diagonal in some basis, equal to a complex diagonal matrix D, then its matrix in any other basis is

$$A = U^\dagger DU, \tag{6.26}$$

where U is the matrix that connects the two bases. This matrix satisfies

$$[A, A^\dagger] \equiv AA^\dagger - A^\dagger A = 0. \tag{6.27}$$

In words, it commutes with its adjoint, and is called a *normal* matrix. The same terminology is used for operators, independent of the basis. For any operator,

$$A = \frac{1}{2}[A + A^\dagger] + i\frac{1}{2}[(-i)(A - A^\dagger)]. \tag{6.28}$$

The two operators in square brackets are each equal to their own Hermitian conjugate, and are called Hermitian. For normal operators, these two Hermitian pieces, analogous to the real and imaginary parts of a complex number, commute with each other, and can be diagonalized simultaneously. This means that the study of normal operators reduces to that of Hermitian operators. It is common to state that all detectables in QM are Hermitian operators. Such a claim would not allow us, e.g., the freedom to take square roots of observable quantities. It is an arbitrary and overly rigid rule.

We will generally abuse language and use the words *operator* and *matrix* interchangeably. When we do, you will understand that we mean the matrix of an operator in some particular basis, which may or may not have been specified. We will also occasionally use the word *representation* to refer to a particular choice of orthonormal basis. For example, in speaking of particles, we often refer to the position representation or the momentum representation. We have also used this word in a rather special context, referring to the Schrödinger and Heisenberg forms of the equations of motion of QM. This context is so special that we usually use the term Schrödinger and Heisenberg *picture*, instead of representation. Later we will also discuss the Dirac or interaction representation of dynamics. The general idea is that abstract operators are the intrinsic dynamical variables/detectable quantities in QM and that a particular choice of basis *represents* the abstract operator as a different matrix. Every such choice is related to every other one by a unitary transformation.

The *spectral theorem* is the statement that *every* operator, which satisfies $[A, A^\dagger] = 0$, is conjugate to a diagonal matrix via unitary transformation, $A = U^\dagger D U$. For a finite dimensional space, counting of parameters in the two equations would seem to indicate that this is true (Exercise 6.3). A rigorous proof in finite dimensions goes along the following lines. The spectrum of an operator A is the set of complex numbers λ such that $A - \lambda$ is not invertible. In finite dimensions, this is equivalent to the statement that there is an eigenstate $|\lambda\rangle$ of A, with eigenvalue λ. Let $d(\lambda)$ be the degeneracy or number of independent eigenstates with the same eigenvalue, and let $P(\lambda)$ be the projection operator on the subspace of the Hilbert space spanned by these eigenstates. In Dirac notation,

$$P(\lambda) = \sum_i |\lambda\,(i)\,\rangle\langle\lambda\,(i)|, \tag{6.29}$$

where

$$A|\lambda\,(i)\,\rangle = \lambda|\lambda\,(i)\,\rangle. \tag{6.30}$$

Then, $A(1 - P(\lambda))$ is a normal operator (prove it) on the Hilbert space perpendicular to the subspace on which $P(\lambda) = 1$. We now repeat the above procedure, until we have written A as a linear combination of a complete set of commuting orthogonal projection operators

$$A = \sum \lambda_i P_i, \tag{6.31}$$

$$P_i P_j = \delta_{ij} P_j \quad \sum_i P_i = 1. \tag{6.32}$$

This is called the *spectral theorem*, and it is one of the most important results in all of mathematics.

6.5 CONTINUUM EIGENVALUES AND UNBOUNDED OPERATORS

There are two subtleties that are encountered in extending this analysis to the case of infinite dimensional separable Hilbert spaces. The first is the phenomenon of continuous spectra, which we have encountered when discussing the single particle position operator x. That is, there can be values of λ for which $A - \lambda$ is not invertible, which do not correspond to normalized eigenvectors of A. The eigenvectors are singular limits of square integrable functions, like the Dirac delta function. The mathematically rigorous way of dealing with this phenomenon is to note that an object defined formally by

$$P(\epsilon, \lambda) = \int_{\lambda-\epsilon}^{\lambda+\epsilon} da \, |a\rangle\langle a| \tag{6.33}$$

is a well-defined projection operator for any choice of λ in the spectrum of A, and any positive number ϵ. Mathematicians speak of a projection valued measure $dP(\lambda)$ and write a general normal operator A as

$$A = \int d\lambda \, \lambda dP(\lambda). \tag{6.34}$$

The integral runs over the spectrum of A in the complex plane. Physicists generally ignore this subtlety, and treat the continuum "eigenstates" as if they were normalized vectors in the Hilbert space, taking care only that none of the expressions they actually use give infinite answers.

The other subtlety has to do with the fact that not all of the operators physicists discuss are actually well-defined operators on the entire Hilbert space. For example, differential operators only make sense when applied to differentiable functions, but most square integrable functions are not differentiable. An even simpler example is the position operator x itself. Acting on an L^2 function of x it gives $xf(x)$, which might not be square integrable. Square integrability of $f(x)$ requires only that it fall faster than $x^{-\frac{1}{2}-\epsilon}$ for small positive ϵ, while square integrability of $xf(x)$ requires a faster falloff. Actually this is not an independent example, because the Fourier transform takes multiplication into differentiation.

Operators that are well defined on each vector in some complete basis of the Hilbert space but not on every vector in the Hilbert space are called unbounded operators. In all physical cases, they arise when we have a one parameter set $B(z)$, of bounded normal operators[3] and we define the derivative of it with respect to the parameter $A = \frac{dB}{dz}$. In this case, the spectral theory of A is derived from the spectral theory of the family. Whenever we have one parameter groups of symmetries, the infinitesimal generators of the symmetry group are constructed in this way.

However, we are often given an explicit form for the operator A, and asked to derive the family $B(z)$ from it. For example, we are given the Hamiltonian, and are asked to construct the time evolution operator $e^{-\frac{i}{\hbar}Ht}$. Then, in principle, we have to prove the existence of

---

[3] Bounded operators have the property that $\|B\psi)\| \leq \|B\|\|\psi\rangle\|$ for a fixed positive number $\|B\|$ and any vector in the Hilbert space. For normal operators, $\|B\|$ is the largest absolute value of any point in the spectrum of B. See Exercise 6.4

bounded functions of A, from the properties of A alone. Very occasionally, in physics, there are subtleties that require us to be a little more careful about the math of unbounded operators than usual. We will encounter one such subtlety for zero angular momentum wave functions in spherical coordinates, in Chapter 7.

6.6   SUMMARY

Let us end this section by summarizing the properties of operators in separable Hilbert spaces, using Dirac notation. Given a linear operator A and a basis $|e_n\rangle$ the matrix elements of the operator in that basis are

$$A_{mn} = \langle e_m | A | e_n \rangle. \tag{6.35}$$

Using the resolution of the identity $1 = \sum_k |e_k\rangle\langle e_k|$, we have

$$\langle e_m | AB | e_n \rangle = \sum_k \langle e_m | A | e_k \rangle\langle e_k | B | e_n \rangle, \tag{6.36}$$

which is the conventional rule for matrix multiplication. We can rewrite the operator in terms of its matrix and the dyadic operators $|e_k\rangle\langle e_l|$ as

$$A = \sum_{kl} A_{kl} |e_k\rangle\langle e_l|. \tag{6.37}$$

In particular, if A is a normal operator ($[A, A^\dagger] = 0$, see the next paragraph), its form in the basis given by its eigenvectors is

$$A = \sum_k a_k |a_k\rangle\langle a_k|. \tag{6.38}$$

The adjoint of an operator A, denoted $A^\dagger$ is defined by

$$\langle e_m | A^\dagger | e_n \rangle \equiv \langle e_n | A | e_m \rangle^*. \tag{6.39}$$

This equation is true in any basis (Exercise 6.5). An operator is called *normal* if $[A, A^\dagger] = 0$. This implies (this is the spectral theorem) in particular that if there is a basis where the matrix of A is diagonal, then $A^\dagger$ is also diagonal in that basis. Particular examples of normal operators are the unitary operators, for which $UU^\dagger = U^\dagger U = 1$. The matrix of a unitary operator is a unitary matrix, which is the same as the statement that both the rows and columns of the matrix are an orthonormal basis for the Hilbert space. In fact, the unitarity equation implies that, given an orthonormal basis $|e_n\rangle$, the vectors $|f_n\rangle = U|e_n\rangle$ are another orthonormal basis for the space.

The spectral theorem states that any normal operator can be written as

$$A = \int da\mu(a)|a\rangle\langle a|. \tag{6.40}$$

The spectral weight $\mu(a)$ is a measure, which means that it can have delta function contributions $\delta(a - a_n)$, as well as a continuous piece. For the discrete spectrum (the delta function contributions), the corresponding symbol $|a_n\rangle$ is a genuine normalized eigenstate of A. In the continuous spectrum, we think of $|a\rangle$ as a linear function on states, defined by the value of the wave function $\langle a|\psi\rangle \equiv \psi(a)$ for each normalizable state $|\psi\rangle$. The "continuous eigenstates" obey delta function normalization

$$\langle a|b\rangle = \frac{1}{\mu(a)}\delta(a - b).$$

It is conventional to use the fact that every normal operator can be written as the sum of two commuting Hermitian operators and to write the above formula in terms of integrals over a real parameter a. The spectral density $\mu(a)$ then has real and imaginary parts.

When A has continuous spectrum, and we express operators in terms of their action "in the basis where A is diagonal," then they become *integral operators*, also called *integral kernels*. For simplicity, we assume that the discrete and continuous spectrum of A are disjoint, and we work in the subspace of the Hilbert space orthogonal to the discrete eigenvectors. We write states in this subspace in terms of their wave functions $\psi(a)$ in the A basis. The completeness relation takes the form

$$1 = \int |a\rangle\mu(a)da \, \langle a|,$$

and we have

$$\langle a|a'\rangle = \delta(a - a').$$

It is convenient to absorb the square root of the nonsingular spectral density $\mu(a)$ into the definition of the wave function:

$$\psi(a) \equiv \sqrt{\mu(a)}\langle a|\psi\rangle. \tag{6.41}$$

A general linear operator takes the form

$$\langle a|B|\psi\rangle = \int B(a, a')\psi(a'). \tag{6.42}$$

Note that differential operators can also be written in this form, with kernels like

$$D(x, y) = \frac{d}{dx}\delta(x - y).$$

6.7   DIRECT SUMS AND TENSOR PRODUCTS

Given a pair of Hilbert spaces, $\mathcal{H}_e$ and $\mathcal{H}_f$, we can define two new Hilbert spaces, called the direct sum, $\mathcal{H}_e \oplus \mathcal{H}_f$, and direct or tensor product, $\mathcal{H}_e \otimes \mathcal{H}_f$. It is simplest to do this in terms of a choice of bases $|e_n\rangle$ in $\mathcal{H}_e$ and $|f_N\rangle$ in $\mathcal{H}_f$. The direct sum is the set of complex linear combinations $\sum a_n|e_n\rangle + \sum b_N|f_N\rangle$. It has dimension $D_e + D_f$. The tensor product space is spanned by vectors $\sum a_{nN}|e_n\rangle \otimes |f_N\rangle$. It has dimension $D_e D_f$. We often drop the $\otimes$ sign when talking about vectors in a tensor product. A straightforward argument, which is left to the reader, shows that the resulting spaces are independent of the choice of bases.

The notion of the direct sum underlies the spectral theorem. A given normal operator decomposes the Hilbert space into a direct sum of eigenspaces, in each of which it takes on a fixed numerical value. In the case of continuous spectra, the direct sum is replaced by something called a direct integral, which is the rigorous definition of some of the expressions we have written in Dirac notation.

In physics, the tensor product arises when discussing the joint states of independent systems. For example, a *pair* of ammonia molecules, labeled A and B, at low energy, has states

$$|+\rangle_{3A} \otimes |+\rangle_{3B}, \quad |+\rangle_{3A} \otimes |-\rangle_{3B}, \quad |-\rangle_{3A} \otimes |+\rangle_{3B}, \quad |-\rangle_{3A} \otimes |-\rangle_{3B}. \tag{6.43}$$

There are two general theorems about operators that act separately on each factor of the tensor product. Such operators are denoted $1 \otimes A_2$ or $A_1 \otimes 1$, and we have

$$(1 \otimes A_2)(A_1 \otimes 1) = (1 \otimes A_1)(A_2 \otimes 1) = A_1 \otimes A_2 = A_2 \otimes A_1. \tag{6.44}$$

We also have

$$\text{Tr}\,[A_1 \otimes A_2] = \text{Tr}_1\,[A_1]\text{Tr}_2\,[A_2], \tag{6.45}$$

where the traces with subscripts are traces over the individual factors in the tensor product.

6.8 THE GENERALIZED UNCERTAINTY RELATION

We have seen that if two operators commute, then they are simultaneously diagonalizable, which means that there are states of the system in which they are both definite at the same time. In this section, we want to make a quantitative estimate of the mutual uncertainty of two operators that do not commute. Given any probability distribution, any quantity A will have an expectation value $\langle A \rangle$. We define a quantity $\bar{A} \equiv A - \langle A \rangle$, which has zero expectation value. If the values of A had a Gaussian, or normal distribution, then the width of the Gaussian would be determined by $\langle \bar{A}^2 \rangle$. A wider Gaussian, with more uncertainty, would correspond to a larger value for this mean square average. We therefore define the uncertainty of A, ΔA, for *any* probability distribution by

$$\Delta A \equiv \sqrt{\langle \bar{A}^2 \rangle}. \tag{6.46}$$

Let us now apply this definition to the probability distributions of two noncommuting Hermitian operators, A and B, in a pure quantum state $|\psi\rangle$. We first observe that

$$\langle\psi|[\bar{A}, \bar{B}]|\psi\rangle = i\text{Im}\langle\psi|\bar{A}\bar{B}|\psi\rangle. \tag{6.47}$$

This means that

$$|\langle\psi|[\bar{A},\bar{B}]|\psi\rangle|^2 \leq |\langle\psi|\bar{A}\bar{B}|\psi\rangle|^2 \leq |\langle\psi|\bar{A}^2|\psi\rangle||\langle\psi|\bar{B}^2|\psi\rangle|. \tag{6.48}$$

In the last inequality, we have used the Schwarz inequality

$$|\langle\chi|\phi\rangle| \leq \||\chi\rangle\|\||\phi\| \tag{6.49}$$

applied to the vectors $|\phi\rangle = B|\psi\rangle$ and $|\chi\rangle = A|\psi\rangle$. We end up with the generalized Heisenberg uncertainty relation

$$\Delta A \Delta B \geq |\langle\psi|[A,B]|\psi\rangle|. \tag{6.50}$$

If we apply this to the position and momentum of a particle, which satisfy $[x,p] = i\hbar$, the right-hand side is independent of the state and we get

$$\Delta x \Delta p \geq \hbar. \tag{6.51}$$

This is Heisenberg's famous uncertainty relation between position and momentum. The careful reader will note that we have implicitly used the fact that $[\bar{A},\bar{B}] = [A,B]$, which is true because the barred quantities differ from the unbarred ones by an additive multiple of the identity, and the identity commutes with everything.

6.9 SYMMETRIES AND CONSERVATION LAWS

In QM, a symmetry transformation must be an invertible operation S, which takes one vector in the Hilbert space into another. It cannot change the probability of finding that something is true, so $S(|v\rangle)$ must have a scalar product with $S(|w\rangle)$, which has the same magnitude as $\langle w|v\rangle$. Unitary operators have this property, because they preserve the entire scalar product, including its phase. There is a class of nonlinear operators which also satisfy the weaker condition of preserving probability. An operator is called conjugate linear if

$$C(a|v\rangle + b|w\rangle) = a^*C(|v\rangle) + b^*C(|w\rangle). \tag{6.52}$$

A conjugate linear operator can also have the property that the scalar product of $C(|v\rangle)$ and $C(|w\rangle)$ is equal to $\langle v|w\rangle$, the complex conjugate of the original scalar product. Such operators are called antiunitary. The product of two antiunitary operators is a linear unitary operator (Exercise 6.6), so any two antiunitaries differ by multiplication with a unitary. We will see below that the operation of *time reversal* must be taken to be antiunitary. If time reversal, T, is a symmetry operation of a given system, then any other symmetry is either represented by a unitary operator or by TU where U is unitary.

Symmetry operations must conserve probability, but that is not sufficient. In order to be a symmetry of a given system, an operation must also preserve its equations of motion. We can write the equations of motion of QM in either the Schrödinger or Heisenberg pictures

$$i\hbar\partial_t|\psi\rangle = H|\psi\rangle, \tag{6.53}$$

or

$$\hbar\partial_t O = i[H, O]. \tag{6.54}$$

If we take

$$|\psi\rangle \to S|\psi\rangle, \tag{6.55}$$

where S is a putative symmetry operation, these equations must be preserved. In the case of the Schrödinger equation, this means $S|\psi\rangle$ should satisfy the same equation as $|\psi\rangle$. Recall, that in the Schrödinger picture, operators are time independent. Acting with S on both sides of the Schrödinger equation we get

$$\pm i\hbar\partial_t S|\psi\rangle = SHS^{-1}S|\psi\rangle, \tag{6.56}$$

where the plus sign corresponds to unitary S, while the minus sign corresponds to antiunitary S. The symmetry criterion, *for symmetries that do not reflect $t \to -t$*, is

$$SHS^{-1} = \pm H, \tag{6.57}$$

while if time reflection is involved then

$$SHS^{-1} = \mp H. \tag{6.58}$$

The upper sign is for a unitary symmetry operation, and the lower for an antiunitary one. If we work in the basis where H is diagonal, we see that the cases with a minus sign are possible only if the spectrum of energy eigenvalues is symmetric around zero. Very few Hamiltonians have this property. In particular, any Hamiltonian for interacting particles, which has the property that particles become free when they are far from each other, *cannot* have a spectrum that is symmetric around zero. The free particle Hamiltonian is bounded from below and its spectrum is continuous. All bound states of the particles, which have energies below the bound of the free Hamiltonian, have a discrete spectrum.[4]

Thus, with Wigner, we argue that symmetries which involve time reflection are antiunitary, while those which do not are unitary. Note that this does not require that time reflection

---

[4] Up till now, our quantum formalism has shared with classical mechanics the principle that the choice of the zero of energy does not effect physics. If we allowed antiunitary symmetries that preserve time orientation, or unitary symmetries that reverse it, the zero of energy would acquire an absolute significance.

itself is a symmetry. Indeed, a variety of experiments on elementary particle physics suggest that it is not. However, there is a rigorous theorem in relativistic quantum field theory, which shows that TCP, the combined operation of time reflection (T), space reflection (P), and the operation (C) which exchanges particles and their antiparticles, is a symmetry of any model of local interactions, which is invariant under space-time translation and Lorentz transformations.[5]

Apart from TCP, we can always think about symmetries, which are linear unitary operations that commute with the Hamiltonian. Choose, among all the symmetry operations of a given model, a maximally commuting set, U_i. The fact that U_i commute with the Hamiltonian implies that we can diagonalize them all, and that the time evolution of the system does not change their joint eigenvalues $e^{i\phi_i}$. That is to say, the phases $e^{i\phi_i}$ are *conserved quantities*, constants of the motion.

We can also see this connection between symmetries and conservation laws by looking at the Heisenberg picture equations of motion. In the Heisenberg picture, states are time independent, and operators whose definition is not explicitly time dependent, vary according to

$$\hbar \partial_t O = i[H, O]. \tag{6.59}$$

For the symmetry operators U_i, this implies that $\partial_t U_i = 0$, which is the Heisenberg picture form of a conservation law.

This connection between symmetries and conservation laws was first discovered in classical mechanics by E. Noether. Noether actually proved the connection only for symmetries that depend on a continuous parameter a. Her proof is reviewed in Appendix C. A continuous family of symmetries in QM, depending on a single parameter, is a family of unitary transformations $U(a)$.[6] We can always choose another parameterization $a \to f(a)$, where f is an invertible continuous function. However, if $[U(a), U(b)] = 0$ for any two different values of the parameter, then there is a natural choice for the parameterization.

First note that $U(a)U(b)$ is a symmetry, if the original unitaries are. It is unitary, and commutes with the Hamiltonian. Let us choose the parameterization such that $U(a)U(b) = U(a + b)$. If that is the case, we can write $U(a) = e^{iQa}$, where Q is a Hermitian operator, if we are willing to make the extra assumption that the dependence on a is differentiable. This assumption is valid for all known continuous symmetry groups that have been used in QM. Q is known as the *infinitesimal generator* of the one parameter group of symmetries $U(a)$. Clearly, $[Q, H] = 0$ and Q is the quantum analog of Noether's conservation law. The Noether connection between symmetries and conservation laws is tighter in QM than in

---

[5] The evolving cosmology in which we live, does not have these symmetries, but they are currently violated by very small amounts, inversely proportional to the age of the universe, so there is a good approximate theory of the world which does have these symmetries.

[6] This discussion should remind the alert reader of our treatment of spatial translations in Chapter 3.

classical Lagrangian physics, because in QM, all operations are also potentially measurable quantities. The idea that an operation is connected to a conserved quantity is natural in this context.

The set of all unitary symmetries of a given quantum system forms a mathematical structure known as a *group*. Every symmetry transformation has an inverse and the product of two symmetries is always a symmetry because we can evaluate its commutator with the Hamiltonian using the Leibniz rule. More precisely, in QM we have a representation of an abstract group of symmetries by a group of unitary transformations in a Hilbert space. The classification of all possible groups and their unitary representations is one of the great problems of mathematics, and it has not been solved. However, mathematicians have found out an enormous amount of information about this problem, much of it generated by examples from physics. We cannot possibly do justice to this material at the level of this course. Appendix D on symmetries summarizes a few of the most important facts and concepts.

Above we discussed a maximal set of commuting symmetry operators U_i for a given system. There can be other symmetries, which do not commute with the U_i. The full set of symmetry operations forms a group. If that group is *nonabelian*, i.e., contains elements that do not commute with each other, then the minimal *faithful*[7] representation of it has dimension greater than 1. A representation of a group is called *irreducible* if the only operators which commute with all the group representation matrices are multiples of the unit matrix. Since the Hamiltonian commutes with all symmetry group elements, it must take on the same value in every state of an irreducible representation. If the group is nonabelian, this leads to degeneracies in the spectrum of the Hamiltonian. Without such a symmetry explanation, the occurrence of such degeneracies would not be expected. A random Hermitian matrix will not have any equal pair of eigenvalues. We will see examples of this below when we discuss rotation invariance and the hydrogen atom.

6.9.1 Dynamical Symmetry Groups

The Hamiltonian itself is a symmetry generator. We introduced it to describe the invariance of quantum dynamics under time translation. Our brief discussion of nonabelian symmetry groups suggests that there could be an expanded concept of symmetry, in which the symmetry group contains operators that do not commute with the Hamiltonian. Indeed, such a concept of symmetry was introduced into physics by Galileo, when he argued that the equations of physics (the language here is a bit anachronistic) looked the same in all inertial reference

---

[7] A faithful representation is one in which the full group multiplication table is represented by nontrivial matrices. An example of a nonfaithful representation is one in which every group element is represented by the unit matrix. A slightly less trivial one is the representation of the group of permutations in which every even permutation is represented by the unit matrix, and every odd permutation is represented by a matrix with eigenvalues ± 1.

frames. Inertial reference frames are systems of Cartesian space coordinates, which are moving with respect to each other at constant velocity, $\mathbf{x} \to \mathbf{x} + \mathbf{v}t$. This transformation is called a *Galilean Boost*. For a particle of mass m, the momentum changes under such a transformation by an amount

$$\mathbf{p} \to \mathbf{p} + m\mathbf{v}, \tag{6.60}$$

and so the particle's kinetic energy changes. The Hamiltonian is not invariant. Nonetheless, the equations of motion in the new frame are identical to those in the old, if there is no external potential. If the system contains multiple particles and the potential depends only on coordinate differences, then the equations of motion still look the same in both frames. Furthermore, only the kinetic energy changes

$$\frac{1}{2} \sum m_i \mathbf{v}_i^2 \to \frac{1}{2} M \mathbf{v}^2 + \mathbf{v} \cdot \mathbf{P}, \tag{6.61}$$

where $M = \sum m_i$ is the total mass and $\mathbf{P} = \sum m_i \mathbf{v}_i$ is the total momentum. In nonrelativistic physics, both the mass and the momentum are conserved if the system is invariant under spatial translation, so this law for the change of the Hamiltonian is universal and involves only conserved quantum numbers. For very small $\mathbf{v} = \delta\mathbf{v}$, the change in the Hamiltonian is

$$\delta H = \delta\mathbf{v} \cdot \mathbf{P}, \tag{6.62}$$

while the change in momentum is

$$\delta\mathbf{P} = M\delta\mathbf{v}. \tag{6.63}$$

If the action of Galilean boosts on a state is

$$|\psi\rangle \to (1 + iN_i \delta v_i)|\psi\rangle, \tag{6.64}$$

where N_i is a Hermitian operator, then the corresponding action on operators is

$$\delta O = -i[N_i \delta v_i, O]. \tag{6.65}$$

Thus,

$$P_i = -i[N_i, H] \tag{6.66}$$

and

$$M\delta_{ij} = -i[K_i, P_j]. \tag{6.67}$$

Similarly

$$t\delta_{ij} = -i[K_i, X_j], \tag{6.68}$$

where $\mathbf{X}$ is the center of mass position coordinate, the canonical conjugate of the total momentum. We can find a solution of these relations by insisting that $H = \frac{\mathbf{P}^2}{2M} + H_{int}$, where

H_{int} depends only on relative coordinates and their conjugate momenta, as well as the spins of the constituent particles, and

$$N_i = \frac{M}{\hbar} X_i - t \frac{P_i}{\hbar}. \tag{6.69}$$

Note that N_i has explicit time dependence. In Exercise 6.7, you will verify that this makes it time independent, despite the fact that it does not commute with the Hamiltonian. This is in accord with Noether's classical theorem, which relates the invariance of the classical action under Galilean boosts, to a conservation law. The form we have written for the Galilean boost operator corresponds precisely to the quantum version of Noether's expression. Those observant readers with long memories will also notice that we have completed the second justification for the form we wrote for the free particle Hamiltonian.

Symmetry groups which include the Hamiltonian, but do not commute with it, are often called Dynamical Symmetries, or Spectrum Generating Symmetries. Galilean boosts tell us that center of mass motion decouples from the internal dynamics, and completely determine the energy momentum relation (also called the *dispersion relation*) for the center of mass motion. The true symmetry of the world, to which Galilean boost invariance is only a slow motion approximation, is the Lorentz group. In QM, this symmetry has profound consequences for dynamics. It forbids the existence of interacting systems in which particle number is conserved, and when combined with the notion of causality/locality (which says that quantum interference should respect the restrictions imposed by a maximal propagation velocity), it puts extremely strong constraints on the properties and interactions of elementary particles. We do not have time to study this in this first course in QM, but there are a number of excellent textbooks on relativistic Quantum Field Theory [17], where you can follow the story further.

There are a few other examples of dynamical symmetries in problems of physical interest. Perhaps the most powerful is the Runge–Lenz–Pauli symmetry of the hydrogen atom, which completely determines the spectrum in terms of group theory. There are also a variety of models in one space and one time dimension with similar spectrum generating symmetries [18].

6.9.2 Projective Representations of Symmetry Groups

Before leaving the topic of symmetry, we should touch on a peculiar aspect of symmetry groups in QM, which has no classical analog. A group is determined by its multiplication law $g_1 g_2 = g_{12}$. In QM, we have a unitary transformation $U(g)$ for each group element and we might have expected that

$$U(g_1)U(g_2) = U(g_1 g_2), \tag{6.70}$$

but this is too strong a requirement. Suppose instead that

$$U(g_1)U(g_2) = U(g_1g_2)e^{i\phi(g_1,g_2)}, \tag{6.71}$$

where ϕ is real. The fundamental predictions of QM are predictions for expectation values of the form $\langle\psi|A|\psi\rangle$, and these are all insensitive to the extra phase in the multiplication law. So such *projective representations* of groups are allowed in QM.

One may ask the question, "Why do we care?" if the choice of phase factor in the projective representation does not affect physical predictions. The answer is that it *does* affect physical predictions through the choice of allowed representations of the group. The most important example of this is the group of three-dimensional rotations $SO(3)$. The most obvious representation of this group is a three-dimensional vector V^i, and one can make products of these $V^i \ldots V^j$ which are called *tensors of rank n*, if there are n factors in the product.

On the other hand, we may recall the 2×2 Pauli matrices, which have a multiplication table

$$\sigma_a\sigma_b = \delta_{ab} + i\epsilon_{abc}\sigma_c, \tag{6.72}$$

which is covariant under rotations if σ_a transforms like a three-vector. Consider the unitary transformation

$$U(\theta, e_3) = e^{i\frac{\theta}{2}\sigma_3} = \cos(\theta/2) + i\sin(\theta/2)\sigma_3, \tag{6.73}$$

and compute

$$U^\dagger(\theta, e_3)\sigma_a U(\theta, e_3). \tag{6.74}$$

It is obvious that this does not change σ_3. Using the Pauli algebra, it gives

$$U^\dagger(\theta, e_3)\sigma_i U(\theta, e_3) = \cos^2(\theta/2) - \sin^2(\theta/2)\sigma_i - i\sin(\theta/2)\cos(\theta/2)(\sigma_3\sigma_i - \sigma_i\sigma_3), \tag{6.75}$$

for $i = 1, 2$. This is

$$\cos(\theta)\sigma_i + \sin(\theta)\epsilon_{ij}\sigma_j, \tag{6.76}$$

where the two-dimensional Levi-Civita symbol is defined in the obvious way.

We see that $U(\theta, e_3)$ is a unitary transformation, which operates on the Pauli matrices like a rotation around the 3 axis, with angle θ. In the Exercises, you will show that $U(\theta, e_a) = e^{i\frac{\theta}{2}e_a\sigma_a}$, where e_a is a three-dimensional unit vector, gives a rotation with angle θ around the e_a axis. So we have a two-dimensional unitary representation of the rotation group.

Well, not quite. Let $\theta = 2\pi$. The corresponding rotation is equal to the unit matrix, and indeed, the Pauli matrices are invariant under this transformation. However,

$$U(\theta, e_3) = e^{i\frac{\theta}{2}\sigma_3} = -1. \tag{6.77}$$

So the multiplication equation $R(\pi, e_3)^2 = 1$ of the rotation group is not satisfied in this representation. Note, however, that this transformation $U = -1$ commutes with all other elements of the group. In Exercise 6.8, you will prove that this implies that we have a projective representation. In this case, the phases are just ± 1.

This representation is actually an ordinary, nonprojective representation of another group, the group of unitary 2×2 matrices with determinant 1, which is called $SU(2)$. Writing a general 2×2 matrix as a sum of the unit matrix and the Pauli matrices:

$$U = e^{i\phi}(n_0 + in_a\sigma_a), \tag{6.78}$$

where n_0 is real, you will show in Exercise 6.10 that the unitarity condition is that n_a is real and $n_0^2 + n_a^2 = 1$, while the determinant condition implies $\phi = 0$. The unitarity condition is solved by $n_0 = \cos(\alpha)$ and $n_a = \sin(\alpha)e_a$, and we recognize both the relation to $3D$ rotations and the fact that the $SU(2)$ group has the geometry of a sphere in 4 Euclidean dimensions.[8] The three-dimensional rotation group is gotten from $SU(2)$ by identifying elements $\pm U$ with each other.

A construction like this is quite general. If G is a group, the set of elements that commute with everything else in the group is called Z, the *center of the group*. We can now define a new group called G/Z (G mod Z) by identifying group elements related by $g_1 = Z_{12}g_2$, where Z_{12} is an element of the center. If G has continuous parameters, we can think of the group as a space with topology. One of the interesting topological questions one can ask about a space is whether there are closed curves, which cannot be contracted to a point. For example, on a sphere, in any dimension, there are no such closed curves. On the other hand, on a two-dimensional torus, the surface of a donut, there are two different kinds of noncontractible closed loops, going around the donut in two orthogonal directions (Figure 6.1). If our group G, like the group $SU(2)$ has *no* noncontractible loops, the group G/Z *will* have contractible loops. Namely, take an open curve in $SU(2)$, which goes from a group element U to $-U$. In $SO(3) = SU(2)/Z_2$, this is a closed loop, but it cannot be contracted because it connects two different points in $SU(2)$.

Figure 6.1 Noncontractible loops on a torus, but not a sphere.

---

[8] One may suspect that there is a *four*-dimensional rotation group hiding somewhere here. Find it in Exercise 6.9.

There is an amusing little experiment you can do, to convince yourself that the rotation group has a noncontractible loop in it. Fill a glass of water half full. Hold the bottom in the palm of your hand, stick your arm out and try to rotate the glass around a full circle, without spilling. You will find that your arm is twisted in a way that you cannot get out of without either undoing the rotation *or* doing *another* 2π rotation. See https://blogs.scientificamerican.com/roots-of-unity/a-few-of-my-favorite-spaces-so-3/ for an illustration of this, and of a similar experiment with belts or long braids of hair. It is the second option that is the surprise. This shows not only that there are objects for which 2π rotations are nontrivial, but also that a 4π rotation is always trivial. We have discovered the hidden $SU(2)$ group in a classical experiment.

Objects that transform under rotations like the two by two matrix representation we have just discussed are called *spinors*. They were first discovered by Pauli, and the mathematician Cartan investigated the mathematical properties of spinors in any space dimension. Given our little example, it might seem that no localized particle could transform like a spinor. Your hand with its glass of water is sensitive to the 2π rotation because your arm is attached to your body. For a point particle, one could imagine something similar happening if there were some kind of infinite string emanating from it. The string would have to extend out to infinity in order to avoid questions about what is attached to the other end. The string could not carry any energy or momentum and other things should be able to go right through it without noticing that it is there. Sounds like magic. Actually we will see in Chapter 15 that we can build mathematical models which behave exactly like particles attached to such magical invisible strings. There are particles that transform like spinors and are sensitive to 2π rotations, and they all behave as if they have magical strings attached, a property which manifests itself in what is called *Fermi-Dirac* statistics.

6.9.3 Examples

Let us go back to the example of the ammonia molecule, with which we began our exploration of QM. We will learn in Chapter 11, that the meaning of picture (Figure 6.2) of this molecule that we drew, is tied to an approximation in which we can think of the nuclei of the atoms composing the molecule as stationary sources of electric field, sitting at points in space.

Figure 6.2 Ammonia in the Born–Oppenheimer approximation.

We know that this is impossible for quantum particles, because of their position and momentum cannot be simultaneously definite. However the commutator of the position and *velocity* of a particle is

$$[X, \dot{X}] = i\frac{\hbar}{M}, \tag{6.79}$$

so if there is some sense in which M is very large, then this uncertainty is small. Thus, a very massive particle can stay in place for a long time, without violating the uncertainty relation. Of course, mass and time have dimensions, so we have to say "long (heavy) compared to what?" In a molecule, the electron mass sets a mass scale and if we have fixed energy, it also sets a time scale. The basic idea of the so-called Born–Oppenheimer approximation is that one can freeze the nuclei at fixed positions over time scales that are long enough that the electrons settle into their ground state in the external potential of the fixed nuclei. This then creates a potential (the Born–Oppenheimer potential) for the nuclei since the ground state energy will depend on their positions. The nuclei sit in harmonic oscillator ground state wave functions localized within a distance that scales like an inverse power of M from the minima of the Born–Oppenheimer potential.

In the ammonia molecule, the positions of the three identical hydrogen atoms form a plane and the symmetry under exchanging these atoms suggest that the minima of the B-O potential should sit at the vertices of an equilateral triangle. The most symmetric configuration for the nitrogen is obviously at the center of the triangle, but there is no reason for it to be in the same plane as the hydrogens, so it probably chooses to be above or below the plane. The Hamiltonian for Coulomb interactions between electrons and nuclei is invariant under spatial reflections, so there must be two equilibrium positions of equal energy.

The full symmetry group of the stationary molecule is thus a combination of cyclic permutation of the three hydrogen atoms, and the reflection in the plane of the hydrogens. These two types of transformation commute with each other. The Z_3 permutation is a discrete rotation in the plane, and rotations in the plane are invariant under reflections through it. Call the transformation on the Hilbert space of states of the nuclei, which performs a $2\pi/3$ rotation U_3 and the reflection U_2. Both operators commute with the Hamiltonian, and with each other. They satisfy

$$U_3^3 = U_2^2 = 1. \tag{6.80}$$

As a consequence, we can consider eigenstates of H such that

$$U_3|k, p\rangle = e^{\frac{2\pi ik}{3}}|k, p\rangle \tag{6.81}$$

$$U_2|k, p\rangle = e^{\pi ip}|k, p\rangle, \tag{6.82}$$

where $k = 0, 1, 2$ and $p = 0, 1$.

These are the only symmetries of the Hamiltonian with nuclei in the Born–Oppenheimer potential, but when we add the electrons back to the system, the U_3 transformation is part of the full rotational symmetry. The states with $k \neq 0$ have nonzero angular momentum and it turns out that this gives them higher energy. You can understand this by remembering the classical dependence of energy on angular momentum, and adding to that the fact that in QM, bounded motions have discrete eigenenergies. We will compute the rough size of rotational levels of molecules in Chapter 11. As a consequence, the lowest energy states are labelled by the two valued quantum number p. In terms of the notation of Chapter 2, these are the states

$$|\pm\rangle_1 = \frac{1}{\sqrt{2}}(|+\rangle_3 \pm |-\rangle_3) \tag{6.83}$$

and the 3 index can now be thought of as the direction perpendicular to the plane of the hydrogen atoms in the ammonia molecule. The reflection symmetry operator is precisely the Pauli operator σ_1, so these are eigenstates of σ_1.

The general theory of symmetry shows us that these two states are eigenstates of the Hamiltonian, because symmetry transformations can always be diagonalized simultaneously with the Hamiltonian and the distinct eigenstates of the symmetry operator σ_1, $|\pm\rangle_1$, form a complete basis for the space. This implies that the Hamiltonian is a function of the symmetry operator

$$H = a + b\sigma_1. \tag{6.84}$$

We have just rederived our theory of the low-energy states of the ammonia molecule, using symmetry conditions alone.

Another simple example of the use of symmetry in QM is the theory of a particle moving on a plane, under the influence of a potential that depends only on the radial distance from the origin. The Hamiltonian is

$$H = \frac{1}{2}(P_1^2 + P_2^2) + V(R), \tag{6.85}$$

where $R \equiv \sqrt{X_1^2 + X_2^2}$. Introduce $Z = \frac{1}{\sqrt{2}}(X_1 + iX_2) \equiv Re^{i\phi}$. The system is invariant under the rotation $Z \to e^{i\alpha} Z$, for any α. Potentials, which preserve this symmetry, are functions only of $Z^\dagger Z$. We can also have potentials of the form

$$V_N = f(Z^N, Z^\dagger Z) + h.c., \tag{6.86}$$

where $h.c.$ refers to the Hermitian conjugate operator. Z^N will be invariant under a rotation by angle α if

$$e^{i\alpha N} = 1, \tag{6.87}$$

which is solved by the N-th roots of unity

$$e^{\frac{2\pi i k}{N}},$$

with $k = 0, \ldots N - 1$. Systems with such a potential are said to be invariant under the group Z_N of N-th roots of unity (Exercise 6.15 asks you to prove that this set of complex numbers forms a group under multiplication). This group is *generated by* the single element $e^{\frac{2\pi i}{N}}$, which means that every element of the group is a power of that element. Such groups are called cyclic groups. They are the building blocks for the set of all finite abelian groups.

Given a group of symmetries in QM, one always asks how vectors in the space of states transform under the group. For the group Z_N, this classification is easy to do. We have an infinite dimensional space of invariant wave functions $\psi(z^N, z^*z)$. (Note that we are using small letters here. When speaking about the potential, which is an operator, we wrote it as a function of the operator Z. Here we are describing states in terms of their wave function in the "basis" where Z is diagonal. z is the eigenvalue of Z.)

We can also have wave functions of the form $\psi_k = z^k \psi(z^N, z^*z)$, which transform as

$$\psi_k \to e^{\frac{2\pi i k}{N}} \psi_k,$$

under the generating element of Z_N. The action under the other transformations is just obtaining by iterating this equation. In Exercise 6.16, you will use Fourier's theorem for periodic functions to prove that this is the most general transformation law allowed. That is, any wave function can be written as a linear combination of wave functions of one of the types listed above. This is an example of a general theorem about QM representations of groups which are *compact*. That term refers to groups for which the set of parameters necessary to specify the most general group element is a closed and bounded subset of a real vector space of finite dimension. Examples of compact groups are groups with finite numbers of elements and groups of rotations or unitary transformations in any number of dimensions.

The action of the group Z_N is also useful for describing the motion of particles on the surface of a cone. You can make a cone by cutting an angular wedge out of a sheet of paper, and then gluing the edges together (Figure 6.3).

A circuit around at fixed distance from the singular apex will traverse an angle less than 2π. When that angle is $2\pi/N$, we can get the same figure by identifying points on the

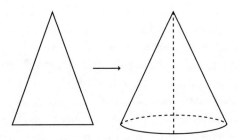

Figure 6.3 Making a cone by gluing.

original sheet under the transformations of Z_N. It follows that, for cones with these opening angles, the physics of the motion is obtained from physics on the plane, by simply restricting attention to quantities invariant under Z_N. That is, the Hilbert space of motion on the cone, is the subspace of the Hilbert space of motions on the plane consisting of those wave functions that are invariant under Z_N.

In this context, the group Z_N is not really a symmetry group, because every state in the physical Hilbert space is invariant under it. Rather the group is there to show us that the description of motion on the cone in terms of motion on the plane is *redundant*. Points on the plane differing by a Z_N transformation are just copies of points in the fundamental wedge with angles between 0 and $2\pi/N$.

The redundant description is useful. If you'd been asked to construct a description of motion on a cone, you might have started by trying to solve the motion on some smooth curved two dimension surface and then taken a limit where the curvature went to zero everywhere except at one point, where it became infinite. For the discrete set of opening angles for which the trick works, it is much simpler to solve the problem on the plane, and then project out the solutions that are not invariant under Z_N. This kind of use of a symmetry group goes by the name of *gauge symmetry* because it is reminiscent of the redundancy familiar from Maxwell's electrodynamics. In electrodynamics, we describe the electric and magnetic fields by

$$\mathbf{E} = \dot{\mathbf{A}} - \nabla\phi, \tag{6.88}$$

$$\mathbf{B} = \nabla \times \mathbf{A}. \tag{6.89}$$

These physical fields are left invariant if we subject the vector potential $\mathbf{A}$ to a *gauge transformation*[9]

$$\mathbf{A} \rightarrow \mathbf{A} + \nabla\chi, \tag{6.90}$$

with $\dot{\chi} = 0$. It is convenient to use the vector potential to solve Maxwell's equations, but all physical quantities are invariant under the gauge transformation. *Gauge Symmetries* have turned out to be the basis of the standard model of particle physics. We will encounter them again in Chapter 15, where we will use a Z_2 gauge theory to give a more profound understanding of Fermi Statistics. Gauge symmetries are also crucial for understanding exotic particles called *anyons* that can only propagate in a plane. Finally, note that if we describe multiparticle states of identical particles in terms of many body wave functions, instead of quantum fields, then we must treat the permutation group S_n, which exchanges particles, as a gauge symmetry or redundancy. Every state must be invariant under the subgroup A_n of even permutations. Fermion states are odd under the Z_2 factor group S_n/Z_n of odd

---

[9] This is not the most general form of gauge transformation. We use the special case to illustrate the analogy.

permutations (every odd permutation can be written as some particular transposition times an even permutation).

As our final example of symmetry groups, we note that many molecules contain a plane[10] which is the form of a regular N-gon, and the Hamiltonian of the model is invariant under the geometrical symmetries of the polygon. To understand what these are, we denote the positions of the vertices as complex numbers $z_k = Re^{\frac{2\pi i k}{N}}$. Then the operations

$$z_i \to e^{\frac{2\pi i k}{N}} z_i = z_{i+k}, \text{for all } i,$$

and

$$z_i \to z_i^*$$

are symmetries of the polygon. The addition is modulo N arithmetic. Geometrically, complex conjugation is a reflection around the horizontal axis in the plane (Figure 6.4). In Exercise 6.17, you will show that these operations form a group with $2N$ elements, called the *dihedral group*, D_N. If we denote the multiplication by the k-th root of unit S_k and complex conjugation by C, then the $2N$ elements are S_k and $R_k = CS_k$. Any product of two of these operations gives another one on the list, and each operation has an inverse.

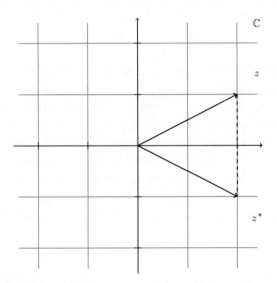

Figure 6.4 Complex conjugation.

---

[10] The idea that a molecule has a fixed shape, is an approximate statement in QM, related to the fact that the masses of nuclei are at least two thousand times the mass of the electron. As a consequence, the uncertainties in nuclear positions are much smaller than those of electrons, and low-energy excitations of the molecule can be treated as small perturbations around a picture where the nuclei are classical particles at rest. We will explore this *Born–Oppenheimer* approximation in more detail in Chapter 11.

In QM, each of these symmetry transformations should be represented by a unitary transformation in a Hilbert space. Given any two such representations, we can form another one by making block diagonal matrices with each block being one of the representations. This is what we called the *direct sum* above. Obviously, the symmetry restrictions on energy eigenvalues will apply separately to each block, so it is sufficient to study only *irreducible* representations, which cannot be decomposed into such blocks. For the dihedral groups, the simplest such representation can be obtained by the following geometrical trick.

If we think of the vertices of the polygon as points in a two-dimensional plane, then the symmetries are represented as two-dimensional real matrices. Let us do the example of D_3, the dihedral group of the equilateral triangle. The vector representations of the vertices are

$$z_0 = \begin{pmatrix} 1 \\ 0 \end{pmatrix},$$

$$z_0 = \begin{pmatrix} \cos(2\pi/3) \\ \sin(2\pi/3) \end{pmatrix},$$

$$z_0 = \begin{pmatrix} \cos(2\pi/3) \\ -\sin(2\pi/3) \end{pmatrix},$$

In Exercise 6.18, you will write the six elements of the dihedral group as two by two real matrices. Since the matrices preserve the length of vectors, they satisfy $M^T M = 1$. Now you can let the same real matrices act on a complex two-dimensional Hilbert space. Since they are real, they satisfy $M^\dagger M = 1$, which means that they are unitary transformations. So we have constructed a two-dimensional unitary representation of the group.

To examine the possibility of more general unitary representations, we note that the transformations $S_k = S_1^k$ all commute, so we can diagonalize all of their unitary representatives by diagonalizing $U(S_1)$. The multiplication law of the group tells us that $U(S_1)^3 = 1$, so its eigenvalues can only be third roots of unity. Now note that the group multiplication law also implies

$$CS_1 = S_{-1}C = S_2 C = S_1^2 C. \tag{6.91}$$

Acting with both sides of this equation on an eigenstate $|\omega\rangle$, we get

$$\omega C|\omega\rangle = S_1^2 C|\omega\rangle, \tag{6.92}$$

so $C|\omega\rangle$ is an eigenstate of S_1^2 with eigenvalue ω. This implies that it is an eigenstate of S_1 with eigenvalue ω^*. The whole Hilbert space is a direct sum of spaces with eigenvalues $1, e^{\pm \frac{2\pi i}{3}}$. The transformation $U(C)$ takes the eigenspace with eigenvalue z_i into that with eigenvalue z_i^*. This means that it commutes with $U(S_1)$ in the z_0 eigenspace and acts like the block matrix

$$\begin{pmatrix} 0 & 1 \\ 1 & 0 \end{pmatrix}, \tag{6.93}$$

mapping the two other eigenspaces (which, we conclude, must be of equal dimension since $U(C)$ is unitary (Exercise 6.19)) into each other. Thus, on the subspace of complex eigenvalues of $U(S)$, we have

$$U(C) = \begin{pmatrix} 0 & 1 \\ 1 & 0 \end{pmatrix} \tag{6.94}$$

and

$$U(S) = \begin{pmatrix} e^{2\pi i/3} & 0 \\ 0 & e^{-2\pi i/3} \end{pmatrix}. \tag{6.95}$$

These are block matrices, with any finite block size B. In Exercise 6.20, you will show that one can rearrange rows and columns so that this is equivalent to a direct sum of B identical copies of the representation

$$U(C) = \sigma_1, \qquad U(S) = e^{\frac{2\pi i \sigma_3}{3}}. \tag{6.96}$$

Rearrangement of rows and columns is implemented by a permutation matrix, which is a special example of a unitary matrix, so we say that every representation with complex eigenvalues is *unitarily equivalent* to a direct sum of copies of the two-dimensional representation. For the subspace where $U(S) = 1$, we can find two solutions of $C^2 = 1$, so we have a number of copies of one-dimensional representations, in each of which $U(S) = 1$ and $U(C) = \pm 1$. However, the representations in which $U(C) = 1$ are not *faithful* representations: two different group elements are represented by the same unitary matrix. To make faithful representations, we combine the one-dimensional representations into two-dimensional representations, where $U(S)$ is the two-dimensional unit matrix and $U(C) = \sigma_3$.

The symmetry constraints on the physics of a system invariant under the group D_3 are simply that the Hamiltonian commute with every symmetry generator. This is equivalent to the statement that it is proportional to the unit matrix in every irreducible representation of the group. In the representations with complex eigenvalues of $U(S)$, where the nonabelian character of the group is preserved, this implies a degeneracy of eigenvalues. Conversely, if a system exhibits an exact degeneracy of eigenvalues, then the group of unitary transformations in the degenerate subspace commutes with the Hamiltonian, and is a symmetry of the problem. The only thing to be wary about in using this general statement is the following. For most systems, the Hamiltonian is constructed in terms of some simple operators, like positions and momenta of particles. It is complicated to find its eigenvalues and eigenstates. Thus, the unitary symmetry operations coming from a degeneracy may not have a simple expression in terms of the fundamental variables. In this case, we would consider the degeneracy accidental, and guess that a small deformation of the Hamiltonian written in terms of position and momentum variables would entirely remove the degeneracy.

A similar remark should be made about the group of symmetries defined, for any Hamiltonian, by unitary operators diagonal in the same basis as the Hamiltonian. For those students who know advanced classical mechanics, these are the analog of the *action and angle variables* of classical mechanics. We have not emphasized this in our discussion of classical mechanics, but one can phrase the Hamilton–Jacobi equation as a technique for finding a complete set of conservation laws of any Hamiltonian system. The angle variables are conjugate (in the Poisson bracket sense) to the conservation laws and basically have a harmonic time dependence. For generic systems, the conservation laws are extremely complicated functions of the original position and momentum coordinates, and finding them is tantamount to finding all the complicated solutions of the equations of motion. Symmetry transformations/conservation laws are only useful in either classical or quantum mechanics, if they are relatively simple functions of the fundamental variables out of which the Hamiltonian is constructed.

6.10 EXERCISES

6.1 Prove the Schwarz inequality $|\langle v|u\rangle| \leq \sqrt{|\langle u|u\rangle\langle v|v\rangle|}$.

6.2 Prove that if the diagonal matrix elements $\langle s|A|s\rangle$ of an operator vanish for every state in the Hilbert space, then the operator is equal to 0.

6.3 If two operators do not commute, then $e^{A+B} \neq e^A e^B$. The corrections to this formula are written in terms of two complementary results, the Baker–Campbell–Hausdorff formula and the Zassenhaus Lemma (the two differ in writing the left-hand side of the equation as the right-hand side plus a series of corrections), or vice versa, and they involve multiple commutators of A and B. Derive the first two terms in the formula $e^{A+B} = e^A e^B e^{C_1} e^{C_2} \ldots$ where C_k involves k fold commutators of A and B.

6.4 Let $A(t)$ be a time-dependent operator. Write the Karplus–Schwinger formula expressing $\frac{d}{dt} e^{A(t)}$ in terms of $\frac{dA}{dt}$. A convenient trick is to introduce a second parameter $W(s,t) = e^{sA(t)}$ and write a differential equation for $\frac{\partial W}{\partial s}$.

6.5 The Gram–Schmidt orthogonalization procedure. Start with a nonorthonormal basis $|a_i\rangle$, with $\langle a_i|a_j\rangle = C_{ij}$. We want to construct an orthonormal basis, with $|e_1\rangle = \frac{|a_1\rangle}{\langle a_1|a_1\rangle^{1/2}}$, one of the basis vectors. The basic idea is to choose a second basis vector via the formula

$$|e_2\rangle \propto |a_2\rangle - |e_1\rangle\langle e_1|a_2\rangle.$$

The proportionality constant is determined so that $|e_2\rangle$ has norm one. We continue in this fashion, subtracting from $|a_k\rangle$ all its projections on $|e_i\rangle$ with $i < k$ and then normalizing the resulting vector. Express the coefficients in the expansion of $|e_i\rangle$ in terms of $|a_i\rangle$ in terms of quantities derived from the matrix C_{ij}.

6.6 In an N-dimensional Hilbert space, consider a unitary operator satisfying $U^N = 1$, with nondegenerate spectrum. Its eigenvalues are $e^{\frac{2\pi i k}{N}}$ for $k = 1 \ldots N$. Denote the corresponding eigenstates by $|k\rangle$. Let V be the cyclic permutation operator that takes $|k\rangle \rightarrow |k+1 \ (mod\ N)\rangle$. Show that V is unitary, and satisfies $V^N = 1$ and $UV = VUe^{\frac{2\pi i}{N}}$. Show that any operator in the space can be written in the form

$$A = \sum_{k,l=0}^{N-1} a_{kl} U^k V^l.$$

6.7 Take the limit $N \rightarrow \infty$ in Exercise 6.6. To be more precise, consider states of the form $|f\rangle = \sum f_k |v_k\rangle$, where $V|v_k\rangle = e^{\frac{2\pi k}{N}} |v_k\rangle$, and $\sum |f_k|^2 = 1$ and use Fourier's theorem on Fourier series to show that these converge to periodic functions on a circle $z = e^{i\theta}$ with $0 \le \theta \le 2\pi$. U converges to the multiplication operator by $e^{i\theta}$ and $V^k \rightarrow e^{i\alpha p_\theta}$, where $\alpha = \frac{k}{N}$ and $p_\theta = \frac{1}{i}\partial_\theta$. Show that the spectrum of p_θ consists of all the integers.

6.8 The circle in the previous exercise had radius 1. θ was a dimensionless angle variable. If we want to understand how things depend on the radius of the circle, we should introduce a variable with dimensions of length. We write $\theta = 2\pi \frac{X}{R}$. Let us choose the range of θ to be $-\pi \le \theta \le \pi$ so that $-\frac{R}{2} \le X \le \frac{R}{2}$. Similarly, we write $p_\theta = KR$. K makes infinitesimal shifts in the variable X. The eigenvalues of K are integers divided by R. In the limit $R \rightarrow \infty$, they become a continuum. Show that in this limit, we recover the Hilbert space of a free particle propagating on an infinite line, with K and X playing their usual role.

6.9 Consider the interval $[-1, 1]$ and a nonnegative function on it, $\mu(x)$, which you can take to be piecewise continuous. Define the moments of μ by $M_n = \int_{-1}^{1} \mu(x)x^n$. Show that you can find polynomials $P_k(x)$ for all nonnegative k, such that

$$\int \mu(x)P_k(x)P_l(x) = \delta_{kl}.$$

Do this explicitly for $k = 0, 1, 2, 3$, and show that the coefficients are expressed in terms of the moments. Argue that the same will be true for all k. This exercise is an infinite dimensional analog of Gram–Schmidt orthogonalization.

6.10 Let x_a, $a = 0 \ldots 3$ be coordinates of a point on the sphere in four dimensions $\sum_a x_a^2 = 1$. Define

$$U = x_0 + i x_a \sigma_a,$$

where σ_i are the Pauli matrices. Show that $U(x_a)$ is unitary and has determinant equal to one. Conversely, show that every 2×2 matrix with these properties has this form (recall that the unit matrix and the Pauli matrices form a basis in the space of all complex two by two matrices).

6.11 Evaluate

$$U^\dagger(x_a) \sigma_i U(x_a) = R_{ij}(x_a) \sigma_j.$$

That is, show that the unit operator does not appear on the right-hand side. Show that the 3×3 matrix R_{ij} is real and satisfies $R^T R = 1$, so that it is a rotation matrix. This is a mapping of the group $SU(2)$ of 2×2 unitary matrices into the group $SO(3)$ of three-dimensional rotations. Show that both $U = \pm 1$ are mapped into the identity rotation.

6.12 Prove that if a Hamiltonian H is bounded from below $H \geq k$, then any diagonal matrix element of the resolvent operator

$$D_s(z) = \langle s | \frac{1}{z - H} | s \rangle,$$

has an integral representation

$$D_s(z) = \int_C \frac{\rho(w)}{2\pi i (z - w)},$$

with $\rho \geq 0$. The contour C encircles the spectrum of H, which lies on the interval $[K, \infty]$, in a counterclockwise manner.

6.13 Schur's Lemma: Consider a subalgebra $\mathcal{A}$ of operators on a finite dimensional Hilbert space, which acts *irreducibly*. That is, there is no proper subspace of the Hilbert space, apart from the zero vector, which is left invariant by the action of $\mathcal{A}$. If C is an operator such that C and $C^\dagger$ commute with every element of $\mathcal{A}$, Schur's lemma states that C is proportional to the unit operator. Alternatively, if the Hermitian conjugate of any operator in $\mathcal{A}$ is in $\mathcal{A}$, then we need only assume that C commutes with $\mathcal{A}$.

6.14 For diagonal operators, the identity

$$\det A = e^{\text{tr} \ln A}$$

is obvious. Prove that it is true for all normal operators, using properties of the determinant and trace. Now let us make a small change $A \to A + \delta A$. Show that

$$\det A \to \det A(1 + \mathrm{tr}\,[\delta A A^{-1}]),$$

to first order in δA.

6.15 Prove that the N-th roots of unity form a group under multiplication, generated by the element with the smallest phase. The axioms for a group are that we have a multiplication rule for any two elements, which gives a third element as their product, that there is an identity element which satisfies $eg = g$ for every group element g, and an inverse such that $gg^{-1} = 1 = g^{-1}g$, for every element g.

6.16 Prove that every function on the plane is a linear combination of functions of the form $z^k f_k$, where f_k is invariant under the group Z_N and $k = 0 \ldots N - 1$.

6.17 Show that the combined operations of multiplying a complex number by an N-th root of unity, and complex conjugation, form a group with $2N$ elements.

6.18 Write the six elements of the dihedral group of the equilateral triangle as 2×2 matrices.

6.19 Show that unitarity of the matrix $U(C)$ in a representation of the group D_3 with complex eigenvalues for $U(S)$ implies that the dimensions of the two eigenspaces with complex conjugate eigenvalues are equal.

6.20 Show that one can exchange rows and columns of the block $2B \times 2B$ matrices, $U(C), U(S)$, representing the group D_3 and write them as a direct sum of B copies of the 2×2 matrices σ_1 and σ_3.

6.21 Explain how the previous comment is related to the equation

$$A \otimes B = B \otimes A$$

for the tensor product of any two matrices. You will have to remember that the first tensor product is defined by multiplying each matrix element of the matrix A, by the matrix B.

6.22 Find, up to unitary equivalence, all unitary irreducible representations of the group D_4.

6.23 Consider a particle of charge q moving on a circle of radius r. If the circle encloses a solenoid of magnetic flux F, show that after setting a bunch of parameters equal to 1, the Lagrangian is

$$L = \frac{1}{2}\dot{\theta}^2 + \alpha\dot{\theta}.$$

Construct the Hamiltonian and find its eigenfunctions and eigenvalues. Show that time reversal symmetry is valid for this problem for two different values of α. One of them is obviously $\alpha = 0$, what is the other? Show that for the nonzero value of α the time reversal invariant problem has a degenerate ground state.

6.24 The symmetry group of the problem in Exercise 6.23, thought of classically, is $U(1) \times T$. Show that when α is nonzero, the $U(1)$ and T operations do not commute. The faithful representations of this nonabelian symmetry group have more than one dimension. Show that the degenerate ground state is a two-dimensional representation of this group.

6.25 Now add a general periodic potential that is a function of 2θ to the Hamiltonian. The problem is no longer exactly soluble. Show that the nonabelian symmetry is still present at the nonzero, T invariant, value of α. This means that the ground state is at least doubly degenerate.

CHAPTER 7

Rotation Invariance and the Hydrogen Atom

7.1 INTRODUCTION

This chapter is devoted to the exact solution of the hydrogen atom, which is the basis for much of our understanding of atomic physics. We first discuss rotationally invariant problems in general. This leads to another Noetherian triumph: angular momentum is the generator of rotations (just as the Hamiltonian and momentum operators generate time and space translations), and the symmetry determines all of its properties in quantum mechanics (QM). In particular, this insight leads to the exact form of Bohr's approximate formula for the quantization of angular momentum. It also shows us that different angular momentum components cannot commute with each other. This has the consequence that energy eigenstates in rotationally invariant theories always come in $2l + 1$ dimensional degenerate multiplets. Finally, rotational invariance completely determines the form of the Schrödinger wave functions, as functions of the angles of a spherical coordinate system, and reduces the problem to an infinite set of one-dimensional Schrödinger equations. For hydrogen, we can solve all of these equations in terms of functions called Laguerre polynomials. The hydrogen spectrum has a degeneracy beyond that implied by angular momentum. This is a consequence of the conservation of the direction of the semimajor axis of elliptical orbits, a quantity called the Laplace–Runge–Lenz vector.

Finally, we will outline the explanation of Mendeleev's periodic table in terms of the hydrogen atom and the Hartree approximation for multielectron atoms.

7.2 UNITS AND SYMMETRIES

The hydrogen atom is modeled by a Hamiltonian

$$H = \frac{\mathbf{P}_N^2}{2m_N} + \frac{\mathbf{P}_e^2}{2m_e} - \frac{e^2 Z}{4\pi\epsilon_0 |\mathbf{X_e} - \mathbf{X_N}|}. \tag{7.1}$$

This Hamiltonian actually describes a whole family of one electron *ions*, with nuclear charge Z and nuclear mass m_N. The hydrogen atom is the case $Z = 1$, and $m_N = m_{proton} \sim 2,000 \ m_e$. The potential depends only on the relative coordinate $\mathbf{R} = \mathbf{X_e} - \mathbf{X_N}$, so the system is invariant under translations and the total momentum $\mathbf{P_T} = \mathbf{P}_N + \mathbf{P}_e$, commutes with the Hamiltonian and is conserved. If we go to the rest frame of the ion, then $\mathbf{P_T} = 0$, and we can rewrite the Hamiltonian in terms of the relative momentum $\mathbf{P} = \mathbf{P}_N - \mathbf{P}_e$, which satisfies

$$[\mathbf{R}_i, \mathbf{P}^j] = i\hbar\delta_i^j. \tag{7.2}$$

The Hamiltonian is

$$H = \frac{\mathbf{P}^2}{2m_R} - \frac{e^2 Z}{4\pi\epsilon_0 |\mathbf{R}|}. \tag{7.3}$$

The reduced mass is given by

$$m_R = \frac{m_e m_N}{m_e + m_N} = m_e + o(m_e/m_N). \tag{7.4}$$

The correction to $m_R = m_e$ is less than 1 part in $2,000$.

If we introduce the Bohr radius by

$$a_B = \frac{4\pi\epsilon_0 \hbar^2}{m_e e^2} \sim 5.29177 \times 10^{-11} m \tag{7.5}$$

and the Rydberg energy by

$$|E_1| = \frac{\hbar^2}{2m_e a_B^2} = 13.6057\text{eV} \equiv E_{Rydberg} \tag{7.6}$$

(the hydrogen ground state energy, like all bound state energies, is conveniently taken to be negative. It is -1 in Rydberg energy units), then we can write a completely dimensionless Schrödinger equation by defining $\mathbf{R} = a_B \mathbf{X}$. We get

$$[-\nabla^2 \psi - \frac{2Z}{r}]\psi = E\psi, \tag{7.7}$$

where E is the energy in Rydberg units and $r = \sqrt{\mathbf{X}^2}$, measured in Bohr radius units

We will begin by analyzing a more general spherically symmetric problem $-\frac{Z}{r} \to V(r)$, and then specialize to the Coulomb problem, which is exactly soluble. As usual, we first use the three-dimensional rotational symmetry, to identify conservation laws, which commute with the Hamiltonian, and then solve the Schrödinger eigenvalue problem with fixed values of the conserved quantities.

Any rotation is a rotation in some plane, by some angle θ. It also has a *sense*: either clockwise or counterclockwise. The set of all counterclockwise rotations in the [12] plane is a one parameter group of symmetries, represented in QM by unitary operators $e^{i\theta K_{12}}$, where $-K_{21} = K_{12} \equiv K_3$ is a Hermitian operator called the infinitesimal generator of the symmetry. θ goes from 0 to 2π. In writing $K_{12} = K_3$, we are using the geometric fact that every plane in three dimensions has a unique axis perpendicular to it. Note that K_3 changes sign if we change the sense of rotation, but not if we do a reflection of all three coordinates. More generally we define

$$K_{ij} = \epsilon_{ijk} K_k, \tag{7.8}$$

as the generator of infinitesimal rotations that take the i-axis into the j-axis. It is obvious from this formula that the three components K_k should transform as a pseudovector under rotations: a vector that does not flip sign under reflection. The fact that $K_{1,2}$ transform as the components of a two-dimensional vector under the rotation generated by K_3 is expressed via the equation

$$e^{-i\theta K_3} K_1 e^{i\theta K_3} = \cos\theta K_1 + \sin\theta K_2. \tag{7.9}$$

For small θ, this says that

$$-i[K_3, K_1] = K_2. \tag{7.10}$$

A generalization of this formula, which works for rotations around any axis is just

$$[K_i, K_j] = i\epsilon_{ijk} K_k, \tag{7.11}$$

where we have used the summation convention for the repeated index k.

It is easy to see intuitively why rotations do not commute with each other. The rotation around the three-axis changes the one-axis into the two-axis, so you do not get the same thing after applying it as you do before applying it. The real surprise in QM is that, for a system invariant under rotations, we must have

$$[K_i, H] = 0, \tag{7.12}$$

which, by the Heisenberg equations of motion implies that the pseudovector $\mathbf{K}$ is a conserved quantity for rotation invariant systems. From classical mechanics, it is familiar that angular momentum is conserved for spherically symmetry systems. Angular momentum is a pseudovector, so it must be proportional to K_i. By dimensional analysis,

$$L_i = \hbar K_i.$$

In principle, there could be a numerically constant in this equation, but as we will see, choosing the constant equal to one gives us precisely Bohr's rule for quantization of angular momentum. In addition, you will prove in Exercise 7.1 that the classical formula

$$\mathbf{L} = \mathbf{X} \times \mathbf{P} \tag{7.13}$$

or

$$L_i = \epsilon_{ijk} X_j P_k \tag{7.14}$$

gives precisely those commutation relations for L_i.

7.3 IRREDUCIBLE REPRESENTATIONS OF THE COMMUTATION RELATIONS

Since the L_i commute with the Hamiltonian, we will simplify the problem of solving the Schrödinger equation by diagonalizing as many of the angular momentum operators as we can before writing the eigenvalue problem. Here we come to one of the weird features of QM. *Since the angular momentum generators do not commute, we can diagonalize at most one of the generators in any given state.* In terms of probabilities, *if we are in a state in which L_3 has a precise value, then $L_{1,2}$ are uncertain.*

The square of the angular momentum

$$\mathbf{L}^2 \equiv L_1^2 + L_2^2 + L_3^2 \tag{7.15}$$

is invariant under rotations, and so must commute with all of the L_i. You will verify this, using the commutation relations, in Exercise 7.2. This means that we can find a complete set of states $|n, l, m\rangle$ of the system in which both $\mathbf{L}^2$ and L_3 are fixed numbers.

$$\mathbf{L}^2 |n, l, m\rangle = \hbar^2 l(l+1)|n, l, m\rangle, \tag{7.16}$$

$$L_3|n, l, m\rangle = \hbar m|n, l, m\rangle, \tag{7.17}$$

The object of the quantum theory of angular momentum is to find the values of l and m that are allowed by the commutation relations. We do this by acting with the operators $K_{1,2}$ on these eigenstates, much as we found the spectrum of the harmonic oscillator. The additional label n refers to the action of operators that commute with the angular momentum. In the problem of a single particle in a spherically symmetric potential, a complete commuting set of such operators consists of functions of the Hamiltonian. So in this problem, n will label the different energy eigenstates for fixed values of l and m. Rotations are a symmetry of a much broader class of problems, involving any number of particles. In these more complicated systems, n might refer to many other conserved quantities, besides the energy. Our analysis of the possible values of l and m will be valid in *all* of these systems.

It will be simpler to work in terms of the dimensionless operators K_i. We have already seen that $K_{1,2}$ transform as a two-dimensional vector under rotations generated by K_3. We can think of the two-dimensional plane as the complex plane, with complex coordinates

$$K_{\pm} = K_1 \pm iK_2, \tag{7.18}$$

$$K_+ = (K_-)^{\dagger}, \tag{7.19}$$

in which case rotations act by $K_{\pm} \to e^{\pm i\theta} K_{\pm}$. Using either this geometric intuition, or directly from the fundamental commutation relations, we see that

$$[K_3, K_{\pm}] = [K_3, K_1] \pm i[K_3, K_2] = i\epsilon_{312}K_2 \mp \epsilon_{321}K_1 = \pm K_{\pm}. \tag{7.20}$$

In other words, $K_{\pm}$ are raising and lowering operators for the eigenvalue of K_3, just as creation and annihilation operators raise and lower the eigenvalue of the harmonic oscillator Hamiltonian.

The parallel with the harmonic oscillator is not exact, because we can compute

$$K_+K_- = K_1^2 + K_2^2 + i[K_2, K_1] = \mathbf{K}^2 + K_3 - K_3^2 \tag{7.21}$$

and

$$K_-K_+ = K_1^2 + K_2^2 - i[K_2, K_1] = \mathbf{K}^2 - K_3 - K_3^2. \tag{7.22}$$

In either order, these are products of an operator with its own adjoint, so the expectation value of both the left- and right-hand sides of both equations, in any state, are positive. Applying this to simultaneous eigenstates of $\mathbf{K}^2$ and K_3, we get

$$l(l+1) \geq m^2 \mp m. \tag{7.23}$$

We know that $K_{\pm}$ raise and lower the eigenvalue, m of K_3 by one unit, without changing l. The only way we can satisfy the inequalities for all states is if there are minimum and maximum values of m such that

$$K_+|l, m_{max}\rangle = 0 = K_-|l, m_{min}\rangle. \tag{7.24}$$

We then have

$$0 = \langle l, m_{min}|K_+K_-|l, m_{min}\rangle = l(l+1) - m_{min}^2 + m_{min}, \tag{7.25}$$

$$0 = \langle l, m_{max}|K_-K_+|l, m_{max}\rangle = l(l+1) - m_{max}^2 - m_{max}. \tag{7.26}$$

Combining these two equations, we find that

$$l(l+1) = m_{max}(m_{max}+1) = m_{min}(m_{min}-1). \tag{7.27}$$

The solution $m_{max} = m_{min} - 1$ of the second equality is incompatible with $m_{max} \geq m_{min}$, so we must have

$$l = m_{max} = -m_{min}. \tag{7.28}$$

Starting from $|l, m_{min}\rangle$ and acting p times with K_+, we raise m to $m_{min} + p$. For consistency, this must stop at m_{max}, so

$$2l = m_{max} - m_{min} = p, \tag{7.29}$$

with p a positive integer. We conclude that l is quantized in half integral units and m runs between $-l$ and l. Angular momentum is thus quantized, as Bohr hypothesized, but he got two things wrong about the quantization rule. First, we find that $\mathbf{L}^2 = \hbar^2 l(l+1)$, rather than l^2, so Bohr's rule is only right in the limit of large l.

The half integer rather than integer quantization can be eliminated, at least for orbital angular momentum, by insisting that a rotation by 2π return every state to itself. The operator $e^{2\pi i K_3}$ must be the unit operator, which means that K_3 must have integer eigenvalues. This argument is too glib, as we will see below. In nonrelativistic physics, where one can make a clear separation between the intrinsic spin of particles and their orbital angular momentum, we can assign the funny behavior under 2π rotations to an intrinsic angular momentum carried by a single particle. This separation is incompatible with relativity. We will sort this confusion out in a little while, when we talk about spin.

In order to find the wave functions of the hydrogen atom, we need to use the representation of the angular momentum operators in coordinate basis. This is usually done in spherical coordinates, and we will record the angular momentum operators in spherical coordinates now. The simplest is L_3, which simply shifts ϕ. It is thus

$$L_3 = \frac{\hbar}{i} \partial_\phi. \tag{7.30}$$

It is convenient to use complex combinations of the coordinates in the 12 plane

$$z = x_1 + ix_2 = r \sin(\theta) e^{i\phi} \tag{7.31}$$

and its complex conjugate

$$z^* = x_1 - ix_2 = r \sin(\theta) e^{-i\phi}. \tag{7.32}$$

The generators $L_\pm$ have to be proportional to $e^{\pm i\phi}$ to satisfy the commutator $[L_3, L_\pm] = \pm \hbar L_\pm$. Thus, they have the form

$$L_+ = x_3 \partial_{z^*} - z \partial_3, \tag{7.33}$$

$$L_- = x_3 \partial_z - z^* \partial_3. \tag{7.34}$$

It is now a simple change of variables to show that

$$L_\pm = \pm\hbar e^{\pm i\phi}(\partial_\theta \pm i\cot(\theta)\partial_\phi). \tag{7.35}$$

We can now compute

$$\mathbf{L}^2 = L_+ L_- + L_3^2 - \hbar L_3. \tag{7.36}$$

In principle, we can compute the angular wave functions, which are called spherical harmonics, by solving the differential equations

$$\mathbf{L}^2 Y_{lm} = \hbar^2 l(l+1)Y_{lm}, \tag{7.37}$$

$$L_3 Y_{lm} = \hbar m Y_{lm}. \tag{7.38}$$

The second equation is solved by

$$Y_{lm} = e^{im\phi}P_{lm}(\theta),$$

and this turns the first equation into

$$\sin(\theta)\partial_\theta[\sin(\theta)\partial_\theta Y_{lm}] + m^2 Y_{lm} = -l(l+1)\sin^2(\theta)Y_{lm}. \tag{7.39}$$

You should verify this as an unofficial exercise. This is called the *associated Legendre equation* and its solutions the associated Legendre functions. The equation came up in the 18th and 19th centuries and the properties of these functions were studied extensively. You can find tables and numerous recursion relations, integral representations, etc., in many online resources.

Rather than following this time honored route, we will get to the Legendre functions by using another representation of the sphere. Instead of thinking about spherical coordinates, we can think about functions on the sphere as functions of a three-dimensional unit vector. It is then obvious how the functions

$$n_{a_1}\ldots n_{a_k} \tag{7.40}$$

transform under rotations. In particular,

$$K_a n_b = i\epsilon_{abc}n_c, \tag{7.41}$$

where we use the summation convention.

There are two ways to use the unit vector representation of the sphere to make Legendre functions, and we will start with the simplest way of constructing all of the Y_{lm} for integer l. The space of all square integrable functions on the sphere is acted on by a unitary representation of the rotation group $U(R)f(n^a) = f(R^{ab}n^b)$, where R is a 3×3 rotation matrix. We have just shown by abstract operator algebra, that there are subsets of functions

$Y_{lm}(\mathbf{n}) = \langle \mathbf{n}|l, m \rangle$ which transform into themselves. For fixed l, there are $2l + 1$ functions so we say that we have a $2l + 1$ *dimensional representation of the rotation group*. Under rotations, its obvious that the subset of functions $\tilde{T}^{a_1 \cdots a_l} \equiv n^{a_1} \cdots n^{a_l} 1$ for fixed l, transform into themselves under rotations. If one contracts two of the indices of $\tilde{T}$ with the Kronecker delta, one gets the $\tilde{T}$ with two fewer indices. This transforms into itself separately. So, for example

$$T^{ab} = \tilde{T}^{ab} - \frac{1}{3}\delta^{ab}, \tag{7.42}$$

transforms into itself. Note that it has five independent components, as we would expect for a spherical harmonic of $l = 2$. The components that have fixed value of $L_3 = \hbar m$ are obtained by taking complex linear combinations of components, $n_{\pm} = n_1 \pm i n_2$. That is T^{++} has $m = 2$, T^{--} has $m = -2$, $T^{3\pm}$ has $m = \pm 1$, and T^{33} has $m = 0$. Now, for any l define, recursively

$$T^{a_1 \cdots a_l} = \tilde{T}^{a_1 \cdots a_l} - C_l[\delta^{a_1 a_2} T^{a_3 \cdots a_l} - \sum \text{combinations}]. \tag{7.43}$$

The combinations are added and the constant C_l chosen to make sure that $\delta^{a_i \, a_j} T^{a_1 \cdots a_l} = 0$ for any pair of indices. The T's are called *traceless symmetric tensors*, and it is easy to verify that the l-th T has $2l + 1$ independent components. Up to normalization, the traceless symmetric tensors *are* the spherical harmonics Y_{lm}, which is to say the "column vector" representatives of the states $|l, m \rangle$ in the basis where n^a are diagonal. Since any function on the sphere can be written as a (in general infinite convergent) linear combination of the $\tilde{T}$'s and therefore of the Y_{lm}, we have found a basis of the whole Hilbert space. Note that each l value occurs exactly once.

An easy way to do the counting of independent Y_{lm} in the traceless tensor representation is to work in terms of complex coordinates, n^3 and $n^{\pm} = n^1 \pm i n^2 = e^{\pm i\phi} \sin(\theta)$. Using the fact that $n_+ n_- = 1 - n_3^2$, we see that there is a complete set of independent components of $n^{a_1} \cdots n^{a_l}$ of the form $Y_{l,\pm m} = n_{\pm}^m f_{lm}(\cos(\theta))$, for $0 \leq m \leq l$. The polynomials f_{lm} can be determined in various ways. One is the orthonormality constraint

$$\int \sin(\theta) d\theta d\phi \, Y_{lm}^* Y_{ln} = 2\pi \int_{-1}^{1} dx (1 - x^2)^m f_{lm}(x) f_{ln}(x) = \delta_{nm}. \tag{7.44}$$

Different people prefer different methods. Use the traceless condition or the orthogonality condition at your own discretion. Of course, nowadays the easiest way to find a formula for spherical harmonics is to look them up on the web. These analytical methods can be reserved for those times when you want to calculate a Y_{lm} after an electromagnetic pulse or cyber attack has disabled the internet.

---

[1] We are raising the indices here to avoid double subscripts.

Now we want to rederive the same results using operator methods. Most people will find these more cumbersome, but they are useful as an introduction to techniques for general symmetry groups, where there are no obvious analogs of the simple functions $\tilde{T}^{a_1\cdots a_l}$, which jump started the discussion of the previous paragraphs. These operator methods are most useful for low values of $|m - l|$.

We decompose K_a into K_3 and $K_\pm$ as in our derivation of the spectrum. The analogous decomposition of $n_a{}^2$ is into n_3 and $n_\pm = n_1 \pm in_2$, and $n_3^2 = 1 - n_+n_-$. $n_\pm$ are simply the variables z and $z*$ introduced above, with the factor of r removed. It is then simple algebra to show that

$$K_\pm n_\pm = 0, \quad K_\pm n_\mp = \pm 2n_3 \tag{7.45}$$

and

$$K_\pm n_3 = \mp n_\pm. \tag{7.46}$$

We can write all monomials in components of n_a in terms of either

$$Z_{pq} \equiv n_+^p n_-^q$$

or

$$n_3 Z_{pq},$$

since even powers of n_3 are just polynomials in n_+n_-. The K_a operators are all first order differential operators, so they satisfy Leibniz' rule when applied to a product of two functions: $K_a(fg) = K_a(f)g + fK_a(g)$. We conclude that

$$K_+(Z_{pq}) = 2qZ_{p(q-1)}n_3, \tag{7.47}$$

$$K_-(Z_{pq}) = -2pZ_{(p-1)q}n_3, \tag{7.48}$$

$$K_+(Z_{pq}n_3) = 2qZ_{p(q-1)}(1 - n_+n_-) - Z_{(p+1)q}, \tag{7.49}$$

$$K_-(Z_{pq}n_3) = -2pZ_{(p-1)q}(1 - n_+n_-) + Z_{p,(q+1)}. \tag{7.50}$$

The action of K_3 on any of these functions, just multiplies it by $p - q$.

We would like to find polynomials, linear combinations of Z_{pq} and $Z_{pq}n_3$, which have fixed values of $\mathbf{K}^2$ and K_3. These functions are the *spherical harmonics* $Y_{lm}(n_a) = Y_{lm}(\theta, \phi)$. They are the coordinate basis representatives of the abstract states, $|lm\rangle$. Recall that

$$n_3 = \cos(\theta),$$

$$n_\pm = \sin(\theta)e^{\pm i\phi}.$$

---

[2] We have switched back to lowered indices here because we no longer have to write expressions that would generate double subscripts.

These are in fact the three components of Y_{1m}, as you can verify by specializing the general analysis below to $l = 1$.

For general l, we start by looking for Y_{ll}, which satisfies $K_+ Y_{ll} = 0$, and $K_3 Y_{ll} = l Y_{ll}$. The obvious solution is

$$Y_{ll} \propto Z_{l0}, \tag{7.51}$$

and it is unique. The proportionality constant can be calculated by integrating $|Z_{l0}|$ over the sphere, since Y_{ll} is normalized to 1:

$$\int d\Omega |Z_{l0}|^2 = 2\pi \int_0^\pi \sin^{2l+1}(\theta) d\theta = 2\pi^2 \frac{(2l)!}{2^{2l}(l!)^2}.$$

Thus,

$$Y_{ll} = [\pi^2 \frac{(2l)!}{2^{2l-1}(l!)^2}]^{-1/2} \sin^l(\theta) e^{il\phi}. \tag{7.52}$$

A similar analysis shows that

$$Y_{l-l} = [\pi^2 \frac{(2l)!}{2^{2l-1}(l!)^2}]^{-1/2} \sin^l(\theta) e^{-il\phi}. \tag{7.53}$$

Our general discussion of angular momentum now tells us that

$$Y_{lm} \propto K_-^{l-m} Y_{ll}, \tag{7.54}$$

and we can carry out the differentiation by using the formulae above for the action of K_- on $Z_{pq} \times (1, n_3)$. When $l - m$ is small, this is quite quick and efficient. On the other hand, when m is close to $-l$ we can instead start from $Y_{l-l} \propto Z_{0l}$, and apply powers of K_+. Either method generates all of the Y_{lm}, but it is less tedious to apply one at the top of the m spectrum and the other at the bottom. Another useful observation that comes from this analysis is that, for any l and $m = 0$, $Z_{pp} = (n_+ n_-)^p = (1 - n_3^2)^p$, so that Y_{l0} is a polynomial in $n_3 = \cos(\theta)$. These polynomials are proportional to the Legendre polynomials [19] and can be computed simply by requiring that they form an orthonormal basis of polynomials on the interval $[-1, 1]$.

The proper normalization of all spherical harmonics can be obtained by computing

$$\langle ll | K_+^{l-m} K_-^{l-m} | ll \rangle. \tag{7.55}$$

This is the squared norm of the state $|K_-^{l-m}|ll\rangle$. The trick for doing this computation is to note that

$$\langle ll | K_+^{l-m} K_-^{l-m} | ll \rangle = \langle ll | K_+^{l-m-1} (\mathbf{K}^2 - K_3^2 + K_3) K_-^{l-m-1} | ll \rangle. \tag{7.56}$$

The state $K_-^{l-m-1}|ll\rangle$ is an eigenstate of $\mathbf{K}^2$ with eigenvalue $l(l+1)$ and of K_3 with eigenvalue $m+1$, so this is equal to

$$(l(l+1) - (m+1)^2 + m + 1)\langle ll|K_+^{l-m-1}K_-^{l-m-1}|ll\rangle. \tag{7.57}$$

We can keep on using this trick until we eliminate all of the raising and lowering operators and get back to the normalized state $|ll\rangle$.

7.4 ADDITION OF ANGULAR MOMENTA

Suppose we have two systems, which have angular momenta j_1 and j_2. The theory of addition of angular momenta tells us which representations of the rotation group there are in the composite system made by combining them. This formula is useful for studying states of two particles, but also for a single particle when we decompose its total angular momentum into spin and orbital parts. The mathematical problem is that we have two commuting copies $J_a^{(1)}$ and $J_a^{(2)}$ of the angular momentum. We can form their sum $J_a = J_a^{(1)} + J_a^{(2)}$, and verify that it also satisfies the angular momentum algebra. We want to know how the full Hilbert space decomposes into irreducible subspaces of dimension $2j + 1$, where $J_a^2 = \hbar^2 j(j+1)$.

The full Hilbert space consists of the states

$$|j_1\ m_1\ j_2\ m_2\rangle; \quad -j_1 \le m_1 \le j_1; \quad -j_2 \le m_2 \le j_2. \tag{7.58}$$

It has dimension $(2j_1 + 1)(2j_2 + 1)$. In the language of Chapter 6, it is the tensor product of the two individual Hilbert spaces. The values of J_3 for each of these states are simply $m_1 + m_2$.

The analysis begins by noting that there is a unique state with $J_3 = j_1 + j_2$. The theory of angular momentum then tells us that there must be a whole multiplet of states with J_3 between $j_1 + j_2$ and $-(j_1 + j_2)$. Now look at states with $J_3 = j_1 + j_2 - 1$. There are two such states $|j_1\ m_1\ j_2\ m_2 - 1\rangle$ and $|j_1\ m_1 - 1\ j_2\ m_2\rangle$. One linear combination of them is the state $|j_1 + j_2\ j_1 + j_2 - 1\rangle$ from the highest angular momentum multiplet. The orthogonal linear combination must be the top of a new multiplet with $j = j_1 + j_2 - 1$. Next we look at $J_3 = j_1 + j_2 - 2$. There are three such states

$$|j_1\ m_1 - 2\ j_2\ m_2\rangle \quad |j_1\ m_1\ j_2\ m_2 - 2\rangle \quad |j_1\ m_1 - 1\ j_2\ m_2 - 1\rangle.$$

Two linear combinations are accounted for by states in the two multiplets we have already found, while the third one indicates the presence of a new multiplet, with $j = j_1 + j_2 - 2$. You should convince yourself that this pattern repeats: at each level $J_3 = j_1 + j_2 - k$ there is always exactly one extra state, which is not contained in the multiplets uncovered at previous levels. The process stops, however, when $k = 2j_{min}$, the smaller of the two angular momenta, because past that point we cannot lower J_3 by lowering m_{min}. Thus, the lowest angular momentum we get to is $j_{min} = |j_1 - j_2|$. Thus, *the addition of angular momenta j_1 and j_2 leads to one copy of total angular momentum for each value of j between $j_1 + j_2$ and $|j_1 - j_2|$.* You should convince yourself that the sum of the dimensions of these representations sum up to the dimension of the tensor product space.

There are now two different bases for the tensor product space, $|j_1 \, m_1 \, j_2 \, m_2\rangle$ and $|j \, m\rangle$. The scalar products between them define the matrix elements of the unitary transformation, which takes one basis into another. They are called Clebsch-Gordon coefficients and you can find tables of them on the World Wide Web.

$$C(j, m; j_1, m_1, j_2, m_2) \equiv \langle j \, m | j_1 \, m_1 \, j_2 \, m_2 - 1\rangle. \tag{7.59}$$

7.5   THE HAMILTONIAN OF SPHERICALLY SYMMETRIC POTENTIALS

Now we want to rewrite the Hamiltonian of a spherically symmetric problem by exploiting our knowledge of angular momentum and rotation invariance. The Hamiltonian commutes with angular momentum, so it must be constructed out of r, ∂_r, and $\mathbf{L}^2$. The potential is simply a function of r. The kinetic term $-\nabla^2$ scales like $\frac{1}{r^2}$, so it must be a linear combination

$$-\nabla^2 = A\partial_r^2 + \frac{B}{r}\partial_r + \frac{f[\mathbf{K}^2]}{r^2}. \tag{7.60}$$

Since it is a second-order differential operator, the function $f[\mathbf{K}^2]$ must be linear, a simple multiple of $\mathbf{K}^2$. Any higher power of $\mathbf{K}^2$ would be a higher order differential operator, and could not come from the Laplacian. We can evaluate A and B by letting the operator act on a function that is constant on the sphere, for which $\mathbf{K}^2\psi = 0$, $\psi = \psi(r)$. Then

$$\partial_i\psi(r) = \frac{x_i}{r}\partial_r\psi. \tag{7.61}$$

It follows that

$$-\nabla^2\psi = -(\partial_r^2 + \frac{2}{r}\partial_r)\psi, \tag{7.62}$$

so $B = -2 = 2A$. To determine the coefficient of the angular momentum term, we apply the Laplacian to the function $\frac{x_3}{r} = \cos(\theta)$. This is a vector under rotations ($l = 1$) and is independent of r so

$$\frac{\mathbf{K}^2}{r^2}\cos(\theta) = \frac{2}{r^2}\cos(\theta).$$

On the other hand,

$$\nabla^2\frac{x_3}{r} = 2\nabla(x_3) \cdot \nabla(\frac{1}{r}),$$

where we have used the fact that $x_3 \nabla^2 \frac{1}{r} = 4\pi x_3 \delta^3(x) = 0$. This gives us

$$\nabla^2 \frac{x_3}{r} = 2\partial_3(\frac{1}{r}) = -\frac{2}{r^2} \cos(\theta).$$

This tells us that the coefficient in the $\mathbf{K}^2$ term is 1.

$$-\nabla^2 = -(\partial_r^2 + \frac{2}{r}\partial_r) + \frac{\mathbf{K}^2}{r^2}. \tag{7.63}$$

That is, we have derived the separation of variables of the Schrödinger equation for a spherically symmetric potential, and found an efficient method to construct the exact angular wave functions Y_{lm}. We also understand why these are simultaneous eigenfunctions of H, the third component of angular momentum, and the square of the angular momentum. The angular momentum eigenvalues are quantized, reproducing Bohr's rule for large l, as we might have expected for wave functions confined to a compact space.

The eigenvalue equation for fixed angular momentum is

$$[-\partial_r^2 - \frac{2}{r}\partial_r + \frac{l(l+1)}{r^2} + V(r)]f_l(r) = E_l f_l(r). \tag{7.64}$$

For single electron ions, the potential is just

$$V(r) = -\frac{2Z}{r}.$$

Energies are expressed in Rydbergs and lengths in Bohr radii. We now exploit the fact that before rescaling everything depended only on the combination Ze^2. So, we can do our computations for hydrogen, and get the answer for ions of higher charge by scaling $e \to \sqrt{Z}e$ in the answer.

The normalization condition for the wave function is

$$\int_0^\infty |f_l(r)|^2 r^2 dr = 1. \tag{7.65}$$

Near $r = 0$, if $l \neq 0$ the wave function behaves like r^a, with

$$-a(a-1) - 2a + l(l+1) = 0. \tag{7.66}$$

The solutions are $a = l$ and $a = -l - 1$. Only $a = l$ is normalizable. For $l = 0$, both solutions are normalizable, but the expectation value of the Coulomb term in the Hamiltonian is logarithmically divergent if we choose $a = -1$. This would mean that if such a behavior were allowed in the Hilbert space the Hamiltonian would be unbounded from below.

This is a place where we have to be a little more careful about the definition of unbounded operators. To prove Hermiticity of the Hamiltonian, we have to integrate the radial derivatives by parts. However, if we choose the singular behavior at the origin for $l = 0$, the surface terms do not vanish. In fact, they diverge. So, functions are in the part of the Hilbert space where H is a Hermitian operator, only if they go to a constant at the origin for $l = 0$. One says that normalizable wave functions with the r^{-1} behavior are not in the domain of definition of the unbounded Hamiltonian, and so cannot be eigenstates.

If we look at $r \to \infty$, then the equation for the logarithm, S of the wave function becomes

$$-\partial_r^2 S - (\partial_r S)^2 - \frac{2}{r}\partial_r S = E - V(r). \tag{7.67}$$

The potential goes to zero at infinity, and we know $E < 0$ if we want a wave function that goes to zero. The leading solutions are $S = \pm\sqrt{-E}r$, and only the negative sign gives something normalizable. Now we know we are in trouble. The boundary condition at the origin fixes the wave function, up to an over all constant, but this means that in general, there will be a nonzero coefficient

$$A(E)e^{\sqrt{-E}r}.$$

The only way to get a normalizable solution is to tune E to one of the zeroes of $A(E)$. This is the origin of energy quantization for bound states.

Notice that for positive energy, the behavior is very different. The two solutions behave like $e^{\pm i\sqrt{E}r}$, which correspond to incoming and outgoing radial waves. These solutions describe scattering of electrons off the nucleus. The electron comes in from infinity, is attracted by the Coulomb potential, but has enough energy to escape to infinity. There is no quantization condition.

Next we note that we can simplify our equation by defining $f_l \equiv u_l/r$. The point is that

$$\partial_r f = \partial_r u/r - u/r^2$$

and

$$\partial_r^2 f = \partial_r^2 u/r - 2\partial_r u/r^2 + 2u/r^3,$$

so that

$$\partial_r^2 f + \frac{2}{r}\partial_r f = \partial_r^2 u/r.$$

In terms of u_l and the rescaled variable $y = r\sqrt{-E}$, the Schrödinger equation is

$$\partial_y^2 u_l = [\frac{l(l+1)}{y^2} - \frac{y_0}{y} + 1]u_l, \tag{7.68}$$

where $y_0 = \frac{2Z}{\sqrt{-E}}$. The normalization condition for u_l is just

$$\int_0^\infty dy|u_l(y)|^2 = 1. \tag{7.69}$$

For the bound states, our analysis of the behavior near 0 and ∞ suggests that we write

$$u_l(y) = y^{l+1}e^{-y}v(y), \tag{7.70}$$

where $v(y) = \sum_{k=0}^{\infty} c_k y^k$ has a power series expansion. The equation for v is

$$y\partial_y^2 v + 2(l + 1 - y)\partial_y v + [y_0 - 2(l + 1)]v = 0. \tag{7.71}$$

When we examine the action of the various terms in this equation on a power y^k, we see that the first term lowers the power by 1, the second has a term that also lowers the power by 1 and a term that leaves the power fixed, and the third term leaves the power fixed. The zeroth order term in the series gives

$$2(l + 1)c_1 = [2(l + 1) - y_0]c_0.$$

More generally, we get

$$c_{k+1} = \frac{(2(k + l + 1) - y_0)}{(k + 1)(k + 2l + 2)}c_k. \tag{7.72}$$

The series terminates for the discrete values $y_0 = 2(k_{max} + l + 1)$. If it does not terminate, its large k behavior is

$$c_{k+1} = \frac{2}{k + 1},$$

or $c_k \sim \frac{2^k}{k!}$, which is the expansion of e^{2y}, giving the bad asymptotic behavior at infinity. If we define $n = k_{max} + l + 1$, then we find that the energy in Rydbergs is quantized as $E_n = -\frac{1}{n^2}$ Rydbergs. Rydberg's original experiments showed that the spectral lines of hydrogen fit the formula $\omega_{mn} = \frac{1}{n^2} - \frac{1}{m^2}$ when measured in units of Rydbergs/$\hbar$.

A curious feature of this formula is that the spectrum has a degeneracy beyond that expected from angular momentum analysis. For any l, we can get the same energy as long as $n \geq l+1$. The degeneracy of the nth level is thus n^2. The origin of this degeneracy is a hidden conservation law of the hydrogen atom. It is well known that the bound classical orbits in a Coulomb potential are ellipses, so that the semimajor axis of the ellipse is a conserved vector throughout the motion. The energy of an orbit does not depend on the direction of the semimajor axis, but it does depend on its magnitude, and is in fact completely determined by it. This conservation law was pointed out by Laplace in the 18th century and Runge and Lenz in the 19th century and is called the Runge–Lenz vector or the Laplace–Runge–Lenz (LRL) vector. Pauli showed that the algebra obtained by combining the quantum mechanical L–R–L vector with angular momentum, completely determines the spectrum of hydrogen. We do not have space here to go into the details. See Exercise 7.4.

The wave functions of the hydrogen atom and one electron ions have the form

$$\psi_{nlm}(r, \theta, \phi) = Y_{lm}(\theta, \phi)\sqrt{\frac{8Z^3(n - 1 - l)!}{2n^4[a_B(n + l)!]^3}}e^{-\frac{y}{2}}y^l L_{n-l-1}^{2l+1}(y), \tag{7.73}$$

with

$$y = \frac{2Zr}{na_B}.$$

The functions $L_{n-l-1}^{2l+1}(y)$ are polynomials called the associated Laguerre polynomials. Our normalization conditions for these polynomials differ from, e.g., the form you will find in the Wikipedia article on Laguerre polynomials.

The recursion relation for the coefficients of these polynomials is

$$c_{k+1} = \frac{2(k+l+1-n)}{(k+1)(k+2l+2)}c_k. \tag{7.74}$$

Recalling that the Gamma function obeys

$$\Gamma(z+1) = z\Gamma(z), \tag{7.75}$$

we see that

$$c_k = \frac{2^k \Gamma(k+l+1-n)}{\Gamma(k+1)}\Gamma(k+2l+2). \tag{7.76}$$

You will explore more properties of Laguerre polynomials in Exercises 7.7–7.9.

Some simple examples are the ground state and first excited state radial wave functions

$$f_{10} = 2(\frac{Z}{a_B})^{3/2}e^{-\frac{Zr}{a_B}}, \tag{7.77}$$

$$f_{20} = (\frac{Z}{2a_B})^{3/2}(2 - \frac{Zr}{a_B})e^{-\frac{Zr}{2a_B}}, \tag{7.78}$$

$$f_{21} = (\frac{Z}{2a_B})^{3/2}\frac{Zr}{\sqrt{3}a_B}e^{-\frac{Zr}{2a_B}}. \tag{7.79}$$

Notable general features are that higher Z implies wave functions closer into the nucleus, while higher l or n have the opposite effect. For angular momentum, this is just a consequence of the repulsive angular momentum barrier, which dominates the Coulomb attraction at small r, while for higher energy states, it is just a consequence of our scaling argument.

The energy differences between different levels can be measured by exciting the hydrogen atom out of its ground state, by subjecting it to an external force. While our simplified treatment finds all of the excited levels of hydrogen to be stable, this is no longer true once we consider corrections to the Coulomb approximation for the effect of the interaction between hydrogen and the quantized electromagnetic field. That interaction includes terms in which the atom can make transitions to a lower level and emit a single photon whose energy, according to the Einstein $\hbar\omega$ rule, corresponds to radiation of a given frequency. Multiphoton emissions can also occur but the probability for them is suppressed by powers of the fine structure constant $\alpha_{em} \sim 10^{-2}$. These processes are allowed by energy conservation, and they are the main processes that occur when the hydrogen atom is in a relatively isolated environment. In such circumstances, since the system of lower energy atom plus emitted photons has more states[3] than the excited atom, the higher excited states will decay. They are only metastable (because the fine structure constant is small, so the probability of decay is small).

The emission spectra from hydrogen and other atoms have an enormous number of both practical and pure science applications. They are a primary tool for identifying materials, as well as the primary tool for probing the distant universe. Indeed, it was Hubble's discovery of a systematic red shift in the spectral lines from distant objects, which led to our modern picture of an expanding universe. The association of energy levels with spectra gives rise to names for the energy levels associated with the visual signatures of the spectra they give rise to. Thus, we have *s(harp)*, *p(rincipal)* and *d(iffuse)* for the spectra associated with $l = 0, 1, 2$. These are followed by *f(undamental)*, *g*, and so on down the alphabet. This terminology for angular momentum states has entered into the colloquial language of physics, in ways that has nothing to do with spectra. Thus, you will often hear particle physicists talking about s or p wave scattering amplitudes (see the next chapter), or condensed matter physicists talking about d wave superconductors. The latter terminology is particularly amusing, because it refers to the behavior of the superconducting condensate, an approximately classical field, in the ground state of certain types of superconducting materials, rather than to the behavior of a quantum wave function.

7.6 THE PERIODIC TABLE

We now want to sketch the QM of multielectron atoms. The key idea is to treat them in an approximation where each electron can be thought of as propagating in a self-consistent field generated by all of the others. This idea goes back to Hartree [20] who proposed to calculate the self-consistent field as the sum of the nuclear Coulomb field and the Coulomb field generated by the electrons, which he wrote as

$$V_{Hartree} = \frac{1}{4\pi\epsilon_0} \int d^3y \, \frac{1}{|(\mathbf{x} - \mathbf{y})|} \sum_i \psi_i^*(\mathbf{y})\psi_i(\mathbf{y}). \tag{7.80}$$

---

[3] In statistical mechanics language, this is just the statement that the final states of atomic ground state plus far away photon have higher entropy than the initial state of an excited atom with no photon. In few body systems, we usually say that "the phase space of the final state is larger." The point is that while transitions from excited atom to less excited atom plus photon, and atom plus photon to excited atom, occur with equal probability when the photon is near the atom, the first of these always occurs, while the latter only occurs if an incoming photon gets close enough to the atom.

The wave functions are the individual electron wave functions, which we take to be the lowest eigenstates of the self-consistent Hamiltonian $-\frac{\hbar^2}{2m_e}\nabla^2 + V_{Hartree}$, consistent with the constraints of Fermi statistics, which we will discuss below.

It turns out that, in the limit that the nuclear mass is $\gg m_e$, the idea of a self-consistent field is exactly valid,[4] but Hartree's approximation to it is not. Much work in the physics of atoms and ions goes into calculating the correct form of the electron density profile, which determines the self-consistent field and the ground state energy. What is remarkable is that the self-consistent field method is valid, even though the idea of a multielectron wave function, which is simply a(n antisymmetrized) product of single electron wave functions is not. The true multielectron wave function for an atom has complicated multiparticle correlations.

Nonetheless, a picture of the atom in terms of single electrons in a self-consistent potential is sufficient for understanding the gross structure of the periodic table, if not the numerical details of atomic ground state energies. It is what we will use. Within such a picture, the two key concepts are screening of the nuclear charge by electrons and Fermi–Dirac statistics. Let us begin by explaining the latter. Imagine that there were no correlation at all between the wave functions of individual electrons. Then we could put each electron into the ground state of the self-consistent potential. We would expect ionization energies to be at least N times the ionization energy for hydrogen for an N electron atom, and apart from that, all atoms would behave in exactly the same way. There would be no explanation for chemistry as we know it. Pauli hypothesized his *exclusion principle* to avoid this theoretical disaster. If no two electrons are allowed to be in the same state, then there can be only two electrons in the single particle ground state, once we include the fact that the electrons have two spin states, and that the self-consistent potential is spin independent. Similarly, there can be two electrons in each of the ψ_{20} and ψ_{21} states. Note that we can retain the hydrogenic labeling for the states, even though the self-consistent potential is not the Coulomb potential, as long as it *is* a spherically symmetrical potential. The label n labels the number of nodes of the radial wave function (Exercise 7.15). We should no longer expect the ψ_{20} and ψ_{21} states to be degenerate, since the potential no longer has the Runge–Lenz conservation law of the Coulomb potential.

As we will see below, Pauli's principle is just what we need to explain chemistry, but what is the physical origin of his oracular declaration? The answer turns on the subtleties of identical particles in QM. We have seen above that photons, the particles of which light is composed, are excitations of a quantized field. As such, it makes no sense to make a distinction between two states which differ by exchanging two photons, since they correspond to the same field configuration. Multiparticle states of photons are automatically invariant under exchange of the labels of individual photon states.

---

[4] This is the fundamental theorem of Density Functional Theory [21]. We will discuss it briefly in Chapter 11.

It is natural to be curious about whether the same could be true for other particles and there is a paradox in classical statistical mechanics which indicates that it must be so. When we open a partition between a box containing hydrogen and a box containing nitrogen molecules, initially at some fixed temperature, T, the entropy of the system increases. This change in entropy is called *the entropy of mixing*. The statistical mechanics argument for the entropy of mixing goes through just as well if the two halves of the box contain identical gases, but this contradicts experimental data. Gibbs concluded from this that one should not count as distinct, states which differed only by the exchange of particle labels.

Fermi–Dirac statistics involves another subtlety, which is purely quantum mechanical. The identical particle gauge symmetry says that each permutation must take any multiparticle state into a physically identical one, but in QM the overall phase of a state cannot be measured. It is a famous statement in group theory that every permutation can be written as a product of transpositions, which simply exchange a particular pair of particles. The square of a transposition is the identity, so if a state picks up a phase under transposition, it must be ± 1. Every permutation will thus multiply the state by either ± 1. The group of all permutations is a complicated object, and we have to ask whether there are consistent ways to assign plus and minus signs to the action of all transpositions in such a way that the transformations indeed satisfy the multiplication table of the group.

One obvious way to make a state invariant up to a phase under all permutations is to assign (-1) to *every* transposition. Then every permutation which is a product of an even number of transpositions gets a plus sign, and every odd product gets a minus sign. The set of all even permutations is a subgroup of the permutation group called *the alternating group A_N*. Every odd permutation has the form $P = T_{ij}A$, where A is in A_N and T_{ij} is a transposition of any fixed pair of particles. It turns out that in space dimensions greater than or equal to three, this assignment of phases is the only one compatible with the statistics gauge symmetry, apart from choosing all phases equal to one, which is Bose–Einstein statistics. We will get a deeper understanding of this when we study the Aharonov–Bohm effect.

The nontrivial even/odd choice is called Fermi–Dirac statistics, and the corresponding particles are called fermions. Fermions obey Pauli's exclusion principle. If we consider a two fermion state, $\psi(x_1, x_2) = -\psi(x_2, x_1)$ and so the state $\psi(x_1)\psi(x_2)$ is not allowed. More generally, for an N fermion state, the simplest thing we can build from N independent one particle wave functions ψ_i is the *Slater determinant*

$$\psi(x_1 \ldots x_N) = \det \psi_i(x^j) = \frac{1}{n!}\epsilon^{(i_1 \ldots i_n)}\epsilon_{(j_1 \ldots j_m)}\psi_{[i_1]}(x^{[j_1]}) \ldots \psi_{[i_n]}(x^{[j_n]}).$$

We have discussed Fermions and Fermi–Dirac statistics in Chapter 5, where we realized it as an alternate way to quantize the creation and annihilation operators of a collection of harmonic oscillators (*aka* a quantum field with a linear field equation). The field theory approach is by far the most efficient way to deal with both bosons and fermions, but in this chapter, we will stick to multiparticle language, since most older textbooks, much of the old literature on QM and the Graduate Record Exam, use this language.

A crude estimate of the self-consistent potential is obtained by writing

$$V_{eff}(\mathbf{x}) = \frac{1}{4\pi\epsilon_0}[-\frac{Z}{|\mathbf{r}|} + \int d^3y \, \frac{Z}{|\mathbf{r}-\mathbf{y}|}n(\mathbf{y})], \qquad (7.81)$$

where

$$n(\mathbf{x}) = \sum \psi_i^*(\mathbf{x})\psi_i(\mathbf{x}), \qquad (7.82)$$

is interpreted as the electron density at the point $\mathbf{x}$. This is called the Hartree Approximation.[5] The potential is called self-consistent because the single particle eigenfunctions, which appear in its definition, are obtained by solving the Schrödinger equation with that potential. One proceeds by starting with an ansatz for the potential, solving the equation, and computing the refined potential, then iterating. The procedure converges fairly rapidly.

As we said above, this crude approximation does not get the details right, but we can use it to understand the qualitative properties of atoms. Fermi–Dirac statistics, in the form of the Pauli principle, says that electrons must be added to higher and higher energy levels. The level structure has bands, corresponding to the degeneracies of the hydrogen atom, but split by the fact that the potential is no longer Coulomb. It *is* spherically symmetric, so the exact angular momentum degeneracies remain, but the accidental degeneracies are split.

We can understand the nature of that splitting, by noting that the Schrödinger equation for angular momentum l has a repulsive $\frac{\hbar^2 l(l+1)}{2m_e r^2}$ term in it, which keeps higher angular momentum states further away from the origin. Also, states of higher n have smaller probability to be near the origin. As we fill in the lower energy levels, the electron density near the origin screens the nuclear Coulomb potential, so the higher n levels, and the higher l levels for fixed n feel a less attractive potential. Their energies are thus higher (they are less bound). Thus, we should expect that, for a given n, the splittings of the exact Coulomb degeneracies will make higher l states have higher energy. Indeed we may expect that some of the bands of levels may overlap, with the higher l states at some value n being higher in energy than the low l states at the next higher value.[6]

Note that all of the wave functions will have exponential tails, though falling at different rates due to the fact that they see a screened value of Z. The last electron added will, at long distance, see a $Z_{eff} = 1$ attractive Coulomb potential, so it will still be bound. The potential will deviate from the Coulomb form near the origin, but the wave function of the last electron (for reasonably large Z) has low probability to be near the origin.

---

[5] See Chapter 18 for an explanation of the relation of the Hartree approximation to variational estimates of the many electron ground state energy. See Chapter 11 on Atomic, Molecular and Condensed Matter physics for the description of Density Functional Theory, which replaces the Hartree approximation.

[6] Recall also that the splitting between the nth and $n + 1$st level of the hydrogen atom goes to zero as n gets large.

The final idea that we need to understand the structure of the periodic table is the fact that electrons actually have two spin states, and that, in the approximation we are using, the energy is independent of the spin. We will talk more about electron spin below, but this is all we need for now. For $Z = 1$, we of course have hydrogen. For $Z = 2$, the self-consistent potential is no longer Coulomb, but it still has a nondegenerate ground state level, and we can put both electrons in this level, by giving them opposite spin. The ionization energy of helium is almost four full Rydbergs, somewhat reduced by the screening effect. This makes helium very stable, which is to say, chemically inert. Atoms make chemical combinations by trading electrons, either "swapping" as in ionic bonding, or "sharing" (this is called covalent bonding). It costs a lot of energy to tear an electron away from a helium atom, less than can be gained by swapping or sharing it with another nucleus.

At $Z = 3$, which is called lithium, one electron must go into the $n = 2$ shell. It will choose the state with $l = 0$, because of the screening effect we have discussed. That is, while the $n = 2$, $l = 0, 1$ levels of single electron ions are degenerate, the actual potential felt by the electrons in the $n = 2$ level of Lithium differs from the Coulomb potential because of the screening due to the electrons in the $n = 1$ shell. The lower angular momentum states have probability distributions closer to the origin, and so have lower energy. Lithium is chemically active because a Li^+ ion puts the extra electron in the same $l = 0$ shell (because of spin degeneracy), which has fairly low energy. It will tend to form ionic bonds with other atoms. $Z = 4$, beryllium puts its fourth electron in that same shell. It is more deeply bound than the Li^+ ion because of the net charge on the latter. Beryllium is not a noble gas like helium, because the $l = 1, n = 2$ shell is not split by very much from the $l = 0, n = 2$ shell, so it is easier to excite Beryllium's electrons out of their ground state. Boron, with $Z = 5$, now has to have an electron in this shell, and because of the combined spin and orbital angular momentum degeneracies, we can proceed up to $Z = 10$ (neon) before the $n = 2$ shell is filled. Neon, like helium, is not terribly reactive, because the next highest energy level is quite far away. On the other hand $Z = 9$, Fluorine is highly reactive because the Fl^- ion has low energy and so is relatively stable. We can begin to see how the chemical activities of atoms, and which ions they like to form, are explained naturally by the independent electron picture of atomic structure.

The next row of the periodic table $n = 3$ is the place we first see dramatic effects of electron–electron repulsion. The $n = 3, l = 0$ states are sodium and magnesium (quite reactive, as expected), followed by six states (aluminum through argon) filling up $n = 3, l = 1$. However, the repulsion has pushed the "expected" $n = 3, l = 2$ states up, so that they are higher than the $n = 4, l = 0$ states. Thus, argon, with a relatively large gap to the next available level, is a noble gas, and ends the third row of the periodic table with only 8 elements.

For the $n = 4$ row, we start by filling the $l = 0$ level with potassium and calcium, then get to the 10 elements corresponding to the $n = 3, l = 2$ levels (scandium through zinc), and finish with gallium through krypton, filling in the $n = 4, l = 1$ levels. Again, the gap to the next set of levels makes krypton a noble gas, and we go on to the next row. You can find the rest of the story in a number of atomic physics texts [22]. Reading those texts, especially the more modern ones, you will find that our detailed quantitative understanding of the physics of multielectron atoms is far from complete, and involves much more sophisticated analysis than the simple model we have presented.

The states of multielectron atoms are described by chemists and atomic physicists using a notation that combines the old spectral names for the different orbital angular momentum states, with the independent particle picture of atoms, which we have just sketched. Thus, for example, the ground state of neon is denoted by

$$(1s)^2(2s)^2(2p)^6. \tag{7.83}$$

This means, "There are two electrons in the $1s$ orbital (opposite spin), two in the $2s$ orbital and six in the three $2p$ orbitals." The word orbital refers to the single particle wave functions that are used in the Slater determinant approximation to the many electron wave function. The labels on the orbitals refer to the fact that they are solutions to the self-consistent single particle Schrödinger equation, with a given value of orbital angular momentum (the $s, p, d, f \ldots$ label) and a given number of radial nodes. Often this is shortened by just writing the symbol for the noble gas preceding the element in the periodic table, followed by the orbital configuration of the electrons above that. Like many archaic notations, this one is not so much a reflection of reality, but of the human desire to catalog things neatly. As such, there are sure to be questions about it on the physics GRE.

7.7 THE SPIN OF THE ELECTRON

The suggestion that electrons have intrinsic angular momentum came from Uhlenbeck and Goudschmidt [23], who invoked it to explain the splitting of levels of atoms in external magnetic fields. As we have seen, the existence of spin, and the fact that the Hamiltonian of nonrelativistic electrons commutes approximately with the spin operator, is crucial to understanding the structure of the periodic table. In this section, we will outline Pauli's theory of the electron spin.

We already have all the mathematical apparatus to do this, since the spin of a single electron is the quintessential example of a two state system. Recall that the Pauli matrices satisfy

$$\sigma_a\sigma_b = \delta_{ab} + i\epsilon_{abc}\sigma_c. \tag{7.84}$$

It follows that

$$[\hbar\frac{\sigma_a}{2}, \hbar\frac{\sigma_b}{2}] = i\epsilon_{abc}\hbar\frac{\sigma_c}{2}, \tag{7.85}$$

which are the angular momentum commutation relations for the generators $J_a = \hbar\frac{\sigma_a}{2}$. The Casimir operator of the representation is

$$\hbar^s j(j+1) \equiv J_a^2 = \hbar^2 \frac{3}{4}, \tag{7.86}$$

so the spin $j = \frac{1}{2}$.

Before proceeding, let us examine more closely what we mean by the spin of a particle, which comes down to a question of what we mean by a particle. Democritus defined atoms as little indivisible objects, out of which everything else is built as a composite. Today, we know that atoms themselves are composites, and even that the proton and neutron in the nucleus of the atom are composites of particles called quarks. We discovered this first by finding that the charge distributions inside atoms and protons and neutrons were not structureless, and later by breaking these objects up into their constituents.[7]

A composite object will have many excited states in which there is orbital angular momentum in the rest frame of the center of mass. One of these states would be the lowest energy level, and it need not have zero total angular momentum. At an energy scale below that of the excited levels, we will see an apparently pointlike object with a fixed value of j. This is the definition of spin of an "elementary" particle. We can never know for sure whether a particle is elementary or composite, because our experiments might not have enough energy to excite the higher rotational levels. So, we define an elementary particle as one in which, at our current level of accuracy, we see states at rest transforming in some particular "spin j" representation of angular momentum.

This is not the end of the story though. Orbital angular momentum is quantized in *integer* units, because wave functions are required to be periodic under 2π rotations. However, we have seen that the mathematics of the rotation group allows for half integer spins. This is allowed, *if we make a rule that no physical operator can change the number of half integral spin particles from even to odd or vice versa*, because quantum expectation values and transition matrix elements *will* be periodic. Note that one cannot obtain half integral spin from internal orbital angular momentum of constituents. As it turns out, the spins of electrons and quarks, those particles which have not yet revealed any inner structure, are half integral. One might be tempted to say that these might be truly elementary, or that at least their structure is at a scale where our notions of space might themselves break down.

Let us leave these heady speculations and come back to the theory of electron spin. Electrons interact with the electromagnetic field because they are charged. They experience Coulomb and Lorentz forces in the presence of external electric and magnetic fields. It was well known to 19th century physicists that even neutral particles can interact with the magnetic

---

[7] In the case of quarks inside nucleons, there is a confining force, a tube of chromoelectric flux, which does not let quarks exist as free particles, but we have done experiments which let the constituent quarks in a nucleon fly far apart from each other.

field by virtue of having *magnetic multipole moments*, the most important of which is the dipole moment. Classically, a magnetic dipole moment can be modeled by a term in the energy of the form

$$\delta H = \mu \cdot \mathbf{B}, \tag{7.87}$$

where μ is the magnetic dipole. It transforms as a vector under proper three-dimensional rotations, and a pseudovector or axial vector (like the magnetic field) under reflections. In QM, μ will become an operator. Throughout most of this book, we will make the assumption that the electromagnetic field is in a coherent state and replace the field operator $\mathbf{B}$ by the classical field that parametrizes that coherent state. We will neglect the small quantum fluctuations of the field around this classical value. Thus, the magnetic dipole term in the Hamiltonian is approximately an operator in the Hilbert space of a single particle. Since it is invariant under Galilean transformations (once we figure out the proper form for the magnetic field in each frame), we can work out what this operator is in the rest frame of the particle.

The states of a particle at rest transform as a representation of the rotation group. As we have said, for an "elementary" particle like an electron we assume that the representation is irreducible, which is to say that it corresponds to a fixed value of the spin, j. The magnetic dipole operator transforms just like the angular momentum. The most general operator in the space can be constructed as a polynomial in the angular momentum operators. Using the commutation relations, and the formula $J_a^2 = \hbar^2 j(j+1)$, we can write the general operator as a sum of totally symmetrized tensors $J_{a_1 \ldots a_k}$, which are traceless in each pair of indices. The space of operators in the spin j Hilbert space can be considered the tensor product of the Hilbert space with itself. The rules for addition of angular momentum tell us that this contains every integer spin between 0 and $2j$, exactly once.

The bottom line is that the only operator transforming like a magnetic moment is the angular momentum generator J_a itself. Thus,

$$\mu_a = \gamma J_a. \tag{7.88}$$

For a model of an electron as classical spinning charge distribution, with charge and mass distributed uniformly around the axis of spin, one derives such a formula with

$$\gamma = -\frac{e}{2m_e}. \tag{7.89}$$

γ is called the *gyromagnetic ratio*. The correct quantum value of γ for an electron is not predicted by this classical formula and is a free parameter in the nonrelativistic quantum theory of the electron, but *is* predicted by quantum electrodynamics (QED), the relativistic quantum field theory of electrons interacting with the quantized electromagnetic field. QED multiplies the classical result by a dimensionless number called the electron g factor, which can be computed in a power series in the fine structure constant. QED gives

$$g = 2(1 + \frac{\alpha}{2\pi} + o((\frac{\alpha}{\pi})^2)),$$

where $\alpha \sim \frac{1}{137}$ is the fine structure constant $\frac{e^2}{\hbar c}$. Many higher order corrections have been computed, for both the electron and the muon and the agreement between theory and experiment is better than any other result in the history of science.

The Hamiltonian we have written for the hydrogen atom commutes with the electron spin. Now let us consider what happens when we subject the hydrogen atom to a small uniform magnetic field, pointing in the three direction. The orbiting electron will experience Lorentz forces due to this field. We will study these in Chapter 9, but they give rise to very small effects compared to the Coulomb attraction of the nucleus. For small field B_3, the dominant effect will be that the previously degenerate spin states are split in energy. We can in fact calculate this splitting exactly because the spin Hamiltonian $H_s = \frac{\gamma}{2}\sigma_3 B_3$ commutes with the Coulomb Hamiltonian. The entire effect (which by the way is called the Zeeman effect) is to add $\pm\frac{\gamma}{2}B_3$ to each energy level, with the sign determined by the 3 component of the spin.

For multielectron atoms, the Zeeman effect is more complicated, because the spin Hamiltonian is $\sum_i \frac{\gamma}{2}\sigma_3(i)B_3$. Thus, for example, in the ground state of Helium, the two electrons are combined into the antisymmetric spin singlet state and the spin Hamiltonian has no effect. The Zeeman effect is usually analyzed by applying the single electron spin Hamiltonian to the *unpaired* electrons in an atom.

In fact, things are even more complicated than this, because the relativistic corrections to the Schrödinger equation do not conserve spin and orbital angular momentum separately, but only the combined rotation generator

$$\mathbf{J} = \mathbf{L} + \mathbf{S}. \tag{7.90}$$

The leading term that is invariant under rotations, but not separate spin rotations, has the form

$$\delta H = k\mathbf{L} \cdot \mathbf{S}. \tag{7.91}$$

This interaction comes from two effects of order $\frac{v^2}{c^2}$. The first, called the Larmor effect, is simply the relativistic transformation of the Coulomb field of the nucleus into the instantaneous rest frame of the orbiting electron. This gives rise to a magnetic field, which interacts with the magnetic moment of the electron. The second effect, first computed by Thomas [24], has to do with the fact that the orbiting electron is accelerating, so the rate of precession of its spin in a given field is not the same as it would be in an inertial frame. The effect of the so-called Thomas precession is to reduce the Larmor energy by a factor of 1/2.

The result of these two relativistic corrections is a coefficient k given by

$$k = \frac{1}{2rm_ec^2}\frac{dU(r)}{dr}, \tag{7.92}$$

where $U(r)$ is the nuclear potential. Using $2\mathbf{L}\cdot\mathbf{S} = \mathbf{J^2} - \mathbf{L^2} - \mathbf{S^2}$, we see that the full Hamiltonian in the absence of an external magnetic field, commutes with $\mathbf{J^2}, \mathbf{L^2}, \mathbf{S^2}$ and J_3, but not S_3 or L_3. Thus, when the $L-S$ coupling Hamiltonian is comparable to or larger than the Zeeman Hamiltonian, the simple description of the Zeeman effect as splitting two degenerate spin states is no longer valid.

7.8   SPIN PRECESSION AND SPIN RESONANCE

Electron spin resonance (ESR) is a technique that uses the simple Zeeman effect to identify materials with unpaired electrons. It has applications throughout chemistry and biology. It is closely related to nuclear magnetic resonance (NMR)[8], where the splitting of nuclear spin levels in an external field is used. Nuclear magnetic moments are inversely proportional to nucleon masses, and are thus $\sim 2,000$ times smaller than the electron moment. For a fixed magnetic field, the energy splittings are thus much smaller in the nuclear case, which means that NMR signals involve electromagnetic radiation of much lower frequency than ESR signals.

The basic idea of ESR or NMR experiments is to subject a sample containing a large number of atoms with the same unpaired electron/nuclear spin to both a source of monochromatic radiation and an external magnetic field B_0. The magnetic field produces a splitting $\Delta E = g\mu_B B_0$ between two spin states (here g is a g-factor, which might be different than the free electron $g_e \sim 2$ because of the environment in which the unpaired electron sits). At finite temperature, the average number of electrons in each state will differ, by a Boltzmann factor $e^{-\frac{\Delta E}{kT}}$ (see Chapter 12 on Statistical Mechanics). One then tunes the field B_0 so that $\Delta E = \hbar\omega$, the photon energy of the monochromatic beam. At this *resonance frequency* the electrons can move between the higher and lower state by (stimulated) emission or absorption of a photon. Since the lower state is more heavily populated, this will give rise to a net absorption of the radiation, which we can monitor by detecting the beam after it is passed through the sample. For microwaves of frequency $\sim 9,390$ MHz, the resonance (for $g = g_e$) occurs at $B_0 \sim .335$ Tesla, while for the same magnetic field, NMR occurs for a frequency of about 14 MHz.

ESR signals can give information about the atomic and molecular structure of the compounds in which the electrons that produce the response are embedded. For example, as we noted above, LS coupling can change the nature of the electronic spin states and correlate them with the orbital state of the electron. This effect leads to the replacement of the free electron g_e factor by a matrix. The i-th component of the magnetic moment is given by

$$\mu_i = \mu_B g_{ij} \frac{\sigma_j}{2}, \tag{7.93}$$

---

[8] NMR is widely used in medical applications, and has been renamed *magnetic resonance imaging* (MRI) because of the widespread public fear of anything involving nuclear physics.

and ESR experiments measure g_{ij}. There are also interesting complications introduced by interaction of the electron spin with nuclear spins, and these can also be exploited to learn about the structure and composition of the materials in which the electron is embedded.

7.9 STERN–GERLACH EXPERIMENTS

For free particles, moving at speeds much smaller than the speed of light, the components of spin are conserved quantum numbers. In principle, that means that we can study states with fixed values of S_3, the component of spin along some given axis. A general pure state of the particle spin (for $\mathbf{S}^2 = \frac{1}{2}\frac{3}{2}$) will have the form

$$|\psi\rangle = a_+|+\rangle_3 + a_-|-\rangle_3. \tag{7.94}$$

A Stern–Gerlach machine is a device for separating out the states of different S_3 component. It exploits the interaction between the magnetic moment of the particle and an external magnetic field. Stern–Gerlach machines only work for neutral particles that are not subject to the much stronger Coulomb forces which could mask the S–G effect. S–G filtering is often used in thought experiments, to prepare pure quantum states with specified properties.

The Hamiltonian for the magnetic moment interaction with a magnetic field pointed in the 3 direction is

$$\Delta H = \mu B_3(\mathbf{X})\sigma_3. \tag{7.95}$$

It modifies the free particle Heisenberg equation of motion from $\dot{\mathbf{P}} = 0$ to

$$\dot{\mathbf{X}} = \frac{\mathbf{P}}{m}, \tag{7.96}$$

$$\dot{\mathbf{P}} = -\mu\nabla B_3\sigma_3. \tag{7.97}$$

The parameter μ is twice the magnetic moment. The full Hamiltonian commutes with σ_3, so we can solve the Heisenberg equations separately for $\sigma_3 = \pm 1$.

Consider a magnetic field

$$B_3 = f(X_1^2 + X_2^2)X_3, \tag{7.98}$$

where f is a smooth function, which vanishes when its argument is larger than r^2, and is approximately equal to 1 for smaller argument. We will neglect effects of the gradient of f. We choose the Heisenberg wave function of the particle to be

$$\psi = e^{-\frac{x_2^2 + x_3^2}{d^2}} e^{-\frac{(x_1 - R)^2}{d^2}} e^{i\frac{px_1}{\hbar}}, \tag{7.99}$$

where $R \gg r$.

The classical picture corresponding to this Hamiltonian is shown in Figure 7.1.

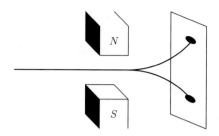

Figure 7.1 The Stern–Gerlach experiment.

There is a magnet centered around the origin of the plane $x_3 = 0$. The initial probability distribution for the particle's position is centered around a point in that plane far from the magnet, but its average momentum sends it toward the region where the magnetic field is nonzero. The momentum is conserved until the expectation value of the coordinate $\mathbf{X(t)}$ is inside the region where $f \neq 0$. At that point $\dot{P}_3 = \pm\mu$ and the expectation value of $X_3(t)$ moves away from $x_3 = 0$ *in a direction that is correlated with the sign of the spin component*. The expectation value of $X_1(t)$ continues to move, and eventually exits the region where $f \neq 0$.

The effect of the S–G apparatus is to entangle the spin state of the particle with the X_3 position of the particle. If we put detectors above and below the plane at positions $x_1 > r$, $x_3 = \pm R$, then each detector will see particles with definite polarization. Running the experiment many times, we can measure the probabilities $|a_\pm|^2$ by counting the numbers of clicks in each detector (we assume we know the sign of μ). More importantly, we now have a source of particles prepared in pure σ_3 eigenstates.

7.10   EXERCISES

7.1 Prove that $L_i = \epsilon_{ijk}R_jP_k$ satisfies the same commutation relations as $\hbar K_i$, where K_i is the generator of rotations.

7.2 Find the form of the operators $K_\pm$ and K_3 in spherical coordinates.

7.3 Find the norm of the state $K_-^{l-m}|ll\rangle$.

7.4 The spherical harmonics have the form $Y_l^m(\theta, \phi) = e^{im\phi}P_l^m(\cos\theta)$. Using the form of the operators K_i derived in Exercise 7.2, find the equation for $P_l^m(w)$ as a function of $w = \cos\theta$. Argue that $P_l^m = P_l^{-m}$. Show that for $m > 0$, the formula

$$P_l^m(w) = (1 - w^2)^{m/2}\frac{d^m P_l^0}{dw^m},$$

solves the equation. The polynomials $P_l^0 \equiv P_l$ are called the Legendre polynomials.

7.5 The classical Kepler problem has the same form as the Coulomb problem, but the coefficient in the inverse square law is GM_1M_2 and the electron mass is replaced by the reduced mass $\mu = \frac{M_1M_2}{M_1+M_2}$. Consider this as a quantum mechanical problem. For the earth sun system, $M_1 \ll M_2$ and so we replace $\frac{Ze^2}{4\pi\epsilon_0}$ by GMm and m_e by m. Calculate the Bohr radius for this system, in centimeters and the Rydberg energy, in joules. Requiring that expectation value of the velocity be less than that of light gives us a constraint on n, the energy level for which the analysis is applicable. The expectation value of $\mathbf{p}^2/2m$ in the n-th state is of order the binding energy

$$\frac{10^{188}}{n^2} \text{ joules,}$$

and this must be $< mc^2 = 5.4 \times 10^{41}$ joules. Thus, $n > 10^{73}$.

7.6 Equate the binding energy for the orbit of the earth, as given by classical physics, to the binding energy given by the Bohr formula, in order to estimate the value of n for the earth's orbit. Does it obey the bound of Exercise 7.5?

7.7 We wrote the solution to the radial equation for hydrogen in terms of associated Laguerre polynomials, $L_q^p(x)$ which are solutions of the equation

$$xL_q^p{}'' + (p+1-x)L_q^p{}' + qL_q^p = 0.$$

Show that the only the equations for $q = 0$ have a constant solution, and only those for $q = 1$ have a linear solution. Find those solutions, each of which should contain an undetermined multiplicative constant.

7.8 A natural conjecture from the result of Exercise 7.7 is that L_q^p is a q-th order polynomial. Mathematicians like to choose the normalization constants so that the polynomials are *monic*: the coefficient of the term with the highest power of x is equal to 1. The L_k^p for $k \le q$ are a basis for all polynomials of order $\le q$. xL_q^p is a new linearly independent polynomial and so L_{q+1}^p must be a linear combination of this and the L_k^p with $k \le q$. It turns out that this relation is rather simple. Show that

$$(A+Bx)L_q^p + CL_{q-1}^p$$

satisfies the equation for L_{q+1}^p for appropriate choice of the constants.

7.9 The polynomials L_q^p for fixed p can be considered eigenfunctions, with different eigenvalue, of the operator

$$x\frac{d^2}{dx^2} + (p+1-x)\frac{d}{dx}.$$

Show that this operator is Hermitian on the space of functions on the interval $[0, \infty]$ with scalar product defined by

$$\langle f|g \rangle = \int_0^\infty ds \, x^p e^{-x} f^*(x) g(x).$$

Show that the Laguerre operator is Hermitian with respect to this scalar product so that the L_q^p for different q are orthogonal. Compute the norm of the monic polynomials. It is easiest to do this by using the explicit form of the polynomials that we derived in the text.

7.10   The Legendre polynomials P_l are eigenfunctions of the operator $\mathcal{D} = \frac{d}{dw}([1 - w^2]\frac{d}{dw})$, with eigenvalues $-l(l + 1)$. Find a scalar product

$$\langle f|g \rangle \equiv \int_{-1}^{1} dw \, \mu(w) f^*(w) g(w),$$

on the space of functions on the interval $[-1, 1]$, such that $\mathcal{D}$ is an Hermitian operator. Argue that the different Legendre polynomials are orthogonal with respect to this scalar product.

7.11   Orbits in the Coulomb potential are ellipses and the semimajor axis of the ellipse stays constant in time. This indicates the existence of a new vector conservation law $\mathbf{E}$, which, unlike angular momentum, changes sign under reflection. To understand how it arises, note that the equation of motion has the form

$$m\frac{d^2\mathbf{r}}{dt^2} = \frac{a}{r^2}\hat{\mathbf{e}},$$

where

$$a \equiv \frac{Ze^2}{4\pi\epsilon_0},$$

and $\hat{\mathbf{e}}$ is the unit vector from the origin to the position of the particle. Recalling that the motion is in a plane with coordinates (r, ϕ), and defining $\mathbf{e}_\phi$ to be the unit vector in the direction of the velocity, we have

$$\dot{\mathbf{e}}_\phi = -\dot{\phi}\,\hat{\mathbf{e}}.$$

This just says that the acceleration is always directed toward the center. Also, the magnitude of the angular momentum is $l = mr^2\dot{\phi}$. Show that we can write the equation of motion as

$$\frac{d\mathbf{v}}{dt} = -\frac{a}{mr^2}\frac{mr^2}{l}\dot{\mathbf{e}}_\phi.$$

Show that this implies a new conservation law. Call the new conserved vector $\mathbf{h}$, for Hamilton, the person who discovered it.

7.12 Construct the first three Legendre polynomials by insisting that they be orthonormal.

7.13 Since $\mathbf{h}$ and $\mathbf{L}$ are conserved, so is $\mathbf{h} \times \mathbf{L}$. Show that for bound orbits this is proportional to the Laplace–Runge–Lenz vector, the vector defined by the semimajor axis of the ellipse.

7.14 Define the generating function of Legendre polynomials by $G(w, s) = \sum_{l=0}^{\infty} s^l P_l(w)$. Show that

$$\frac{d}{dw}\left([1 - w^2]\frac{dG}{dw}\right) = -\frac{d^2(sG)}{ds^2}.$$

Show that a solution of this equation is

$$G(w, s) = (1 - 2sw + s^2)^{-1/2}.$$

7.15 Prove that the nth solution of the radial wave function for hydrogen has n nodes: i.e., n places where the wave function vanishes.

CHAPTER 8

Scattering Electrons on a Nucleus

8.1 INTRODUCTION

This section contains a brief description of the scattering solutions of the hydrogen atom, and a brief introduction to scattering theory. The material is of much less general utility than our discussion of the bound state spectrum, so our treatment is somewhat cursory. More material on scattering theory can be found in Chapter 16.

8.2 POSITIVE ENERGY EIGENFUNCTIONS AND SCATTERING AMPLITUDES

In celestial mechanics, we are familiar with the fact that gravitational attraction can form bound systems like the solar system. We also know about other trajectories, in which two objects, subject to gravitational attraction, suffer a close encounter, but then one of them (in the rest frame of the other) wanders off into space, never to return. In quantum mechanics, the analog of these "close encounters of the second kind" are called scattering states. If you have done all the exercises of Chapter 4, you have already encountered these in a one-dimensional context. In this chapter, we want to study the scattering states of electrons in hydrogen-like ions.

These are solutions of the Schrödinger equation with positive energy:

$$[-\nabla^2\psi - \frac{Z}{r}]\psi = k^2\psi. \qquad (8.1)$$

k^2 is the energy in units of the absolute value of the Rydberg energy, and k is also the wave number in units of the Bohr radius. We can make an angular momentum decomposition of this eigenvalue problem, and the resulting radial equation is

$$[-\partial_r^2 - \frac{2}{r}\partial_r + \frac{l(l+1)}{r^2} - \frac{2Z}{r}]\psi_{lm}(r) = k^2\psi_{lm}(r). \tag{8.2}$$

The Z dependence can be removed by scaling $r \to \frac{r}{2Z}$ and $k \to 2Zk$, just as in the bound state problem.

Let us use these rescaled variables and analyze the behavior of the solutions as $r \to \infty$. Dropping the angular momentum and potential terms, which vanish in this limit, we have

$$[-\partial_r^2 - \frac{2}{r}\partial_r]\psi_{lm}(r) = k^2\psi_{lm}(r). \tag{8.3}$$

We can get rid of the first derivative term by writing $\psi = r^a\phi$ so that

$$[-\partial_r^2 - \frac{2}{r}\partial_r]\phi(r) - a(a-1)r^{-2}\phi(r) - 2(a+1)/r\partial_r\phi(r) - 2a/r^2\phi(r) = k^2\phi(r). \tag{8.4}$$

For $a = -1$, both the first derivative and $1/r^2$ terms cancel and $\phi \sim \frac{e^{\pm ikr}}{r}$. Thus, the radial equations for all angular momenta, have two solutions, which behave at infinity like $\psi_{lm}^{0\ pm} \sim \frac{e^{\pm ikr}}{r}$, where we take k to be positive. The functions $\psi_{lm}^{0\pm}$ are solutions of the free particle Schrödinger equation. When combined with the time-dependent factor $e^{-i\hbar k^2 t}$, they look like incoming and outgoing radial waves, with incoming corresponding to e^{ikr}. They have continuum normalization, with the scalar product of two different values of k, and the same value of the $\pm$ index, being $\delta(k-k')$. We would like to understand what these solutions have to do with the problem of scattering. For the moment, however, note that the general solution of the Schrödinger equation has a singularity at $r = 0$, because both the Coulomb potential and the angular momentum barrier are singular there. The Hamiltonian will only be a Hermitian operator if we restrict attention to the linear combination which is nonsingular at the origin. As a consequence, at infinity, if we normalize the coefficient of $\frac{e^{-ikr}}{r}$ to 1, the solution will behave like

$$\frac{e^{-ikr}}{r} + A(k)\frac{e^{+ikr}}{r}. \tag{8.5}$$

Conceptually, we can formulate the problem of scattering as follows. Start with an incoming wave packet

$$\psi_{in}(\mathbf{x}, t) = \int \frac{d^3k}{(2\pi)^{3/2}}[\psi_{in}(k)Y_{in}(\hat{\mathbf{k}})e^{ik\hat{\mathbf{k}}\cdot(\mathbf{x}-\mathbf{x_0})}e^{-i\frac{\hbar^2 k^2}{2m}(t-t_0)}]. \tag{8.6}$$

We want the initial angular wave function Y_{in} to be concentrated in a small solid angle. It does not change with the free time evolution. For a scattering problem, we want to take $t_0 \to -\infty$ and $|x_0| \to \frac{-\hbar k_0 t_0}{2m}$, where k_0 is the center of the momentum space wave packet

$\psi_{in}(k)$. The idea is that, in this limit, because the potential falls off at infinity,[1] this function is a better approximation to the solution of the full time-dependent Schrödinger equation, when t is large and negative. The problem of scattering is to find the evolution of this solution as $t \to \infty$.

Physical intuition tells us that two possible things can happen to a classical charged particle thrown into a Coulomb field of opposite charge from infinity. It can be captured into a bound orbit, or it can follow a hyperbolic orbit back out to infinity. For a two body system, which of these two things happens is entirely determined by conservation laws. If the energy is positive, the particle scatters off to infinity, and if it is negative, it becomes bound. If more than two particles are involved in the scattering, more complicated things can occur. Only the total energy is conserved, so an incoming scattering state of 3 or more particles can transform into one in which a two particle subsystem has negative energy, but the total is made up by the positive energy of the third. Thus, the complete set of asymptotic states of such a system will include not only the original particles, but bound states of pairs, or more complicated subsystems of them. In the real world, things are even more complex, because of the existence of massless photons. In relativistic quantum mechanics, particles can be created from empty space if sufficient energy is available. A single electron can scatter off a single ion, and be absorbed into a negative energy bound state, with energy conservation guaranteed by the emission of a positive energy photon. However, in our model problem, there are no photons, and only a single electron scattering off an ion. Positive energy in the past then implies that the state in the future is also a scattering state. No bound states appear in the final state.

If we now return to the exact eigenfunctions of the positive energy Coulomb Schrödinger equation, we can state the scattering/bound state problem more clearly. The spectral decomposition of the Hilbert space for fixed l is given by the resolution of the identity

$$1 = \sum_n |n\rangle\langle n| + \int_0^\infty dk |k\rangle\langle k|. \tag{8.7}$$

The continuum states $|k\rangle$ appearing in this decomposition are the eigenstates of the radial Hamiltonian for the appropriate value of l. The Hamiltonian is a second-order ordinary differential operator, with a singularity at $r = 0$. Only one of the two linearly independent solutions gives a normalizable solution in the vicinity of this singularity.[2] As $r \to \infty$ this

---

[1] Actually, the Coulomb potential does not fall off fast enough at infinity to justify this assumption. The formalism of scattering theory requires modification to deal with this problem. We will ignore the issue, but if you are bothered by it, imagine turning the Coulomb potential into a Yukawa potential with a range of 10 billion kilometers. Formal scattering theory, as explained in Chapter 16, will apply to this problem, as all of the issues will have to do with certain quantities that do not have finite limits as the range goes to infinity. For any experiment in a laboratory of size much smaller than the range, these must be irrelevant.

[2] For $l = 0$, as in the bound state problem, both solutions are square integrable near $r = 0$ but only one is sufficiently well behaved that no probability current flows into the singular point.

unique eigenstate wave function is a linear superposition of incoming and outgoing radial waves:

$$\psi_l(k^2, r) \to A(k, l)\frac{e^{ikr}}{r} + B(k, l)\frac{e^{-ikr}}{r}. \tag{8.8}$$

We define the incoming state by setting $A = 1$ and the outgoing state by $B = (-1)^{l+1}$. The two are obviously related by multiplication by a phase

$$S_l(k) = (-1)^{l+1}\frac{B(k, l)}{A(k, l)} \equiv e^{i\delta_l(k)}. \tag{8.9}$$

The prefactor $(-1)^{l+1}$ arises because the amplitude for no scattering is the amplitude to take an incoming wave vector $\mathbf{k}$ to an outgoing wave vector $-\mathbf{k}$, since the orientation of incoming and outgoing vectors on the sphere at infinity is opposite. The prefactor takes into account the transformation of spherical harmonics under this spatial reflection, so that the amplitude for this nontransition is 1 if the *phase shifts* $\delta_l(k)$ vanish.

The amplitudes $A(k, l), B(k, l)$ of course depend on the precise nature of the incoming wave packet. The actual response of the system is encoded in the ratio B/A, which, together with the prefactor, we have called the scattering matrix element $S_l(k)$. The actual physical setup of scattering involves localized incoming and outgoing wave packets, focused around two directions $\mathbf{\Omega_{in/out}}$. The amplitudes for scattering a particular incoming wave packet into a particular outgoing wave packet will be matrix elements of an operator in the space spanned by a complete set of either the incoming or outgoing wave packets. This is the *Scattering Operator* and its matrix is the *S-matrix*. The *S*-operator commutes with the Hamiltonian, the total momentum and the total angular momentum, for any translationally and rotationally invariant system. The fact that for a two body rotationally invariant system, the Schrödinger equation has only one solution for each value of energy and angular momentum shows that the *S*-matrix is diagonal in the basis where energy and angular momentum are diagonal. The phases $e^{i\delta_l(k)}$ are its eigenvalues. In Chapter 16, we will discuss scattering theory for potentials that are not spherically symmetric, and in that case the *S*-matrix will not be diagonal in any easily accessible basis.

8.3   ANALYSIS OF THE COULOMB SCATTERING AMPLITUDES

If you search through a book of special functions or Google "Coulomb Scattering Wave Functions," you will find that the solutions of the positive energy Schrödinger equation are special cases of Confluent Hypergeometric functions. A lot is known about these functions, including exact integral representations. From these, one can extract the Coulomb phase shifts:

$$e^{2i\delta_l(k)} = \frac{\Gamma(l + 1 + \frac{i}{k})}{\Gamma(l + 1 - \frac{i}{k})}, \tag{8.10}$$

where k is the wave number in Bohr radii, which corresponds to the incoming energy.

The Euler Gamma function has poles when its argument is a nonpositive integer. This happens for negative energy, and the poles are precisely where the bound states of the hydrogen atom sit. We have seen this phenomenon for the square well potential in Chapter 4 and we will see in Chapter 16 that this is a general phenomenon: bound states show up as poles in scattering amplitudes when they are analytically continued to "unphysical" regions where the energy variable is incompatible with the kinematics of scattering. The effect of these complex poles on scattering can be dramatic, if the imaginary part of the energy is small.

In a scattering experiment, the detector measures the number of particles per unit time, $N(\mathbf{\Omega})$ that are scattered into a solid angle $\mathbf{\Omega}$ relative to the direction of the incoming particle. The differential cross section is

$$\frac{d\sigma}{d\Omega} = \frac{N(\mathbf{\Omega})}{J}, \tag{8.11}$$

where J is the flux of particles in the incoming beam. J is computed from the probability current $-i\frac{\hbar}{m}\psi^*\nabla\psi + c.c.$ which, when integrated over a small surface area, gives us the total probability per unit time, that a particle with wave function ψ will cross that surface. For a plane wave $\psi = e^{i\mathbf{k}\cdot\mathbf{r}}$, J is just $\frac{\hbar k}{m}$ for a surface orthogonal to $\mathbf{k}$.

The scattering eigenfunction for an initial plane wave can be computed by doing the expansion of the plane wave in spherical coordinates and using the scattering solutions for fixed angular momentum. To compute the cross section, we need only the asymptotic behavior of the scattered wave. You will do this carefully in Exercises 8.1 and 8.2. The asymptotic behavior of the full in state wave function is

$$\psi_{in} = e^{i\mathbf{k}\cdot\mathbf{r}} + f(\theta)\frac{e^{-ikr}}{r}, \tag{8.12}$$

where

$$f(\theta) = \sum_{l=0}^{\infty} \frac{e^{2i\delta_l(k)} - 1}{2ik} P_l(\cos(\theta)). \tag{8.13}$$

There is no dependence on the azimuthal angle ϕ because the potential is rotation invariant and the choice of incoming plane wave is invariant under rotation of ϕ.

The current of outgoing particles can be separated out by going to a nonzero scattering angle θ. It is

$$\mathbf{J}_{out} = \frac{\hbar k}{mr^2}\mathbf{\Omega}|f(\theta)|^2. \tag{8.14}$$

Here we have neglected interference terms between the incoming plane wave and the scattered wave. If, instead of a plane wave, we would chosen a wave packet concentrated around the $\mathbf{k}$

direction these would cancel almost completely. The area element around $\mathbf{\Omega}$ is $dA = r^2 d\Omega$. If we dot the scattered current into $\mathbf{\Omega}$ and compute $N(\mathbf{\Omega})$ we get

$$\frac{d\sigma}{d\Omega} = |f(\theta)|^2. \tag{8.15}$$

To get a feeling for the meaning of these formulae, let us consider scattering of classical particles in a central potential $V(r)$.

In classical scattering, the particle follows a fixed trajectory, which is a hyperbola for the Coulomb potential. The angle of scattering is determined by impact parameter $b(\theta)$, which is the distance of closest approach to the center. θ is the polar angle of a coordinate system whose north pole is determined by the direction of the incoming particles. By rotational invariance, the impact parameter is independent of the azimuthal angle ϕ. If we have a beam of particles, with incoming flux J, the number of particles scattered per unit time between θ and $\theta + d\theta$ is equal to the number of incident particles per unit time between b and $b + db$. The number of particles scattered into solid angle Ω is

$$N(\Omega)d\Omega = 2\pi N \sin(\theta)d\theta = J2\pi b \frac{db}{d\theta}d\theta. \tag{8.16}$$

See Figure 8.1 for the geometry leading to these equations.

The differential cross section is defined to be the ratio between the number of particles per solid angle scattered into the direction Ω and the incident flux. It has dimensions of area, which explains the name *cross section*. Classically we have

$$\frac{d\sigma}{d\Omega} = \frac{b}{\sin(\theta)}\left|\frac{db}{d\theta}\right|. \tag{8.17}$$

In Exercise 8.3, you will verify that for a Coulomb potential

$$\frac{d\sigma}{d\Omega}_{classical} = \frac{Z^2 e^4}{(16\pi\epsilon_0)^2 \sin^4(\theta/2)}. \tag{8.18}$$

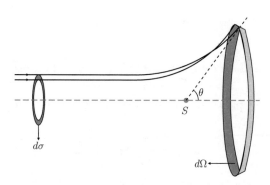

Figure 8.1 The geometry of a scattering experiment.

Quantum mechanically, we have seen that

$$\frac{d\sigma}{d\Omega} = |f(\theta)|^2. \tag{8.19}$$

We can relate $f(\theta)$ to the partial wave phase shifts, by expressing the asymptotic form of the wave function as a sum over spherical harmonics. If we take the direction of the incoming wave $\mathbf{k}$ as the polar angle, then

$$e^{i\mathbf{k}\cdot\mathbf{r}} = e^{ikr\cos(\theta)} = \sum_{l=0}^{\infty} j_l(kr) P_l(\cos(\theta)), \tag{8.20}$$

where

$$\left(-\frac{d^2}{dr^2} - \frac{2}{r}\frac{d}{dr} + \frac{l(l+1)}{r^2}\right) j_l(kr) = k^2 j_l(kr). \tag{8.21}$$

The solution is a function of kr because we can set $k = 1$ by rescaling r. The solution of this equation, which is nonsingular at $r = 0$ is called *the spherical Bessel function of order l*. You will explore some of its properties in Exercise 8.4.

Putting together the results of that exercise with our discussion of the wave function for fixed l, we conclude that

$$f(\theta) = \frac{1}{ik}\sum_{l=0}^{\infty}(2l+1)[e^{2i\delta_l(k)} - 1]P_l(\cos(\theta)). \tag{8.22}$$

For the Coulomb potential this gives

$$f(\theta) = \frac{1}{ik}\sum_{l=0}^{\infty}(2l+1)[\frac{\Gamma(1+l+i\eta)}{\Gamma(1+l-i\eta)} - 1]P_l(\cos(\theta)). \tag{8.23}$$

The Coulomb scattering amplitude is very singular at $\theta = 0$, and the total cross section

$$\sigma = \int d\Omega \, \frac{d\sigma}{d\Omega}, \tag{8.24}$$

is infinite. The total cross section is the probability per unit initial flux that particles are scattered by the potential. The reason it diverges is the long range nature of the Coulomb potential. No matter how large the impact parameter, there will be a nonzero deflection of the charged particle trajectory.

The infinite total cross section is a symptom of an even more serious problem that arises when we take into account the quantization of the electromagnetic field, of which the classical Coulomb potential is just the classical, nonrelativistic remnant. Classical charged particles radiate when they accelerate, and the fact that scattering amplitudes do not fall off with impact parameter means that particles are always radiating photons, no matter how far apart they are. Moreover, the classical radiation field of an accelerated charge is a coherent state, which has an amplitude to consist of an arbitrarily large number of photons. These facts make the essential hypothesis of scattering theory that particles become free when far apart, false.

This problem has little practical importance, because most of the photons radiated have very low energy and we can just define *inclusive probabilities*, where we sum over states of photons with energy less than some arbitrary cutoff and define a density matrix for other particles. These probabilities are all that an experimentalist with limited energy resolution can ever measure. Nonetheless, the problem is not completely solved at the conceptual level. A similar problem arises for gravitons, the hypothetical particle connected to Einstein's gravitational field, and it may lie at the heart of the nature of the quantum theory of gravity.

8.4 EXERCISES

8.1 The equation for positive energy Coulomb wave functions is

$$[-\partial_r^2 - \frac{2}{r}\partial_r + \frac{l(l+1)}{r^2} - \frac{1}{r}]\psi_l(r) = k^2\psi_l(r).$$

Show that the substitution $\psi_l = r^l e^{\pm ikr}\phi_l$ takes this into an equation of the form

$$[\mp r\partial_r^2 + (b \pm ir)\partial_r + a]\phi_l = 0.$$

What are the values of a and b?

8.2 The fact that r appears linearly in the equation for ϕ_l in the previous exercise, motivates us to write

$$\phi_l(r) = \int_c^d dz \ e^{zr} f_l(z).$$

The integral is taken over some open contour in the complex plane, with the indicated endpoints. Derivatives become powers of z and linear terms in r can be converted, via integration by parts, into first derivatives of f_l with respect to z. If we choose the endpoints to be places where $f_l(c) = f_l(d) = 0$, the surface terms in the partial integration vanish. Find the appropriate contour and solve for f_l. Show that the endpoints are finite points in the plane and that f_l is nonsingular on the contour of integration. Prove that ϕ_l is nonsingular at finite r, and use this integral formula to extract its behavior as $r \to \infty$ and thus the phase shifts.

8.3 Show that the classical formula for the Coulomb scattering cross section (first derived by Rutherford) is

$$\frac{d\sigma}{d\Omega} = \frac{Z^2 e^4}{(16\pi\epsilon_0)^2 \sin^4(\theta/2)}.$$

8.4 In terms of the variable $z = kr$, the spherical Bessel equation is

$$(-\frac{d^2}{dz^2} - \frac{2}{z}\frac{d}{dz} + \frac{l(l+1)}{z^2})j_l(z) = j_l(z).$$

Show that near $r = 0$, the right-hand side is negligible and the solution has the power law form z^a. Find the values of a and argue that only one of them is acceptable. As $r \to \infty$, show that only the first term on the left-hand side, and the term on the right-hand side are important, so that the solutions are $e^{\pm iz}$. Eliminating one of the powers at the origin implies that we get a specified linear combination of the two exponentials. The equation is invariant under $z \to -z$ so we can always choose even and odd combinations. Show that the value of l determines the combination $e^{iz} \pm e^{-iz}$.

8.5 Solve for the power series expansion of $j_l(z)$.

CHAPTER 9

Charged Particles in a Magnetic Field

9.1 INTRODUCTION

This chapter discusses Landau's solution of the motion of a charged spinless particle in a constant magnetic field. This classic problem has become more and more important as time passed and really became a key topic in condensed matter physics with the discovery and explanation of the Fractionally Quantized Hall Effect.

The basic feature of Landau's solution is that the eigenstates are grouped into discrete bands with infinite degeneracy. This is directly related to the guiding center solutions of the corresponding classical problem: charged particles spiral around a fixed position in the plane perpendicular to the field. In the limit of large fields, all but the lowest Landau level get very large energy, but the physics of that degenerate level, once electron interactions are taken into account is intricate and fascinating.

9.2 THE LORENTZ FORCE AND LANDAU'S SOLUTION

The Lorentz force on a charged particle moving in a time-independent magnetic field is

$$m\ddot{\mathbf{x}} = q\dot{\mathbf{x}} \times \mathbf{B}. \tag{9.1}$$

In order to write these in the Euler–Lagrange form, we must introduce the vector potential $\mathbf{B} = \nabla \times \mathbf{A}$. The classical Lagrangian coupling a charged particle to a static magnetic field with vector potential $\mathbf{A}(\mathbf{x})$ is

$$L = \frac{1}{2}m\dot{\mathbf{x}}^2 + q\dot{\mathbf{x}} \cdot \mathbf{A}(\mathbf{x}). \tag{9.2}$$

In Exercise 9.1, you will show that the Euler–Lagrange equations of this Lagrangian are simply the Lorentz force equations. Applying the rules of analytical mechanics to the above Lagrangian leads to a Hamiltonian

$$H = \frac{1}{2m}(\mathbf{p} - q\mathbf{A}(\mathbf{x}))^2. \tag{9.3}$$

In classical mechanics, only the equations of motion matter, and these are the usual Lorentz force equations, which contain only the magnetic field $\mathbf{B} = \nabla \times \mathbf{A}$, so classical mechanics is invariant under gauge transformations $\mathbf{A} \to \mathbf{A} + \nabla\Lambda$. In quantum mechanics (QM), this is less obvious because $\mathbf{A}$ appears in the Schrödinger equation

$$\frac{(-i\hbar\nabla - q\mathbf{A})^2}{2m}\psi = i\hbar\partial_t\psi. \tag{9.4}$$

The key to understanding what is going on is the realization that the overall phase of the wave function has no physical meaning in QM. The transformation $\psi \to e^{iq\Lambda}\psi$, with constant Λ, is a redundancy or *gauge symmetry* of ordinary QM. Without a magnetic field, or if the particle has no charge, this is *not* true of phase transformations with variable Λ. However, it is easy to see (Exercise 9.2) that if we combine a variable phase change $\Lambda(\mathbf{x})$ with the gauge transformation $\mathbf{A} \to \mathbf{A} + \nabla\Lambda$, then the Schrödinger equation *is* invariant.[1] This is equivalent to the statement that under the combined transformation,

$$\mathbf{D}\psi \equiv (\nabla - i\frac{q}{\hbar\mathbf{A}})\psi \to e^{iq\lambda/\hbar}\mathbf{D}\psi \tag{9.5}$$

$\mathbf{D}$ is called the *covariant derivative operator*.

For multiparticle systems, the corresponding transformation of the multiparticle wave function is

$$\psi(\mathbf{x_1}\ldots\mathbf{x_n}) \to e^{i\sum_i \frac{q_i}{\hbar}\Lambda(\mathbf{x_i})}\psi(\mathbf{x_1}\ldots\mathbf{x_n}), \tag{9.6}$$

where q_i is the charge of the i-th particle. Note that the gauge function and vector potential depend only on three space coordinates, while the wave function depends on the $3N$ coordinates of all the particles. In the formalism of *Quantum Field Theory*, which we have alluded to in Chapter 5, and will discuss a bit more in Chapters 11 and 12, we have operator valued quantum fields, which are functions of a single position coordinate, but which describe the entire collection of any number of identical particles with fixed quantum numbers.[2] The Hamiltonian is built from these field operators and gauge invariance is implemented by replacing derivatives by covariant derivatives for each field operator. The full gauge invariance of

---

[1] The gauge function Λ must be independent of time, unless we introduce a scalar potential to make a covariant time derivative. In the gauge where the scalar potential is zero, time-dependent gauge transformations of $\mathbf{A}$ generate physical electric fields $\mathbf{E} = \dot{\mathbf{A}}$.

[2] In relativistic quantum field theory, a single field describes both particles and antiparticles.

Maxwell's equations is incorporated into QM by adding the scalar potential $\Phi(\mathbf{x}, t)$ and letting the vector potential depend on time as well. The coupling of particles to Φ is dictated by replacing time derivatives by covariant time derivatives

$$D_t = \partial_t - iq\Phi.$$

In this formalism, if one wants to describe the full QM of the electromagnetic field interacting with charged particles, then all one has to do is add the Hamiltonian of Maxwell's theory and quantize the electromagnetic field according to canonical rules. The quantization is a bit subtle, because gauge invariance tells us that not all the fields are physical variables, but there are various ways of dealing with this [17].

Let us return to single particle QM, and solve the simplest problem involving magnetic fields, namely the constant $\mathbf{B}$ field pointing in the 3 direction. The most convenient gauge for the vector potential is to take $A_3 = 0$ and $A_1 = \frac{1}{2}x_2 B, A_2 = -\frac{1}{2}x_1 B$. The Schrödinger equation is

$$\frac{-\hbar^2}{2m}[\partial_3^2 + (\partial_2 + i\frac{qB}{2\hbar}x_1)^2 + (\partial_1 - i\frac{qB}{2\hbar}x_2)^2]\psi = E\psi. \tag{9.7}$$

Define $x_i = \sqrt{\frac{\hbar}{2B}}w_i$ and $E = \frac{\hbar^2 qB}{2mc}\epsilon$. In terms of the new variables, we have

$$-[\partial_3^2 + (\partial_1 - i\frac{w_2}{2})^2 + (\partial_2 + i\frac{w_1}{2})^2]\psi = \epsilon\psi. \tag{9.8}$$

The equation is obviously invariant under translation in the w_3 direction, so there are solutions that are simultaneous eigenfunctions of $K_3 = -i\partial_3$, which have the form $e^{ik_3 w_3}\chi(w_1, w_2)$, where χ satisfies

$$-[(\partial_1 - i\frac{w_2}{2})^2 + (\partial_2 + i\frac{w_1}{2})^2]\chi = [\epsilon - k_3^2]\chi. \tag{9.9}$$

The covariant derivative operators $D_i \equiv \partial_i - i\epsilon_{ij}w_j$ satisfy

$$[iD_1, iD_2] = i, \tag{9.10}$$

just like canonical position and momentum operators $iD_1 = Q$, $iD_2 = P$, so we can write the Schrödinger equation in operator form as

$$(P^2 + Q^2)\chi = [\epsilon - k_3^2]\chi, \tag{9.11}$$

which is the equation for a harmonic oscillator. Remember Feynman's dictum: the same equations have the same solutions. There are no springs attached here, but we can use our knowledge of oscillators to solve the equations. The equations of motion (classical or quantum) are

$$\ddot{Q} = -4Q, \tag{9.12}$$

so that the dimensionless frequency (which is the same as the dimensionless energy) is $\omega = 2$. Thus, the quantized energy levels are

$$\epsilon = 2n + 1 + k_3^2, \tag{9.13}$$

where n is a nonnegative integer and k_3 is a real number. The wave functions in terms of Q are just the familiar Hermite functions, concentrated around $Q = 0$. In terms of the original variables w_i, the equation $\sqrt{2}a|\chi\rangle = Q + iP|\chi\rangle = 0$ reads

$$[i\partial_1 - \partial_2 + \frac{1}{2}(-w_2 + iw_1)]\chi(w_1, w_2) = 0. \tag{9.14}$$

There are an infinite number of degenerate solutions to this equation, which have the form

$$f(z)e^{-\frac{1}{4}(w_1^2 + w_2^2)} = f(z)e^{-zz^*}, \tag{9.15}$$

where f is an analytic function of the complex variable $z = \frac{1}{2}(w_1 + iw_2)$. A basis of normalizable (but not orthogonal) solutions to this equation is gotten by choosing $f(z) = z^n$. These functions have fixed angular momentum around the point $w = 0$ in the plane. On the other hand, if we look at $f(u, z) = e^{2zu^* - uu^*}$, then we obtain a solution which is not localized at the origin. We can get a better idea of the properties of this wave function by multiplying it by the gauge transformation $e^{z^*u - zu^*}$ to get $e^{-(z-u)^*(z-u)}$. The probability density is

$$P(z, z^*) \propto e^{zu^* + z^*u - zz^*}, \tag{9.16}$$

and is peaked at

$$z = u. \tag{9.17}$$

The ground state wave functions centered at $z = u$ are called *guiding center solutions*.

We can understand the infinite degeneracy we have encountered by thinking about the translation symmetry of the problem. A constant magnetic field is translation invariant in all three directions in space. This is not apparent in the vector potential, which must depend on w in order to generate a nonzero field. We chose a potential which was manifestly rotation covariant around the point $w = 0$, but there are other gauges, where the same field is generated by a potential whose rotation symmetry is around the point $z = u$ instead. The Schrödinger equation in a fixed gauge is not gauge invariant. What we showed above is that by multiplying the wave function in one gauge by a position dependent phase factor, we obtain the wave function in another gauge.

The different guiding center solutions are not, however, gauge transformations of each other. They are physically different solutions of the same equation, in a fixed gauge. They are not, individually, translation invariant, but they do have the same energy eigenvalue. This is, in general, all QM tells us about symmetries. The translation operator commutes with the

Hamiltonian, but this just tells us that it takes any state into a state with the same energy. Of course, we can try to superpose all of these different states, to obtain another degenerate state which *is* translation invariant. However, if the symmetry in question is translation, we run into a problem. The functions $f(u,z)e^{-\bar{z}z}$ are all normalizable. The unique translation invariant superposition of them

$$\int d^2u \; f(u,z)$$

is *not* normalizable (Exercise 9.3), and so is not an allowed state in the Hilbert space.

The real question is, why are there localized normalizable[3] eigenfunctions for a particle moving in a constant field? The answer can be found by combining the classical physics of this system with the uncertainty principle. Classically, if we start a particle off with nonzero velocity, the Lorentz force accelerates it perpendicular to that velocity. This means, in particular, that the work done on the particle by the field is zero, so that its kinetic energy remains unchanged. The particle will move in a circle, called a Larmor orbit, whose radius is determined by the equation

$$\frac{mv^2}{r} = qBv,$$

which says that the centripetal acceleration necessary to the circular motion comes from the Lorentz force. Thus, particles in a constant magnetic field perform helical motions at constant energy, moving freely in the x_3 direction while performing Larmor orbits in the plane perpendicular to the field. The center of the Larmor orbit can be located anywhere in the plane.

In classical physics, the velocity is a continuous parameter and can be taken to zero. In QM, because of the uncertainty principle, a particle bound in a finite region of the plane cannot have a completely certain velocity. Indeed, the Heisenberg equations of motion tell us that the velocity v_i is proportional to $p_i - qA_i$, so that its two planar components do not commute with each other in a constant B field! As usual, confinement of the motion to a finite region (normalizability of the wave functions) leads to quantization of the energy. The guiding center states that we have found are the quantum analogs of Larmor orbits and the reason for the absence of a translationally invariant ground state is now clear.

This problem was first worked out by Landau [25] and the quantized degenerate states for fixed k_3 are called *Landau levels*. In particular, when B is large and the gap between levels grows, most of the physics is described by the lowest Landau level.

We have now gone about as far as we can go with single particle QM. We will still use it to understand a variety of approximation methods, but much of the physics to which QM applies involves multiple particles in an important way. We have already seen this in our

---

[3] None of the actual eigenfunctions ψ is actually normalizable, because of the free motion in the x_3 direction. We are really talking about normalizability of the wave function χ for fixed k_3.

discussion of Fermi statistics and the periodic table. Before starting in on the discussion of atomic, molecular and solid state physics though, we should take a step back to discuss some of the conceptual issues surrounding probability, measurement, and the nature of classical reality, which we touched on but did not resolve in the first chapter. Students of a practical cast of mind may want to skip the next chapter on first reading, and go on to real physics. It is worth coming back and finishing this "philosophical" chapter at some point. Too many physicists, who know how to use QM, still feel uncomfortable with its conceptual basis. This discomfort is unnecessary.

9.3 EXERCISES

9.1 Show that the Euler–Lagrange equations of the Lagrangian

$$L = \frac{1}{2}m\dot{\mathbf{x}}^2 + q\dot{\mathbf{x}} \cdot \mathbf{A}(\mathbf{x})$$

are the Lorentz force equations.

9.2 The Schrödinger equation for a charged particle in a magnetic field is

$$\frac{\hbar^2}{2m}(\nabla - iq\mathbf{A}(\mathbf{x}))^2\psi + (V - E)\psi = 0.$$

Show that if we transform $\psi \to e^{iq\Lambda(\mathbf{x})}\psi$, then this becomes

$$\frac{\hbar^2}{2m}(\nabla + iq\nabla\Lambda - iq\mathbf{A}(\mathbf{x}))^2\psi + (V - E)\psi = 0.$$

Thus, a position dependent phase is equivalent to a gauge transform of the vector potential, and the equation is invariant under the combined operation of multiplying by a phase and doing a gauge transformation.

9.3 Show that the translation invariant superposition of guiding center solutions is not normalizable.

9.4 Solve the problem of a particle in a constant magnetic field using the gauge $A_1 = 0$ for the vector potential.

9.5 The natural form of the solutions to Exercise 9.4 are states that are eigenstates of the translation operator P_2. Show that the infinite degeneracy of the eigenstates is now understood in terms of translations of the x_2 coordinate in this gauge, and interpret the nonnormalizability of translation invariant states in this gauge.

9.6 Show that the lowest Landau level eigenstates discussed in the text are instead eigenstates of the angular momentum operator $X_1 P_2 - X_2 P_1$. Write the angular momentum eigenstates as linear combinations of the gauge transformations of the P_2 eigenstates. Be careful to do the proper gauge transformation to convert the wave functions in one gauge into those in the other.

CHAPTER 10

The Meaning of Quantum Measurement and the Emergence of a Classical Reality

10.1 INTRODUCTION

In this chapter, we will outline the way in which a classical world can emerge as an approximation to quantum mechanics (QM), and assess the likelihood of measuring the quantum corrections to the classical behavior of macroscopic objects. We will see that these corrections are extraordinarily small in typical states of these objects, a fact which accounts for the difficulty we have with coming to an intuitive understanding of quantum rules. Those of you who want to get on with the mathematical formalism of QM, and its use in atomic and condensed matter physics, can skip this chapter on a first reading, but it is worth your while to come back to it.

The key word in the previous paragraph is *macroscopic*, and we begin by defining precisely what we mean by that. We have seen that the mass of the electron and the strength of the electromagnetic coupling define a natural length scale for atomic structures, called the Bohr radius. Its size is about 10^{-8} cm. A cubic sample of solid material, slightly less than 0.1 cm on a side thus has about $N > 10^{20}$ "unit cells" of atomic dimensions. In gaseous or liquid states of matter, the atoms are further apart so the same number of degrees of freedom take up somewhat more space. We will talk here about models of solids but similar estimates work for liquids and gases as well.

The interactions between the N unit cells are local, each cell is strongly coupled to only a few nearest neighbors, because long range Coulomb forces are screened. We will see that in typical local models of a large number of variables, the number of states with almost degenerate energies, in energy bands far above the ground state, increases like e^N. We will also see that *collective coordinates*, averages over of order N variables, have uncertainties that scale like $N^{-1/2}$ and the number of states with a fixed expectation value of those coordinates is also exponential in N. As a consequence, we will show that QM defines, with exponential precision, probabilities for histories of the collective coordinates, which satisfy Bayes' rule and all of our expectations for a classical probability theory. In other words, there is a small subset of variables in any large quantum system with local interactions, which behave exactly as we would expect a classical system subject to small random perturbations to behave. Since all of our coarse grained observations of the world are observations of collective coordinates, QM can explain the apparently classical nature of the world. In particular, the above results allow us to condition predictions for the future of the entire world, on a particular history of some of the collective coordinates. This is the procedure which has become known as "collapse of the wave function."

We will also talk briefly about the phenomenon of unhappening, in which the disintegration of some macrosystem into elementary particles forces us to give up predictions based on a particular history of its collective coordinates (unless some other macroscopic system has made a macroscopic record of that history), and return to the original uncollapsed wave function to make predictions about the future.

Finally, we will discuss Bell's theorem and further results which show that any attempt to find a more classical explanation of quantum probabilities will have to have truly bizarre features.

10.2 COUNTING STATES AND COLLECTIVE COORDINATES

The simplest possible model, which suffices to see how the counting works, assigns a two-dimensional quantum Hilbert space to each point of a lattice with a spacing of order the atomic scale. For simplicity of visualization, take the lattice to be cubic, though we will not really use that in our estimates. A complete set of variables for this system is a set of Pauli matrices $\sigma(i)$, one for each lattice site. The Pauli matrices on different sites commute with each other. A simple, soluble Hamiltonian for this system is

$$H = -J \sum_{<i,j>} \sigma_3(i)\sigma_3(j) + B \sum_i \sigma_3(i). \tag{10.1}$$

J and B are positive constants. The symbol $< i, j >$ means that we sum over all pairs of the nearest neighbors on the lattice. We will call $\sigma_3(i)$ the local spin of the system. We can see by inspection that the first term likes all of the "spins" to be aligned, while the second term prefers them to be negative. Low lying excitations of this ground state are gotten by flipping a single spin, at a cost in energy of $Jc + B$, where c is the number of different nearest neighbors. If there are N different lattice points, then there are N different states of this type, depending on which spin we flip (use periodic boundary conditions to avoid edge effects at the "end" of the lattice).

Next we can flip two spins. If they are not nearest neighbors, then the energy cost is twice that of flipping a single spin, while if they are, we get an energy lower by J. There are $N(N-1)$ excitations of this type, most of them exactly degenerate. Similarly, there are $\frac{N!}{k!(N-k)!}$ states with k spins flipped, again mostly degenerate in energy. For large N and $1 \ll k \ll N$, the degeneracy is of order e^{kN}. It is this exponential behavior of the number of states with the volume N, which is responsible both for the success of the methods of statistical mechanics and the emergence of classical behavior.

Now imagine perturbing the Hamiltonian by a small term

$$\delta H = \epsilon \sum_{<i,j>} \sigma_1(i)\sigma_1(j). \tag{10.2}$$

The problem is no longer exactly soluble but when $\epsilon \ll J, B$, we can see the qualitative nature of the effect on the spectrum fairly easily. We will systematize this kind of QM perturbation theory in Chapter 13, but for the time being we need only qualitative results. For states between which the unperturbed problem gives a gap in energy of order J or B, the gap will only be changed a little. By contrast, for the exactly degenerate states, the leading order correction to the energy is gotten by diagonalizing the perturbation δH in each degenerate subspace. This gives a matrix whose size is order the degeneracy and whose eigenvalues are all within something of order $\pm\epsilon$ of the original degenerate energy level. If we start in some particular state, for example one of the original degenerate eigenstates of $\sigma_3(i)$, it will evolve under the action of the perturbed Hamiltonian around the subspace of erstwhile degenerate states, in a fairly random manner.

It is difficult to make mathematically rigorous statements, which quantify the degree of randomness in this evolution, but the assumption of randomness is the key hypothesis in both the explanation of the thermal properties of matter[1] and the emergence of classical physics.

10.3 COLLECTIVE COORDINATES

The second important concept in the explanation of classical behavior is the notion of a *collective coordinate* of a macroscopic system. In our model, these are variables like

$$\Sigma \equiv \frac{1}{N} \sum \sigma(i). \tag{10.3}$$

---

[1] Unfortunately, this explanation is far beyond the scope of the current elementary textbook.

They are sums over the lattice, or a reasonably large fraction of the lattice points, of terms involving only a few lattice sites, close to each other. These particular collective coordinates satisfy commutation relations

$$[\Sigma_a, \Sigma_b] = \frac{i}{N} \epsilon_{abc} \Sigma_c \tag{10.4}$$

and equations of motion (we set $\delta H = 0$ for this equation).

$$\dot{\Sigma}_a = B \epsilon_{a3b} \Sigma_b + J \sum_{<i,j>} \epsilon_{ab3} \sigma_b(i) \sigma_3(j). \tag{10.5}$$

The equation of motion relates macroscopic operators to other macroscopic operators, with N independent coefficients. The commutators of macroscopic operators are very small, so that their quantum fluctuations are of order $\frac{1}{\sqrt{N}}$. This follows in a very general way for operators that are defined as sums over all lattice points of products of local spins at only a few neighboring points.

A further general property is that these macroscopic operators take on the same values for an exponentially large number, e^{cN} of the (large k) eigenstates of the unperturbed Hamiltonian. This is easy to see for the operators Σ, but is true for some large class of operators. It is easy to see that our estimates did not depend on the detailed nature of the system, but only on the following two properties

- The system consists of a large number N of mutually commuting variables, which live on some kind of lattice (which need not be regular as long as it is regular when averaged over large enough sizes). The Hamiltonian is a sum over the whole lattice, of *local* operators; operators that depend on only one site and a few nearest neighbors. For high enough excitation energy, with $E - E_0$ scaling like N, the number of states with energy between E and $E \pm \delta$ scales like e^{cN}.

- Collective coordinates C_i are operators which are *averages* of local operators over a large fraction of the lattice. The generalized uncertainty relations imply that the uncertainties in collective coordinates scale like $\frac{1}{\sqrt{N}}$. The Heisenberg equations of motion for the $C_i(t)$, close on the C_i themselves, because of the local form of the Hamiltonian. For a given initial state, which is a minimal uncertainty state for a set of C_i satisfying a closed set of equations of motion, if we write

$$C_i(t) = \langle \psi | C_i(t) | \psi \rangle I + \Delta_i(t),$$

with I the identity operator, then $\Delta_i(t)$ is an operator whose matrix elements remain small for $t \ll \sqrt{N}$. The expectation values $c_i(t)$ then satisfy an approximately closed system of classical equations over these time scales. These define a classical history for the system. It is important that the number of variables appearing in these classical equations is finite as $N \to \infty$. It is also clear that the number of states that share the same classical history, in a band with high enough energy, scales exponentially with N.

As an example, we can consider the center of mass coordinate conjugate to the total momentum of a bound system consisting of a large number of atoms. This satisfies

$$[X_i, \dot{X}_j] = \frac{i}{M\hbar} \delta_{ij},$$

$$M\ddot{X}_i = -\frac{\partial V}{\partial X_i}.$$

The equation of motion for the expectation value of X_i is just the classical Newton equation. Since $M \sim N$ and the potential is a sum of terms acting on individual atoms, the classical motion has a time scale of order 1 and minimal uncertainties of order $N^{-1/2}$.

For bits of matter made of even the lightest elements, we have

$$\delta X \delta V \geq \frac{10^{-4} cm^2}{N sec}.$$

A solid cube of side 10^{-1} cm will have $N \sim 10^{20}$. Even if our initial state exceeds the minimal uncertainty by four orders of magnitude, we can have a position uncertainty of 10^{-6} cm and a velocity uncertainty of 10^{-14} cm/s in the initial state. The position uncertainty will grow to the size of the object in a time of order 10^{13} s, if the bit of matter moves without the influence of external forces. As we have seen, the uncertainty will grow even more slowly in the presence of a potential.

Quantum uncertainties are thus quite small even for quite small macroscopic objects, but don't seem beyond the realm of measurement. The real problem is proving that the fluctuations that we see are quantum mechanical in nature. Indeed, the first direct experimental verification of the theory of atoms came from observations of *Brownian motion*; the small fluctuations of the positions of dust particles suspended in a fluid. A perfectly adequate classical theory of these fluctuations was developed independently by Einstein and Smoluchowski. It satisfies Bayes' rule for conditional probabilities.

The classical theory of Brownian motion is "wrong," since it treats atoms and molecules as classical particles, but it is a valid emergent theory, because the quantum treatment of the same problem reproduces the classical theory with incredible accuracy. To distinguish between the two, we would have to observe not just uncertainty, but interference phenomena, and the violation of Bayes' rule, for the collective coordinates of the dust mote.

10.4 INTERFERENCE OF CLASSICAL HISTORIES

In the energy regime which we are discussing, interference is suppressed by factors of order e^{-cN}, rather than the mere power laws by which quantum uncertainty is concealed. The suppression is related to the large number of independent quantum states that have the same classical history. By arguments just like those we used to count the degeneracy of energy eigenstates of high enough energy, the number of linearly independent states that have the same classical initial conditions is exponentially large.

Now let us consider the time evolution of the system, starting from some generic state in some large energy eigenspace of the unperturbed Hamiltonian. Label the collective coordinates by C_i. Their expectation values will satisfy

$$\frac{d}{dt}\langle\psi(0)|e^{i(H+\delta H)t}C_i e^{-i(H+\delta H)t}|\psi(0)\rangle = \langle\psi(0)|f_i(C_j(t))|\psi(0)\rangle, \qquad (10.6)$$

and involve only functions of the other collective coordinates. $C_i(t)$ is the Heisenberg operator equal to C_i at $t = 0$. Because the quantum fluctuations of the C_i are small, these equations are, with accuracy $\frac{1}{\sqrt{N}}$, classical equations, which relate the expectation values to themselves. This defines a classical history of the collective coordinates. For a given classical history of the collective coordinates, the time for which fluctuations falsify the predictions of the classical equations is of order $\sqrt{N}t_0$. The natural time scale of atomic physics is about 10^{-10} seconds, so this is a time scale of order seconds for a piece of solid material 0.1 cm on a side. These are very rough estimates, and for some collective coordinates, the time is much longer. However, as we will see, the real parameter that determines *classicality* is not the size of the fluctuations. Indeed, we know from the classical theory of Brownian motion that even a classical physicist studying the interaction of a macroscopic variable with a system composed of many atoms expects to see statistical fluctuations, coming from the unknown state of the atomic system. To distinguish the quantum fluctuations from those predicted by such a classical model, we have to observe interference phenomena between different classical histories.

In the $N \to \infty$ limit, one can argue that there cannot be such interference. The equation

$$\langle\psi(0)|C_i(t)|\psi(0)\rangle = C_i^{cl}(t) \qquad (10.7)$$

is quadratic in $|\psi(0)\rangle$ and so does not define a subspace of the Hilbert space. However, when $N \to \infty$, one can define orthogonal subspaces of the Hilbert space in which the $C_i(t)$ are all simultaneously diagonal and equal to their classical values.

The claim is that when N is finite and large, a typical state with some classical history has overlap e^{-N} with one that has another history. There are three different kinds of arguments for this, the last of which predicts an even smaller overlap.

- The first argument is pretty rigorous, but involves an assumption. It is easy to argue that the number of linearly independent states with a given history is of order e^{cN}.

For example, in our spin model, the constraint $\frac{1}{N}\langle\Sigma_3\rangle = M$ is one equation on the 2^N eigenstates of $\sigma_3(i)$. We have seen that constraining the energy in a band high enough above the ground state still leaves an exponential number of states satisfying the energy and magnetization constraints. As long as the classical equations that determine the history, effectively close on a finite number of expectation values $\ll N$, this counting continues to work.

- Two typical states with the same expectation value $c_i(t)$, defining a classical history, will have exponentially small scalar product. This is a simple consequence of the fact that the space of states with the same history has an exponentially large dimension. The scalar product of two randomly chosen vectors is thus exponentially small. Consider some collective coordinate C_i whose local form involves only k sites, with $k \ll N$. There is a basis of states, which have the form[2]

$$\otimes_{p=1}^{N/k}|\psi_p\rangle, \qquad (10.8)$$

where each state is acted on only by the operators in one of N/k disjoint clusters of k sites. C_i is a sum of terms that act only within a cluster, and some that couple different clusters. Our argument will ignore the coupling terms and is therefore not rigorous unless $k = 1$. To change the expectation value of C_i by an amount of order 1, we have to change each of the states $|\psi_p\rangle$ into $|\psi_p'\rangle$ with $|\langle\psi_p'|\psi_p\rangle| = c_p < 1$. The overlap of the two product states is then of order $\prod_{p=1}^{N/k} c_p \sim e^{-bN}$. Roughly speaking, this argument says that two states with different classical histories will *always* have exponentially small scalar product.

- The time scale of microscopic change of the quantum state of our system ranges from much shorter than the time scale for the classical motions $c_i(t)$ to much longer. Time scales in QM are determined by energy differences and in atomic systems, these range from thousands of electron volts to $e^{-cN}1,000$ eV (because so many states have to fit into an energy band of typical atomic scale). This has two consequences. First, it is almost never the case that the classical motion is so slow that it doesn't affect the microstate (so that the second item above becomes relevant). Second, the interference terms in a QM probability calculation are time dependent, and average to zero over the shortest time scale of the classical motion.

- Finally we want to mention the notion of environmental decoherence. We have been talking about a small piece of solid matter, with $N \sim 10^{20}$. Such systems are rarely isolated from their environment very well, and coupling to the environment makes the

---

[2] Recall the notion of tensor product from the chapter on Hilbert space.

effective value of N in the above estimates much larger. This is true even if we imagine that the system discussed above was the needle on a dial on a device whose scale is of order 10 cm and imagine that the device itself is sealed inside a completely shielded laboratory, in a vacuum. When the needle moves from one position to another, it is interacting with a different part of the device. So in most cases, taking $N \sim 10^{20}$ in the estimates above, grossly overstates how big the interference terms are for different histories of the expectation values of collective coordinates. Many discussions of decoherence emphasize the vastly more important effect of environmental decoherence, and do not treat the decoherence of the collective coordinates of even tiny bits of matter, due to their own internal structure. We prefer to emphasize the relatively small $e^{-10^{20}}$ effect of internal structure because it shows how remote quantum interference effects are, even for quite small systems.

10.5   SCHRÖDINGER'S CAT AND SCHRÖDINGER'S BOMB

We can now use these remarks to understand the mysterious "collapse of the wave function" and "Schrödinger's cat paradox" that haunt so much of the literature on the interpretation of QM. Consider a microsystem like the two state Ammonia Molecule of Chapter 1, in a superposition of states with the two orientations of the electric dipole

$$|\psi\rangle_{Ammonia} = a|+\rangle_3 + b|-\rangle_3. \tag{10.9}$$

The dipole moment is uncertain in this state, and if we evolve the system with the Hamiltonian $H = \epsilon\sigma_1$, we cannot describe the time-dependent probability distributions in terms of probabilities for histories of the value of σ_3. This is because of interference terms. The matrix elements of products of Pauli matrices in this state depend on $a^*b + b^*a$ as well as the probabilities $|a|^2 = 1 - |b|^2$.

Von Neumann [26] was the first to realize that a model of the measurement procedure was the coupling of this state to a macroscopic system, via perfectly unitary QM, to obtain what is called an *entangled* state

$$a|+\rangle_3 \otimes |C_+\rangle + b|-\rangle_3 \otimes |C_-\rangle, \tag{10.10}$$

where $C_\pm$ refer to two positions of a macroscopic needle on a measuring apparatus. The work on Decoherence theory [27] in the 1970s and 1980s made the crucial observation that there were an exponentially large number of linearly independent states in the ensemble of states that have the same expectation value for the needle position. As a consequence, for the reasons sketched above, once the entanglement between the state of the ammonia dipole and the apparatus has occurred, no future measurement will be able to detect the interference between the two eigenstates of the dipole position. The ammonia Hamiltonian still acts, but it is negligible compared to the interaction between the molecule and the apparatus. It cannot change the position of the needle.

In other words, once a microsystem becomes entangled with a macrosystem in a way that correlates its state with the expectation value of a collective coordinate, then Bayes' rule and the emergent notion of probabilities for histories become applicable to the particular ammonia molecule that was measured in a particular run of the experiment. QM predicts that the needle will go up with probability $|a|^2$ and we can decide to condition our predictions for future experiments on the outcome of one particular experiment where the needle went up. You can decide to get married as a consequence of the needle going up in a particular run of the experiment to verify the quantum predictions for the ammonia molecule. The quantum description of this sounds weird if you think of QM telling you about what goes on in particular experiments. You are in a superposition of being married and not married.

This is a correct description of the quantum state of the world, but the quantum state is nothing but a probability distribution. We have to ask whether there is anything about the superposition of married/not married that is more disturbing than if you decided whether or not to get married depending on whether or not Hurricane Katrina hit New Orleans (which, if you lived in the Big Easy on the day of the hurricane, might have been a very relevant factor in your decision about getting married). And since your marriage is in fact correlated with the position of a macroscopic needle on a dial, Bayes' rule is valid with incredible accuracy. In addition, the position of the needle is predicted to be pretty close to its expectation value. The fluctuations in the position of the needle are $\ll 10^{-10}$ of the central value, and because of the validity of Bayes' rule, they are identical to those of a theory that attributes the fluctuations to unmeasured classical "hidden variables."

From a practical point of view, you can treat the quantum prediction of your marital state as you would the predictions of the weather equations for hurricanes: there are probabilities for histories, and once we do an experiment, we know which of the histories "really occurred" and we can throw away our probabilistic prediction for things that "did not happen," and renormalize our probability distribution so that, e.g., $a = 0$ and $b = 1$, if you did not get married. The way we do this in QM is called "collapse of the wave function." Classical probabilists exploit the linearity of their equations for probability distributions to prove Bayes' rule and replace the probability distribution $P_1 + P_2$, after a measurement consistent with P_1 but not P_2, by $P_1(1 - \int P_2)^{-1}$. In QM, we replace $a\psi_1 + b\psi_2$ by ψ_1 (recall that $|a|^2 + |b|^2 = 1$) to achieve the same end. And that end is to *compute the probabilities for future events conditioned on the result of some particular event that affected a collective coordinate of a macroscopic object.*

The philosophical stance behind the use of Bayes' rule in classical probability theory is, however, very different than that which underlies the collapse of the wave function in QM. In classical physics, we subscribe to the belief that we *could* have predicted the precise behavior of the system, if only we had been able to determine the initial state of all the microscopic variables. The theory is supposed to predict the exact history of the system, in principle. In this context, the use of Bayes' rule is supposed to represent a refinement of our knowledge about that exact history, and the act of throwing away the part of the probability distribution that disagrees with what actually happened in a particular experiment is the correct thing to do. It represents what we would have done at the outset, if we had known enough about all the original initial conditions to make the correct prediction in the first place.

In QM, the conceptual role of Bayes' rule/collapse of the wave function is quite different. There are no exact histories. Probability is intrinsic and inescapable, and does not result from our inability to measure all the relevant variables. It results from the fact that there are *no* states of the system in which all the variables that appear in the equations of motion have definite values. There are, however, macroscopic collective coordinates for which the probability predictions of QM obey Bayes' rule with fantastic accuracy. Quantum predictions for probabilities conditioned on values of the collective coordinates are the same as those in a classical statistical theory, up to corrections that are, in principle, too small to be measured. Furthermore, the statistical uncertainties in these variables are small, so that their expectation values determine approximate histories. Thus, we can use Bayes' rule and collapse the wave function, as long as we are always talking about the behavior of these almost-classical variables.

There is a little thought experiment, which is a sort of combination of the double slit experiment and the Schrödinger's cat experiment, which illustrates how careful one must be about the use of Bayes' rule in QM. Consider an experiment, taking place inside a small isolated laboratory out in intergalactic space. Like Schrödinger's cat experiment, this one consists of the correlation of the dipole moment of an ammonia molecule, with a macroscopic collective coordinate, this time the minute hand on a macroscopic clock. After the measurement, the state of the system is

$$a|+\rangle_3 \otimes |3:00\rangle + b|-\rangle_3 \otimes |3:30\rangle. \tag{10.11}$$

As before, there is an exponentially large number of clock microstates compatible with each position of the minute hand. Decoherence makes it impossible to see interference between the two parts of the wave function, and would encourage us, as in the classical argument for the double slit experiment, to conclude that we can make arguments about predictions to the future of this measurement, by saying, "Either the clock reads 3 or 3 : 30. The probability of something happening in the future is the sum of the probabilities for what would have happened given one of those two exclusive alternatives."

However, what we have not yet told you is that the clock is actually the timer on an explosive device, which is set to go off at 4 o'clock. It is a chemical explosion, which will blow the entire laboratory into its constituent atoms, leaving no macroscopic trace. However, it is not powerful enough to give nuclei in the debris relativistic velocities.

In order to understand the actual predictions of QM in this system, it is convenient to use the language of Feynman diagrams. We have seen in the case of ammonia, that atomic systems, consisting of charged particles, can emit photons. In the theory of *quantum electrodynamics*, there is a probability amplitude for a charged particle like an electron, propagating through space, to emit a photon, which is then absorbed at a later time by another charged particle, let us say a proton. This amplitude is small, and the probability for all possible things a pair of charged particles can do is approximately calculable in terms of this single elementary pair of processes, with which Feynman associated a space-time diagram, and a set of rules converting each element in the diagram to a part of the calculation of the probability amplitude. The diagram is shown in Figure 10.1.

The electron propagates from its original position to a space-time point labeled by a four-vector x, where it emits a photon. The electron then propagates to its final space-time position, and the photon propagates to a point y, where it is absorbed by the proton. The proton then propagates to its own final position. The points x and y are arbitrary, and part of Feynman's rule is that one must integrate the amplitude over all possible values of both x and y. In particular, we integrate over the times the emission and absorption events could have happened. Feynman's prescription is backed up by a host of solid theoretical arguments, and leads to a theory that provides the most precise agreement with experiment in the history of science.

Let us apply Feynman's prescription to our little thought experiment, asking for the probability, far in the future, of an event in which a proton originating in the explosion encounters a photon emitted in the explosion and scatters from it. The process of proton (or electron) colliding with a photon is called Compton scattering. If there were only one time at which the particles could have been emitted, then we would predict that this event could never occur far in the future. The photons propagate away from the space-time event of the explosion, at the speed of light, while the protons are much slower. At very late times, all protons are separated from all photons by a large spatial distance.

However, Feynman's prescription, assuming the initial state was the entangled state above, with nonzero amplitudes for both positions of the hand of the clock, says that we must sum the probability *amplitudes* over the two possible times for the explosion and then

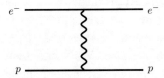

Figure 10.1 Feynman diagram for virtual photon emission and absorption.

square to get the probability. There is an amplitude for a proton to be emitted if the explosion occurred earlier, and an amplitude for a photon to be emitted in the later version, and therefore there is an interference term in the amplitude for all possible things that can occur, in which the early proton, scatters off the late photon. That amplitude would be proportional to *a*b + b*a and would vanish if we assumed that the wave function had actually collapsed via some real physical process, at the time of the measurement.*

In the statistical interpretation of QM that we have advocated in this book, there is no real problem with interpreting this result in a way that is consistent with both our classical intuition and the correct prediction about photon proton scattering probabilities. Given the initial state of the ammonia molecule in a superposition of two directions for the dipole, we get an unambiguous prediction of a nonzero scattering amplitude. However, we also see that if we ask the question: what is the probability of proton photon scattering, given that the explosion occurs at a particular time, we find the answer is zero. What this apparent contradiction means, is that, after both explosions have definitely occurred, there is no way to condition measurements on the explosion having occurred at a particular time.

If we insert another correlation with a macroscopic object into the system, then we have a new set of variables to which we can apply the rules of Bayesian conditioning. That is, if the explosion debris makes macroscopic changes in some distant piece of matter, then we can again define conditional probabilities based on the time at which those changes occurred. Amplitudes projected in this way will show no photon proton scattering amplitude at late times, but this is because they are answering a different question: not, "what is the unconstrained amplitude for photon proton scattering at late times given the initial state of the ammonia molecule," but "what is the amplitude constrained by the condition that the distant piece of matter registers a macroscopic effect of the explosion at a particular time."

If we interpret QM as saying something about definite results of particular experiments, all of this begins to seem problematic. If we interpret it statistically then, as long as we specify precisely what we are talking about, nothing weird occurs. The theory is tested by doing multiple runs of the experiment and comparing the results to carefully phrased theoretical calculations. We simply cannot use classical logic, and the idea that one of two macroscopic things definitely happened in any one of the runs of the experiment to conclude what the theory predicts will happen after all macroscopic trace of those things has disappeared. Or rather, we can use such logic, but it pertains to uncheckable claims. The fact that a particular run of the experiment led the clock to go off at a later time is something that we can use to condition predictions about future experiments. It is sensible to use it as long as some macroscopic record, for which the usual notions of history and conditional probabilities for particular events in a history make sense. We could insist on continuing to use only the branch of the wave function that predicted that particular sequence of macroscopic events even after all macroscopic record of those events has disintegrated into microscopic particles. It will predict correctly, that if there was a chain of macroevents in the past, in which the explosion took place at a particular time, then, no matter which time it was, there is a vanishing amplitude for any proton photon scattering long after the explosion.

If, on the other hand, we ask about the amplitude for postexplosion Compton scattering of debris, with no conditions about what happened to macroscopic objects at intermediate times, then there is a finite amplitude for the scattering to occur. The complexity of the intermediate state, and the fact that our description of it in terms of collective coordinates is very coarse grained, is not relevant to this computation. The complexity of the intermediate state is reflected in correlations between the single photon and proton states, which participate in the scattering, and the multiparticle wave function of all of the other debris. These correlations imply, with probability $(1 - e^{-cN})$ that the initial states of the proton and photon are maximally uncertain, consistent with the fact that they collide,[3] but still give a probability of collision proportional to $a * b + b * a$.

This kind of experiment is called *unhappening*, a term which dramatizes the fact that our assumption that the equations of physics, tell us in principle about things that are "really happening" in a given run of an experiment *is wrong*. There can be no interpretation of QM which retains the notion of probabilities for histories,[4] except as an emergent concept, applicable to the collective coordinates of macroscopic objects as long as those objects exist. Given an initial state during the period such objects exist one can use Bayes' rule to collapse the wave function according to the observed behavior of the collective coordinates.

Once the collective coordinates disappear then predictions about the future must revert to the initial uncollapsed wave function. This is completely bizarre, acausal and nonlocal if one insists on thinking about the wave function as a real object, or even as a probability distribution for histories of something. As long as one thinks about it as a probability distribution for instantaneous values of things that cannot all be certain at the same time, no logical contradictions or bizarre behavior occurs. Things are never in two places at the same time, they merely have probabilities of being in two places at the same time. As we have seen, if one is too glib about the description of experiments, without acknowledging that experiments always involve macroscopic objects, and paying careful attention to the differences between probabilities conditioned on some macroscopic behavior and probabilities which are not so conditioned, then one gets into conceptual trouble. It is our macroscopic conditioning; thought processes evolved in a world where everything of importance seemed to be a macroscopic collective coordinate, that gets in our way of having an intuitive understanding of quantum phenomena.

---

[3] This means that, with probability of $o(1)$, with no exponential suppression, the proton and photon are in wave packets that collide with each other, but the exact form of the packet and the spin states of the initial particles, are completely uncertain.

[4] Unless, with the followers of Bohmian mechanics, one abandons Bayes' rule for probabilities of histories. This is a stance I find logically contradictory—a mere playing with words. See Appendix A on Interpretation for a more detailed discussion.

In some distant future, if the dreams of both quantum computation and artificial intelligence can be realized, the human race might find itself in conversation with sentient beings who have an intuitive understanding of the rules of QM. Perhaps they would be able to explain things better than I have. On the other hand, they might find themselves confronted with the same problem we encounter, if we contemplate teaching calculus to a chimpanzee. I will reserve further comments about the interpretation of QM to an appendix.

10.6 THE EINSTEIN–PODOLSKY–ROSEN (EPR) PARADOX, BELL'S THEOREM, AND OTHER ODDITIES

In relativistic QM, elementary particles can decay into others because there is no separate conservation law of mass. In particular, a massive particle at rest can decay into two light particles with equal and opposite momenta because we can have

$$mc^2 = 2\sqrt{p^2c^2 + m_e^2c^4}, \tag{10.12}$$

where we have called the light particle mass m_e because we want to take them to be electrons. We will also make an assumption about the production mechanism of the electron pair, namely that they are emitted with zero orbital angular momentum w.r.t. to the point of emission, in the rest frame of the decaying particle. It follows from conservation of angular momentum that the spin state of the electron pair is a singlet, if the decaying particle has spin zero. The singlet state can be written in the basis where the 3 components of the spins are diagonal as

$$|\psi_0\rangle = \frac{1}{\sqrt{2}}(|+-\rangle - |-+\rangle). \tag{10.13}$$

The electron and positron are in wave packets traveling in opposite directions at a speed $s = \frac{pc}{\sqrt{p^2 + m_e^2 c^2}}$. They will travel a distance $d = st$ in a time that the width of their packets spreads by an amount of order $\frac{\hbar d}{sm_e\Delta} \ll d$. The fact that the pair has zero orbital angular momentum means that there is an equally likely chance for them to be traveling in any particular direction. We will be studying many decays, so we simply place a pair of Stern–Gerlach machines at positions $\pm d$ along the 1 axis and do our measurements only on those pairs which happen to pass through the machines. The thought-experimental setup is shown schematically in Figure 10.2.

Recall from Chapter 7 that a Stern–Gerlach machine scatters incoming particles with different spin components into different directions in space. Let us add to each machine an absorber such that only electrons or positrons with $\mathbf{n} \cdot \sigma = 1$ get to our detector. When we choose the unit vector $\mathbf{n}$ in the three direction we find the following interesting anticorrelation:

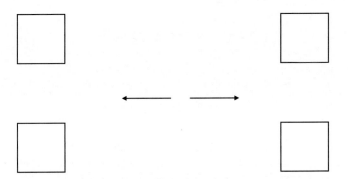

Figure 10.2 Experimental setup for Bell's inequality thought experiment.

in each incident of decay[5] only one of the detectors goes off, although conservation laws tell us that particles entered both detectors. Which detector goes off is completely random, like the flip of a fair coin. Einstein Rosen and Podolsky [30] considered this experiment to somehow violate the principle of locality, since measurement of the spin of the electron immediately determines what the spin of the positron was a space-like distance $2d$ away. According to the principle of relativity, no signal can propagate between the two detectors in the arbitrarily small time between the two measurements. In fact, we can duplicate the results of this experiment with a purely macroscopic system, obeying the rules of classical statistics with arbitrary accuracy. We simply place one black and one white ball in a sequence of boxes. The box contains a random spinner, which rotates the two balls around inside and then opens its $\pm\mathbf{e_1}$ faces and ejects whichever ball happens to be sitting there in the appropriate direction. The ball ejected in the plus direction will be black, with probability $1/2$ and there will be an exact anticorrelation between the colors of the balls received at distant detectors in each run of the experiment.

J.S. Bell pointed out what appears to be a more serious apparent violation of locality by studying what happens when the unit vector $\mathbf{n}$ of one of the detectors points in an arbitrary direction. The projection operator

$$P \equiv \frac{1}{2}(1 + \mathbf{n} \cdot \sigma) \tag{10.14}$$

is the quantity that determines whether the particle going through the rotated detector hits the screen or not. Without loss of generality, we let the detector detecting the particle whose spin appears to the right be the rotated one. There are now four possible outcomes: both particles are detected, neither is detected, the unrotated detector detects a particle and the rotated one does not, and vice versa. The corresponding projection operators are

---

[5] We ignore the fact that the decay probability is isotropic in this part of the analysis. It can be taken into account, with a lot of words, and no extra enlightenment.

$$\frac{1}{4}(1 \pm \sigma_3) \otimes (1 \pm \mathbf{n} \cdot \sigma) \equiv P_{\pm\,\pm}.$$

Their expectation values are

$$P_{+;+\theta} = P_{-;-\theta} = \frac{1}{2}\sin^2(\frac{\theta}{2}), \tag{10.15}$$

$$P_{+;-\theta} = P_{-;+\theta} = \frac{1}{2} - \frac{1}{2}\sin^2(\frac{\theta}{2}), \tag{10.16}$$

where θ is the angle by which the detector is rotated. By rotation invariance, we can think of these probabilities as those for two detectors rotated into direction $\mathbf{n}_{1,2}$ with $\mathbf{n_1} \cdot \mathbf{n_2} = \cos(\theta)$. In QM, one considers these probabilities as the probabilities that pairs of particles created in decays of the heavy particle will register in two detectors oriented in the given directions.

This calculation is inconsistent with the assumption that one can assign a probability to the history of every component of the spin of the electron for a given initial state. Suppose we assumed that in every run of this experiment, the electron had some definite value of $\mathbf{n} \cdot \sigma$, for every choice of the unit vector, which the theory could not predict, but to which it assigns probability. We could classify each run as definitely having the value ± 1, *for every value of θ*. Then we would have probabilities $P(\pm; \pm\,\theta)$ to take on each of these values, given the value of σ_3 for the other particle.

Now consider, according to the rules of classical probability theory, the probability $P(+; +\theta = \pi/2)$ that the first particle can pass the unrotated detector *and* the second can pass at $\theta = \pi/2$. According to classical statistical thinking, we can divide all runs of the experiment up according to the value of the spin of the particles in directions with $\theta = \pi/4$, *even when we do not have an apparatus setup to measure the spin at this angle*. Now according to classical reasoning, we should have

$$P(+\theta_1 = \pi/4; +\theta_2 = \pi/2) + P(+\theta_1 = 0; +\theta_2 = \pi/4) \geq P(+\theta_1 = 0; +\theta_2 = \pi/2). \tag{10.17}$$

Here are the assumptions in the argument. In each run of the experiment, the spin component of each particle in each direction has some definite value, with the two values anticorrelated by angular momentum conservation. We are interested in runs that have the second particle spin up in the $\pi/2$ direction, while the first is up in the 0 direction. Among runs satisfying the former condition, the spin of the first particle in the $\pi/4$ direction could be either up or down and the spin of the second is always anticorrelated with it. These particles have a large space-like separation at times less than but of order d/s. But the probability $P(+\theta_1 = \pi/4; +\theta_2 = \pi/2)$ might have nonzero contributions when the spin in the $\theta_1 = 0$ direction is not up, and $P(+\theta_1 = 0; +\theta_2 = \pi/4)$ might have contributions when the spin in the θ_2 direction is not up. So, the inequality follows, since every run that contributes to the probability on the right-hand side is included in one of the probabilities on the left.

If we plug in the quantum formulae for these probabilities, using the rule $P(\pm\theta_1; \pm\theta_2) \equiv P_{\pm\theta_1; \pm\theta_2}$ we get

$$P_{+\theta_1=0; +\theta_2=\pi/4} + P_{+\theta_1=\pi/4; +\theta_2=\pi/2} = \sin^2(\pi/8) = .1464, \qquad (10.18)$$

while the left-hand side is

$$P_{+\theta_1=0; +\theta_2=\pi/2} = \frac{1}{2}\sin^2(\pi/4) = .2500. \qquad (10.19)$$

The inequality is clearly violated. Since it is violated by an amount of order 1, no quibbling about imprecision of measurements will save us from this conclusion. If we insist on sticking to a formalism in which there are probabilities for histories, one in which the statistical fluctuations in the observed properties of particles are due to interactions with some "hidden variables" whose state we do not observe, then those hidden variables must mediate interactions between particles at space-like separation, in order to reproduce the quantum results. At the time that Bell first announced these inequalities, experiments where the particles were sufficiently space-like separated had not yet been done. Brilliant work by Aspect and others [31] soon closed this loop hole. Quantum predictions are correct, and their interpretation in terms of some sort of classical statistical theory must involve bizarre nonlocal interactions.

Anyone who has followed the exposition of QM in this book knows what is wrong with the classical statistical reasoning that led to Bell's inequality. QM produces probability distributions for every normal operator in Hilbert space. But it is impossible to find a state in which all operators have definite values. Operators that do not commute cannot take on definite values in the same state. This is the content of the generalized uncertainty relation. Our computations were done in a state where the operator $\sigma_3^{(1)}\sigma_3^{(2)} = -1$ and the operators $\sigma^{(1)} + \sigma^{(2)} = 0$. Other combinations of Pauli matrices, such as the spins at various angles, have only a probability distribution. The meaning of those distributions is that if we set up a correlation between eigenvalues of one of those uncertain operators, and the collective coordinates of a macroscopic system, then, by doing repeated runs of the experiment and keeping track of the history of the macrosystem (which has an emergent meaning, married only by terms in the quantum predictions that are exponentially small in the number of atoms in the system), the quantum distribution reflects the frequency with which the macrodetector will respond in a certain manner. The basic assumption of the classical analysis, that we can assign values to all variables, independent of their coupling to a macrosystem, is simply wrong. QM is an intrinsically statistical formalism, which only allows for the classical statistical notion of probabilities for histories as an emergent concept valid in limited circumstances, which we have outlined in the early parts of this chapter.

Bell's inequality shows that any attempt to duplicate the results of QM with some sort of classical statistical framework will require the introduction of interactions that are nonlocal in space and violate the principle that signals cannot travel faster than light. One might ask whether Bell's result implies some sort of nonlocality in quantum theory itself, and many popular accounts claim that that is the case. This is simply wrong. Although we cannot explore it in this book, the formalism of quantum field theory allows for a straightforward generalization to take into account the constraints of special relativity and causality [17]. Roughly speaking, all one has to do is to take field equations covariant under Lorentz transformations, and quantize them following the prescriptions in this book. The fundamental postulate of these relativistic theories is that operators can be localized in regions of spacetime, and that the operators in two space-like separate regions commute with each other. This implies that one can never set up EPR correlations, which will allow for superluminal propagation of information.

There is an even simpler exposition of the basic quantum vs. classical conundrum revealed by Bell's inequality. This is due to Greenberg Horn and Zeilinger (GHZ), though the version given here is due to David Mermin [32]. Extensive references to the literature on these examples can be found in Mermin's article. The example is a simplification of earlier constructions by Kochen and Specker, and by Peres. Consider three independent sets of Pauli spins, which we can consider to be the spins of a widely separated triplet of spin 1/2 particles. We want to ask the question of whether there could be some classical statistical explanation found for the apparently random values found in experiments measuring various components of the spin. So, we imagine that there are some other "hidden variables," and that if we knew the values of those variables, we would have a definite prediction for the numerical value of every normal operator on the eight-dimensional spin space. We will call $v(A)$ the value of an operator A without labeling it by the values of the hidden variables. Our argument will work for any choice of hidden variables.

QM predicts that mutually commuting operators can all have simultaneous values. Let us consider the following group of 10 operators (we label the vector components of the spins by x, y, z, and the particle labels by $1, 2, 3$.

$$\sigma_x^1 \sigma_y^2 \sigma_y^3, \ \sigma_x^1, \sigma_y^2, \ \sigma_y^3. \tag{10.20}$$

$$\sigma_y^1 \sigma_x^2 \sigma_y^3, \ \sigma_y^1, \sigma_x^2, \ \sigma_y^3. \tag{10.21}$$

$$\sigma_y^1 \sigma_y^2 \sigma_x^3, \ \sigma_y^1, \sigma_y^2, \ \sigma_x^3. \tag{10.22}$$

$$\sigma_x^1 \sigma_x^2 \sigma_x^3, \ \sigma_x^1, \sigma_x^2, \ \sigma_x^3. \tag{10.23}$$

We have written 16 operators above but there are duplications and you can check that there are only 10 independent ones. The reason for arraying them as above is that each row consists of four mutually commuting operators. QM insists that these can all have simultaneous values, and furthermore, that the product of the values in each row is 1. Any assignment of values to the 10 operators, for fixed values of the hidden variables must satisfy this constraint, or disagree with QM about the results of actual experiments. QM asserts in addition that each of these 10 operators can take on only the values ± 1.

On the other hand, the four triple products of spins all commute with each other as well and their product is -1. Thus, we must have

$$-1 = v(\sigma_x^1 \sigma_y^2 \sigma_y^3) v(\sigma_y^1 \sigma_x^2 \sigma_y^3) v(\sigma_y^1 \sigma_y^2 \sigma_x^3) v(\sigma_x^1 \sigma_x^2 \sigma_x^3) \tag{10.24}$$

$$= v(\sigma_x^1) v(\sigma_y^2) v(\sigma_y^3) v(\sigma_y^1) v(\sigma_x^2) v(\sigma_y^3) v(\sigma_y^1) v(\sigma_y^2) v(\sigma_x^3) v(\sigma_x^1) v(\sigma_x^2) v(\sigma_x^3) = 1. \tag{10.25}$$

The last equality follows because the value of each individual spin appears twice. This contradiction is avoided in QM because the different mutually commuting sets do not commute with each other, and so cannot have simultaneous values.

In a hidden variable theory, one can predict probabilities for histories of the combined quantum and hidden variables. In each of those histories, *all* QM operators are supposed to take on values compatible with their allowed values in QM, and satisfying some of the relations that QM predicts. Since QM has states in which all operators in a commuting set take on definite values, it seems like a minimal requirement that the values assigned to these operators at some time, in a particular history, satisfy relations like those above, which are true in *any state*, for each possible collection of commuting operators. We have just seen that this is impossible in a Hilbert space with eight states,[6] and so it is impossible in any larger Hilbert space.

The EPR/Bell arguments are based on a somewhat different criterion of what a minimal set of requirements for a sensible hidden variable theory must be. They emphasize the requirement that one's theory should not allow information to be transferred faster than light. Bell's inequalities show that this is impossible, and the three spin example above demonstrate this in an even simpler manner, if we assume that the three spins are far removed from each other in space. Let us assume we are in a state such that

$$\sigma_x^1 \sigma_y^2 \sigma_y^3 |\psi\rangle = \sigma_y^1 \sigma_x^2 \sigma_y^3 |\psi\rangle = \sigma_y^1 \sigma_y^2 \sigma_x^3 |\psi\rangle = |\psi\rangle. \tag{10.26}$$

This is possible because the three operators commute and each has eigenvalues ± 1. The product of the three operators is $-\sigma_x^1 \sigma_x^2 \sigma_x^3$, so $|\psi\rangle$ is also an eigenstate of this operator, with eigenvalue -1. Now imagine that the statistical ensemble defined by $|\psi\rangle$ allowed for the concept of histories of all of the operators $\sigma_a^i(t)$ plus some hidden variables $H_A(t)$. The probabilities are obtained by averaging over the hidden variables.

The state $|\psi\rangle$ has correlations between the spins of the distant particles. One knows the value of the y component of any one particle's spin if one knows the x component of one of the

---

[6] In fact it is already impossible for three states, though the proof is much harder. There is a simple four state analog of what we have presented [32], but that example does not lead to a Bell/EPR paradox with locality.

far away particles and the y component of the other. Such nonlocal correlations in space can only be created, if there is no superluminal propagation of information, by events in the past history of the system, and in a hidden variable theory, each complete past history gives us definite values for all of the components of all three spins. The apparent quantum fluctuations are supposed to come from averaging over the unmeasured behavior of the hidden variables.

The values of the spin components of each individual particle can be established by local measurements, unlike, for example, the operator $\sigma_x^1 \sigma_y^2 \sigma_y^3$. Call these values s_a^i. Furthermore, these values, in any particular history which contributes to the statistical ensemble defined by $|\psi\rangle$, must satisfy all of the exact relations between mutually commuting variables that the quantum variables satisfy. Thus,

$$s_x^1 s_y^2 s_y^3 = s_y^1 s_x^2 s_y^3 = s_y^1 s_y^2 s_x^3 = 1 \tag{10.27}$$

and

$$s_x^1 s_x^2 s_x^3 = -1. \tag{10.28}$$

These are inconsistent, because the product of the four triplets of spins is equal to one since each spin is ± 1.

The conclusion of discussions of Bell's theorem is usually stated as "there can be no local hidden variable theory," implicitly suggesting that some kind of nonlocal hidden variable theory, consistent with the idea of probabilities for histories of microscopic variables might be found in the future. The GHZ form of Bell's theorem seems to rule out this possibility. It says that even if we restrict attention to variables that we know can be measured, the spin components of localized particles, we cannot consistently assign values to those variables in any particular history, which are consistent with the values that QM assigns to collections of mutually commuting (but nonlocal) operators, which we can construct as products of the local variables. We have demonstrated this contradiction in a particular state, and there are a variety of other states where the same contradiction exists. In more general states, one must resort to inequalities of Bell's type to find a contradiction.

In an Appendix, we will examine a variety of interpretations of QM, which purport to make the theory more compatible with our intuition. None of them resolve the paradoxes above.

10.7   THE RELATION BETWEEN THEORY AND EXPERIMENT IN QM

In our classical theory of the low energy ammonia molecule, the theory of measurement is so trivial as not to require discussion. The system has only two states, which are completely determined by the value of the quantity σ_3. Even if we have some uncertainty about the initial state, the probability distribution $p(\sigma_3)$ is completely determined by doing repeated measurements of σ_3. On the other hand, in the quantum theory, the general pure state is of the form

$$\alpha_1|+\rangle_3 + \alpha_2|-\rangle_3, \tag{10.29}$$

where (α_1, α_2) is a complex unit vector in two dimensions. In general, a choice of initial state does not make a prediction for the value of σ_3. Repeated measurements of identically prepared states, measuring only the value of σ_3, only serves to determine $|\alpha_i|^2$ and do not determine the relative phase. The reason for this, as we have just discussed, is that the term "measurement of σ_3" means entanglement of eigenstates of σ_3 with the values of pointer variables of a macroscopic system. By definition, pointer variables are shared by large ensembles of states, with the property that the overlap between states with different values of the pointer variable are doubly exponentially small. In the entangled state, it is, in principle, impossible to recover the phase difference between α_1 and α_2. Thus, in order to determine experimentally what the quantum state is, one must not only do repeated experiments, but also measure different noncommuting variables. Remarkably, this is a problem even if the initial state is almost an eigenstate of σ_3. Consider the state

$$\epsilon|+\rangle_3 + \sqrt{1 - |\epsilon|^2}|-\rangle_3, \tag{10.30}$$

where ϵ is a very small complex number. Measurements of σ_3 are expected to give -1, with probability $1 - |\epsilon|^2$. That means that we have to do of order $(1 - |\epsilon|^2)^{-1}$ measurements of σ_3 in order to determine that $\epsilon \neq 0$. On the other hand, measurements of σ_1 will give a nonzero average of order $|\epsilon|$, *which depends on the phase of ϵ*. Thus, it is almost never possible to actually determine the quantum state of a system by measuring only a single variable.

This discussion might make one skeptical of the possibility of ever preparing an initial quantum state in a reproducible way. In fact, for quantities like the electric dipole of the ammonia molecule, this is not an issue, because of rotation invariance. Given the assumption that one can build an apparatus that can entangle the σ_3 eigenstates with macroscopic pointers, rotation invariance implies that we can do the same for every operator $\mathbf{n} \cdot \sigma$. This then gives us a method for preparing eigenstates of $\mathbf{n} \cdot \sigma$ for any direction in space, by simply rotating our measuring apparatus, and throwing away runs of the experiment in which the wrong value appears. Such *polarizers* have been known for photons since the mid-19th century. The Stern–Gerlach experiments described in Chapter 7 show us how to prepare quantum spin states of neutral molecules. Modern experiments in high energy physics have managed to prepare quantum states of the spin of the electron.

Theoretical physicists often talk as if we can do the same thing for every Hermitian operator in an arbitrarily complicated quantum system. This is highly unlikely to be true. Nonetheless, it is tautologically true for any operator that *can* be measured. To say this more precisely, A is an operator in some quantum system, and that system interacts with a macrosystem in such a way that the joint system evolves from $\sum \alpha_i |a_i\rangle \otimes |Ready\rangle$ to

$$\sum \alpha_i |a_i\rangle \otimes |P_i >, \tag{10.31}$$

where $|P_i\rangle$ are states in the ensembles with values P_i for collective coordinates of the macrosystem. We have seen that the quantum predictions for the probability distributions of such collective coordinates are, with incredible precision, consistent with Bayes' rule and the notion of probabilities for histories of the P_i. As a consequence, it is entirely consistent with QM to consider a given run of an experiment, and define probabilities for the future of that run conditioned on the value of P_i that appeared in that run. This procedure prepares the state of the microsystem in the eigenstate $|a_i\rangle$ of A.

Note that while we have used the word "measured" in the previous paragraph, no actual laboratory, and certainly no conscious observer, is necessary to such state preparation. All that is needed is a *Schrödinger's cat* correlation between a property of the microsystem, and the values of a collective coordinate of some macrosystem. A lot of the frustration of critics of QM with the standard Copenhagen interpretation of the subject[7] has more to do with the use by the founders of the subject, of words like measurement, and observable rather than with the actual physical content of that interpretation. Readers wishing to find a more learned discussion of these issues along the lines presented here, but with a wealth of extra detail and many references, would profit from the recent review article by Hollowood [33]. For the purposes of this textbook, it is now time to leave these interpretational issues and return to the use of QM to describe phenomena in the world in which we live.

10.8 EXERCISES

10.1 Analyze Feynman's description of the double slit experiment from the point of view of this chapter. Is the wave phenomenon shown pictorially in Feynman's lectures a real wave, or a summary of the results of repeated double slit experiments with identical initial conditions? Give a careful explanation of why the interference pattern disappears when one, in each of those experiments, places a detector in the system with the resolving power to tell which slit the electron went through.

10.2 Hyperion is one of the satellites of the planet Saturn. It is an irregular shaped rock of size about 140 km, and its intrinsic rotational motion is chaotic. The tidal forces exerted by Saturn and its other moons make the motion unpredictable. A simplified model Hamiltonian for the motion of Hyperion was given by Wisdom et al. [35]:

$$H = \frac{L^2}{2I_3} - \frac{3\pi^2}{T^2}\left(\frac{a}{r(t)}\right)^3 (I_2 - I_1)\cos(2\phi - 2\theta(t)). \tag{10.32}$$

---

[7] I feel that the account of QM in this textbook follows the Copenhagen interpretation, with additions by the inventors of the theory of decoherence.

The I_i are the moments of inertia, with $I_2 - I_1 > 0$, a is the semimajor axis of the orbit around Saturn and $r(t), \theta(t)$ describe the orbit, and are considered to be fixed functions, determined from the observed orbit of the satellite. L is the third component of the internal angular momentum of Hyperion. If R is the average radius of Hyperion and we define $x = R\phi$ and $p = \frac{L}{R}$, then because R is large (~ 300 km), p is almost continuous, even in QM.

Chaos of the classical motion implies that if we take a disk in x, p space, it turns into an "amoeba" of equal area in a time of order t_c. The length of the boundary of the disk grows like $e^{\frac{t}{t_c}}$. For Hyperion, the orbital period T is a few days and t_c is about 100 days. In the classical theory, an initial probability distribution becomes very stretched out in the x variable in a time of order t_c.

In QM, we can define coherent states of width δ

$$|x, p\rangle = N \sum_{m=-\infty}^{\infty} e^{im\phi} e^{-\frac{\delta^2}{\hbar^2}(p - \frac{m\hbar}{R})^2}. \tag{10.33}$$

If δ is small, these states have small position uncertainty. N is a normalization constant, which guarantees that these are normalized states. Compute it.

If we take the momentum uncertainty in this quantum state to be of order the classical thermal fluctuations in momentum on the surface of Hyperion, then $\delta = \frac{\hbar}{\sqrt{mT}}$. For Hyperion, $m \sim 10^{19}$kg $\sim 10^{55}$ eV, and $T \sim 10^{-2}$ eV. Thus, the position uncertainty in such a state is tiny, while the momentum uncertainty is what one might have expected from classical physics. Such a state is quasiclassical, while a state with large δ has large quantum fluctuations of the position coordinate.

Define a phase space density for any quantum wave function by

$$\rho_{ph} = |\langle x, p|\psi\rangle|^2. \tag{10.34}$$

Compute this density for $|\psi\rangle = |x', p'\rangle$, where both states have the same δ.

If the quantum motion stayed semiclassical for all times, as happens for harmonic oscillator quantum states, then we would remain in a coherent state with small δ. We could view the initial phase space distribution as representing classical uncertainty, i.e., lack of knowledge about which coherent state the system was in, which would then propagate forward in time using the classical equations for the coherent state parameters. For a chaotic system, this is not consistent. Compute the time t_q for the classical uncertainty of the position to grow to size R, its maximal allowable value. At that time, either the classical and quantum phase space densities are very different, or the system is no longer in a coherent state with small δ. Since the experimentally observed distribution does indeed spread, we conclude that Hyperion must be in a superposition of many small δ coherent states at time t_q. Compute t_q and show that it is short enough that the actual satellite we observe, must in fact be in such a quantum superposition. Argue that, because the phase space variables x, p are collective coordinates of a very large object, the quantum predictions of the superposition of coherent states are almost identical to those of a classical statistical model. Estimate the size of the interference corrections to those predictions. More details on this discussion of the dynamics of Hyperion can be found in [33] and the references cited there.

CHAPTER 11

Sketch of Atomic, Molecular, and Condensed Matter Physics

11.1 INTRODUCTION

In this chapter, we will sketch the most important applications of quantum mechanics (QM), to atomic and molecular physics, and the theory of complex materials, also called condensed matter physics. These are huge subjects, and most of the literature on them is devoted to either sophisticated approximation schemes or advanced theoretical discussions of special corners of the field like high-temperature superconductivity or the fractional quantum hall effect. In an elementary text, we can do no more than outline some major topics and ideas, without much computational or theoretical depth.

We will begin by introducing the basic Bohr–Rydberg units, which control the energies and length scales of all of these problems. Once we have written everything in terms of dimensionless variables, the only large parameters in the system are the ratios between the nuclear masses and the electron masses, as well as the charges of any large Z nuclei in the material. In the limit where the latter are large, it is intuitively clear that only the last few energy levels of the atom, where the electrons see only a screened Coulomb field, can participate in any sort of collective low-energy excitation of the system. These are called the *valence* electrons.

The Born–Oppenheimer approximation takes into account the lowest order in an expansion of the energy in powers of the ratio of the masses of the electron and the nuclei of atoms, numbers which are all less than 10^{-3}. At leading order, the nuclei are frozen into fixed positions, which are determined by minimizing the sum of their mutual Coulomb repulsion and the ground state energy of electrons for fixed nuclear positions. This gives rise to the notion of classical shapes for molecules and crystals. Many low-lying excitations of the system can be understood as rotations of the entire molecule or vibrations of the nuclear positions around their minimum values. These have quantized energy levels with spacings of order $\frac{m_e}{m_N}$ and $\sqrt{\frac{m_e}{m_N}}$, in Bohr–Rydberg units, respectively. In crystals, the vibrational excitations of long wavelength can have much lower energy.

Solving the problem of interacting electrons in an external potential is hard, and a number of sophisticated approximation methods have been invented to solve it. The most commonly used method, which is amenable to explanation at the level of this book, is Density Functional Theory (DFT). We will give a rough sketch of DFT. We will also introduce a simple model to explain the band structure of crystalline solids and the distinction between conductors, insulators, and semiconductors. Finally, we will give the reader a hint of Landau's Fermi Liquid theory, a simple model with a remarkably wide range of applicability.

11.2   THE BORN–OPPENHEIMER APPROXIMATION

In the nonrelativistic limit, the Hamiltonian for charged point particles interacting via electromagnetic fields is given by

$$\sum_i \frac{\mathbf{P_i^2}}{2m_i} + \frac{e^2}{8\pi\epsilon_0} \sum_{i \neq j} \frac{Z_i Z_j}{|\mathbf{R_i} - \mathbf{R_j}|}, \tag{11.1}$$

where $-Z_i$ is the charge of the i-th particle in units of the electron charge, and m_i its mass. In atomic, molecular, and condensed matter physics, the relevant particles are the nuclei of the atoms in question, and the electron.

The simplest system of this type is the hydrogen atom, and we have seen that it defines a characteristic energy scale, the Rydberg $|E_1| = \frac{\hbar^2}{2m_e a_B^2} = 13.6057$ eV, and a characteristic length scale, the Bohr radius $a_B = \frac{4\pi\epsilon_0 \hbar^2}{m_e e^2} = 5.29177 \times 10^{-11}$ m. e is the charge of the electron in SI units. Using these scales, we can rewrite the Schrödinger equation for any condensed matter system in terms of dimensionless variables, which we denote by $\mathbf{r_i}$. The dimensionless canonical momenta satisfy

$$[r_a, p_b] = i\delta_{ab}. \tag{11.2}$$

The rescaled Hamiltonian operator is

$$H = \sum \frac{m_e}{m_i} \mathbf{p_i}^2 + \sum_{i \neq j} \frac{2Z_i Z_j}{|\mathbf{r_i} - \mathbf{r_j}|}. \tag{11.3}$$

Its eigenvalues are energies in Rydberg units.

In a neutral condensed matter system, with N electrons, we have $\sum Z_I = N$, where Z_I are the nuclear charges. The kinetic terms of the nuclei are suppressed by the small numbers $\frac{m_e}{m_I}$. This suggests the following approximation, due to Born and Oppenheimer [36]:

- Fix the nuclear positions r_I and solve the Schrödinger equation for the electrons, finding the ground state energy for fixed positions $E(r_I)$.

- Minimize $E(r_I)$ w.r.t. the positions of the nuclei plus the Coulomb repulsion of the nuclei.

- Expand $E(r_I)$ around the minimum, obtaining a system of coupled harmonic oscillators. The kinetic terms of the oscillators are of order m_e/m_I and the potential terms are $o(1)$. Therefore the frequencies and the quantized energy levels of the nuclear motion are of order $\sqrt{m_e/m_I} < 2 \times 10^{-2}$ in Bohr/Rydberg units. This is a small perturbation to the value of $E(r_I)$ at the minimum.

- We also find that the wave functions of the nuclei are Gaussian, centered around the equilibrium positions, with width $\sim \sqrt{m_e/m_I}$, so that we can think of the nuclear positions as roughly fixed, giving the system a "classical shape." This is the origin of the little ball and stick models of molecules and solids, which used to populate chemistry labs. Recall how these models led us to our first view of QM in the physics of the ammonia molecule.

- Depending on the shape of the atom, molecule, or solid,[1] there may be other energy levels whose energy is much lower than $\sqrt{m_e/m_I}$. These levels are related to global rotations and translations of the system.

We see that much of the quantum physics of nonrelativistic systems, interacting solely via electromagnetism, is related to solving the problem of electrons interacting with themselves, and with an external potential generated by the nuclei, viewed (approximately) as point sources. To first approximation, we evaluate the ground state energy of the electrons for fixed nuclear positions and minimize the resulting *Born–Oppenheimer potential*, $V_{BO}(\mathbf{r_I} - \mathbf{r_J})$ with respect to the nuclear positions. This results in a "classical shape" for the system. Corrections to this are calculated by expanding the Born–Oppenheimer potential around its minimum, leading to system of coupled oscillators with a Hamiltonian

$$H_{nuc} = \sum_I (m_e/m_I)\mathbf{p_I}^2 + \sum_{I,J} K_{Ia,Jb}(r_I - r_I^*)^a (r_J - r_J^*)^b, \qquad (11.4)$$

where the stars indicate the positions of the minima. a, b are the spatial components of the vectors in parentheses. The frequencies of classical motion of the normal modes of these oscillators are obtained by diagonalizing the matrix

$$[\Omega^2]_{Ia,Jb} = \left(\frac{m_e}{m_I}\right)^{1/2} K_{Ia,Jb} \left(\frac{m_e}{m_J}\right)^{1/2}, \qquad (11.5)$$

---

[1] We are talking about the ground state of the system, for which almost all systems of large numbers of atoms are in the solid phase.

and are of order $\omega_{typical} \sim (\frac{m_e}{m_I})^{1/2}$ in Bohr/Rydberg units. The quantized energy levels are half integer multiples of these frequencies. The ratio $(\frac{m_e}{m_I})^{1/2}$ varies between .022 for Hydrogen and $\sim .0014$ for U_{238}.

Atomic energy differences, of order a Rydberg, correspond to emission of light in the visible to ultraviolet part of the spectrum and can range into the X-ray region for atoms with fairly large nuclear charge.[2] This is the reason that we are able to identify elements, even on distant stars, in terms of their visible spectra. It also indicates that the physical mechanism responsible for vision in most animals must involve excitations of electronic energy levels in the atoms of the visual receptors.

In contrast, the vibrational spectra of molecules are smaller by a factor $\sim .01$ and correspond to infrared (IR) light, which we feel as heat. This makes molecules potent agents in the heat balance of planetary atmospheres. Solar radiation striking the rocky surface of the inner planets is reradiated into the atmosphere, mostly as IR radiation because higher energy photons are efficiently absorbed in solids that are not transparent (this is the fundamental physics behind the colloquial notion of transparency). That IR radiation will either propagate out into space, or be absorbed by excitation of the vibrational levels of gases in the atmosphere. It turns out that certain molecules, like CO_2 and methane, have vibrational level spacings that correspond to much of the IR radiation from the planet's surface, and so act as potent *green house gases*. Changes in the atmospheric concentrations of these green house molecules will thus have dramatic effects on the energy content of the atmosphere, which determines both the average temperature on the surface and the amount of energy available for the creation of storm systems.

11.3 COLLECTIVE COORDINATES, ROTATIONAL LEVELS, AND PHONONS

Some of the oscillation frequencies of molecules and atoms are in fact identically equal to zero if the system is isolated in infinite space. The original Hamiltonian of interacting electrons and nuclei is invariant under simultaneous rotations and translations of all particle coordinates. The Born–Oppenheimer potential, as a consequence, shares these symmetries. That is, the potential is constant in the multinucleon configuration space, along the directions where all the nuclear positions are translated or rotated simultaneously. These flat directions correspond to zero frequency oscillations in the expansion of the potential around its minimum.

Viewed more globally, beyond the small oscillation approximation, the motions along the flat direction of the potential are free motion and free rotation of the entire system. The energies associated with collective translational and rotational motion come from the kinetic term $\sum (m_e/m_I)\mathbf{p_I}^2$. The translational energy can be set to zero, by making a Galilean boost

---

[2] Recall that binding energy of the ground state of a heavy atom scales like Z^2 in Rydberg units.

to the rest frame of the system. The rotational motions come from $H_R = (m_e/m_I)I_{ab}^{-1}L_aL_b$, where I_{ab} is the moment of inertia tensor calculated from the classical Born–Oppenheimer shape of the system and L_a the components of total angular momentum in the rest frame. The energy levels are quantized, and we can understand the scale of the quantization by examining the symmetric case where $I_{ab} \propto \delta_{ab}$. In this case, the rotational Hamiltonian is proportional to the square of the angular momentum operator, whose eigenvalues are $l(l+1)$. Thus, the rotational splittings are down by another factor of $\sqrt{m_e/m_I}$ compared to the vibrational levels. For many molecules, including water at temperatures close to room temperature, this gives photon frequencies in the microwave energy regime. The ubiquitous microwave ovens, which inhabit most of our kitchens, operate primarily by exciting the rotational levels of the water in various foods.

11.3.1 Water

The water molecule H_2O contains 10 electrons, so solving its Schrödinger equation is highly nontrivial, even in the Born–Oppenheimer approximation. There are many sophisticated calculations, which demonstrate that the B–O shape of the molecule lies in a plane, with a bond angle between 110 and 100 degrees, in good agreement with experimental data.

The rough scale of rotational energy levels of water is $\frac{m_e}{m_{H_2O}} \times 10$ eV, or about 10^{-4} eV, which is in the microwave range. Water vapor indeed accounts for much of the atmospheric absorption of electromagnetic radiation in this range. Absorption of microwaves by water is also the basic physical principle behind the microwave oven. Most of our foods contain a lot of water and the ovens excite the rotational levels of those water molecules, uniformly in the food sample. The dissipation of that rotational energy is what cooks the food.

Many of the most important properties of water, such as the fact that its solid form is less dense than the liquid, near the freezing point, or its solvent properties, have to do with the interactions between water molecules and are not understood on a quantitative level. However, it is clear that the relatively weak binding of hydrogen to oxygen in individual molecules is crucial. We know this because *heavy water*, where hydrogen nuclei are replaced by deuterium behaves very differently. The larger mass of the deuterium atoms means that their nuclei are more tightly bound to the oxygen atom. Indeed, there is experimental evidence that the bond lengths in heavy water molecules are shorter than those in ordinary water [41]. Note that in the Born–Oppenheimer approximation, the bond lengths are independent of nuclear masses, as long as they are large. It is possible that the correct explanation of these observations will involve a study of the effect of rotational levels of water on the scattering experiments described in [41], rather than a breakdown of the Born–Oppenheimer approximation. The size of the effect looks like something of order $\sqrt{m_e/m_{proton}}$. These experiments also show that the properties of the liquid state of heavy water are different than those of light water. Clearly, even for molecules as simple as water there is much to be done.

Indeed, replacing 25%–50% of a multicelled animals water content with heavy water leads to a variety of toxic effects, from sterility to breakdown of numerous important reactions essential to life. The effects are undoubtedly a result of the tighter binding, but detailed mechanisms have not yet been worked out. Remarkably, single celled organisms like bacteria seem to tolerate up to 90% replacement of their water supply by heavy water.

11.3.2 Phonons

For an isolated molecule, the translational collective coordinate is not very interesting. It just describes the free motion of the molecule. We can always use the Galilean symmetry of nonrelativistic physics, to study the system at rest, and obtain its properties in any other reference frame by using Galilean transformations. The internal energies are unaffected by the boost. However, there is a type of system for which excitations associated with translational invariance are important and interesting. These are solids, which are the *almost* ubiquitous state of matter at low enough temperature.

Consider a system of volume V in Bohr units, which contains N nuclei. We want to study the ground state of this system in the limit $V \to \infty$, with $\rho \equiv N/V$ fixed. The nuclei repel each other, but the electrons cancel that repulsion. The cancellation cannot be exact, because we have seen that nuclear wave functions are localized at length scales of order $\sqrt{m_e/m_I}$ in Bohr units, while electron wave functions around low Z nuclei are localized only to within Bohr radii. Even around high Z nuclei, only the innermost electrons are localized within $1/Z$ of a Bohr radius.

The rough picture of the ground state of such systems is that the electrons mostly cluster around individual nuclei, forming atoms or ions with small net charge, and the residual attractive forces between these atomic/ionic constituents form a stable structure. The simplest example is an ionic solid, like sodium chloride, in which ions of charge ± 1 form a regular cubic array. Thinking of the ions as classical point charges, this is clearly a minimum of the energy. We will discuss the Density Functional Formalism, which can estimate the corrections to this naive picture, in the next section.

This sort of crystalline structure exemplifies an interesting phenomenon in the limit $V \to \infty$ at fixed density. In that limit, the underlying Hamiltonian is invariant under continuous translations, *but the crystalline ground state is not.* If the crystal has a finite size, $V \to \infty$ at fixed but large N, then it has a translation collective coordinate, the position of its center of mass, and the true energy eigenstates are states where the entire crystal has fixed momentum. However, for the infinite crystal, there is no meaning to moving the system around. It takes up all of space, and has infinite mass.

Instead, the infinite system has a degenerate set of decoupled ground states, in which the position of any one nucleus is translated, with the entire lattice kept rigid. These states are decoupled in the sense that every localized excitation of one of them is orthogonal to all of the others. This phenomenon, in which a symmetry of the Hamiltonian of a system, is realized as a transformation which moves one to a decoupled ground state in a Hilbert space orthogonal to all local excitations of the original ground state, is called *spontaneous breaking of symmetry*. It is quite common and pervades much of both condensed matter and particle physics.

Now, let us think about the modes of oscillation of the nuclei around the stable crystal lattice. They are described by displacement fields $\mathbf{\Delta}(\mathbf{x}_I)$. $\mathbf{\Delta}(\mathbf{x}_I)$ is the amount by which the coordinate of the I-th nucleus is displaced from its lattice equilibrium position. If all the $\mathbf{\Delta}$'s are equal, then the energy of the displacement is zero, because this is just the translation to a new degenerate ground state. It follows that if $\mathbf{\Delta}$ varies very slowly over the lattice, these excitations must have an energy that goes to zero. This spectrum of low-energy vibrational excitations of a crystal is called the phonon spectrum. $\mathbf{\Delta}$, since it is small, will satisfy a linear equation of motion, which is invariant under lattice translations. We can Fourier transform the Heisenberg equation of motion for $\mathbf{\Delta}$ and get a harmonic oscillator equation, with a frequency $\omega(\mathbf{k})$ that depends on the wave number, and goes to zero as the wave vector goes to zero. This *dispersion relation* for the frequency has only the symmetries of the lattice. Thus, $\mathbf{\Delta}$ behaves like a quantized field. Phonons, the eigenstates of this field Hamiltonian, behave like particles, in the same way that photons do, and are our first example of what condensed matter physics call *quasiparticles*. Our system is made up of "fundamental" particles called electrons and nuclei, but it also contains collective excitations with particle like behavior, which do not have independent existence outside of the material whose low-energy behavior they characterize.

11.4 THE HYDROGEN MOLECULAR ION

We now turn to a description of the simplest system to which our general remarks about atoms and molecules apply: the hydrogen molecular ion. This consists of two protons and a single electron. In our theoretical treatment, they interact only via Coulomb forces. There are simple variations of this system in which one or both of the protons are replaced by a deuterium nucleus.

The Born–Oppenheimer potential for the hydrogen molecular ion is invariant under simultaneous translation or rotation of the positions of the two protons. As a consequence, it depends only on the distance between them. There is no potential energy for either the center of mass coordinate or the orientation of the relative position vector. In the exercises, you will show that the Hamiltonian for these variables is (in Bohr/Rydberg units)

$$H_{coll} = \frac{m_e \mathbf{P}_{cm}^2}{2m_P} + \frac{m_e}{m_P} \mathbf{L}^2. \tag{11.6}$$

The Born–Oppenheimer potential (which we will define to include the internuclear Coulomb repulsion) will depend only on r_p.

To compute the Born–Oppenheimer potential, we must solve the problem of an electron in the potential

$$V(\mathbf{x}) = \frac{-1}{\mathbf{x} - d\hat{\mathbf{z}}} + \frac{-1}{\mathbf{x} + d\hat{\mathbf{z}}}, \tag{11.7}$$

and find the ground state energy as a function of $r_p = 2d$. Note that we have chosen the separation between the protons to lie along the z axis. The electron Hamiltonian is invariant under rotations around the z axis and we can diagonalize the z component of orbital angular momentum L_3. It is also invariant under electron spin rotation. The energy is completely independent of the spin, and depends on L_3 through an additive term proportional to L_3^2, with a positive coefficient. The ground state is thus doubly degenerate in spin, and has $L_3 = 0$. The Hamiltonian reduces to

$$H = p_r^2 + p_z^2 - \frac{1}{\sqrt{r^2 + (z-d)^2}} - \frac{1}{\sqrt{r^2 + (z-d)^2}}. \tag{11.8}$$

In the two limits, $d \to 0$ and $d \to \infty$, this problem becomes exactly soluble. In the first limit, it is just the single electron ion problem, with a nucleus of charge 2. The $d \to \infty$ limit is a little more subtle. If we consider an electron wave function localized near $z = d$, then the problem reduces to that of the hydrogen atom in this limit. The same can be said for a wave function localized near $z = -d$. The correct limiting wave function for the ground state is the superposition of the two, with positive sign (see Exercise 11.1). The splitting between the positive and negative superpositions is exponentially small at large d.

We conclude that the electronic contribution to the Born–Oppenheimer potential ranges between -4 Rydbergs at $d = 0$ and -1 Rydberg at $d = \infty$. You will show in Exercise 11.3 that the increase with increasing d is monotonic. Thus, the electrons produce a net attraction between the protons. The argument in Exercise 11.3 depends on a theorem due to Feynman and Hellman, which we will prove in Chapter 13. This is an equation for the change of an energy eigenvalue under a change in some parameter λ in the Hamiltonian.

$$\partial_\lambda E(\lambda) = \langle E(\lambda) | \partial_\lambda H(\lambda) | E(\lambda) \rangle. \tag{11.9}$$

You will calculate the leading correction to the -1 Rydberg at large d in Exercise 11.2. The attraction cancels the protons' Coulomb repulsion only partially, leading to a repulsive force that falls off like $1/d^4$. The repulsion also dominates at very small d, but it turns out there is an intermediate regime where the attraction due to the electrons dominates, and there is a local minimum at a point d^*. We will estimate d^* in the exercises.

Thus, in the Born–Oppenheimer approximation, the system of two protons and an electron has a bound state, in which the distance between the two protons is d^* and the electron wave function is localized in an ellipsoidal cloud whose semimajor axis is of order a few times d^*. The protons bind into a hydrogen molecular ion by virtue of "sharing" the electron in order to lower the system's energy. Chemists call this kind of binding a *covalent bond*.

We can make a crude model of molecular hydrogen by simply placing two electrons in the same position space wave function as that found in the molecular ion, with their spins in the antisymmetric spin-singlet state $\frac{1}{\sqrt{2}}(|+-\rangle + |-+\rangle)$. This model neglects the effect of the electron–electron repulsion, on the shape of the electron wave function. A better model might be $\psi_{ion}(x_1)\psi_{ion}(x_2)f(\frac{|\mathbf{x_1}-\mathbf{x_2}|}{2a_{Bohr}})$, where ψ_{ion} is the single electron wave function in the hydrogen molecular ion, and f is a smooth function vanishing when the distance between the electrons is less than a few ionic Bohr radii, and is otherwise equal to 1. We will learn how to optimize the choice of unknown functions in an ansatz like this in the chapter on the Variational Principal.

11.5 DENSITY FUNCTIONAL THEORY

This section is meant to be a short summary of DFT rather than even an introductory formal treatment of it. It is intended to outline the logic of the modern approach to atoms, molecules, and solids. The Born–Oppenheimer approximation reduces nonrelativistic atomic, molecular, and condensed matter physics to the quantum problem of electrons interacting via the Coulomb potential, in a background external potential, which screens out the total electronic charge. We can view this as the Homogeneous Electron Gas—an artificial system with a constant background positive screening charge density, in an external potential. In the real world, that external potential is the sum of nuclear Coulomb potentials, but it is convenient to let it be a general function of position. If one can solve this problem for every $V(\mathbf{x})$, and calculate the ground state energy $E[V(\mathbf{x})]$, then one can calculate V_{BO} for the nuclei as a special case, and begin the process of minimizing and expanding about the minimum, which we sketched above.

No one knows how to solve the homogeneous electron gas exactly. For calculating V_{BO}, the perturbation series techniques we will develop in Chapter 13 show us that we only need to know how to calculate ground state expectation values of products of density operators

$$\langle \psi_0 | N(\mathbf{x_1}) \dots N(\mathbf{x_1}) | \psi_0 \rangle. \tag{11.10}$$

Here $|\psi_0\rangle$ is the ground state of the homogeneous electron gas with no external potential. The density operator is defined by the equation

$$N(\mathbf{x}) = \sum_p \delta(\mathbf{x} - \mathbf{x_p}), \tag{11.11}$$

where the sum is over all electrons. This operator measures the electron density at the point $\mathbf{x}$ in any state. Expectation values of products of it in the ground state of the homogeneous gas tell us about the expected value of the local electron density, as well as its fluctuations.

In two remarkable papers written in the early 1960s, Hohenberg et al. [21] showed that one could reformulate the calculation in terms of a functional of the expectation value of the density operator. To understand them, we will need to borrow some wisdom from Chapter 18. The variational principle discussed in that chapter shows that $E[V]$ is equal to the minimum, over all normalized states, of the expectation value of the Hamiltonian

$$H = H_{HEG} + \int d^3x \, \langle \psi | N(\mathbf{x})(V - \bar{V})(\mathbf{x}) | \psi_0 \rangle. \tag{11.12}$$

Here $\bar{V}$ is the integral of V over all space, and H_{HEG} includes electron kinetic energy, electron–electron Coulomb repulsion, and $\bar{V}$, a constant potential which cancels off the total electron charge. H_{HEG} treats the nuclear charge distribution as a smeared out homogeneous positive charge density.

The key observation now is that the term in the expectation value of the *energy*, which depends on $V - \bar{V}$, is sensitive to the quantum state of the electrons only through the expectation value of $N(\mathbf{x})$,

$$n(\mathbf{x}) \equiv \langle \psi | N(\mathbf{x}) | \psi \rangle, \tag{11.13}$$

in the test state $|\psi\rangle$.

This motivates a two step procedure for finding the minimum energy. First look only at states that have the same expectation value, $n(\mathbf{x})$ for $N(\mathbf{x})$ and minimize the expectation value of H_{HEG} among those states for fixed $n(\mathbf{x})$. This calculation defines the *density functional* $F[n(\mathbf{x})]$. Then the expectation value of the energy is

$$E[n(\mathbf{x})] = F[n(\mathbf{x})] + \int d^3x \, n(\mathbf{x})(V - \bar{V})(\mathbf{x}). \tag{11.14}$$

The ground state energy $E[V]$ is just obtained by minimizing this functional over all possible densities $n(\mathbf{x})$. This constrained search for the minimum reduces the problem of atomic, molecular, and solid state physics to a classical variational problem, once the density functional $F[n(\mathbf{x})]$ is known. What is remarkable about this result is that F is a quantity that can be calculated in the homogeneous electron gas, without reference to particular choices of the nuclei, which make up the substance of interest.

Of course, the problem of calculating F is not easily soluble. It was first approached by Kohn and Sham, using the following sequence of approximations. First one writes a term $F_0[n]$, which is the value of the functional F for a gas of electrons with no Coulomb repulsion. This term incorporates the constraints on the density that arise from the Heisenberg uncertainty principle. The high cost in electron kinetic energy suppresses densities that vary on short wavelengths. The second term in the K–S approximation to the density functional is called the Hartree term. It has the form

$$F_{Hartree}[n] = \int d^3x \, d^3y \, \frac{n(\mathbf{x})n(\mathbf{y})}{|\mathbf{x} - \mathbf{y}|}. \tag{11.15}$$

This term approximates the Coulomb repulsion between the electron density *operators* $N(\mathbf{x})$ by that between the expectation values, $n(\mathbf{x})$ of those operators in the lowest energy state at fixed $n(\mathbf{x})$. The Hartree term can be justified in certain variational approximations to the multielectron ground state wave function, which we will discuss in Chapter 18. The Hartree approximation was widely used at the time K–S wrote their paper, and computer codes had been written to explore its consequences. In a previous section, we showed how it gave qualitative understanding of the structure of atoms and solids. It was never a huge quantitative success.

K–S proposed to improve the Hartree approximation by adding a term

$$F_{LDA} = \int d^3x \; \epsilon(n(\mathbf{x})), \tag{11.16}$$

to the density functional. The function $\epsilon(n)$ is the ground state energy of the homogeneous electron gas for constant density n. It is not calculable analytically, but both high and low density expansions of it were known. Over the years, extensive numerical calculation [37] has given us very reliable estimates of this function over the whole range of densities. The initials LDA stand for *local density approximation*. F_{LDA} would be the entire density functional if spatial variation of the density did not exist in real substances. Roughly speaking, the K–S idea was that the dependence on spatial variations captured by F_0 and $F_{Hartree}$ was enough, together with F_{LDA} to get a good approximation to the density functional for all materials.

To facilitate the calculation of F_0, as well as to exploit existing computer codes for the Hartree approximation, Kohn and Sham wrote the density as

$$n(\mathbf{x}) = \sum_i \psi_i^*(\mathbf{x})\psi_i(\mathbf{x}).$$

The summation index i runs over the number of electrons, K, in the system and ψ_i are chosen to be K orthonormal single electron wave functions. This ansatz is motivated by the Hartree approximation, in which the multielectron ground state is approximated by a product of single electron wave functions. The ψ_i are called Kohn–Sham orbitals. One can show [38] that if we write

$$F_0[n] = \sum_i \psi_i^*(-\frac{\nabla^2}{2m} - \mu)\psi_i, \tag{11.17}$$

then the minimum of $F_0[n] + F_{Hartree}[n] + F_{LDA}[n] + \int V(\mathbf{x})n(\mathbf{x})$ with respect to the K–S orbitals is the same number and gives rise to the same minimizing density, as direct variation of the density functional w.r.t. to $n(\mathbf{x})$. On the other hand, as you will verify in the exercises, the variational equations with respect to the K–S orbitals are the same as the Schrödinger equation for the lowest lying states in a self-consistent potential. Only the equation for the self-consistent potential is different than the equation we wrote in Chapter 7, based on the Hartree approximation.

It turns out that the K–S equations do not, in most cases, give results accurate enough for the needs of chemistry and materials science. In the 1990s, a number of "semiempirical" terms were added to the K–S expression for $F[n]$ [39], which achieved the required accuracy. These terms were motivated by theoretical considerations, but contain a variety of free parameters. Those parameters are fixed by fitting of order 10 well-studied materials, and the resulting density functionals then give excellent results for a host of other systems. A systematic review can be found in [37].

DFT is a beautiful idea which, after empirically fit improvements, has become a powerful tool in the study of materials. We still do not have a systematic way of calculating the density functional from first principles, so there is a lot of both fundamental and practical work to be done in this field.

In recent years, a different approximation scheme, called Dynamical Mean Field Theory (DMFT), has challenged the DFT approach to condensed matter systems. Unfortunately, the basic concepts underlying DMFT would take up too much time and require a level of sophistication not expected of most students reading this book. Nonetheless, anyone interested in the application of QM to the physics of materials should be aware of the literature on DMFT [40].

11.6 ELEMENTS OF THE THEORY OF CRYSTALLINE SOLIDS

If we study a very large sample of a homogeneous collection of molecules,[3] we might expect the ground state of the system to be approximately invariant under spatial translation. Translation invariance is broken by the walls of the box containing the sample but if the size of the box is much larger than the Bohr radius, this is a very small effect. Nonetheless, most systems do not have translation invariant ground states.

Recall the separation of the electrons in such a system into valence electrons, and electrons bound to the nuclei making up each molecule. In the Born–Oppenheimer approximation, the nuclear positions are frozen. Each molecule is in some position. The energy is lowered by amounts of order the volume of the system by allowing all of the molecules to be bound together. One way to do this is to have different molecular ions with alternating positive and negative charges. These are called ionic solids, and ordinary table salt is a common example. If the separation between the positive and negative ions is too small, they will bind to form neutral clusters and we will not get a solid. If it is too large, we do not get much lowering of energy. The compromise is to have the ions sitting on some kind of regular lattice.

In our discussion of the hydrogen molecular ion, we have also encountered *covalent bonding*, a quantum phenomenon in which two molecules are bound together because an electron

---

[3] In this section, we will let the word molecule stand for either a bound state of some collection of atoms, or a single atom.

is in a superposition of states close to each molecule. Again, there is a characteristic separation, of typical atomic scale at which the molecules prefer to sit, in order to lower their energy. If we have a large collection of molecules, filling a box of volume V, the Born–Oppenheimer potential will be minimized for a lattice configuration of the molecules.

What is a "lattice"? The Hamiltonian of the electrons and nuclei is invariant under translations and rotations, which combine to form the three-dimensional Euclidean group. A lattice is a set of points in space, invariant under some discrete subgroup of the Euclidean group. A crystalline solid is a system of molecules whose ground state is invariant under such a discrete subgroup. It is clear that the first step in the study of crystalline solids is the mathematical problem of classifying all such discrete subgroups. There are 230 different types of such *space groups*, which are symmetries of possible crystal lattices in three dimensions. One of the primary applications of DFT and DMFT has been to determine the preferred symmetry group for particular substances.

In order to determine the Born–Oppenheimer potential, whose minimization determines the space group, one must solve the problem of valence electrons propagating in Coulomb field of nuclei sitting at the minima of the Born-Oppenheimer (BO) potential. This is, in principle, done by DFT/DMFT, but one also needs to understand that dynamics in order to understand the nature of low-lying excitations in the system. These determine the crucial *transport properties* of the substance: thermal and electrical conductivity, specific heat, *etc.*

The numerical cost and lack of intuitive insight that is involved in full scale DFT/DMFT determination of transport properties leads one to search for shortcuts. For a given lattice, there is a set of useful approximations, mostly developed prior to modern implementations of DFT/DMFT codes, which gives a determination of electron propagation sufficiently accurate to study transport properties. One first studies noninteracting electrons, propagating in an assumed potential with the right symmetries. This leads, as we will see, to something called the *band structure* of the material. The sophisticated numerical methods are necessary to find the details of the band structure, but we can understand its qualitative nature in simpler models. The energy level spectrum contains bands of states where the electrons are free to propagate throughout the material and other bands where they are locked near particular molecules. Depending on where the Fermi surface lies, the material will be a conductor or an insulator or a semiconductor.

For a given material, if it were exactly pure, the number of valence electrons would be completely determined by the geometry of the lattice of molecules, and the number of valence electrons per molecule. However, we can vary the density of valence electrons by a procedure known as doping.[4] One substitutes a density of other molecules at a sublattice of sites. This changes the average density of valence electrons, and gives us a slightly different material which may exhibit sharply different transport characteristics. In theoretical models,

---

[4] Which has *no* connection to the practice common in modern sports and horse racing.

one incorporates doping by adding a chemical potential term to the Hamiltonian for the valence electrons. One can also induce interesting phase transitions by varying the doping sites between a regular array and a random selection of lattice points.

Once one has determined the band structure, one uses what is known as the *tight binding approximation*. In crystals formed by covalent bonding, the electron wave functions are superpositions of states localized near particular molecules. One introduces a lattice approximation to the quantum field theory of electrons, in which there are creation and annihilation operators for electrons at fixed lattice sites, and one replaces the kinetic energy operator by

$$\int \psi^\dagger (-\nabla^2)\psi \to \sum \psi_i^\dagger K_{ij} \psi_j. \tag{11.18}$$

The *hopping matrix* K_{ij} has the symmetries of the lattice. Now one also reintroduces interactions. The most important of these is the remnant of the Coulomb repulsion between the electrons

$$H_{Hubbard} = \lambda \sum (\psi_i^\dagger \psi_i)^2. \tag{11.19}$$

The idea of this term, first introduced by Hubbard, is that the Coulomb repulsion between electrons separated by a lattice site or more, is mostly screened, but that Coulomb repulsion wants to keep electrons from occupying the same site. Note that the Hubbard interaction would vanish if the electron did not have spin, and that for spin one half electrons there are no higher order single site terms that we can add. In order to model the magnetic properties of materials, one often adds nearest neighbor interactions between electron spin densities. These are called Heisenberg interactions, and can explain ferromagnetism and antiferromagnetism.

Hamiltonians of the Hubbard type are known as *effective low-energy field theories*. They aim to characterize universality classes of low-energy thermodynamic and transport behavior. In principle, one would hope to derive them, or at least determine the values of parameters like λ for a particular substance, by calculations based on more ambitious methods like DFT/DMFT. In practice, much condensed matter physics is done by fitting parameters in Hubbard-like effective field theories.

Although it is important to understand the context of these ideas, much of the technical detail of the analysis of solids lies far beyond the scope of this book. We will be able to give just a smidgin of the techniques involved in calculating band structure, and nothing of the technicalities involved in analyzing the Hubbard model.

11.7   BAND STRUCTURE

Let us start with an intuitive picture that helps us to understand the existence of band structure in crystalline solids.[5] Consider a crystalline solid whose unit cells contain some

---

[5] I would like to thank S. Shastry for explaining this to me.

kind of molecule, and imagine introducing an external potential that can rescale the crystal so that the intermolecular spacing becomes larger or smaller. This is a mathematical device and does not correspond to something we can arrange using electrical forces in the real world. When the intermolecular spacing is made much larger than the Bohr radius, it is intuitively clear that the electrons will want to clump into separate neutral molecules. The system will be an insulator. Now bring the spacing down to something of order the Bohr radius. If we consider just two molecules we know, from our discussion of the covalent bonding of atoms into molecules, that there might be a possibility that the energy is lowered by putting some of the electrons into linear combinations of wave functions localized near the different molecules.

If we have a whole lattice of molecules, we can lower the energy even more[6] by using superpositions of wave functions localized near all of the molecules. That is, we look at wave functions of the form

$$\psi(\mathbf{k}) = \frac{1}{\sqrt{N}} \sum_{x_n} e^{i\mathbf{k}\cdot\mathbf{x_k}} \psi(\mathbf{x_k}). \tag{11.20}$$

The position space wave function $\psi(\mathbf{x_k})$ is the bound state wave function of the highest lying single electron orbital in the molecule. It is bound to the center of mass of the molecule and falls off exponentially with distance from the center. In principle, the allowed values of $\mathbf{k}$ are restricted only by the boundary conditions at the edges of the large crystal, so they form a quasi-continuum. The expectation value of the electron current in such an extended state can be nonzero, if we allow boundary conditions which allow current to escape from the sample. For some values of $\mathbf{k}$, such extended eigenstates of the single electron problem will exist, and for others they may not. However, since the $\mathbf{k}$ form an almost continuous set, small changes in $\mathbf{k}$ will *typically* not change whether or not the extended states exist. Small variations in $\mathbf{k}$ will give small changes in energy, so the extended states will lie in quasi-continuous energy *bands* separated by quasi-continuous *gaps* in energy, where no extended states exist.

Turning off our artificial external potential, the same statements are true for the single electron eigenstates in the self-consistent potential calculated from (some approximate form of) DFT. Note that there is nothing in the above argument, which requires the potential to be perfectly periodic, so the existence of band structure is robust, and will be valid in crystals doped with impurities, where we can vary the electron density. As we do so, we change the Fermi level and the crucial question is whether the Fermi level lies within a band, within a gap, or near the edge between a band and a gap.

Materials where the Fermi surface lies in a gap are insulators, while those where it lies in a band are conductors. In the latter, any small external electric field will induce a flow of current, because there are a large set of extended states with nonzero current, which lie close to the ground state and so can be excited by small external perturbations. Something interesting happens when the Fermi surface lies just above or just below the line separating

---

[6] Here we are implicitly invoking the variational principle.

a band from a gap. Near such a band edge, it is easy to move the Fermi level between conducting and insulating states by doping. These materials have high electrical resistance, but allow weak electrical currents to flow. These materials are called *semiconductors*.

If, in a semiconductor, the Fermi surface is just above a gap, then the number of available extended states is small. This means that we can think of the excitations of the ground state as a very dilute gas of individual electrons. The Pauli exclusion principle has little effect and a classical model of nonrelativistic electron propagation captures the transport properties very well. Note that this is very different from what happens for a metal, where the Fermi surface is deep inside the conduction band. There, the charge carrying excitations (see the section on Fermi liquids below) have only a single effective, almost continuous, momentum component, perpendicular to the Fermi surface, and their energy is linear in this momentum. For a semiconductor, the charge carriers are like single free electrons, except that the effective value of the mass does not have to be equal to the electron mass in empty space (the electron kinetic energy term is modified by interactions with the other electrons). This sort of semiconductor is called *n-type* because the charge carriers carry negative charge.

Surprisingly, when the Fermi surface is at the top of a conduction band, almost identical physics occurs. Although "all of the ground state electrons are in extended states," the ground state is stationary and no current flows. Current is carried by excitations of the ground state. At the top of a conduction band, there are no low-energy excitations, which carry negative charge. However, if we dope the material so that the Fermi surface is a little lower, we no longer have enough electrons to fill the entire conduction band. Instead, we have a small density of "missing electrons" or *holes*, which have very low energy, and can be in extended states. These will also behave like nonrelativistic particles, but will carry positive charge. This kind of material is called a *p*-type semiconductor.

Small amounts of doping can change the electrical conductivity of a semiconductor by factors of order 10^3–10^6 and this sensitivity makes them useful for all sorts of devices. Essentially, though this is a simplification, these materials can act as on–off switches for the flow of current. All of modern electronics is based on the properties of semiconductors [42].

Although our intuitive argument for the existence of band structure makes it clear that exact periodicity of the potential is not essential, much of the early literature on the subject is based on a mathematical theorem discovered by Felix Bloch. The group of symmetries of the lattice includes an abelian subgroup of translations. The unitary transformations implementing these symmetries on the Hilbert space of a single electron in a potential invariant under the symmetries have the form $e^{i\mathbf{K}\cdot\mathbf{a_n}}$, where $\mathbf{a_n}$ are the discrete translations that leave the lattice invariant. $\mathbf{K}$ is the wave number operator of Chapter 3. These operators can be diagonalized along with the Hamiltonian and so we can study the spectrum in each eigenstate independently. That is, we can restrict attention to wave functions satisfying

$$\psi(\mathbf{x} + \mathbf{a_n}) = e^{i\mathbf{k}\cdot\mathbf{a_n}}\psi(\mathbf{x}), \qquad (11.21)$$

where $\mathbf{k}$ is a fixed number, called the Bloch wave vector. The set of all $\mathbf{k}$ forms a three-dimensional torus, because if we shift $\mathbf{k} \to \mathbf{k} + \mathbf{r}$ where $\mathbf{r} \cdot \mathbf{a_n} = 2\pi m$ for any integer m, then the phase is left invariant. The discrete set of $\mathbf{r}$ satisfying this condition is called *the reciprocal lattice* and the torus of allowed wave vectors is called the (first) Brillouin zone.

11.7.1 A Simple Model

To get a feel for how band structure works, we will work out the Dirac Comb model, introduced by Griffiths [43], which captures the essential phenomenon. The original toy model for band structure, the Kronig–Penney model [44] is somewhat more complicated. We consider a one-dimensional problem, on a circle of circumference Na in Bohr units and a potential for the dimensionless coordinate $0 \le x \le N$, of the form

$$V(x) = \alpha \sum_{j=0}^{N-1} \delta(x - ja). \qquad (11.22)$$

The wave function satisfies the periodicity condition $\psi(x + Na) = \psi(x)$, which defines the Hilbert space on which the Schrödinger operator acts. a is a number of order 1 and $N \sim 10^{23}$, as appropriate for a macroscopically large crystal. The solutions of the free particle Schrödinger equation are e^{ikx}, with $k = \frac{2\pi n}{Na}$, where n is any integer. They form an almost continuous band in wave number space. We will see that the effect of the potential is to open up gaps in this spectrum.

In the region $0 < x < a$, we have the general solution to the free Schrödinger equation

$$\psi(x) = A\sin(kx) + B\cos(kx), \qquad (11.23)$$

where the eigenvalue in Rydberg units is $E = \frac{k^2}{2}$. For $-a < x < 0$, Bloch's theorem tells us that

$$\psi(x) = e^{-iKa}[A\sin(kx + ka) + B\cos(kx + ka)]. \qquad (11.24)$$

The Schrödinger equation tells us that the second derivative of the wave function is a delta function at $x = 0$, which means that the first derivative is a Heaviside step function and the wave function itself is continuous. Thus, we get two conditions

$$B = e^{-iKa}[A\sin(ka) + B\cos(ka)] \qquad (11.25)$$

and

$$2\alpha B = k(A - e^{-iKa})[-B\sin(ka) + A\cos(ka)]. \qquad (11.26)$$

We solve the first equation by eliminating A in terms of B

$$A\sin(ka) = [e^{iKa} - \cos(ka)]B. \qquad (11.27)$$

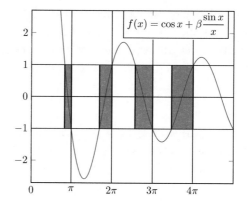

Figure 11.1 The eigenvalue equation for the Dirac Comb.

After substituting this into the second equation, we see that there is a common factor of B in every term, because the wave equation is homogeneous. Thus, we get a consistency condition relating the other parameters. After a bit of algebra, this simplifies to

$$\cos(Ka) = \cos(ka) + \frac{\alpha}{k}\sin(ka). \qquad (11.28)$$

The periodicity condition on the wave function is solved by $K = \frac{2\pi n}{N}$, where n is any integer $\leq N$. Thus, the allowed Bloch wave numbers practically form a continuum.

Now consider the solution of the eigenvalue condition for k given a fixed K. A typical value of n will be $\ll 2\pi N$, so the left-hand side of the equation is close to 1, and is certainly below 1 in absolute value. On the other hand, the function on the right-hand side is not bounded by 1. One can see the nature of the solution graphically in Figure 11.1.

As shown in Figure 11.2, this will cause gaps in k space where there are no eigenvalues. These are interspersed with *bands* of almost continuous eigenvalues. We described in the previous section how the positioning of the Fermi level of the material with respect to these bands leads to a qualitative explanation of the variations in the electrical conductivity of materials.

11.8 THE FERMI LIQUID THEORY OF CONDUCTORS

The first quantum theory of metals treated the electrons as a free gas enclosed in a box of size L, neglecting both the nuclei and the Coulomb repulsion between the electrons. The ground state is found by solving the free particle Schrödinger equation in a box, with energy levels $E = \frac{\hbar^2}{2m}(k_1^2 + k_2^2 + k_3^2)$ where each k_i runs between 0 and ∞ in units of $\frac{2\pi}{L}$. The restriction to positive values arises from the fact that we impose vanishing boundary conditions on the edges of the box, so that the solutions are sine waves. If we have N electrons with $N \gg 1$,

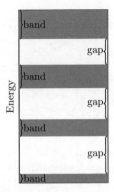

Figure 11.2 The energy spectrum for the Dirac Comb.

then the set of almost degenerate ground states fills up a volume in **k** space. Each allowed set of wave vectors sits at a point in a three-dimensional lattice and we can associate the cube in continuous **k** space formed by the three negative unit vectors emanating from that point. The volume of this cube is $\frac{\pi^3}{L^3}$. For large N, the volume filled up by the states occupied by all fermions is approximately an octant of a sphere of radius k_F with the Fermi wave vector given by

$$\frac{\pi k_F^3}{6} = \frac{Nq\pi^3}{2L^3},\tag{11.29}$$

where the factor of 2 is inserted in the denominator of the right-hand side to account for the fact that two electron spin states can fit in each state. q is the average number of valence electrons per atom, a number of order 1. The Fermi wave number is given by

$$k_F = (3\rho\pi^2)^{1/3},\tag{11.30}$$

where $\rho = \frac{Nq}{L^3}$ is the electron density. The corresponding *Fermi energy* is

$$E_F = \frac{\hbar^2}{2m}(3\rho\pi^2)^{2/3}.\tag{11.31}$$

The total ground state energy is given by integrating $\frac{\hbar^2 k^2}{2m}$ over the octant of the Fermi sphere (Exercise 11.8), resulting in

$$E = \frac{\hbar^2(3\pi^2 Nq)^{5/3}}{10\pi^2 m}L^{-2}.\tag{11.32}$$

The fact that this depends on the volume, and decreases as the volume increases, means that the gas exerts a *pressure* on the walls of the box, which can be computed via the first law of thermodynamics (we are in the ground state, so the temperature is zero): $dE = PdV$. In Exercise 11.9, you will verify that

$$P = \rho^{5/3} \frac{\hbar^2 (3\pi^2)^{2/3}}{5m}. \tag{11.33}$$

This is called *degeneracy pressure*. Its origin is quantum mechanical, a combination of the uncertainty principle allowing smaller electron energy quanta in a larger box, and the fact that there are more states below a given energy to put the fermions in when the box is larger. This degeneracy pressure is crucial to understanding the bounds on stability of white dwarf stars [45] (where however the motion of the electrons is relativistic), and is part of the story of why ordinary matter is stable.

In a more realistic treatment of conductors, one should apply the above reasoning to the extended electron states in a conduction band of the periodic Born–Oppenheimer potential. The formula for individual electron energies will be different, but we still label the states by an almost continuous wave vector $\mathbf{k}$ and an energy $\epsilon(\mathbf{k})$. The function $\epsilon(\mathbf{k})$ need have only the symmetries of the relevant lattice. If, as a function of $|\mathbf{k}|$ at fixed angle, ϵ is monotonically increasing, then there will be a Fermi surface $|\mathbf{k_F}|(\mathbf{\Omega})$, analogous to the Fermi sphere of the free electron system, delineating the maximum wave number at that angle, corresponding to a filled single particle fermion state in the multiparticle ground state.

The symmetry constraints on ϵ are weak, so we might imagine that shape of the Fermi surface is quite nonuniversal. Landau [46] argued, however, that many properties of the system depended only on the existence of the Fermi surface, and not on its shape. In particular, the low-energy excitations around the ground state are characterized by a single parameter, the distance from the Fermi surface, and the energy momentum relation for those excitations is (generically) linear in this parameter. One can extract a lot of information about the low temperature thermodynamics and the transport properties of the system from this fact alone. We will compute the low temperature specific heat of a general Fermi liquid in the next chapter. The "single particle excitations" of Landau–Fermi liquid theory should not be thought of as single electrons. They are really collective excitations of the entire solid, which have the quantum number of the electron. Landau gave these excitations the name *quasiparticle*.

Landau–Fermi liquid theory has turned out to work with extraordinary universality. It can be applied to any system build out of fermions, including those whose constituents are neutral fermionic atoms or molecules. In recent years a much better theoretical understanding of *why* Fermi liquid theory works so well [47] has been developed. An analysis too complicated to summarize in an introductory textbook, shows that almost all possible interaction corrections to a Fermi liquid picture, become weaker as one approaches the Fermi surface. The single exception is an attractive two body interactions between quasiparticles. As first shown by Cooper [48], such interactions lead to a bound pair of quasiparticles, which behaves very much like a boson, and Bardeen, Cooper, and Schrieffer (BCS) [49] showed that the fluid of these Cooper pairs undergoes Bose condensation. If the original quasiparticles are charged, like electrons, this leads to superconductivity. The new, superconducting ground state has an energy gap and is stable.[7] The elegant BCS theory of superconductivity is one of the triumphs of the application of QM to condensed matter systems.

In recent years, experimental condensed matter physicists have discovered more and more new materials, where the Landau–Fermi liquid paradigm fails. The attempt to discover a more general theory of the ground states of these exotic systems is still in its early stages. There is experimental evidence of new classes of universal low-energy behavior but no comprehensive understanding of the correct models, nor even of how many different universality classes might exist. This is one of the most exciting areas in theoretical physics. An entry point to the literature can be found in [50].

11.9 EXERCISES

11.1 Show that the ground state of the hydrogen molecular ion, in the Born–Oppenheimer approximation, for large d is an even superposition of hydrogen wave functions localized around the individual protons.

11.2 Calculate the leading large d correction to the Born–Oppenheimer potential for the hydrogen molecular ion. Show that it leads to attraction (the energy is lowered by making d smaller) and determine how it behaves as a function of d.

11.3 Prove that the ground state energy of electrons for fixed proton separation d is monotonic in d.

11.4 In terms of the Born–Oppenheimer potential $V_{BO}(d)$ (assumed known), and the equilibrium position d^* estimated in Exercise 11.7, calculate the spectrum of oscillations of the hydrogen molecular ion around its ground state.

11.5 Estimate the leading large d dependence of the Born–Oppenheimer force on the protons in the hydrogen molecular ion.

11.6 Given the Hamiltonian

$$H_{ion} = \frac{P_1^2}{2m_{proton}} + \frac{P_2^2}{2m_{proton}} + \frac{e^2}{4\pi\epsilon_0|\mathbf{X_1} - \mathbf{X_2}|} + E_R V_{BO}([X_1^z - X_2^z]/a_B),$$

---

[7] If the underlying quasiparticles are neutral, there is a gapless bosonic excitation, but a nonzero energy gap for fermionic excitations.

where E_R is the absolute value of the Rydberg energy and a_B the Bohr radius, identify the terms in the Hamiltonian that describe rotation of the molecule around its center of mass and estimate the quantum mechanical rotation energies.

11.7 Show that the balance between Coulomb repulsion and Born–Oppenheimer attraction leads to a minimum in the full potential for the protons.

11.8 Integrate $\frac{\hbar^2 k^2}{2m}$ over an octant of the Fermi sphere to show that the ground state energy of the free electron gas is

$$E = \frac{\hbar^2 (3\pi^2 Na)^{5/3}}{10\pi^2 m} L^{-2}.$$

11.9 Show that the pressure of the free electron gas is

$$P = \rho^{5/3} \frac{\hbar^2 (3\pi^2)^{2/3}}{5m}.$$

CHAPTER 12

Quantum Statistical Mechanics

12.1 INTRODUCTION

The key difference between classical and quantum statistical mechanics of particles has to do with identity of particles. In both classical and quantum mechanics (QM), one can write a Hamiltonian for an N particle system, which has an S_N permutation symmetry exchanging the particles. In both cases, we can consider states that are invariant under the symmetry, as well as states that are not. There is no fundamental principle, enunciated in our description of QM, which requires us to reject states that are not invariant under the symmetry group. Rather, the experimental facts of the world we live in require this rejection.

The first indication of this was found by Gibbs, in his discussion of the entropy of mixing between two boxes of gas. If a partition is removed between two boxes containing different gases, then an elementary calculation shows that the entropy increases. There are more states available. However, if we treat the molecules of a single gas as distinguishable particles, which is to say particles whose states[1] need not be symmetric, even though the Hamiltonian is, then exactly the same calculation shows that the entropy increases when we remove a partition between two identical boxes of gas. This does not agree with experimental observations on the thermodynamics of uniform gases.

The solution, in both classical and quantum mechanics, is to view particles as localized excitations of fields. In this interpretation, fields are fundamental, particles are special states of such fields, and the identity of particle states under permutation is a consequence of the fact that permutation of the particles gives the same field configuration. The details of how this works in classical and quantum mechanics are quite different. In classical mechanics, particle-like behavior implies that the state is eternally localized in space, and only special nonlinear field equations have such *soliton* solutions. In QM, however, we have seen that eigenstates

---

[1] Recall that in classical mechanics the state of a system is labeled by a point in its phase space, the space of all solutions of the equations of motion.

of the quantized field Hamiltonian, for linear field theories, have a particle interpretation. The field is an operator which adds or subtracts a particle to/from the system and the field configuration associated with a particle is interpreted as the matrix element of the field $\phi_s(x) = \langle 0|\phi(x)|s\rangle$ between the state $|0\rangle$ with no particles and a single particle state $|s\rangle$. For linear field theories, this is equal to the single particle wave function of the particle state, and the spreading of wave packets is understandable in terms of the fact that free particle states are not localized eternally, because of the Heisenberg uncertainty relations. $\phi_s(x)$ cannot be interpreted as a classical observable of the system. By contrast, coherent states, which are superpositions of states with different particle numbers, DO behave classically in the fully quantized theory.[2]

The lesson is that ALL particles in quantum field theory (QFT) are indistinguishable, and experiment shows that the particle-like objects we detect in the laboratory behave like the identical particles of QFT. From the field theory point of view, the S_N symmetry of the first quantized particle description of particles is a redundancy or *gauge symmetry*, morally similar to the Maxwell gauge symmetry of electrodynamics. The profound difference between ordinary and gauge symmetry transformations is illustrated by the difference between motion on the plane with a potential invariant under rotations by the discrete angle $2\pi/N$, and motion on a cone, with that opening angle. The cone is, mathematically, just the plane with points related by $2\pi/N$ rotations identified. We can study motion on the cone by studying motion on the plane, but we must be careful to choose only solutions of the equations (either classical or quantum) which are *invariant* under these discrete rotations. Similarly, if we consider perturbations of the free motion, then if we are working on the plane we can add an arbitrary potential, whether or not it obeys the symmetry, whereas on the cone, consistent perturbations *must* be invariant. Similarly, an arbitrary perturbation of a free QFT, will always produce multiparticle states of *identical* particles, which are invariant (except for fermionic minus signs) under permutations.

The parenthesis above reminds us that we have already discussed the other odd QM twist on field behavior, the existence of fermionic fields. The multiparticle states of particles created by those fields are NOT invariant under permutations of particles, but pick up a minus sign under odd permutations. This is allowable in QM if all *physical* operators (those which might

---

[2] Some nonlinear field equations have localized classical solutions that are static or periodic in time. In Chapter 17, we will learn about the JWKB approximation: if the Lagrangian of a classical system is proportional to g^{-2} with $g \ll 1$, then one can expand the logarithm of the matrix element of the time evolution operator between two coordinate eigenstates, in a power series in g, the leading term of this expansion is the classical action. In the context of the small g^2 expansion, the localized solutions of the field equations can be interpreted as very heavy particles, if the semiclassical expansion parameter g is small. They remain localized because in the $g \to 0$ limit, they are infinitely heavy and their wave packets do not spread. The discussion of these so-called *soliton* solutions lies far beyond the scope of this book.

appear as a perturbation of the Hamiltonian) are even functions of fermion fields. In that case, the minus sign is undetectable in any matrix elements of physical operators.

Fermi statistics is one of the most important features of the world we live in. It is responsible, as we have seen, for the existence of a myriad of atomic nuclei, and for the periodic table of the elements, which lies at the basis of chemistry. We will learn a little bit more about the fundamental mechanism of Fermi statistics when we discuss the nonlocal phases discovered by Aharonov and Bohm, in Chapter 16.

The point of this long introduction was to convince you that we should do the quantum statistical mechanics of particles that have Bose and Fermi statistics, rather than that which would follow the classical treatment of Boltzmann and Gibbs, who thought of identical particles as limits of nonidentical particles.

12.2   QUANTUM FIELD THEORY OF FERMIONS

We have understood the proper treatment of bosons via the quantization of fields. A system of noninteracting bosons with single particle energy eigenstates ϵ_p has a Hamiltonian

$$H = \sum_p n_p \epsilon_p. \tag{12.1}$$

This formula generalizes the one we wrote for the solution of the quantum mechanical D'Alembert equation. In that case, the labels p stood for three-dimensional momenta, $\mathbf{p}$ and the energies were $\epsilon_{\mathbf{p}} = |\mathbf{p}|$. n_p is the number operator of the p-th harmonic oscillator. It takes integer values between 0 and ∞ and can be written

$$n_p = a_p^\dagger a_p, \tag{12.2}$$

where

$$[a_p, a_q^\dagger] = \delta_{pq}, \tag{12.3}$$

and

$$[a_p, a_q] = 0. \tag{12.4}$$

The energy eigenstates are

$$|p_1 \dots p_n\rangle = a_{p_1}^\dagger \dots a_{p_n}^\dagger |0\rangle, \tag{12.5}$$

where

$$a_p|0\rangle = 0. \tag{12.6}$$

For fermions, we want to obey the Pauli exclusion principle, restricting the values of the number operator to $n_p = 0, 1$ and we want the multiparticle states to be antisymmetric under permutations. The latter requirement is easily satisfied, by replacing commutators by anticommutators. The anticommutator of two operators is $AB+BA \equiv [A, B]_+$. The equation

$$[a^\dagger(p), a^\dagger(q)]_+ = 0 \tag{12.7}$$

clearly implies antisymmetry of multiparticle states. It also implies $a^2(p) = 0 = (a^\dagger)^2(p)$. This is the Pauli exclusion principle.

Consistency with the last pair of equation implies that the commutator $[a(p), (a^\dagger)^2(p)] = 0$, which is inconsistent with the commutator between creation and annihilation operators that we used for bosons, as one can see using the Leibniz rule. The obvious replacement is that

$$[a(p), a^\dagger(q)]_+ = \delta pq. \tag{12.8}$$

The number operators $n(p) = a^\dagger(p)a(p)$ still satisfy

$$[n(p), a(q)] = -\delta_{pq}a(q), \tag{12.9}$$

$$[n(p), a^\dagger(q)] = \delta_{pq}a(q), \tag{12.10}$$

so the Hamiltonian

$$H = \sum_p \epsilon_p n(p), \tag{12.11}$$

describes the states of noninteracting fermions, with single particle energies ϵ_p.

12.3 STATISTICAL MECHANICS OF BOSONS AND FERMIONS

The most general system of noninteracting particles will have a variety of single particle energy eigenstates, labeled by a parameter p, which can be discrete or continuous and range over a finite or infinite number of values. The space of p values might also have a geometrical structure corresponding to some number of spatial dimensions. As an example, p could label both the three vector momentum $\mathbf{P}$ and the J_3 component of the spin σ of a nonrelativistic particle of any spin j. The momentum would be discrete if space were a torus, and would live on a torus, the first Brillouin zone, if space were a lattice. Denote the energy of the p-th single particle state by ϵ_p.

The N-particle states are labeled by N copies of the quantum number $p_1, \ldots p_N$, which are either symmetric or totally antisymmetric under permutations of the copies. The key to statistical mechanics is the realization that a nonredundant labeling of the states is just to give the number of particles $n(p)$ occupying each single particle state p. For bosons, $n(p)$ can be any nonnegative integer, while for fermions, it is either zero or one. For either type of statistics, we have

$$[n(p), n(q)] = 0, \tag{12.12}$$

which means that the Hilbert space breaks up into a tensor product $\mathcal{H} = \otimes_p \mathcal{H}_p$, where $n(p)$ acts as $1 \otimes \ldots n(p) \otimes 1 \ldots \otimes 1$. That is, $n(p)$ acts in a nontrivial way only on the factor $\mathcal{H}_p$. Furthermore, because p is a full set of single particle quantum numbers, the states in $\mathcal{H}_p$ are completely characterized by the eigenvalue of $n(p)$. For fermions, $\mathcal{H}_p$ is two dimensional, while for bosons, it is infinite dimensional.

In nonrelativistic physics, the numbers of each type of stable elementary particle (electrons, protons, neutrons inside nuclei which do not undergo beta decay) are conserved quantum numbers. From a relativistic perspective, this is because we never consider transitions between energy levels differing by more than about 10^{-4} of the electron mass. For the purposes of this chapter, we will consider only a single type of elementary particle, so that there is only one such conserved quantum number

$$N = \sum n(p). \tag{12.13}$$

Conservation means that $[N, H] = 0$, which is evident from

$$H = \sum \epsilon_p n(p). \tag{12.14}$$

12.4 THE PARTITION FUNCTION

Macroscopic systems tend to come to thermodynamic equilibrium: given an initial macrostate, the system quickly settles down into a universal state characterized by a few parameters like temperature and pressure. The basic principle of Boltzmann–Gibbs statistical mechanics is that many macroscopic observables of a system composed of a large number of elementary particles take on the same expectation values in all states which have the same values of a few macroscopic conservation laws. We have discussed some of the mathematics behind that in Chapter 10 on quantum measurement, but the derivation of the laws of statistical mechanics from those of QM is far from complete. Some guides to the literature can be found in [51]. We will take the Boltzmann–Gibbs hypothesis as a true statement, but it is worth outlining the basic lines of argument.

Consider a quantum system like the lattice system of Chapter 10. Its Hilbert space is a tensor product of local Hilbert spaces, and the Hamiltonian operator and other collective coordinates are sums of density operators localized near points in space. We showed that in such systems, once we are sufficiently high above the ground state, the number of states in a small energy band, of order the typical scales E_{typ} in the Hamiltonian, becomes exponentially large in the volume of the system measured in atomic units. Consider an initial state $|\psi\rangle = \sum c_n |E_n\rangle$, where the coefficients are nonzero only in some energy band of order the typical scale or a bit larger, in a region where the density of states $n(E) = e^{S(E)}$ is exponentially large. The time evolution of the density matrix is

$$\rho(t) = \sum_{m,n} e^{-i(\omega_m - \omega_n)t} c_m c_n^* |E_m\rangle\langle E_n|, \tag{12.15}$$

where $\omega_m = E_m/\hbar$. The time average of this density matrix over some interval τ is also a density matrix. If we take $\tau \gg \omega_{typ}^{-1}$, then each individual term averages to $(\Delta\omega_{mn}\tau)^{-1}$. So terms with $(\Delta\omega_{mn}\tau) \gg 1$ become negligible. On the other hand, terms with $(\Delta\omega_{mn}\tau) \ll 1$ have almost no time evolution on time scales of order τ.

Recall that the extreme quasi degeneracy of levels, which leads to $S(E) \sim V/a_B^3$, originated from the fact that we could make local changes of the state and raise the energy by amounts independent of the volume. Collective coordinate operators, by their definition, are insensitive to such local changes, with the sensitivity going to zero as the volume over which the collective coordinate density is averaged goes to infinity. Let us define coarse grained energy bands by the projection operators $P(E) = \int_{E-\frac{\hbar}{\tau}}^{E-\frac{\hbar}{\tau}} de \, |e\rangle\langle e|$. These bands are much smaller than the full range of energies contained in the initial state. The essential hypothesis of quantum statistical mechanics is that collective coordinate operators act like the unit operator within each band, up to corrections of order a_B^3/V. It follows that the time averaged expectation values of products of these operators approach constants, at least on time scales $\tau < t < e^{S(E)}\tau$. The system is said to approach equilibrium in this time interval. Statistical mechanics is the study of these equilibrium expectation values. We generally do computations in statistical mechanics by computing expectation values of local fields. This is a well defined mathematical procedure, and it is usually used to compute only expectation values of a product of a small number of fields. Although local field operators are not collective coordinates, products of a few of them are insensitive to changes of the state of the system far from the points where the operators act, so should also act, approximately, like the unit operator in the subspaces corresponding to coarse grained energy bands.

What then is the equilibrium density matrix, to which the system relaxes for the long time interval $\tau < t < e^S \tau$. Since our argument did not depend on the details of the coefficients c_n, one might expect that the resulting state was also independent of those details. This is the Boltzmann–Gibbs hypothesis. Of course, if the system has conservation laws, then the values of those conserved quantities cannot be changed by time evolution. The generic conservation law in a quantum system with time independent Hamiltonian is the projection operator on an energy eigenstate. We are interested in expectation values of operators that do not make fine-grained distinctions between states in a given energy band with projector $P(E)$. Those expectation values are thus independent of the values of most of the conservation laws. Furthermore, the initial state has been chosen to be a typical superposition of states with different values of those fine-grained conserved quantities. Thus, one might imagine that the density matrix depends only on the values of conserved quantities which are themselves collective coordinates, namely energy, momentum, and angular momentum.

We are typically interested in systems at rest in the laboratory, which are enclosed in a box, so rotation and translation invariance are broken and the only conservation laws are energy and the number of each species of particle, if those particles are in a nonrelativistic energy regime, where particle creation can be neglected. We can also consider electric charge, baryon number, and lepton number to be conserved, even in the relativistic regime.[3] Boltzmann and Gibbs tell us that we should evaluate equilibrium properties of the system by averaging the expectation value over all states with fixed energy and particle number.[4] In quantum mechanical language, the Boltzmann–Gibbs hypothesis is that the density matrix of an equilibrium system is

$$\rho_{eq} = \delta(H - E)\delta_{Nn}, \tag{12.16}$$

where E is a real number and n a positive integer. This is called the *microcanonical ensemble*.

One simple way to explain why a system might be in such a state is to assume that it is maximally entangled with a much larger system. If we label a basis of states of the small system by i and those of the large one by I, then a typical state of the combined system (the tensor product of the individual Hilbert spaces) is

$$|\psi_{typ}\rangle = \sum c_{iI}|i, I\rangle, \tag{12.17}$$

where c_{iI} is a maximal rank complex matrix satisfying

$$\sum_{i,I} |c_{iI}|^2 = 1. \tag{12.18}$$

We have $1 \le i \le n$ and $1 \le I \le N$, with $N \gg n$. If we make no measurements on the larger system, the state of the smaller system is uncertain and its density matrix is

$$\rho_{ij} = \sum_I c_{iI}^* c_{jI}. \tag{12.19}$$

In Exercise 12.1, you will verify this is a positive Hermitian matrix with trace equal to 1.

Now consider the equation

$$Uc^* = c^* V_U, \tag{12.20}$$

where U is an $n \times n$ unitary matrix and V_U an $N \times N$ unitary matrix. For each choice of U, the number of equations for V_U is much smaller than the number of unknowns, so we can always satisfy them. Therefore,

$$U\rho U^\dagger = c^* V_U V_U^\dagger c^T = \rho. \tag{12.21}$$

---

[3] It is believed that baryon and lepton numbers are violated by small effects. The equations of the standard model of particle physics then imply that the sum of baryon and lepton numbers is violated. Many models that attempt to explain the small mass of neutrinos violate the difference, $B - L$, as well.

[4] For simplicity of exposition, we will assume that our system consists of a single type of nonrelativistic particle. The generalization to multispecies models should be self evident.

Since ρ commutes with every unitary matrix it is proportional to the unit matrix, which means that the probability distribution for states of the smaller system is maximally uncertain.

If we choose a random state in the full Hilbert space then the probability that c_{iI} is not of maximal rank is the probability that $N \gg n$ vectors of dimension n span a subspace of dimension $< n$. This goes to zero like a power of n/N, depending on how small a dimension we ask for. For macroscopic systems, n is exponentially large in the volume of the system measured in microscopic units like the Bohr radius. Thus, the expected deviations from a maximally uncertain density matrix are exponentially small, of order $e^{-(V_N - V_n)}$, where V is the volume of the indicated system in atomic units.

We have just sketched a proof of Page's theorem [52], which tells us that a small system that has become entangled with a larger system is exponentially likely to be in a maximally uncertain state consistent with macroscopic conservation laws. We have noted, however, that most of the conservation laws are broken by the macroscopic box that we put the system in, in order to isolate it from its environment. Energy conservation is not destroyed by the box, and energy appears to be conserved on all time scales much shorter than the age of the universe.[5] However, the small and large systems are entangled either by a boundary condition at the beginning of time, or, more reasonably, via some interaction between them. This means that the energy of the small system itself is not conserved, so that the microcanonical ensemble is at best an approximate concept.

Boltzmann was the first to study, within classical mechanics, this situation of a system coupled to a *heat bath*; a much larger system with only weak interaction with the system of interest. The reason for the weak interaction can be thought of as a combination of geometrical and mechanical factors. The two systems interact only at the boundary of the box in which the substance of interest is contained. Furthermore, one can imagine designing a box for which energy transport through the relatively thick walls is much less efficient than processes which reemit the energy from the walls, into the system of interest. Let us write the total Hamiltonian as

$$H = H_{sys} + H_{int} + H_{HB}. \qquad (12.22)$$

An eigenstate of the total Hamiltonian with eigenvalue E_0 can be written as an entangled state of different eigenstates of H_{sys}, and by Page's theorem this is extremely likely to be the maximally uncertain density matrix on the Hilbert space of the *system*, constrained only by the total energy of *system plus heat bath*. The heat bath has a much larger energy than the system, so the probability of finding the system with energy E is zero above the eigenvalue and is a rapidly decreasing function of E as this bound is approached. Indeed, the total energy of the heat bath is much larger than the system energy E, and its number of states

---

[5] In making this statement, we are taking the origin of time somewhere in the current slowly expanding period of universal history.

close to that energy is also much larger. Thus, we can write the logarithm of the number of states at energy E_0, each of which is approximately[6] the tensor product of an eigenstate of the heat bath with energy $E_0 - E$, with a particular eigenstate of the system with energy E as

$$l(E) = L(E_0) - E\frac{dL(E_0)}{dE_0} \equiv L(E_0) - \frac{E}{kT}. \qquad (12.23)$$

Here $L(E_0)$ is the logarithm of the number of states of the heat bath with energy E_0 and we have assumed that it increases with energy. This formula is the statistical mechanics definition of the thermodynamic variable called temperature. k is Boltzmann's constant, which is equal to one if we use energy units for temperature, rather than conventional degrees. Thus, the probability of being in a particular eigenstate of the system, with energy $E \ll E_0$ varies as $e^{-\frac{E}{kT}}$ which is Boltzmann's famous statistical law.

One the other hand, as we have seen in Chapter 10 on quantum measurement, the number of states of the *system* with energy E rises rapidly with E. We have also seen that the total energy of a macroscopic system in a typical excited state is of order the volume V of the system. Dimensional analysis says that this is multiplied by an energy density, ϵ, which has dimensions of energy per unit volume. The entropy, or logarithm of the number of states of the system with fixed energy and volume, is also proportional to the volume. If we are in an energy regime above the scales that appear in the Hamiltonian of the system, then the entropy density must scale like $\epsilon^{3/4}$ so the number of states in volume V behaves like

$$n(E) \sim e^{V(\frac{\epsilon}{\hbar v})^{3/4}} e^{V^{1/4}(\frac{E}{\hbar v})^{3/4}}.$$

The velocity v is the maximum velocity at which signals can propagate in the system. It appears, along with $\hbar$, in order to get a dimensionally correct formula independent of most system parameters, which is what we expect in this energy regime. At lower energies, we expect similar exponential growth with a fractional power of the energy. These expectations are based on the spin model of Chapter 10 and many similar models. The falling exponential Boltzmann weight combines with the rising density of states to make a function sharply peaked at an energy of order VT^3 (we have set $\hbar = v = k = 1$ here). It can be shown that statistical averages with this probability distribution give results identical to those in the microcanonical ensemble at that energy, in the *thermodynamic limit* in which V goes to infinity. This new *canonical ensemble* is much more convenient to use, and much closer to the actual state of most physical systems. The density matrix associated with it is $\beta \equiv (kT)^{-1}$.

$$\rho_{can} = \frac{e^{-\beta H}}{Z}, \qquad (12.24)$$

---

[6] This approximation is valid because of the weak coupling between the system and the heat bath.

where $Z = \text{Tr } e^{-\beta H}$, is called the partition function. It turns out to be a convenient tool for quickly calculating thermal averages. For example,

$$\langle H \rangle = -\partial_\beta \ln Z. \tag{12.25}$$

The canonical partition function for an M particle system is

$$Z_M(T) = \text{Tr } e^{-\beta H} P_M(N). \tag{12.26}$$

Here $\beta^{-1} = kT$, where k is Boltzmann's constant. This formula is written in what is called the Grand Canonical Hilbert space in which each of the operators $n(p)$ is allowed to take on any of its allowed values, so that the number operator N can be any nonnegative integer (it will have a maximum for fermions if there are only a finite number of values of p). The operator $P_M(N) = \delta_{0,\ N-M}$ (Kronecker delta) projects on the subspace on which $N = M$.

The calculation of $Z_M(T)$ is a bit complicated because the projection operator couples together all of the tensor factors. For large systems, it is also calculating a quantity that is hard to measure. We never have microscopic control over the particle number, just as we never have microscopic control over the total energy of a large system. In both cases, we introduce a control parameter which weighs different energy or particle number sectors in such a way that the statistical fluctuations away from the mean are small. This is the mathematical meaning of temperature in the case of energy fluctuations, and corresponds to replacing the microcanonical by the canonical ensemble. The corresponding parameter for particle number is called *chemical potential*. It is denoted by μ and is defined by the *Grand Canonical Partition Function*:

$$Z(\mu, T) = \text{Tr } e^{-\beta(H-\mu N)}. \tag{12.27}$$

This is simple to compute, because the trace of a tensor product of operators is the product of their traces (Exercise 12.2). Thus,

$$Z(\mu, T) = \prod_p \text{Tr } e^{-\beta(\epsilon_p - \mu)n(p)}. \tag{12.28}$$

$$Z(\mu, T) = \prod_p (1 \pm e^{-\beta(\epsilon_p - \mu)})^{\pm 1}. \tag{12.29}$$

The plus sign refers to Fermi–Dirac statistics and the minus sign to Bose–Einstein. A useful mnemonic for remembering the sign is that it is the same sign that appears in the commutation relations

$$aa^\dagger \pm a^\dagger a = 1 \tag{12.30}$$

for the creation and annihilation operators for fermions and bosons.

The equation fixing the expectation value of the total number of particles is

$$\langle N \rangle = \beta^{-1} \frac{\partial \ln Z}{\partial \mu} = \sum_p \frac{e^{-\beta(\epsilon_p - \mu)}}{1 \pm e^{-\beta(\epsilon_p - \mu)}}, \tag{12.31}$$

while that fixing the energy expectation value is

$$\langle E \rangle = \frac{\partial \ln Z}{\partial \beta} = \sum_p \epsilon_p \frac{e^{-\beta(\epsilon_p - \mu)}}{1 \pm e^{-\beta(\epsilon_p - \mu)}}. \tag{12.32}$$

Note that, in the limit of high temperature, when $\beta \to 0$, the term in the denominator, which differentiates between bosons and fermions, is negligible. In this limit, we get formulae that are the same as those Boltzmann derived in classical statistical mechanics, except that we are instructed to treat configurations which differ only by the exchange of identical particles as the same configuration. This inserts a factor of $\frac{1}{M!}$ in the partition function for M particles, and corrects the wrong prediction of an entropy of mixing for identical particles.

12.5 THE LOW TEMPERATURE LIMIT

In the low temperature limit, noninteracting bosons and fermions behave quite differently. For fermions, the formula for the expectation value of n_p in the Grand canonical ensemble is

$$\langle n_p \rangle = \frac{e^{-\beta(\epsilon_p - \mu)}}{1 + e^{-\beta(\epsilon_p - \mu)}}. \tag{12.33}$$

Thus, when $\beta \to \infty$, the expected occupation number for every level with $\epsilon - \mu > 0$ goes to zero. Note that μ is chosen to fix the expectation value of N, the number of fermions. In the low temperature limit, every state below μ has expectation value 1 for n_p. This is the maximum value that operator can take, so we are in a state where that level is filled. Thus, the fermion ground state, which is unique if the eigenvalues ϵ_p are nondegenerate, is the state in which the lowest M single particle states are occupied and all others have no fermions in them. This is a pure state. The maximum occupied eigenstate is called the *Fermi Level*. The concept of Fermi level and the nature of the low lying excitations just above the Fermi level is among the most important in the physics of materials. Remarkably, it is valid in a wide range of materials, even though electrons in solids are far from noninteracting. The underlying reason for this is Landau's Fermi Liquid theory [46] [47]: one can show that the low lying excitation spectrum of a system of fermions is determined to a large extent by the shape of a surface $\epsilon(\mathbf{p})$ in momentum space. Fermionic excitations near this surface are called quasiparticles. They are *not* individual electrons, but a sort of fermionic collective excitation of the system of interacting electrons, and one can show that the interactions of these quasiparticles near the Fermi surface are very weak. Unfortunately, this topic is too advanced for a first course in QM, but serious students of the subject will want to learn about Fermi liquid theory as soon as they can.

More recently, we have discovered classes of materials, including the superconductors with the highest known transition temperatures, which are not well described by Fermi liquid theory. This is an area of intense current interest [50] and it is fair to say that we do not yet have a comprehensive theory of these materials. The exploration of their properties is one of the most interesting areas of modern physics.

The behavior of noninteracting Bosons at low temperature is even more remarkable than that of Fermions. The expectation values of the number operators are

$$\langle n_p \rangle = \frac{e^{-\beta(\epsilon_p - \mu)}}{1 - e^{-\beta(\epsilon_p - \mu)}}. \tag{12.34}$$

In this case, if $(\epsilon_p - \mu) < 0$ the denominator can have poles, and the expectation values blow up. Of course, as the temperature is lowered at fixed chemical potential, this will happen first for the single particle ground state.

Of course, if the expectation value of the total number operator, $N = \sum n_p$ is fixed and finite, we cannot have a divergence in any single occupation number. In real systems, the number of particles in a finite volume is always finite, and $\langle N \rangle = \infty$ is only realized in the thermodynamic limit of infinite volume. This sort of singularity in the thermodynamic limit is called a *phase transition*, and this particular phase transition is called *Bose–Einstein condensation* (BEC). Its real meaning for a finite system is that a macroscopic fraction of the particles in the system are in their single particle ground state. That is $\langle n_0 \rangle \sim V/a^3$, where a is an atomic length scale of order the Bohr radius.

For many years after the theoretical discovery of BEC in the noninteracting Bose gas, the only experimental model of the phenomenon was the superfluid behavior of liquid He^4. He^4 is in fact a fairly dense fluid, rather than a dilute gas, and the noninteracting model is not really a good guide to its detailed properties. In 1995, Cornell, Weiman, and Kepperle [53] succeeded in constructing apparatus which could trap systems of cold dilute atoms, whose properties *are* quite close to those of the noninteracting Bose gas. They received the Nobel Prize in 2001. Since then, many other experiments like this have been performed, and these Bose condensed systems have been used to construct systems that behave like a large collection of idealized textbook problems, and can test rather delicate predictions of QM.

It is worth our while then to study the ideal Bose gas of spinless particles with single particle Hamiltonian $H = \frac{\mathbf{p}^2}{2m}$. The formulae for the expectation values of the number operators and the energy are:

$$\langle N \rangle = V \int \frac{d^3 k}{2\pi^3} \frac{e^{-\frac{\beta(\hbar^2 k^2 - 2m\mu)}{2m}}}{1 - e^{-\frac{\beta(\hbar^2 k^2 - 2m\mu)}{2m}}} + n_0 \frac{e^{\beta\mu}}{1 - e^{\beta\mu}}. \tag{12.35}$$

$$\langle E \rangle = V \int \frac{d^3 k}{2\pi^3} \frac{\hbar^2 k^2}{2m} \frac{e^{-\frac{\beta(\hbar^2 k^2 - 2m\mu)}{2m}}}{1 - e^{-\frac{\beta(\hbar^2 k^2 - 2m\mu)}{2m}}}. \tag{12.36}$$

Note that we have allowed for the possibility of macroscopic occupation of the discrete zero momentum mode of the finite volume system, in writing this large V limit of the formulae. The mass m, $\hbar$, and the temperature can be combined to make a parameter λ, called the thermal de Broglie wavelength because it has dimensions of a length:

$$\lambda \equiv \sqrt{\frac{2\pi\beta\hbar^2}{m}}. \tag{12.37}$$

Now define $\mathbf{k} \equiv \sqrt{4\pi}\lambda^{-1}\mathbf{\Omega}x$, where $\mathbf{\Omega}$ is a unit vector. Then we can do the angular integrals, getting a factor of 4π since the integrand is spherically symmetric. We end up with equations for the number density n and the energy density ϵ:

$$n = \frac{n_0}{V} + \lambda^{-3} 4\pi^{-1/2} \int_0^\infty dx \, \frac{x^2}{ze^{x^2} - 1}. \tag{12.38}$$

$$\epsilon = \frac{16\hbar^2 \pi^{1/2}}{2m\lambda^5} \int_0^\infty dx \, \frac{x^4}{ze^{x^2} - 1}. \tag{12.39}$$

We have defined $z \equiv e^{-\beta\mu}$, which is called the *fugacity*. If $z < 1$, the integrands have poles and these expressions are ill-defined, so physical systems have $z \geq 1$. On the other hand, as a function of z, the integral appearing in the expression for n is bounded by its finite value at $z = 1$. Thus, the integral term appearing in the expression for n is monotonically decreasing with temperature (because of the factor involving the thermal wavelength) for low enough temperature. For any given temperature, there is a density above which we must invoke a finite value of n_0/V in order to solve the equations. For any given density, there is a temperature below which there is a finite value of n_0/V. The line in the (n, T) plane below which this condensate appears is called the phase transition to BEC (Figure 12.1).

What is the Bose–Einstein condensate state? It is best to think of a very large but finite box, and certainly this will be the correct description of any real experiment. The momenta

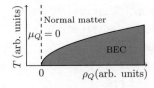

Figure 12.1   Phase diagram of Bose–Einstein condensation.

are discrete. For particles of nonzero momentum, it is simply an ordinary thermal state. For the discrete zero mode, it is a density matrix of maximal entropy constrained by the equation

$$Tr\rho a^\dagger(0)a(0) = n_0, \tag{12.40}$$

where $n_0 \propto V\lambda^{-3}$ is the density we need to fit the thermodynamic equations above for some values of n and T below the transition line. In Exercise 12.3, you will show that this is

$$\rho = \frac{e^{La^\dagger(0)a(0)}}{\text{Tr}\left[e^{La^\dagger(0)a(0)}\right]}, \tag{12.41}$$

where

$$\frac{e^L}{1 - e^L} = n_0. \tag{12.42}$$

As the volume goes to infinity, n_0 goes to infinity and $L \to 0$. The density matrix approaches something proportional to the unit matrix and the entropy becomes infinite.

This is certainly the result in the Grand Canonical ensemble. For the Canonical ensemble though, where the particle number is fixed, it cannot be right. For the system in a box, with a fixed finite number of particles $N \propto V$, the result of lowering the temperature is that more and more of the particles go into the ground state. The Grand Canonical calculation of the limiting density of particles in states with nonzero momentum is still valid for large V (in de Broglie wavelength units), so there is still a Bose–Einstein condensate below a certain temperature, but the density matrix for the zero momentum creation and annihilation operators is quite different. We can use the equation

$$n = \frac{n_0}{V} + \lambda^{-3}4\pi^{-1/2}\int_0^\infty dx\, \frac{x^2}{e^{x^2} - 1} \equiv \frac{n_0}{V} + \rho_>, \tag{12.43}$$

to calculate n_0, the number of zero momentum particles below the transition temperature. However, the term proportional to V in this number is no longer an expectation value. We have fixed the total number of particles as an operator constraint. Below the transition temperature, the term of order V in the number of nonzero momentum particles is $\rho_> V$, and the fluctuations of this operator are suppressed relative to this by $V^{-1/2}$. This means that n_0 is fixed in the canonical density matrix, up to terms of relative order $V^{-1/2}$. The full multiparticle system is mostly in its ground state, and the entropy is all carried by the relic particles of nonzero momentum. In the strict $T \to 0$ limit, the term in the entropy proportional to the volume goes to zero.

For a system of noninteracting bosons confined to a box by a potential $V(\mathbf{x})$, the N-particle ground state wave function is just

$$\Psi_0(\mathbf{x_1}\ldots\mathbf{x_N}) = \psi(\mathbf{x_1})\ldots\psi(\mathbf{x_N}), \tag{12.44}$$

where $\psi(\mathbf{x})$ is the single particle ground state. If we add a repulsive short range two body interaction of the form

$$\delta H = \frac{\lambda}{2} \sum_{i \neq j} \delta^3(\mathbf{x_i} - \mathbf{x_j}), \tag{12.45}$$

then it is plausible, at least at low density, that the ground state is still approximately a product of single particle wave functions (this is the Hartree approximation—see Chapter 11, as well as Chapter 18 on the variational principle). If we calculate the expectation value of the energy in states of this form, and minimize with respect to ψ^*, we get the Gross–Pitaevski equation [54]:

$$(-\frac{\hbar^2}{2m}\nabla^2 + V(x))\psi(\mathbf{x}) = \lambda\psi(\mathbf{x})\psi^*\psi(\mathbf{x}). \tag{12.46}$$

The G–P equation is one of the main tools in the modern theory of Bose–Einstein condensates.

Apart from its resolution of Gibbs' entropy of mixing paradox, quantum statistical mechanics is primarily useful in the study of systems at very low temperatures. It turns out that the variety of peculiar states of matter at low temperature is extremely large and the field promises to remain fascinating into the indefinite future. Landau's Fermi liquid theory and the experimental discovery of dilute Bose–Einstein condensates have made the very simple calculations we have done in this chapter much more useful than one might have thought. However, the real future of the field is probably the study of more exotic states of matter, for which we do not yet have a complete set of theoretical tools.

12.6 STATISTICAL MECHANICS OF A FERMI LIQUID

As we have said, a Fermi liquid is an accurate description of the low lying excitations near the ground state of a wide variety of systems whose underlying constituents are fermions. Those low lying excitations are almost noninteracting fermions. Their single particle energies have the form

$$\epsilon(p, \Omega) = v_F(\Omega)p, \tag{12.47}$$

where v_F is called the *Fermi velocity*. p is a continuous parameter, which describes the distance in momentum space above the Fermi surface, a surface in momentum space, defined by an equation of the form $E(\mathbf{p}) = E_F$. This equation defines a two-dimensional surface in three-dimensional momentum space. Ω is a solid angle (we have written it as a three-dimensional unit vector), which tells us where the excitation is on the surface. For a crystalline solid with a particular lattice, the equation for the Fermi surface will have only the symmetries of the lattice.

The underlying microscopic physics is a complicated interacting problem, and it is not easy to calculate the ground state energy. The miracle of Fermi liquid theory is that the physics of excitations near the ground state is completely characterized by the function $v_F(\Omega)$. The grand canonical partition function of a Fermi liquid is given by the general prescription above

$$\ln Z = p_F^{-1} \int_0^\Lambda dp \int d^2\Omega \, [\ln (1 + e^{-\beta(pv_F(\Omega) - \mu)})]. \qquad (12.48)$$

The cutoff Λ on the momentum integral reminds us that the Fermi liquid approximation is only valid near the ground state. Both Λ and the scale p_F might be expected to be of order $\frac{\hbar}{a_B}$ for a system consisting of nonrelativistic electrons and nuclei interacting via the Coulomb potential.

As long as $v_F(\Omega)$ does not have any zeroes, the contribution to the integrand from momenta near the cutoff is of order $e^{-\beta\Lambda} = e^{-\frac{\Lambda}{kT}}$, so as long as $kT \ll \Lambda$, this is negligible. If we formally take $\Lambda \to \infty$, the integral converges, but we should recognize that it only corresponds to real physics when this temperature inequality is satisfied. In the $\Lambda \to \infty$ limit, we can extract the temperature dependence by defining $\beta p = y$. y has units of inverse velocity. Then,

$$\ln Z = (\beta p_F^{-1}) \int_0^\infty dy \int d^2\Omega \, [\ln (1 + ze^{-yv_F(\Omega)})]. \qquad (12.49)$$

In this equation, $z = e^{\beta\mu}$ is the fugacity.

One can show [47] that to leading order at low temperature, thermodynamic quantities do not depend on the interactions between Landau's quasiparticles. Thus, we should really only use these simple formulae to lowest order in T, because higher order terms will have interaction corrections. We should emphasize that we assume that we are dealing with a normal metal. Bardeen Cooper and Schriefer showed that if the quasiparticle interactions are attractive, then the Fermi liquid ground state is unstable to a superconducting ground state and the thermodynamics is completely different. For large β, the fugacity z is small (the constraint of fixed average density determines the chemical potential to be negative). Thus, we can expand out the logarithm and the leading order result depends on the angle dependent Fermi velocity only through $\int \frac{d\Omega}{v(\Omega)}$. Thus, since the noninteracting Fermi gas is a particular example, the leading low temperature behavior of an arbitrary Fermi liquid is the same as that of a free gas, up to a multiplicative constant. Thus, for example, the specific heat (per particle) is given by Exercise 12.7:

$$C_V = \frac{\pi^2}{2} k \frac{T}{T_F}, \qquad (12.50)$$

where k is Boltzmann's constant. For a free electron gas, the "Fermi temperature" T_F is given by

$$T_F = \frac{\hbar^2}{2mk} (3\pi^2 \rho)^{2/3}, \qquad (12.51)$$

where ρ is the fermion density. For metals, the free electron models gives $m = m_e$. Fermi liquid theory tells us that we get the same low temperature formula with $m_e \to m^*$, the "effective mass" of the quasiparticle. It is important to realize that unlike the case of semiconductors, where the charge carrying excitations have a standard Galilean invariant dispersion relation at small wave number, Landau's quasiparticles *do not* behave like nonrelativistic electrons with an effective mass.

12.7 PLANCK'S RADIATION LAW

The origin of QM was Planck's resolution of the *ultraviolet catastrophe* in the statistical mechanics of the electromagnetic field. A field is just a collection of oscillators with wave numbers $\mathbf{k_n}$, where the $\mathbf{k_n}$ are the allowed wave numbers with (e.g., periodic) boundary conditions on a cubical box of side L, $\mathbf{k_N} = \frac{2\pi}{L}(\mathbf{N})$, where $\mathbf{N}$ is a vector of integers. Each oscillator has frequency

$$\omega(\mathbf{k_N}) = c\frac{2\pi}{L}|\mathbf{N}|,$$

where c is the speed of light. The logarithm of the partition function is just the sum of the logs of the partition functions of these oscillators

$$\ln Z = -2\sum_{\mathbf{k}} \ln\left[1 - e^{-\beta\hbar\omega(\mathbf{k_N})}\right]. \tag{12.52}$$

The factor of two counts the two polarization states of light waves. Because the oscillations are not coupled, we actually have a separate energy conservation law for each frequency, so we can calculate the expectation value of the energy for each frequency by having a different inverse temperature β_k for each value of $\omega(\mathbf{k_N})$. Taking the limit of large L turns sums into integrals. $\mathbf{k}$ becomes a continuous variable. We can think of the allowed values of $\mathbf{k}$ as forming an infinite lattice in three-dimensional space. Each lattice point can be associated with the unit cube in this space, which has that point as its largest vertex in all three directions. Each of these cubes has volume $(\frac{2\pi}{L})^3$. In the limit of large L, the surface in $\mathbf{k}$ space with fixed ω becomes a sphere of radius ω/c in k space.

Using the fact that we have two polarization states for each elementary cube, it is easy to work out that the *density of energy* per unit frequency $d\omega$ is

$$\rho(\omega) = \frac{\hbar\omega^3}{\pi^2 c^3}(e^{\hbar\omega\beta} - 1)^{-1}.$$

This is Planck's famous radiation law, the blackbody spectrum. The ultraviolet catastrophe of classical physics comes from extrapolating the low frequency limit of this formula $\rho = k_B T\frac{\omega^2}{\pi^2 c^3}$ to high frequencies. The physical reason for it is that classical physics assumes that one can get arbitrarily low intensity electromagnetic waves, which cost very little energy, at any frequency. There are more modes at higher frequency, so the statistical sum at any temperature should be dominated by very low energy, very high-frequency electromagnetic waves. The solution of the quantum oscillator problem shows that there is a minimum energy $\hbar\omega$ of any excitation of the quantized electromagnetic field. This minimum energy excitation is called a single *photon* and this connection between particles and fields is the one we pointed out in Chapter 5. At any finite temperature, the high-frequency modes will be Boltzmann suppressed because even one photon carries a very high energy.

Since photons are bosons, one might have thought that one could have a Bose condensate of photons at sufficiently low temperature. The reason that this does not happen under normal circumstances is that photons in a box are constantly being absorbed and reemitted by the atoms in the walls of the box, so that photon number is not conserved. For nonrelativistic particles, energy and momentum conservation normally forbid interactions which change the number of particles. Thus, one has two conservation laws, particle number and energy. Recently, a theory of thermalization processes which could conserve photon number [55] has been developed, and observations of Bose condensed photon fluids have been reported [56].

12.8 EXERCISES

12.1 Show that the matrix $\rho_{ij} = \sum_I c_{iI}^* c_{Ij}$ has the properties of a density matrix.

12.2 Show that the trace of a tensor product of matrices is the product of the traces of the individual matrices.

12.3 Show that the maximal entropy density matrix for the ground state of a system of Bose particles with only the average number of particles constrained to be n_0 is given by

$$\rho = \frac{e^{La^\dagger(0)a(0)}}{\text{Tr}\left[e^{La^\dagger(0)a(0)}\right]},$$

where

$$\frac{e^L}{1 - e^L} = n_0.$$

12.4 Diamagnetism is a term describing the first-order response of a system to an external magnetic field. The material gets a magnetic moment. If the magnetic field generated by that moment points in the direction opposite to the applied field, the system is said to be diamagnetic. If it points in the same direction, it is called paramagnetic. Show that in a system composed of classical charged point particles with no intrinsic magnetic moment, the thermal average of the induced magnetic moment vanishes. This is called van Leuwen's theorem.

12.5 Consider a system of N atoms, each with an intrinsic magnetic moment μ. Treat the motion of the particles, as well as the energetics of the dipoles, classically. Using the result of Exercise 12.4, show that the statistical mechanics of the magnetization in a constant magnetic field is given entirely by the Hamiltonian $H_B = -\mu B \sum_{i=1}^{N} \cos(\theta_i)$, where B is the magnitude of the field and θ_i the angle between the direction of the field and the dipole of the i-th atom. Evaluate the magnetization and the magnetic susceptibility of this system in the canonical ensemble at inverse temperature $\beta = (kT)^{-1}$. The susceptibility is defined as the derivative of the magnetization with respect to B. Show that

$$M \equiv \beta^{-1} \partial_B (\ln Z) = N\mu(\coth(\alpha) - \alpha^{-1}),$$

where $\alpha \equiv \mu B \beta$.

$$\chi = \beta \mu^2 [\alpha^{-2} - \text{csch}^2(\alpha)].$$

χ is the susceptibility per atom. Show that $\chi \sim \beta$ for small β (Curie's law) and evaluate the proportionality constant, which is called Curie's constant.

12.6 In QM, van Leuwen's theorem is no longer valid. We saw in Chapter 9 that for free particles, the motion perpendicular to the field has quantized energy levels with spectrum $\frac{e\hbar B}{mc}(n+1/2)$. Each of these levels was infinitely degenerate due to the translation symmetry. If we put the system in a box, that symmetry is broken and each level has a finite degeneracy g. One can show that at large volume $g = L^2 \frac{eB}{2\pi\hbar c}$, where L is the size of the box. The motion along the field has energies $\frac{p_z^2}{2m}$, where $p_z = \frac{2\pi\hbar}{L}$. Write an integral formula for both the logarithm of the grand canonical partition function and the average particle number for noninteracting charged fermions in a constant magnetic field. Evaluate the formulae in the high temperature limit, compute the susceptibility per unit volume and show that it is nonzero and obeys Curie's law.

12.7 Derive the low temperature specific heat for a free fermion gas with single particle energies $\frac{\hbar^2 \mathbf{k}^2}{2m}$.

CHAPTER 13

Perturbation Theory: Time Independent

13.1 INTRODUCTION

Much of the rest of this book will be devoted to approximation methods. The simplest of these is time-independent perturbation theory. Most interesting problems in physics cannot be solved exactly but many are close to exactly solvable problems and one can write the solution as a power series expansion in some small parameter. This is called *perturbation theory*. The higher-order terms in the series become more and more complicated, but one can often organize each of them into a sum of individual terms, each associated with a picture or diagram. Even when there is no small perturbation parameter in the real physical problem, one can introduce one artificially and try to get insight into more sophisticated approximation schemes by resumming the perturbation theory, keeping only a certain set of simple diagrams at each order.

This chapter will give a general set of rules for computing the perturbation theory for the eigenvalues of a Hamiltonian of the form $H_0 + gV$. These methods are applicable to the discrete spectrum. The discussion of perturbation theory for the continuous spectrum can be found in the next chapter and Chapter 16 on scattering theory.

13.2 BRILLOUIN–WIGNER PERTURBATION THEORY FOR NONDEGENERATE LEVELS

We have seen that there are several quantum mechanical problems, which can be solved exactly. Let H_0 be such a Hamiltonian, and consider

$$H = H_0 + gV, \tag{13.1}$$

where g is a small parameter. The Schrödinger equation can be written

$$(H_0 + gV)(|\psi_0\rangle + |\phi\rangle) = (E_0 + \Delta)(|\psi_0\rangle + |\phi\rangle). \tag{13.2}$$

$|\psi_0\rangle$ is an eigenstate of H_0 with eigenvalue E_0. The projection of the perturbed vector $|\phi\rangle$ on $|\psi_0\rangle$ is ambiguous, because the equation is homogeneous. We can choose it so that the eigenstate is normalized, obtaining a form of perturbation theory first invented by Rayleigh [57] in his study of sound waves, and adapted to quantum mechanics (QM) by Schrödinger [58]. Instead, we will adopt a procedure invented by Brilliouin and Wigner [59] and choose

$$\langle\phi|\psi_0\rangle = 0. \tag{13.3}$$

The resulting state has a squared norm $1 + \langle\phi|\phi\rangle$, and we must remember to divide by the norm when computing expectation values.

The eigenstate of H_0 with eigenvalue E_0 might be degenerate, but we will first consider the case when it is isolated. Define

$$P = 1 - |\psi_0\rangle\langle\psi_0|, \tag{13.4}$$

the projector onto the subspace orthogonal to the unperturbed eigenvalue, and write the two equations obtained by left multiplying the Schrödinger equation by P and $1 - P$.

$$PgV[|\psi_0\rangle + |\phi\rangle] + PH_0|\phi\rangle = PE|\phi\rangle. \tag{13.5}$$

$$|\psi_0\rangle\langle\psi_0|gV(|\psi_0\rangle + |\phi\rangle) = (E - E_0)|\psi_0\rangle. \tag{13.6}$$

The second equation is just an evaluation of the perturbed energy:

$$\langle\psi_0|gV|(|\psi_0\rangle + |\phi\rangle) = E - E_0. \tag{13.7}$$

The first equation can be rewritten as

$$|\phi\rangle = g\frac{P}{E - H_0}V(|\psi_0\rangle + |\phi\rangle). \tag{13.8}$$

We have used the fact that P commutes with H_0. We see the $|\phi\rangle$ is indeed small when g is small, and has the power series expansion

$$|\phi\rangle = \sum_{k=1}^{\infty}[\frac{P}{E - H_0}(gV)]^k|\psi_0\rangle. \tag{13.9}$$

This is, implicitly, a power series in g, because the equation for $E - E_0$ shows that it begins with a term

$$E_1 = g\langle\psi_0|V|\psi_0\rangle, \tag{13.10}$$

with corrections of order g^2. Thus,

$$E = \sum_{k=0}^{\infty} g^k E_k. \tag{13.11}$$

Note also that the dependence on $E - E_0$ in the term with $k = 1$ is an illusion, because the projection operator kills it. Our relatively simple equation for $|\phi\rangle$ hides some of the g dependence. To get an explicit power series in g, we have to expand out $E = E_0 + \sum_{k=1}^{\infty} g^k E_k$.

To get explicit expressions for both $|\phi\rangle$ and $\Delta \equiv E - E_0$, we proceed systematically. The expression for Δ at order g^k requires the knowledge of $|\phi\rangle$ at order g^{k-1}. The expression for $|\phi\rangle_k$ in the expansion $|\phi\rangle = \sum_{k=1}^{\infty} g^k |\phi\rangle_k$ involves all the coefficients E_p up to E_{k-1}. So we get

$$E_1 = \langle \psi_0 | V | \psi_0 \rangle. \tag{13.12}$$

$$E_2 = \langle \psi_0 | V \frac{P}{E_0 - H_0} V | \psi_0 \rangle. \tag{13.13}$$

$$|\phi\rangle_1 = \frac{P}{E_0 - H_0} V | \psi_0 \rangle. \tag{13.14}$$

$$|\phi\rangle_2 = \frac{P}{E_0 - H_0} (V - E_1) \frac{P}{E_0 - H_0} V | \psi_0 \rangle. \tag{13.15}$$

$$|\phi\rangle_3 = \frac{P}{E_0 - H_0} (V - E_1) \frac{P}{E_0 - H_0} (V - E_1) \frac{P}{E_0 - H_0} V | \psi_0 \rangle + \frac{P}{E_0 - H_0} (E_2) \frac{P}{E_0 - H_0} V | \psi_0 \rangle. \tag{13.16}$$

And so it goes...

The terms get more and more complicated as k gets larger. For certain simple systems, notably when H_0 is a collection of harmonic oscillators, the perturbation theory can be simplified further and pictorial representations of individual terms, called (time-ordered) Feynman diagrams, allow one to compute to fairly high orders. In actual fact, the time-dependent perturbation theory we will study in the next chapter, or the associated Feynman path integral method, provide more convenient expressions for computing higher-order terms. However, they do not produce direct expressions for perturbed energy levels.

Note that the operator $\frac{P}{E_0 - H_0}$, which appears everywhere in these formulae, is well defined because the eigenvalue E_0 is nondegenerate. However, it is also obvious that if there are eigenvalues very close to E_0, the higher-order terms in the perturbation series will be very large, unless g is very small. We will see that the methods of degenerate perturbation theory also give us a clue about how to deal with situations like this.

The expression for E_2 in case $|\psi_0\rangle$ is the ground state of H_0, is always negative, and this exemplifies a general principle called the Variational Principle, which we will explore further in Chapter 18. We have already used it in our discussion of the hydrogen molecular ion and the Gross–Pitaevski equation. Consider the expectation value of $H = H_0 + gV$ in any normalized state $\psi\rangle$. Expanding $\psi\rangle$ in eigenstates of H

$$|\psi\rangle = \sum c_n|E_n\rangle, \qquad H|E_n\rangle = E_n|E_n\rangle, \tag{13.17}$$

we obtain

$$\langle\psi|H|\psi\rangle = \sum|c_n|^2 E_n \geq E_{ground}\sum|c_n|^2 = E_{ground}. \tag{13.18}$$

That is, *the ground state energy is the minimum expectation value of H among all normalized states.*

In particular, $\langle\psi_0|H|\psi_0\rangle = E_0 + gE_1 \geq E_{ground}$. On the other hand, perturbation theory tells us that for small enough g, $E_{ground} \approx E_0 + gE_1 + g^2E_2$. The inequality then tells us that E_2 must be negative, and this agrees with our explicit formula.

The most straightforward way to evaluate the higher-order terms in perturbation theory is to insert complete sets of H_0 eigenstates and obtain formulae like

$$\Delta_2 = \sum_{n\neq 0}|\langle\psi_0|V|\psi_n\rangle|^2 \frac{1}{E_0 - E_n}. \tag{13.19}$$

With the notable exception of the important case where H_0 is the Hamiltonian of noninteracting quantized fields, where this strategy leads to Feynman diagrams, these infinite summations can be very tedious. In some cases, there is an alternative, in low orders of perturbation theory, invented by Dalgarno and Lewis [60]. These authors observed that if one can find an operator Ω_2 satisfying

$$V|\psi_0\rangle = [H_0, \Omega_2]|\psi_0\rangle, \tag{13.20}$$

then

$$\frac{P}{E_0 - H_0}V|\psi_0\rangle = \frac{P}{E_0 - H_0}[H_0 - E_0, \Omega_2]|\psi_0\rangle = P\Omega_2|\psi_0\rangle, \tag{13.21}$$

so that

$$E_2 = \langle\psi_0|V\Omega_2|\psi_0\rangle - \langle\psi_0|V|\psi_0\rangle\langle\psi_0|\Omega_2|\psi_0\rangle. \tag{13.22}$$

If Ω is "simple," then the latter formula is easy to evaluate. We will see some examples of this below. To extend the Dalgarno–Lewis (D–L) method to next order, we would have to find an operator satisfying

$$V\Omega_2|\psi_0\rangle = [H_0, \Omega_3]|\psi_0\rangle. \tag{13.23}$$

This is only possible if $E_2 = 0$ (prove it!). So the D–L trick is limited to second-order perturbation theory. However, in that context it can be very useful and it deserves to be explored more than it has been in the literature.

13.2.1 Relation to Rayleigh–Schrödinger Perturbation Theory

The Rayleigh–Schrödinger version of perturbation theory determines the part of the perturbed state parallel to $|\psi_0\rangle$, by insisting that $|\psi_0\rangle + |\phi\rangle_{RS}$ have length one. Thus,

$$|\phi\rangle_{RS} = \frac{|\phi\rangle}{\sqrt{1 + |\langle\phi|\phi\rangle}}, \tag{13.24}$$

where $|\phi\rangle$ is the BW state. It is a lot easier to compute the normalized eigenstate from this formula than to apply the RS rules directly.

There is another aspect of the difference between the two methods of computing the perturbation expansion which is exposed by the above formula. A perturbative calculation of $|\phi\rangle$ gives an infinite number of powers of g in the normalized wave function. Similarly, if we write the expression for the energy in BW perturbation theory

$$(E - E_0) = \langle\psi_0|gV \sum_{k=0}^{\infty} [\frac{P}{E - H_0}(gV)]^k|\psi_0\rangle, \tag{13.25}$$

then at a finite order in the expansion, we get a nonlinear equation for E, whose exact solution contains an infinite number of powers of g. There has not been much exploration of whether the resummations of perturbation theory implicit in the BW formalism lead to better approximations at finite values of g.

13.3 DEGENERATE PERTURBATION THEORY

A major virtue of the Brillouin–Wigner approach to perturbation theory is the ease with which the formalism generalizes to the case where the energy level E_0 is degenerate. The only change is that the projection operator P is replaced by

$$P = 1 - \sum_d |\psi_0^d\rangle\langle\psi_0^d|, \tag{13.26}$$

the projector on the degenerate *subspace*. As in the nondegenerate case, the projector P makes the inverse operator $P(E_0 - H_0)^{-1} = (E_0 - H_0)^{-1}P$ well defined, and the higher-order terms in the perturbation series are formally small for small g.[1]

---

[1] The question of *convergence* of the perturbation series is much more involved. It is always an *asymptotic series* for the actual eigenvalue as a function of g. This means that the difference between the exact answer and the first n terms vanishes like g^{n+1} as $g \to 0$. This is much less than convergence. Convergence would imply that the eigenvalue existed as an analytic function on a disk surrounding the origin in the complex g plane. It is often easy to see that such analyticity violates physical sense. For example, if we perturb a harmonic oscillator by gx^4, then it is manifest that the Hamiltonian is not bounded from below for negative g, so the eigenvalues cannot be analytic functions. We will say a bit more about this in our chapter on the JWKB approximation.

If $|\psi_0^b\rangle$ is any state in the degenerate subspace, we write the perturbed state as

$$|\psi^b\rangle = |\psi_0^b\rangle + |\phi^b\rangle, \tag{13.27}$$

where $|\phi^b\rangle$ is orthogonal to the subspace. The Schrödinger equation takes the form

$$(H_0 + gV)|\psi_0^b\rangle + \phi^b\rangle = E^b(|\psi_0^d\rangle + \phi^d\rangle). \tag{13.28}$$

Then we can solve for $|\phi^b\rangle$ as before

$$|\phi^b\rangle = \sum_{k=1}^{\infty} (\frac{gP}{E^b - H_0} V)^k |\psi_0^b\rangle. \tag{13.29}$$

Now take the scalar product of the Schrödinger equation with $|\psi_0^a\rangle$.

$$\langle \psi_0^a | H_0 + gV | \psi_0^b \rangle + \langle \psi_0^a | gV \sum_{k=1}^{\infty} (\frac{gP}{E^b - H_0} V)^k | \psi_0^b \rangle = E^b \delta_{ba}. \tag{13.30}$$

This equation is only consistent if we choose the basis in the degenerate subspace to be one in which the matrix

$$\langle \psi_0^a | H_0 + gV | \psi_0^b \rangle + \langle \psi_0^a | gV \sum_{k=1}^{\infty} (\frac{gP}{E^b - H_0} V)^k | \psi_0^b \rangle, \tag{13.31}$$

is diagonal. The now split eigenvalues of H are the eigenvalues of this matrix.

This form of degenerate perturbation theory can also handle cases where the level E_0 is not exactly degenerate, but the level splittings in its neighborhood are very small, so that the perturbation series is a bad approximation unless g is extremely small. One simply defines the projector P so that it projects on the space orthogonal to all of the closely spaced levels, and then proceeds as in the degenerate case. The number E_0 then becomes a diagonal matrix in the quasidegenerate subspace, the image of $(1 - P)$.

The typical case in which such quasidegeneracy arises is for large systems with a volume $V \gg 1$ in Bohr units. As we have seen, phonons and other long-wavelength excitations, corresponding to flows of conserved currents, have energies as low as $V^{-1/3}$ in Bohr units, so the ground state is quasidegenerate. In principle, the Brillouin–Wigner method enables one to construct an energy-dependent *effective Hamiltonian*, which describes only the dynamics of these low-energy modes. It turns out, however, that the Feynman path integral formalism provides a much more efficient way to construct the dynamics of these long-wavelength, low-energy excitations.

13.4 THE FEYNMAN–HELLMANN THEOREM

Feynman and Hellmann [61] proved a simple theorem about derivatives of energy levels with respect to parameters, which is often helpful in evaluating the expectation values required for computing first-order perturbation theory. Let $H(\lambda)$ be a one parameter set of Hamiltonians, $E_n(\lambda)$ one of its discrete eigenvalues and $|\psi_n(\lambda)\rangle$ the corresponding normalized eigenstate. Assume that we are working at a value of λ where the eigenstate is nondegenerate. The norm of $|\psi_n(\lambda)\rangle$ is independent of λ so

$$\langle\psi_n(\lambda)|\frac{d}{d\lambda}|\psi_n(\lambda)\rangle + c.c. = 0. \tag{13.32}$$

If we write

$$E_n(\lambda) = \langle\psi_n(\lambda)|H(\lambda)|\psi_n(\lambda)\rangle, \tag{13.33}$$

and use the previous equation, we find the Feynman–Hellmann result

$$\frac{d}{d\lambda}E_n(\lambda) = \langle\psi_n(\lambda)|\frac{dH}{d\lambda}|\psi_n(\lambda)\rangle. \tag{13.34}$$

This can be viewed as a generalization of the first-order perturbation theory formula to Hamiltonians that depend on nonlinear functions of λ.

13.5 EXAMPLES

13.5.1 The Stark Effect

Let us now apply these general and abstract ideas to a simple example: the hydrogen atom in external constant electric or magnetic fields. Let us do the electric field first. The perturbation has the form $V = -eEX_3$, where we have used the rotation invariance of the unperturbed problem to choose the electric field in the x_3 direction. The first-order perturbation theory formula for the perturbed levels is

$$E_{nlm}^{(1)} = E_n - eE\langle n\,l\,m|X_3|n\,l\,m\rangle, \tag{13.35}$$

where E_n is the nth Rydberg energy.

The operator $X_3 = r\cos(\theta)$ is invariant under rotations around the three direction. The unperturbed eigenfunctions have the form $R_{nl}(r)Y_{lm}(\theta,\phi)$, so the matrix element of X_3 is a product of a matrix element of r between two radial wave functions, and a matrix element of $\cos(\theta)$ between a pair of spherical harmonics. Noting that $\cos(\theta)$ is just $\sqrt{\frac{4\pi}{3}}Y_{10}$, we see that computing the angular matrix element reduces to the problem of addition of angular momentum. That is, the wave functions $Y_{10}Y_{lm}$ are just products of spin 1 and spin l eigenstates, which can be expanded in a complete basis of spherical harmonics. The general rules of addition of angular momenta tell us that we get a linear combination of states with spin l and $|l \pm 1|$. Furthermore, we get only states with $L_z = \hbar m$.

More generally, we have

$$Y_{kn}Y_{lm} = c(l', m'; k, n\ l, m)Y_{l'm'}. \tag{13.36}$$

The coefficients $c(l', m'; k, n\ l, m)$ are called Clebsch–Gordon coefficients, and are entirely determined by the group theory of the rotation group. The required matrix element of $\cos(\theta)$ is thus,

$$\langle l'm'|\cos(\theta)|lm\rangle = \sqrt{\frac{4\pi}{3}}c(l', m; 1, 0\ l, m), \tag{13.37}$$

where l' ranges between $|l-1|$ and $l+1$. You will find tables of the Clebsch–Gordon coefficients on the World Wide Web.

The above discussion is a special case of what is known as the Wigner–Eckart theorem [62]. In any rotation invariant system, the rotation generators act on the space of operators via

$$O \to U^\dagger(R)OU(R). \tag{13.38}$$

This formula is analogous to the relation between Schrödinger picture and Heisenberg picture operators. It gives us the version of an operator appropriate in a rotated reference frame. Since the space of operators is a linear space, we can decompose it into irreducible representations of the rotations. Thus, any operator will have a representation

$$O = \sum O_l o_{lm}, \tag{13.39}$$

where the operators O_l commute with rotations, and the operators o_{lm} transform under rotations like the spherical harmonics Y_{lm}. The matrix elements of o_{lm} between unperturbed eigenstates of the rotation invariant system, will be Clebsch–Gordon coefficients, and we will need only the matrix elements of the invariant operators O_l. The appendix on Group Theory contains a short proof of the Wigner–Eckart theorem.

For the ground state of the hydrogen atom, first-order perturbation theory gives zero for the Stark Effect, because of the angular momentum selection rules. The second order energy is

$$-\langle\psi_0|V\frac{P}{E - H_0}V|\psi_0\rangle. \tag{13.40}$$

This is an example where the Dalgarno–Lewis method proves useful. The D–L equation is

$$[H_0, \Omega_2]|\psi_0\rangle = -eEX_3|\psi_0\rangle. \tag{13.41}$$

Since H_0 and the ground state wave function are rotation invariant, Ω_2 transforms like X_3 under rotations. If Ω_2 commutes with the coordinates it has the form

$$\Omega_2 = f(r)\cos(\theta), \tag{13.42}$$

where

$$-(f''(r) + 2f'(r)\partial_r)\psi_0(r) + \frac{2}{r}[f(r) - f'(r)]\psi_0(r) = -\frac{2meE}{\hbar^2}r\psi_0(r). \tag{13.43}$$

Dividing through by ψ_0 and using $\partial_r(\ln \psi_0) = -\frac{1}{a_B}$, we find

$$f(r) = \frac{meE}{\hbar^2}(a_B^2 r + \frac{a_B r^2}{2}). \tag{13.44}$$

The second order energy is thus,

$$E_2 = \langle\psi_0|(-eEX_3)f(r)\cos(\theta)|\psi_0\rangle. \tag{13.45}$$

which is a simple integral. The result is

$$E_2 = -(eE)^2\frac{9a_B^2}{4}. \tag{13.46}$$

In Exercises 13.1–13.2, you will show how to generalize this result in a variety of ways.

13.5.2 The Zeeman Effect

The Zeeman effect is the shift in hydrogen energy levels due to a weak constant magnetic field. It is fairly straightforward to evaluate it, and we will do so in a moment, but first we must point out that the question of exactly how weak the field is, becomes crucially important because of relativistic corrections to the Schrödinger equation, which also have to do with magnetic fields. We have modeled the electromagnetic field of the nucleus by a Coulomb field, and this is valid for electrons at rest, but the electron in the hydrogen atom is moving, albeit at a velocity much less than that of light. As a consequence, it also sees a magnetic field, which is weak because the electron velocity is $\ll c$. The qualitative features of the Zeeman effect depend on whether the external field is larger than or smaller than the effective magnetic field seen by the moving electron. We will first study the effect where the external field is stronger, where it is also known as the Paschen–Back effect.

An external magnetic field acts on both the spin of the electron, and via the *minimal substitution* $\mathbf{P} \to \mathbf{P} - e\mathbf{A}$, where $\mathbf{A}$ is the magnetic vector potential. For a constant magnetic field B in the three direction, we can choose the vector potential to be

$$A_i = -\frac{B}{2}\epsilon_{ij3}x_j. \tag{13.47}$$

To first-order in B we have

$$\frac{(\mathbf{P} - e\mathbf{A})^2}{2m} = eB\frac{P_i\epsilon_{ij3}x_j}{2m} = \frac{eB}{2m}L_3, \tag{13.48}$$

where L_3 is the third component of the orbital angular momentum.

By definition, the interaction with the spin is given by a Hamiltonian

$$-\mu_3 B = -\frac{eg}{2}\sigma_3 B, \tag{13.49}$$

where g is the gyromagnetic ratio, relating the electron magnetic moment to its intrinsic spin. Dirac's relativistic theory of the electron gives $g = 2$. The full theory of Quantum Electrodynamics gives corrections to this, which are smaller by powers of $\frac{\alpha}{\pi}$, where $\alpha \sim \frac{1}{137}$ is the fine structure constant.

The full first-order perturbation is thus,

$$H_{PB} = \frac{e\hbar}{2m_e}B(K_3 + \sigma_3), \tag{13.50}$$

where K_3 is the orbital angular momentum in $\hbar$ units. If we consider the hydrogen eigenstates labelled by (n, l, m, s), with n fixed, this perturbation splits the $(2l + 1)$ degenerate orbital states and the two degenerate spin states. The $m = 0$ state splits into two states with energies $E_n \pm \frac{e\hbar}{2m}B$. More generally, we get states with energies $E_n + \frac{e\hbar}{2m}B(m \pm 1)$.

13.5.3   Fine Structure and the Weak-Field Zeeman Effect

For weaker fields, we have to compute two other (relativistic) perturbations, which compete with the external field. Normally, to first order, we can simply add up the effect of different perturbations, but in a degenerate system, one perturbation may be diagonalized in a different basis than the other. The correct procedure is to diagonalize the sum of all perturbations which are of the same size. We will concentrate on the weak-field regime, where the relativistic perturbations dominate over the external magnetic field. They leave over a degeneracy, which is split by even a tiny external field. The relativistic corrections, which exist even when the external field vanishes, give rise to the *fine structure corrections* to the hydrogen spectrum.

The first fine structure correction comes simply from the relativistic correction to the relation between energy and momentum for the electron. The correct relativistic formula is

$$E^2 - (\mathbf{P}c)^2 = m^2c^4. \tag{13.51}$$

This says that energy and momentum sit on a three-dimensional hyperboloid in a four-dimensional space. This relation is obviously invariant under hyperbolic rotations, which are Lorentz transformations.

In QM, $\mathbf{P}$ is still ($\hbar$ times) the translation operator, so we can write the correction to the Schrödinger equation when $|\mathbf{P}| \ll \mathbf{mc}$ as a term

$$\delta H = -\frac{(\mathbf{P}^2)^2}{4m_e^3c^2} = -\left(\frac{\mathbf{P}^2}{2m_e}\right)^2\frac{1}{m_ec^2}. \tag{13.52}$$

The second expression is $\frac{1}{m_e c^2}(H - V)^2$, for any potential, and when evaluating its matrix elements in any unperturbed eigenstate, we can replace H by the corresponding eigenvalue. The correction is rotation invariant, and commutes with the spin, and so will not break the angular momentum and spin degeneracies, but it will lift the degeneracies between different l states with the same n, because the expectation values of r^{-1} and r^{-2} depend on l. The perturbation is already diagonalized in the angular momentum basis. Thus, just like the Stark effect, this contribution to the fine structure involves matrix elements of r^p between radial Coulomb wave functions.

The second contribution to the fine structure is more complicated, and involves a term proportional to $\mathbf{L} \cdot \mathbf{S}$. Roughly speaking, the orbital motion of the electron produces a magnetic field in its rest frame,[2] and the electron dipole moment interacts with that field. Naively, one could evaluate the magnetic field by saying that in the rest frame of the electron, the proton is moving, thus producing an electric current, which produces a magnetic field. The problem with such a calculation is that the electron rest frame is an accelerated frame, so that the simple Lorentz transformation rules of electric and magnetic fields between inertial frames do not give the correct answer.

L.H. Thomas pointed out that the transformation to an accelerated frame could be viewed as a time-dependent Lorentz boost transformation. We do not have the space here to elaborate on this argument, but will simply record that Thomas' answer was simple: it multiplies the naive field by a factor of $1/2$. The proton current in the Lorentz frame which coincides instantaneously with the electron rest frame is

$$\mathbf{J}(\mathbf{X}) = -e\mathbf{v}\delta^3(\mathbf{X}) = -e\frac{\mathbf{P}}{m_e}\delta^3(\mathbf{X}). \tag{13.53}$$

Solving Maxwell's equation

$$\nabla \times \mathbf{B} = \frac{1}{\epsilon_0 c^2}, \tag{13.54}$$

we get

$$\mathbf{B} = \frac{1}{4\pi\epsilon_0}\frac{e}{mc^2 R^3}\mathbf{L}, \tag{13.55}$$

using

$$\mathbf{L} = \mathbf{R} \times \mathbf{P}. \tag{13.56}$$

Thomas' result reduces this by a factor of $1/2$.

The interaction of the magnetic dipole with this field is

$$\delta H = -\mu \cdot \mathbf{B}, \tag{13.57}$$

---

[2] The field is purely electric in the proton rest frame.

and the magnetic dipole moment of the electron is

$$\mu = g\frac{-e\mathbf{S}}{2m_e} = \frac{-e\mathbf{S}}{m_e}. \tag{13.58}$$

We have used the fact that Quantum Electrodynamics tells us that $g = 2$, up to a small correction. The final result is

$$\delta H = \frac{e^2}{8\pi\epsilon_0 m_e^2 c^2 R^3}\mathbf{S}\cdot\mathbf{L}. \tag{13.59}$$

Now

$$2\mathbf{S}\cdot\mathbf{L} = (\mathbf{L}+\mathbf{S})^2 - \mathbf{L}^2 - \mathbf{S}^2 = \hbar^2[j(j+1) - l(l+1) - 3/4]. \tag{13.60}$$

R^{-3} also commutes with $\mathbf{L}^2, \mathbf{J}^2$ so the states of fixed j and l are those in which this perturbation is diagonal. The allowed values of j for a given l are $l\pm 1/2$, so this perturbation splits the $2(2l+1)$ states in a degenerate spin/angular momentum multiplet, into two groups of degenerate states of size $2l+2$ and $2l$. Note that the external field perturbation, proportional to $B_{ext}S_3$ does not commute with J^2. However, when the field is weak, so that the energy shifts it induces are small compared to the fine structure, all we have to do is evaluate the matrix of S_3 in each of the subspaces with fixed j l and varying j_3. You will do this in Exercise 13.3 and see that the matrix elements are Clebsch–Gordon coefficients.

13.5.4   Coulomb Expectation Values of Powers of R

All of the perturbations of the hydrogen atom that we have studied, involve the expectation values of powers of R in Coulomb wave functions, or matrix elements of such powers between Coulomb wave functions. For R^{-p} with $p = 1,2$ we can evaluate these easily using the Feynman–Hellmann theorem. For other powers, we will need a relation, called Kramers' relation, between the expectation values of any three consecutive powers of r in any unperturbed hydrogen wave function. One writes the radial equation as

$$u_{nl}'' = [\frac{l(l+1)}{r^2} - \frac{2}{ra_B} + \frac{1}{n^2 a_B^2}]u_{nl}, \tag{13.61}$$

to write $\int dr\, ur^p u''$ in two ways. In the first one uses the equation directly, while in the second one integrates by parts before using it. Since Coulomb wave functions are real the method relates expectation values of different powers of r. You will prove this relation in Exercise 13.4.

Kramers' relation is

$$\frac{p+1}{n^2}\langle u_{nl}|r^p|u_{nl}\rangle - (2p+1)a_B\langle u_{nl}|r^{p-1}|u_{nl}\rangle + \frac{p}{4}[(2l+1)^2 - p^2]a_B^2\langle u_{nl}|r^{p-2}|u_{nl}\rangle = 0. \tag{13.62}$$

In a moment, we will evaluate the cases $p = -1, -2$ using the Feynman–Hellman theorem, and Kramers' relation does the rest of the work.

The effective Hamiltonian for the radial wave function $u = rR$ is

$$H = -\hbar^2 2m_e [\partial_r^2 - \frac{l(l+1)}{r^2}] - \frac{e^2}{4\pi\epsilon_0 r}. \tag{13.63}$$

Derivatives of the Hamiltonian with respect to l and e give the requisite expectation values of the first two inverse powers of r. We must be careful in taking the derivative w.r.t. l, because of the degeneracy of l levels. The principle quantum number n is $j_{max} + l + 1$ where j_{max} is the highest power in the Laguerre polynomial. If we varied l with n fixed, we would be making drastic changes in the wave function. Instead, we want to vary l with j_{max} fixed. The energies are

$$E_n = -\frac{m_e e^4}{32\pi^2\epsilon_0^2\hbar^2(j_{max}+l+1)^2}. \tag{13.64}$$

Then the F–H theorem tells us that

$$\langle\frac{1}{r}\rangle = -4\pi\epsilon_0\frac{\partial E_n}{\partial e^2} = \frac{m_e e^2}{4\pi\epsilon_0\hbar^2 n^2}. \tag{13.65}$$

$$\langle\frac{1}{r^2}\rangle = 2m_e\hbar^2(2l+1)\frac{\partial E_n}{\partial l} = \frac{m_e^2 e^4}{(2l+1)8\pi^2\epsilon_0^2\hbar^4 n^3}. \tag{13.66}$$

Putting everything together, we get a formula for the fine structure of hydrogen energy levels

$$E_{nj}^{fs} = \frac{\alpha^2 E_n^2}{2mc^2}(3 - \frac{4n}{j+1/2}). \tag{13.67}$$

Here j is the integer determining the square of the total angular momentum $\mathbf{J} = \mathbf{L} + \mathbf{S}$. α is the fine structure constant $\sim 1/137$. The full second order formula for the energy levels is

$$E_{nj} = -\frac{13.6\text{eV}}{n^2}(1 + \frac{\alpha^2}{n^2}[\frac{4n-3j-3/2}{4j+2}]). \tag{13.68}$$

If we now put the system in a weak magnetic field, the dominant effect is to split levels which are degenerate according to the formula just derived for E_{nj}. These are the states with different m_j values at fixed j. The perturbing Hamiltonian for the weak-field Zeeman effect is still $\frac{e}{2m_e}B(J_3 + S_3)$, but now we have to evaluate its expectation value in the states obtained by diagonalizing the spin-orbit plus fine structure perturbations. These are labeled by the eigenvalues of the operators that commute with the spin-orbit Hamiltonian, which are $\mathbf{J}^2, J_3, \mathbf{L}^2$. Note since the Hamiltonian without the external field is rotation invariant, we are free to choose the component of $\mathbf{J}$ that we diagonalize to be the one in the direction of the external field. For expectation values in states of fixed J_3 we can set

$$\langle m_J | S_3 | m_J \rangle = \langle m_J | \frac{\mathbf{S} \cdot \mathbf{J}}{J^2} J_3 | m_J \rangle, \tag{13.69}$$

because the other components of $\mathbf{S} \cdot \mathbf{J}$ have vanishing expectation values. Noting that

$$\mathbf{S} \cdot \mathbf{J} = \frac{1}{2}(J^2 + S^2 - L^2),$$

we can now evaluate the weak-field Zeeman perturbation as

$$\Delta E_{Zeeman} = \frac{e\hbar}{2m_e}[1 + \frac{j(j+1) - l(l+1) + 3/4}{2j(j+1)}]m_J. \tag{13.70}$$

As promised, the degenerate m_J levels are split. The term in square brackets is called the Lande g-factor. The prefactor $\frac{e\hbar}{2m_e} \equiv \mu_B$ is called the Bohr magneton. In SI units, it is 5.788×10^{-5} eV/T. T stands for Tesla, the unit of magnetic field.

13.5.5   A Three-Dimensional Example

Consider the matrix

$$\begin{pmatrix} E_1 & a_1 & a_2 \\ a_1^* & E_2 & a_3 \\ a_2^* & a_3^* & E_3 \end{pmatrix}. \tag{13.71}$$

If the a_i are small, we can use perturbation theory. The first-order perturbation vanishes because the perturbation is a purely off diagonal matrix in the basis where the unperturbed Hamiltonian is diagonal. Assuming there is no degeneracy, the operator whose expectation value we must evaluate to compute the second-order correction to the ground state E_1 is

$$H_2 = V \frac{P}{E - H_0} V = \begin{pmatrix} 0 & a_1 & a_2 \\ a_1^* & 0 & a_3 \\ a_2^* & a_3^* & 0 \end{pmatrix} \begin{pmatrix} 0 & 0 & a_1 \\ 0 & (E - E_2)^{-1} & a_2 \\ a_1^* & a_2^* & (E - E_3)^{-1} \end{pmatrix} \begin{pmatrix} 0 & a_1 & a_2 \\ a_1^* & 0 & a_3 \\ a_2^* & a_3^* & 0 \end{pmatrix}. \tag{13.72}$$

To compute the expectation value, we only need the $1, 1$ matrix element of this product matrix, which is $\frac{|a_1|^2}{(E-E_2)} + \frac{|a_2|^2}{(E-E_3)}$. Thus, to second order, the implicit B–W equation is

$$E = E_1 + \frac{|a_1|^2}{(E - E_2)} + \frac{|a_2|^2}{(E - E_3)}, \tag{13.73}$$

whose solution through this order is

$$E = E_1 + \frac{|a_1|^2}{(E_1 - E_2)} + \frac{|a_2|^2}{(E_1 - E_3)}. \tag{13.74}$$

There are similar equations for the other two eigenvalues. There are obvious problems when $|E_1 - E_i| \sim |a_{i-1}|^2$ or smaller.

Let us suppose the quasidegenerate pair are $E_{1,2}$. Then, the B–W prescription is to consider the matrix elements of

$$\begin{pmatrix} 0 & a_1 & a_2 \\ a_1^* & 0 & a_3 \\ a_2^* & a_3^* & 0 \end{pmatrix} \begin{pmatrix} 0 & 0 & a_1 \\ 0 & 0 & a_2 \\ a_1^* & a_2^* & (E - E_3)^{-1} \end{pmatrix} \begin{pmatrix} 0 & a_1 & a_2 \\ a_1^* & 0 & a_3 \\ a_2^* & a_3^* & 0 \end{pmatrix}, \tag{13.75}$$

in the two-dimensional subspace. This is relatively easy to do, because the middle matrix is proportional to a one-dimensional projector. The answer is

$$H_2 = \frac{1}{E - E_3} \begin{pmatrix} |a_2|^2 & a_2 a_3^* \\ a_2^* a_3 & |a_3|^2. \end{pmatrix} \tag{13.76}$$

The implicit equation for the quasidegenerate eigenvalues is

$$(E - E_a)(E - E_3) = \alpha_a, \tag{13.77}$$

where $a = 1, 2$ and the α_a are the (properly ordered) eigenvalues of the 2×2 matrix of the previous equation. The leading order contribution gives

$$E = E_a + \frac{\alpha_a}{E_a - E_3}. \tag{13.78}$$

13.5.6 Degenerate Perturbation Theory in a Macroscopic System

We end this discussion with a rather sophisticated example. In our discussion of the physics of solids, we mentioned briefly the Hubbard model. It consists of fermion operators $\psi(i)$ sitting on the points of a regular lattice, with anticommutation relations

$$[\psi_a(i), \psi_b^\dagger(j)]_+ = \delta_{ij}\delta_{ab}, \tag{13.79}$$

$$[\psi_a(i), \psi_b(j)]_+ = 0. \tag{13.80}$$

The fermions have spin $1/2$, and the subscript a labels the spin value $\pm\hbar/2$. The Hamiltonian is

$$H = \sum_{ij} k_{ij}\psi_a^\dagger(i)\psi_a(j) + g^2 \sum_i (\psi_a^\dagger(i)\psi_a(i))^2. \tag{13.81}$$

k_{ij} contains only nearest neighbor terms, and the first term in the Hamiltonian describes processes in which a fermion is destroyed on one site and another created on a nearest neighbor site. Repeated application of this term allows fermions to propagate throughout the lattice. Hermiticity of the Hamiltonian implies that $k_{ij} = k_{ji}^*$. The second term is a repulsion, which tries to forbid two fermions from occupying the same site. They are allowed to do so, consistent with Fermi statistics, because they have two spin states.

Solution of the Hubbard model consists of finding the ground state, for each value of the conserved fermion number $N = \sum_i [\psi_a^\dagger(i)\psi_a(i) - 1]$. We want to study the problem in the limit of large g^2, and for $N = V$ the total number of lattice sites. This is called the problem with a half filled band, because, in principle, we could accommodate twice as many fermions. For $g^2 \to \infty$, states with two fermions on a site are absolutely forbidden. The constraint on the total number of fermions tells us that there must be exactly one fermion on each site. Neither of these two constraints tells us what the fermion spin is on each site, so we actually have a degenerate ground state. Any state, with a fermion on each site, is a ground state, but the spin state of the fermion on each site is undetermined. The operators $\Sigma_m(i) = \psi_a^\dagger(i)(\sigma_m)_{ab}\psi_b(i)$ are the local spin operators, which act on the degenerate subspace of the full Hilbert space.

We now want to consider g^2 large but finite, and construct a perturbation theory with the hopping term $V = \sum_{ij} k_{ij}\psi_a^\dagger(i)\psi_a(j)$ as the perturbation. A single action of V on any state in the degenerate ground state subspace takes us to a superposition of states, each of which has two fermions on some lattice site, and none on one of its nearest neighbors. It is thus orthogonal to the degenerate subspace. All of the matrix elements of V in the degenerate subspace vanish, and there is no contribution to the energy shift in first order in $\frac{1}{g^2}$.

The second-order operator

$$V_2 = (1 - P)V\frac{P}{E - H_0}V(1 - P), \qquad (13.82)$$

is nonvanishing. In order for it to have a nonzero degenerate matrix element, the second action of V must create a fermion where the first action annihilated one, and annihilate a fermion on the doubly occupied site created by the first action. The intermediate state has an energy of order g^2 above the ground state, so the operator V_2 has eigenvalues of order $1/g^2$, if E is small compared to g^2. In Brillouin–Wigner perturbation theory, E is just the total energy of the state in the degenerate subspace, so it *is* negligible.[3] It follows that, to order $1/g^2$

$$V_2 = -\frac{c}{g^2}\sum_{ij} k_{ij}\psi_a^\dagger(i)\psi_a(j)\psi_b^\dagger(j)\psi_b(i)k_{ji}. \qquad (13.83)$$

We can move $\psi_b(i)$ through to the left, since it commutes with $\psi_a(j)\psi_b^\dagger(j)$ (remember that i and j are distinct points). Moving $\psi_a(j)$ to the right of $\psi_b^\dagger(j)$, we pick up a minus sign, plus a term from the nonvanishing anticommutator.

---

[3] Actually, if you have followed the derivation above, E is actually an operator in the degenerate subspace, but it is still true that it is negligible compared to the leading order energies of the excited states.

This results in

$$V_2 = \frac{c}{g^2} [\sum_{ij} |k_{ij}|^2 \psi_a^\dagger(i)\psi_b(i)(\psi_b^\dagger(j)\psi_a(j) - \delta_{ab})].$$ (13.84)

The second term in round brackets is a sum of local density operators $N(x) \equiv \psi_a^\dagger(i)\psi_a(i)$, but these are all equal to 1 on every state in the degenerate subspace, so this term just gives a constant negative shift in the ground state energy, proportional to $\frac{V}{g^2}$, and does not remove the degeneracy.

Now write

$$\psi_a^\dagger(i)\psi_b(i) = \frac{1}{2}[\psi^\dagger \sigma^n \psi (\sigma^n)_{ab} + \psi^\dagger \psi \delta_{ab}],$$ (13.85)

which one can verify by using

$$\text{tr}(\sigma^n) = 0,$$ (13.86)

$$\text{tr}(\sigma^n \sigma^m) = 2\delta^{mn}.$$ (13.87)

In the preceding equations, we have written the labels on the Pauli matrices as superscripts in order to distinguish them from the matrix indices a, b.

Putting these equations together, we find that the second-order Hamiltonian is, apart from the negative constant discussed above

$$V_2 \frac{c}{g^2} [\sum_{ij} |k_{ij}|^2 (N(i)N(j) + \Sigma^n(i)\Sigma^n(j))].$$ (13.88)

The first term is another constant on the degenerate subspace, while the second gives a nearest neighbor interaction between local spins, called the Heisenberg Hamiltonian [63]. The sign of the interaction is such that, if the spins were classical, one would lower the energy by having nearest neighbor pairs to point in opposite directions. This is called an *antiferromagnetic* interaction, and the Hamiltonian is called the Heisenberg antiferromagnet. The actual behavior of its quantum ground state is quite interesting, and depends both on the dimensionality of space and the type of lattice involved. There are still situations where the qualitative nature of the ground state is not understood.

It is clear that all of these systems are insulators, called Mott insulators [64]. The combination of the half filling constraint and the large g^2 limit prevents the transfer of charge through the lattice. On the other hand, for small g^2, Fermi liquid theory applies, and the system is a conductor. The Hubbard model has fascinatingly complex behavior as one varies the coupling, and the nature of the lattice, and it can reproduce the qualitative behavior of many of the states of condensed matter found in nature.

This concludes our discussion of time-independent perturbation theory for bound states. We will discuss the application of perturbation theory to the continuum eigenspectrum in the next chapter.

13.6 EXERCISES

13.1 Determine the energy levels of the anharmonic oscillator

$$H = \frac{P^2}{2m} + \frac{\omega^2 m}{2} X^2 + aX^3 + bX^4,$$

through second order in perturbation theory.

13.2 Consider the Hamiltonian of Exercise 13.1 with $a = 0$. Show that one can solve the Schrödinger equation for the ground state, order by order in a power series in b by an ansatz of the form

$$\psi(x) = \sum_{k=0}^{\infty} b^k P_k(x) e^{-\frac{m\omega}{2\hbar} x^2},$$

where P_k is a polynomial. What is the order of P_k? Show that the Schrödinger equation leads to a difference equation for the coefficients of the polynomials and the perturbed energy level $E = \hbar\omega + \sum_{k=1}^{\infty} b^k E_k$. Verify that this method of solution agrees with the results of Exercise 13.1 for the level E_2.

13.3 For a normal matrix, the equation $P(a) = \det (A - a)$ is a polynomial in a whose roots are the eigenvalues a_k. Now suppose $A = A_0 + \lambda A_1$, where A_i are normal. Argue that the roots of the polynomial $P(a)$ are analytic functions of λ except for branch points where one or more roots coincide.

13.4 Consider a perturbation $V = f(r)\mathbf{L} \cdot \mathbf{S}$ of the hydrogen atom. Consider all those states with a fixed value of the principal quantum number n. Describe how V breaks the degeneracy of those states. Is there any degeneracy left?

13.5 We found the eigenvalues and eigenstates of the harmonic oscillator for any value of the mass and the spring constant. Now consider an oscillator with a different spring constant, which is a small perturbation of the original one. Use perturbation theory to compute the first two corrections to the energy levels and verify that they agree with the exact formula.

13.6 Repeat Exercise 13.5, for a small perturbation of the mass, rather than the spring constant. You will calculate different matrix elements, but come to the same conclusion

13.7 Consider a harmonic oscillator perturbed by a constant force, with potential $V = -FX$. Solve this problem exactly. Then solve it by perturbation theory, to the first nontrivial order. You might want to write X in terms of creation and annihilation operators. Show that the expansion of the exact answer agrees with that calculated by perturbation theory.

13.8 Consider the nucleus of a heavy atom in a molecule, whose other constituents are much lighter. The low-lying vibrational excitations of the molecule consist of oscillations of that heavy atom around its equilibrium position. with frequencies $\omega_i = \sqrt{\frac{k_i}{m}}$, where k_i are the spring constants in the three principle directions of oscillation. Evaluate the effect of an external electric field $\mathbf{E} \cdot \mathbf{X}$ on these levels, to second order in perturbation theory in the electric field strength.

13.9 When an atom, molecule, or nucleus interacts with an external constant electric field, the first-order term in perturbation theory typically vanishes. This is due to a symmetry called *parity* under which all components of $\mathbf{X}$ are reflected. The electric field is a polar vector, and is odd under parity. If the ground state of the system is nondegenerate, then it must be an eigenstate of the parity operator $U(P)$, since that operator commutes with the Hamiltonian. Show that this means the expectation value of the electric field perturbation vanishes.

13.10 In Exercise 13.9, suppose the ground state is degenerate. Since $U^2(P) = 1$, we might have some linear combinations of degenerate states with $U(P) = 1$ and others with $U(P) = -1$. In this case, the electric field can have nonzero matrix elements between states of different parity. Show that if there are k more states of one parity than the other, then the electric field matrix in the degenerate subspace has k zero eigenvalues. Show that in the subspace orthogonal to these zero modes, the electric field operator takes the form $(\mathcal{E}_1 \sigma_1 \otimes K_N + \mathcal{E}_2 \sigma_2 \otimes M_N$, where M_N and K_N are $N \times N$ matrices, and N equal to half the dimension of the orthogonal subspace. Show that, in general, the ground state energy of this system depends on the electric field, to first order in perturbation theory in the electric field. Since the energy has a term linear in $\mathcal{E}$, we say that the atom, molecule, or nucleus has a *permanent electric dipole moment*.

13.11 In our discussion of the Born–Oppenheimer approximation, we showed that molecules had a definite "shape," which is not invariant under rotations. This means that the ground state has a lot of low-lying rotational levels, with energies of order $\frac{m_e}{m_{mol}}$ in Rydberg units. As we saw for ammonia, when the shape is not invariant under parity, the smallest transition probability between the two parity-reversed shapes led to a ground state that was a definite eigenstate of parity. Prove that this means that the expectation value of the electric dipole moment is zero. Thus, the situation outlined in the previous problem is very special. How is it then that molecules have permanent electric dipole moments? The answer lies in the large number of low-lying rotational levels. The typical energies at which we observe molecules are those in which a large number of these rotational levels can be excited. These levels are calculated by fixing the orientation of the molecule (e.g., whether the ammonia molecule has the nitrogen atom above or below the hydrogen plane in some fixed coordinate system) and then quantizing the rotational motion. The actual state we observe is a time-dependent superposition of rotational levels. If we consider a general state, allowing for superpositions of the two orientations of the molecule, there is no reason that the superposition of rotational levels should be the same for the two orientations. Now consider a small perturbation of size ϵ mixing the rotational subspaces for the two different orientations. Show that if we assume random superpositions for the two orientations, the matrix element of the perturbation is at most of order ϵ/N, where N is the dimension of the space of allowed rotational levels. Show further that these matrix elements are time dependent and that their time average over times long compared to $\hbar/\Delta E$ is of order $\hbar/(\Delta E t)$. Here ΔE is a typical energy splitting between rotational levels.

13.12 The typical scale of rotational energy levels is $10^{-3}/A$ in Rydberg units, where A is the total number of protons plus neutrons in the molecule. For ammonia, how high does the temperature have to be before we have to take these levels into account in our description of the physics of the molecule. Recall that room temperature is 273 K, which is about $1/40$ electron volts or roughly 2×10^{-3} Rydberg. Do the same estimate for water molecules.

13.13 Consider an infinite square well confining a particle to $-L \leq x \leq L$ and add a harmonic term $V = \frac{1}{2}kx^2$. Assume that the oscillator frequency $\omega = \sqrt{k/m}$ is much smaller than the characteristic frequency of the particle in the well $\omega_w = \frac{\hbar}{2mL^2}$. Treat the harmonic term as a perturbation and calculate the first order perturbed energies of the ground state and first excited state of the well.

13.14 The problem in Exercise 13.13 can be solved exactly. The solutions of the Schrödinger equation are *parabolic cylinder functions* (consult your favorite online math oracle) and the condition for eigenfunctions corresponds to finding zeroes of these functions. Use the integral representation or power series solution to show that the exact and approximate solutions for the eigenvalues exist.

13.15 Repeat Exercise 13.13 for the bound states in a finite square well

$$V(x) = -\theta(L - x)\theta(x + L)[V_0 - \frac{k}{2}x^2],$$

to lowest order in k.

13.16 Consider a harmonic oscillator on the full real line, perturbed by a finite square well. Calculate the first-order perturbation of the ground state energy and express the answer in terms of error functions.

13.17 Evaluate the weak-field Zeeman effect for the states of the hydrogen atom with $n = 2$.

CHAPTER 14

Perturbation Theory: Time Dependent

14.1 INTRODUCTION

Time-dependent perturbation theory is used to study the response of quantum mechanical systems to time-dependent external fields. However, as a consequence of a trick introduced by Dirac, it is also a simple way to set up a simple perturbation theory for scattering amplitudes. Dyson introduced a perturbative solution for the evolution operator of a general time-dependent system. The formulae involve the fundamental concept of a time-ordered product of operators. The more general concept of a product of operators ordered along some path in a multidimensional space or space-time is one of the most important notions in modern theoretical physics.

Time-dependent perturbation theory is the tool of choice for understanding the excitation and decay of excited atomic or nuclear states by external fields, which is the way we explain spectral lines. We will derive Dyson's and Dirac's formulae, and apply them to a variety of simple problems.

14.2 DYSON'S FORMULA

Time-dependent Hamiltonians arise in a variety of different ways in quantum mechanics (QM), most commonly when one is studying a system coupled to another one whose dynamics is not computed explicitly, but approximated by some external classical time-dependent couplings in the system Hamiltonian. The fundamental formula for solving time-dependent problems is a formal solution of the time-dependent Schrödinger equation first written by Dyson [65]:

$$i\hbar\partial_t U(t, t_0) = H(t)U(t, t_0). \tag{14.1}$$

Here $U(t, t_0)$ is the time evolution operator, the unitary operator which maps the initial state $|\psi(t_0)\rangle$ into the final state at time t

$$|\psi(t)\rangle = U(t, t_0)|\psi(t_0)\rangle. \tag{14.2}$$

We also have the obvious boundary condition $U(t_0, t_0) = 1$, which allows us to write the Schrödinger equation as an integral equation

$$U(t, t_0) = 1 - \frac{i}{\hbar} \int_{t_0}^{t} ds H(s) U(s, s_0). \tag{14.3}$$

This may be formally solved by iteration

$$U(t, t_0) = 1 - \frac{i}{\hbar} \int_{t_0}^{t} ds H(s) - \frac{1}{\hbar^2} \int_{t_0}^{t} ds_1 \int_{t_0}^{s_1} ds_2 \; H(s_1) H(s_2) U(s_2, s_0), \tag{14.4}$$

and so on. We can continue to use the integral equation inside the last integral, to expose more and more explicit powers of $H(s)$. Formally, the answer is written as the sum of an infinite series of terms, with the n-th term involving integrals over an n-dimensional region characterized by the inequalities

$$t \geq s_1 \geq s_2 \ldots \geq s_n \geq t_0. \tag{14.5}$$

Note that the inequalities ensure that the action of the Hamiltonian operators are such that operator order reflects time order.

We can write a more symmetric form of this formula by introducing the notion of a time-ordered product of operators. The time-ordered product is defined by

$$TH(s_1) \ldots H(s_n) = \sum_P \theta(s_{P(1)} - s_{P(2)}) \ldots \theta(s_{P(n-1)} - s_{P(n)}) H(s_{P(1)}) \ldots H(s_{P(n)}). \tag{14.6}$$

$P(k)$ is a permutation of the integers $1 \ldots n$ and we sum over all such permutations. The n-dimensional hypercube, $\mathcal{H}$, defined by $t \geq s_i \geq t_0$ is completely covered by $n!$ regions in which the inequality $t \geq s_1 \geq s_2 \ldots \geq s_n \geq t_0$ is replaced by $t \geq s_{P(1)} \geq s_{P(2)} \ldots \geq s_{P(n)} \geq t_0$. Thus, if $\mathcal{T}$ is the hypertriangular region defined by the first inequality, then

$$\int_{\mathcal{T}} d^n s \; H(s_1) \ldots H(s_n) = \frac{1}{n!} \int_{\mathcal{H}} d^n s \; TH(s_1) \ldots H(s_n). \tag{14.7}$$

The formal solution of the Schrödinger equation can thus be written

$$U(t, t_0) = T exp(-\frac{i}{\hbar} \int_{t_0}^{t} ds/H(s)), \tag{14.8}$$

where the time-ordered exponential is just the sum of the time-ordered integrals of products of $H(s_i)$ over the n-dimensional hypercube, multiplied by $\frac{(-i/\hbar)^n}{n!}$.

14.3   THE DIRAC PICTURE

Now consider a Hamiltonian $H = H_0 + V$, where the perturbation V may depend explicitly on time. We write the Schrödinger picture time evolution operator as

$$U(t, t_0) = e^{-\frac{i}{\hbar} H_0 (t - t_0)} W(t, t_0). \tag{14.9}$$

Rewrite the Schrödinger equation as an equation for W

$$e^{-\frac{i}{\hbar} H_0 t} i\hbar \partial_t W e^{\frac{i}{\hbar} H_0 t_0} = V e^{-\frac{i}{\hbar} H_0 t} W(t, t_0) e^{+\frac{i}{\hbar} H_0 t_0}. \tag{14.10}$$

$$i\hbar \partial_t W = e^{+\frac{i}{\hbar} H_0} V e^{-\frac{i}{\hbar} H_0 t} W(t, t_0) \equiv V_I(t). \tag{14.11}$$

$V_I(t)$ is the operator V, with each Schrödinger picture operator replaced by *the corresponding Heisenberg picture operator evolved with the Hamiltonian H_0.* $V_I(t)$ is called the interaction picture version of the perturbation V and is time dependent even if the original perturbation is time independent. The results of the previous section tell us that

$$W(t, t_0) = T exp\left(-\frac{i}{\hbar} \int_{t_0}^{t} ds \, V_I(s)\right). \tag{14.12}$$

14.4   TRANSITION AMPLITUDES

The quantities of interest in time-dependent perturbation theory are *transition amplitudes*; probability amplitudes for one energy eigenstate of H_0 to convert into another under the action of the perturbation. A typical situation is the decay of an excited state of an atom into its ground state under the influence of a time-dependent perturbation. We will calculate transition amplitudes to first order in the perturbation. To that order, the amplitude is

$$T_{ab}(t) = -\frac{i}{\hbar} \langle E_a | e^{-\frac{i}{\hbar} H_0 (t - t_0)} \int_{t_0}^{t} ds \, V_I(s) | E_b \rangle = e^{\frac{i}{\hbar} E_a (t - t_0)} \int_{t_0}^{t} ds \, \langle E_a | e^{-\frac{i}{\hbar} (E_b - E_a) s} V(s) | E_b \rangle. \tag{14.13}$$

$V(s)$ is now the time-dependent perturbation in the Schrödinger picture. If we assume that the time-dependent perturbation vanishes outside of a finite interval, then we can send the limits of the s integration to $\pm\infty$, whenever $V(s)$ vanishes for $s > t$ and $s < t_0$. The transition amplitude is then given by the matrix element of the Fourier transform of $V(s)$ between initial and final states, evaluated at the energy difference between the initial and final states.

   An important special case is a monochromatic driving force $V(s) = \cos(\omega s) F(s)$, where $F(s)$ is a smooth function which approximates $\theta(a - s)\theta(a + s)$ with $a \gg \omega^{-1}$. If we set $F = 1$, the Fourier transform has poles at $\omega = \pm \frac{E_a - E_b}{\hbar} \equiv \pm \omega_0$. When F vanishes outside the interval $[-a, a]$, those poles are displaced into the complex plane so the result is peaked around the *resonance frequency* ω_0. Setting $F = 1$ and choosing ω close to ω_0, we can write the transition probability, the absolute square of the transition amplitude, approximately as

$$P_{a \to b} \approx \frac{|\langle E_a | V | E_b \rangle|^2}{\hbar^2} \frac{\sin^2[(\omega_0 - \omega)t/2]}{(\omega_0 - \omega)^2}. \qquad (14.14)$$

Here we have taken $t_0 = 0$ and allowed t to be anything. Note that this expression can exceed 1 and cannot be correct for ω too close to ω_0 except when t is near a zero of the sine. Higher order perturbation theory becomes important at other times, as you will explore in Exercise 14.1, and ensures that the transition probability is always less than one.

For finite times, the transition probability oscillates so there are intervals of time over which the transition is very probable, interspersed with intervals where it is unlikely to occur.

14.5 ELECTROMAGNETIC RADIATION FROM ATOMS

The proper framework for treating the interaction of radiation with atoms is the quantum field theory called Quantum Electrodynamics or QED. In that theory, the Coulomb term in the Hamiltonian of atoms, which we have written down is supplemented by an interaction which allows electrons and protons to change their energy by emitting photons, quantized excitations of the electromagnetic field. The excited states of atoms are no longer exact eigenstates of the full Hamiltonian, because they are degenerate in energy with states where the atom has relaxed to its ground state and one or more photons have been emitted to preserve the energy balance. Since there are many more states of the second type, and since the emitted photons quickly propagate far from the atom, the excited state decays. We can only resuscitate it briefly, by shining light on the atom. Indeed, this is how the spectral lines associated with excited states are observed. The excited energy levels still have an approximate meaning, because they are *metastable*: the probability of a photon emission is proportional to the *fine structure constant* $\alpha \sim 1/137$ and it takes a long time, on atomic time scales, for the decay to occur.

In this section, we will describe an approximate scheme for calculating radiative transition probabilities, which is based on treating the electromagnetic field as a time-dependent classical perturbation. Unlike most of the approximations discussed in this book, this semi-classical radiation theory is not the first term in a systematic expansion of the exact transition amplitudes. Nonetheless, it was of great historical importance, and many of the concepts we will introduce survive in the systematic treatment of these amplitudes in QED.

The typical spatial extent of an atom is given by the Bohr radius $\sim 10^{-8}$ cm. The low lying electrons in an atom of high Z are closer to the nucleus than this, by a factor $\sim 1/Z$, but radiative transitions in such atoms involve the electrons in outer shells, which feel only a screened Coulomb potential. On the other hand, the typical energy difference between atomic levels is $\sim 10 eV$. This corresponds to a wave length for light emitted in the transition that is of order 10^{-5} cm, so that the spatial variation of the electromagnetic field is negligible over the size of the atom. This motivates the *dipole approximation* in which one considers the electric field of the emitted or absorbed light to be constant in space

$$\mathbf{E}(t, \mathbf{x}) = E_0 \cos(\omega t)\hat{\mathbf{e}}, \qquad (14.15)$$

where $\hat{\mathbf{e}}$ is the unit vector describing the direction of the field. Taking this to be in the three direction, we can write the electromagnetic scalar potential

$$\Phi(t, \mathbf{x}) = -E_0 x_3 \cos(\omega t). \qquad (14.16)$$

This leads to a perturbation of the Hamiltonian of a particle of charge q:

$$V = -q E_0 x_3 \cos(\omega t). \qquad (14.17)$$

Note that we are neglecting the interaction of the electron with the magnetic field of the electromagnetic wave. This is a consequence of the nonrelativistic kinematics relevant to atomic transitions. The magnetic field interaction is smaller by a factor of $v/c \sim \frac{\hbar}{a_B m_e c} \sim \alpha \sim \frac{1}{137}$ than the electric effect that we are studying.

The transition amplitude is then proportional to the matrix element of the third component of the electric dipole operator

$$\mathbf{d} = q\mathbf{x}, \qquad (14.18)$$

between initial and final states. Call that matrix element d_{ab}. We should note that for X-ray transitions, where the emitted energy is of order 0.1–100 keV, the dipole approximation is not reliable.

Our general results on periodic perturbations give us a dipole transition probability

$$P_{a \to b}(t) = \left(\frac{|d_{ab}| E_0}{\hbar} \right)^2 \frac{\sin^2[(\omega - \omega_0)t/2]}{(\omega - \omega_0)^2}. \qquad (14.19)$$

This result is the same whether one is thinking of a transition from a lower energy state to a higher one or vice versa. The first process, which is called *absorption* (of light, or of a photon), is easy to understand intuitively: the electromagnetic field provides the energy to bump the electron up to a higher state. The second process is there because Hermiticity of the Hamiltonian forced us to include complex waves with both positive and negative frequencies $e^{\pm i\omega t}$, so that we get two possible poles in the transition amplitude.[1]

The second transition is called stimulated emission, and its existence was first pointed out by Einstein, as well as the fact that the probabilities for absorption and stimulated

---

[1] Never forget that the poles are an artifact of letting the wave exist forever, neglected the damping factor $F(t)$, which converts the poles into finite enhancements of the transition probability.

emission are equal. At a deeper level, the reason for the connection between these two phenomena stems from fact that the electromagnetic field is quantized, and quantized with Bose statistics (Chapters 5 and 12). The positive frequency $e^{-i\omega t}$ terms in the field are multiplied by photon annihilation operators, and are responsible for absorption processes, while the negative frequency terms multiply photon creation operators. The creation operator terms are responsible for both the stimulated emission probability we have just discussed and the *spontaneous emission* process in which an atom makes a transition from an excited state to the ground state (the excited state decays into the ground state plus a photon). The probabilities for all three processes are the same, because the creation and annihilation terms are related by Hermiticity, and so give rise to amplitudes which are complex conjugates of each other. Remarkably, Einstein understood all of this [66] in 1917, 8 years before the Schrödinger equation was invented and even longer before the invention of QED. His arguments[2] used only general notions of probability theory and the Planck distribution for thermal photons, which was discovered in 1900. The derivation of Planck's distribution requires QED.

The process of stimulated emission is the fundamental principle underlying lasers and masers (light or microwave amplification by stimulated emission of radiation). Recall, from the elementary theory of the harmonic oscillator, that a creation operator acting on a state with n photons gives $\sqrt{n}$ times the normalized state with $n+1$ photons. Now suppose we have somehow introduced a *population inversion* into a distribution of identical atoms, so that, contrary to the expectations of Boltzmann's statistical mechanics, there are more atoms in the excited state $|E_a\rangle$ than in the ground state $|E_b\rangle$. The presence of a low amplitude external field, whose frequency is tuned near the transition frequency ω_0 produces both absorption of photons, and stimulated emission, with equal probability. Since the population is inverted, we end up with more photons of the frequency ω_0 than we started with, but this means a stronger field, which enhances stimulated emission, etc. We end up with a final state having a very strong field, all at the frequency ω_0.

14.6 INCOHERENT PERTURBATIONS AND RADIATIVE DECAY

So far we have studied electromagnetic perturbations of fixed frequency, polarization, and direction of propagation.[3] We now ask how the formulae change for incoherent radiation, such as one might encounter in a thermal bath. The first step is to write E_0^2 in our formulae for transition rates in terms of the energy density u in the electromagnetic field.

$$E_0^2 = \frac{2}{\epsilon_0} u. \tag{14.20}$$

---

[2] You can find a clear explanation of Einstein's arguments in Griffiths' popular textbook [67].

[3] The direction of propagation is transverse to the polarization. It did not appear explicitly because we made the approximation of a field with no spatial variation.

We are working in SI units and ϵ_0 is the permittivity of the vacuum [68]. In a situation where the electromagnetic field is an incoherent sum of waves of different frequencies, the energy density is written as

$$u = \int \rho(\omega)d\omega, \tag{14.21}$$

and it is plausible, and turns out to be correct, simply to put the transition rate we have calculated at fixed ω under the integral sign

$$P_{b\to a}(t) = \frac{2}{\epsilon_0 \hbar^2}|d_{ab}|^2 \int_0^\infty \rho(\omega)\frac{\sin^2[(\omega-\omega_0)t/2]}{(\omega-\omega_0)^2}d\omega. \tag{14.22}$$

If $\rho(\omega)$ is a slowly varying, smooth function, and the other term in the integral is sharply peaked around $\omega = \omega_0$, we can pull $\rho(\omega_0)$ out of the integral. Then, introducing $x = (\omega_0 - \omega)t/2$, we can evaluate the transition rate approximately (Exercise 14.1) as

$$P_{b\to a}(t) \approx \frac{\pi|d_{ab}|^2}{\epsilon_0 \hbar^2}\rho(\omega_0)t. \tag{14.23}$$

Note that this gives a constant transition *rate* rather than the oscillatory behavior found for the idealized case of monochrome radiation. This formula is a special case of a very general result, called *Fermi's Golden Rule: a transition rate is the product of a squared matrix element, and the density of final states.* In many examples, there are *many* final states and we must sum or integrate this result over all of them to get what is called the inclusive rate for the transition. This is what is relevant if we do not measure the detailed properties of the final state. This is commonly the case for decays of excited states. A common catchphrase for this sum over possible final states is "integrating over phase space."

The above formula still assumes fixed directions of propagation and polarization for the electromagnetic field. The quantity d_{ab} is the matrix element of the dot product of the unit vector of polarization $\hat{\mathbf{e}}$ with the dipole operator $\mathbf{x}$. Averaging over polarizations[4] of the quantity $|\hat{\mathbf{e}} \cdot d_{ab}|^2$ gives us

$$\sum_i |\hat{\mathbf{e}}_i \cdot d_{ab}|^2 = d_{ab}^{*\,k}\Pi_{kl}d_{ab}^l = 1 - |\hat{\mathbf{n}} \cdot d_{ab}|^2. \tag{14.24}$$

Here Π_{kl} is the two by two projection matrix on the subspace of three-dimensional space orthogonal to the direction of propagation $\hat{\mathbf{n}}$. We now average this over the direction of propagation by doing the integral

$$I = \frac{1}{4\pi}\int \sin(\theta)d\theta d\phi \sin^2(\theta) = \frac{1}{3}. \tag{14.25}$$

[4] In the current context, where we are discussing absorption or stimulated emission, we do this average because we assume we are in a state with equal probabilities for the polarizations of the impinging radiation. This is logically different than the *sum* over final states we would do, by Fermi's Golden Rule, if we did not observe the final polarization.

The final result for the transition rate is

$$R_{b \to a} = \frac{\pi}{3\epsilon_0 \hbar^2} |d_{ab}|^2 \rho(\omega_0). \tag{14.26}$$

As an example, we can calculate the transition rates between states of a charged harmonic oscillator, and in Exercise 14.1, you will do similar calculations for transitions in the hydrogen atom. A charged oscillator is a crude model of a radio transmitter or other device for producing electromagnetic waves. In this case, the dipole operator is a linear combination of a creation and an annihilation operator (of the oscillator, not the electromagnetic field!) so it has matrix elements only between neighboring states.

$$\langle n|\mathbf{d}|m \rangle = q\sqrt{\frac{m\hbar}{2m\omega}} \delta_{n,m-1} \hat{\mathbf{e}}, \tag{14.27}$$

where we have only taken the matrix element to the lower state because we are discussing emission of radiation in the decay of the excited state. The resonance frequency ω_0 is of course just ω so the transition rate is

$$R_m = \frac{mq^2\omega^2}{6\pi\epsilon_0 m_e c^3}. \tag{14.28}$$

For constant transition rate, if we start with N_m atoms in the state m, the number left after time t is e^{-t/R_m} and we define the lifetime of a state to be the time at which only N_m/e of the atoms are left in the state m, so the lifetime is R_m^{-1} the inverse of the rate. The half-life, defined as the time at which half the excited states have decayed away, is used in discussions of nuclear decays, and is given by $R_m^{-1}\ln 2$.

The energy radiated in this decay is $\hbar\omega$ and the power radiated is this energy multiplied by the decay rate

$$P = \frac{q^2\omega^2}{6\pi\epsilon_0 m_e c^3}\left(E_m - \frac{1}{2}\hbar\omega\right). \tag{14.29}$$

The quantum rate of energy emission is of course zero for the ground state. You will calculate the corresponding power for a classical oscillator in Exercise 14.2. That power is of course proportional to the acceleration of the charge, and the total power radiated in one cycle of classical oscillation can be calculated using the equations of motion. It is

$$P_{cl} = \frac{q^2\omega^2}{6\pi\epsilon_0 m_e c^3} E, \tag{14.30}$$

and it is the same as the quantum rate when $\hbar = 0$.

14.7 SELECTION RULES

The rule that oscillator states decay only to the next lowest state is an example of a *selection rule*. Selection rules follow generally from symmetries of the unperturbed problem, which are broken by the perturbation. The transformation properties of the perturbation under the symmetry of the unperturbed system, then constrain the allowed transitions to lowest order in perturbation theory. Generally, most of these constraints are lifted in second-order perturbation theory and none of them are exact (unless of course there is some exact residual symmetry). In the case of the oscillator, the symmetry is the unitary transformation generated by the Hamiltonian. The dipole perturbation does not commute with this, but it transforms simply, being the sum of an energy raising and an energy lowering operation.

A general situation in which selection rules show up is a small perturbation of a rotation invariant system. The perturbation V can be expanded in terms of operators which transform as components of some integer spin irreducible representation of the rotation group, which is to say, like some spherical harmonic. To first order in perturbation theory, we can treat each of these irreducible components separately. Let us denote the components of an irreducible perturbation as V_A. A runs from $-j_V$ to j_V, where J_V is the spin of the representation under which V_A transforms. Then an irreducible perturbation has the form, $V_{irr} = g^A V_A$. Now consider the action of the angular momentum operators J_a on a subspace of states given by $V_A|j, m\rangle$, where $|j, m\rangle$ is some collection of degenerate eigenstates of the unperturbed Hamiltonian which transform in the representation with spin j. Then

$$J^a V_A|j, m\rangle = [J^a, V_A]|j, m\rangle + V_A (J^{(j)\,a})_{mk}|j, k\rangle. \tag{14.31}$$

The matrix $J^{(j)\,a}_{mk}$ is the spin j representation of angular momentum. Now use the fact that

$$[J^a, V_A] = i(J^a)_{AB} V_B, \tag{14.32}$$

to conclude that this subspace of states transforms under rotations like the states of a pair of distinct particles, one with spin j and the other with the spin, call it j_V, of the V_A representation. The rules of addition of angular momentum tell us that this contains every angular momentum between $j_V + j$ and $|j_V - j|$ exactly once. Thus, if we compute the matrix elements

$$\langle J, M|V_A|j, m\rangle, \tag{14.33}$$

which induce transitions between eigenstates of the unperturbed Hamiltonian under the influence of the perturbation, then these matrix elements vanish unless $j_V + j \geq J \leq |j_V - j|$. In addition, if the perturbation $g^A V_A$ only contains certain of the J_3 eigenstates in the representation spanned by V_A, then only changes in J_3 corresponding to those values, are allowed.

These remarks are formalized in the Wigner–Eckart theorem [62] which states that

$$\langle J, M | V_A | j, m \rangle = c_{M,A,m}^{J,j_V,j} R(J, j_V, j).\tag{14.34}$$

The coefficients $c_{M,A,m}^{J,j_V,j}$ are completely determined by the group theory of angular momentum (they are called Wigner 3j symbols or Clebsch–Gordon coefficients), while the *reduced matrix elements* $R(J, j_V, j)$ depend on the angular momentum only in the indicated fashion, but are specific to the particular system and perturbation. You will find a quick proof of this theorem in Appendix D on group theory.

Let us apply these rules to first-order electromagnetic transitions in the dipole approximation. The perturbation is a vector and carries $m = \pm 1$ in some direction perpendicular to the direction of propagation of the perturbing wave.[5] Thus, dipole transitions only occur between states whose j values differ by 1 or 0. Furthermore, the value of m *must* change by ± 1 in any transition.

It is an unfortunate fact that transitions that do not obey these rules were called forbidden transitions by the early practitioners of atomic QM. They are forbidden only in the electric dipole approximation. Variation of the electric field across the atom is only down by a factor of 100 to 1,000, as are magnetic effects and corrections coming from higher orders in time-dependent perturbation theory. Thus, forbidden transitions actually occur, albeit at suppressed rates.

14.8 EXERCISES

14.1 Calculate the rate of electric dipole transitions between two states of the hydrogen atom.

14.2 Calculate the final state for the dipole transition of a charged oscillator assuming the initial state is a coherent state $|z\rangle$. Do this by expressing the coherent state as a sum of eigenstates, and using the formulae in the text.

14.3 Consider a time-dependent Hamiltonian

$$H(t) = \hbar\omega(t)a^\dagger a,$$

where a is the usual annihilation operator. Find the selection rules for transitions between oscillator states if $\omega(t) = \omega + \delta\omega(t)$, where the time-dependent piece is small. What is the relationship between this Hamiltonian and a harmonic oscillator with time-dependent frequency?

---

[5] The fact that photons cannot have a longitudinal polarization cannot be understood in nonrelativistic QM, because it is a special property of representations of the Lorentz group describing massless particles with spin. However, this fact *is* encoded in the classical Maxwell equations, so it was understood before a full appreciation of the properties of QED.

14.4 Consider a hydrogen atom in its ground state, in the presence of a time-dependent vector potential (**Warning**: for the rest of the problems in this chapter, we will use Gaussian units for electromagnetism. You can convert to the SI units used in the rest of the book by replacing $E_0 \to \sqrt{4\pi\epsilon_0} E_0$.)

$$\mathbf{A}(\mathbf{r},t) = c\frac{\mathbf{E_0}}{\omega}\cos(\mathbf{k}\cdot\mathbf{r}-\omega t),$$

where ω is positive. c is the speed of light. We want to calculate the probability that this perturbation ejects the electron from the atom into one of the Coulomb scattering states. This problem will take up the next few exercises. We have broken it into bite size pieces, but you should do all of them, eventually. In principle, what we are doing is calculating the matrix element of this time-dependent perturbation between the ground state of hydrogen and a scattering state with some outgoing momentum $\mathbf{p}$. We are going to make an approximation based on the fact that for large enough momentum we expect the outgoing electron to spend most of its time far from the atom, so that the scattering state can be approximated by a plane wave. To leading order in perturbation theory in E_0, show that the time-dependent Hamiltonian is

$$V(t) = \frac{e}{2m\omega}(e^{i(\mathbf{k}\cdot\mathbf{X}-\omega t)} + h.c.)\mathbf{E_0}\cdot\mathbf{P}.$$

In writing this formula, you must use the Coulomb or radiation gauge for the vector potential. In this gauge, $\nabla\cdot\mathbf{A} = 0$. Show that the matrix element between initial and final states of the term shown explicitly gives a delta function $\delta(E_f - E_i - \hbar\omega)$ when integrated over time. Show that the complex conjugate term gives a delta function that vanishes when $E_f > E_i$. The complex conjugate term would describe extraction of energy (spontaneous emission) from the atom, which is impossible for the ground state. The fact that the two appear with equal strength in the Hamiltonian, a consequence of Hermiticity, is the fundamental fact underlying the equality of Einstein's A and B coefficients.

14.5 In the previous exercise, $\hbar\mathbf{k}$ is the momentum transferred to the atom (which is mostly carried by the outgoing electron because the nucleus is so heavy) by the space-time-dependent field. The absolute value of $\mathbf{k}$ is equal to ω/c because the field satisfies Maxwell's equations. $\hbar\omega = E_f - E_i$ which is R Rydbergs with $R > 1$ for a liberated electron. Thus, in Rydbergs, $\hbar k = R/c$. On the other hand, the typical scale of momentum in the ground state wave function of hydrogen is $\hbar/a_B$. Show that the ratio of the momentum transfer to the typical momentum is of order $\alpha_{em}R = \frac{e^2 R}{\hbar c} \sim \frac{R}{137}$. Argue that this means that we can have a fast moving outgoing electron but still neglect the term proportional to $\mathbf{k}\cdot\mathbf{X}$ in $V(t)$. This is called the *dipole approximation*. Show that it means that we can treat the electric and magnetic fields as constant in this approximation.

14.6 Show that putting all of these approximations together, the amplitude for the transition is

$$A_{fi} = 2\pi\delta\left(\frac{E_f - E_i}{\hbar} - \omega\right) N \int d^3x \; e^{-i\mathbf{k_f}\cdot\mathbf{x}} \frac{\mathbf{E_0}}{\omega}\cdot(-i\hbar\nabla e^{-r/a_B}),$$

where

$$N = \frac{e}{2m}\left(\frac{1}{2\pi\hbar}\right)^{3/2}\left(\frac{1}{\pi a_B^3}\right)^{1/2}.$$

By integrating by parts show that this is just

$$A_{fi} = N\frac{\mathbf{E_0}\cdot\mathbf{k_f}}{\omega}\sqrt{2\pi}\delta\left(\frac{E_f - E_i}{\hbar} - \omega\right)\int d^3x \; e^{-i\mathbf{k_f}\cdot\mathbf{x}}e^{-r/a_B}.$$

14.7 To do the integral in the previous exercise, we use spherical coordinates

$$\int d^3x \; e^{-i\mathbf{k_f}\cdot\mathbf{x}}e^{-r/a_B} = 2\pi\int_0^\pi d\theta\int_0^\infty drr^2\sin(\theta)e^{i|p_f|r\cos(\theta)}e^{-r/a_B}.$$

Do the remaining integrals.

14.8 The total transition probability is given by the square of the expression for A_{fi} in Exercise 14.6, after doing the integral. Notice that this includes the square of the energy delta function, which is infinite. By going back to Exercise 14.4, argue that this infinity comes from an integral over time, and is therefore due to the fact that we assumed the perturbation was a plane wave. Argue that if the time is finite but very long we should view the infinity as simply the length of time over which the wave interacted with the atom. Show that the transition probability per unit time, which is called the *rate* is

$$R_{fi} = \frac{4a_B^3e^2}{m^2\hbar^4\pi\omega^2}|\mathbf{E_0}\cdot\mathbf{k_f}|^2[1 + (k_fa_B/\hbar)^2]^{-4}\delta\left(\frac{k_f^2}{2m} - E_i - \hbar\omega\right).$$

This formula is an example of Fermi's Golden Rule.

14.9 The delta function means that this rate is singular (though again this is because we assumed the time interval was infinite). In a realistic experiment, the detector has a finite opening angle, and energy resolution, so we should integrate this rate over a small region between k_f and $k_f + dk_f$ and multiply it by $d\Omega$, the detector resolution in solid angle. Do the integral over momentum and solid angle, to find the total rate. Show that it is $p_f = \hbar k_f$

$$R_{fi}^{tot} = \frac{16a_B^3e^2p_f^3E_0^2}{3m\hbar^4\omega^2}[[1 + (p_fa_B/\hbar)^2]^{-4}.$$

4.10 The previous Exercise calculated the rate of ionization of hydrogen. The rate of energy absorption from the beam of light is

$$\frac{dE_{ab}}{dt} = \hbar\omega R_{fi}.$$

A plane wave has infinite total energy because it is spread over all of space. Show that if we put up a perfectly absorbing screen of area σ transverse to the beam, it will absorb energy at a rate $\frac{dE_{ab}^{screen}}{dt} = \frac{\sigma c E_0^2}{8\pi}$. We therefore define the *absorption cross section* σ_{ab} for ionization of hydrogen by the ratio of the real absorption rate to that of a perfect absorber. Show that it is given by

$$\sigma_{ab} = \frac{128 a_B^3 \pi e^2 p_f^3}{3 m \hbar^3 \omega c [1 + p_f^2 a_B^2/\hbar^2]^4}.$$

The cross section is a useful object because it is independent of the characteristics of the beam. A real experiment would involve a collimated beam of light rather than a plane wave. As long as the beam is constant over the size of the atom, the same cross section will be found.

14.11 Calculate the cross section for ionization of hydrogen when the emitted electron has $5, 10, 50$ Rydbergs of kinetic energy. Compare these cross sections to the geometrical size of the atom, which is defined by the region where the probability distribution for the electron is not exponentially small, and is of order $4\pi a_B^2$.

14.12 The cross section falls off rapidly for large p_f. The electron inside the atom does not have a fixed p_f because of the uncertainty principle. Argue that the rapid falloff is due to the fact that it is improbable to find a large electron momentum in the unperturbed atom. Note that the cross section does not have as fast a falloff as the probability distribution because the interaction Hamiltonian is proportional to p_f and because the number of final states with momentum p_f grows with p_f.

14.13 If we repeat the ionization calculation for an ion of charge Z, the cross section scales like Z^2. Go back over the derivation and estimate the range of momenta over which this calculation is valid.

CHAPTER 15

The Adiabatic Approximation, Aharonov–Bohm, and Berry Phases

15.1 INTRODUCTION

The adiabatic approximation deals with time-dependent Hamiltonians whose variation is slow compared to the oscillation frequencies in the quantum state. In a typical situation, the gap between the ground state and the first excited state is much larger than $\hbar\omega$ where ω is any frequency in the Fourier spectrum of the time dependence. The adiabatic theorem says that the solution of the time-dependent Schrödinger equation is then a phase times the time-dependent ground state. Michael Berry showed that an important part of the phase has topological properties. Berry's phase is responsible for a number of the most bizarre phenomena in quantum theory. The most famous is the Aharonov–Bohm effect, where the quantum phase of the wave function of a charged particle can measure the magnetic field of a solenoid, even though the particle is never in a region with nonzero field strength.

15.2 ADIABATIC ENERGY EIGENSTATES

We have seen that the effect of a time-dependent perturbation on an eigenstate of the system is predominantly to cause transitions between levels whose energy difference is of order the frequency of the time-dependent field. In this chapter, we will consider time-dependent Hamiltonians whose frequencies are much smaller than the energy difference between a pair of levels. The most common situation to which such an approximation is applicable is systems where the ground state is separated from the first excited state(s) by a relatively large gap.

If the Hamiltonian is time dependent, and is not a small perturbation of a time-independent system, what we mean in the previous paragraph by the "levels" and the "ground state" are the eigenstates of the time-dependent Hamiltonian

$$H(t)|E_i(t)\rangle = E_i(t)|E_i(t)\rangle. \qquad (15.1)$$

It is the gap in this time-dependent spectrum to which we were referring. For a general time-dependent situation, these states exist but are not terribly interesting, but if $H = H(\omega t)$, with $\hbar\omega \ll |E_i - E_j|$, then the *adiabatic theorem* shows us that they are the right states to consider. For large systems, it is virtually impossible to satisfy this inequality for all pairs of states. If the characteristic energy scale of the problem is E_{char}, then in a large system, we will have energy splittings of order $e^{-s(E)}E_{char}$, where $s(E)$ is the entropy or logarithm of the number of states with energy $\sim E$. However, the gap between the ground state and the first excited level is of order E_{char}, so the adiabatic theorem will be valid whenever $\hbar\omega \ll E_{char}$. We will state the adiabatic theorem for the ground state, and hope that the interested reader will see that it can apply in more general contexts.

The adiabatic theorem is the statement that if we begin in the adiabatic ground state $|E_0(t=0)\rangle$ then the system, to a good approximation, will evolve into

$$e^{\langle E_0|\partial_t|E_0\rangle - i\int_0^t ds\, \frac{E_0(s)}{\hbar}}|E_0(t)\rangle.$$

To understand how the state $|E_0(t)\rangle$ actually evolves, we differentiate the eigenvalue equation:

$$\dot{H}(t)|E_0(t)\rangle + H(t)d|E_0(t)\rangle/dt = \dot{E}_0(t)|E_0(t)\rangle + E_0(t)d|E_0(t)\rangle/dt. \qquad (15.2)$$

If we differentiate the normalization equation for $|E_0(t)\rangle$, we get

$$0 = \langle E_0(t)|\partial_t|E_0(t)\rangle + \partial_t(\langle E_0(t)|)|E_0(t).\rangle \qquad (15.3)$$

This shows that the component of the time derivative of the adiabatic eigenstate, along the eigenstate itself, is purely imaginary.

Introducing the projection operator $P(t)$ on the subspace orthogonal to $|E_0(t)\rangle$, we can write

$$\partial_t|E_0(t)\rangle = \langle E_0|\partial_t|E_0\rangle|E_0(t)\rangle + P(t)\partial_t|E_0(t)\rangle \qquad (15.4)$$

and we get

$$P(t)\partial_t|E_0(t)\rangle = \frac{1}{E_0(t) - H(t)}P(t)\dot{H}(t)(1 - P(t)) - (1 - P(t))\dot{H}(t)\frac{1}{E_0(t) - H(t)}P(t)|E_0(t)\rangle.$$
$$(15.5)$$

Note that the second term annihilates $|E_0'(t)\rangle$. We add it to make the evolution operator δH, defined below, a Hermitian operator.[1] The component of $\dot{H}|E_0(t)\rangle$ in the $|E_0(t)\rangle$ direction is, by the Feynman–Hellman theorem, equal to $\dot{E}_0|E_0(t)\rangle$ and so cancels from the above equations.

Now let us try to solve the Schrödinger equation with an ansatz

$$|\psi(t)\rangle = e^{-i\phi}|E_0(t)\rangle. \tag{15.6}$$

The result is

$$(\dot{\phi} - i\langle E_0(t)|\partial_t|E_0(t)\rangle)|E_0(t)\rangle = \frac{E_0(t)}{\hbar}|E_0(t)\rangle + \delta H|E_0(t)\rangle, \tag{15.7}$$

where

$$\delta H = i\hbar[P\frac{1}{E_0 - H}\dot{H}(1 - P) - (1 - P)\dot{H}\frac{1}{E_0 - H}P], \tag{15.8}$$

and we've used the eigenvalue equation $[H(t) - E_0(t)]|E_0\rangle = 0$. This equation is inconsistent, unless the term involving δH is negligible. When δH *is* negligible, it is the proof of the adiabatic theorem.

If $H(t)$ is a smooth function of ωt with $\hbar\omega \ll E_{gap}$, then δH is of order $\hbar\omega/E_{gap}$. Note also that at large eigenvalue, a region we might worry about if the spectrum of the Hamiltonian is unbounded, this operator is still bounded by something of order ω. Indeed, we can write

$$H(t) = \sum \alpha_i(\omega t)H_i,$$

and at large eigenvalue

$$\delta H = \omega \frac{P\sum \alpha_i'H_i}{\sum \alpha_i H_i},$$

where the prime denotes derivative with respect to the argument of the function. The operator multiplying ω is bounded by something of order 1 in the large eigenvalue region. For example, if we had $H(t) = \frac{p^2}{2m(t)} + V(x, t)$, then the large eigenvalue region is dominated by the kinetic term and we can approximate $\frac{\dot{H}}{H} \sim -\frac{\dot{m}}{m} \sim \omega$. It should be clear that nothing in this argument actually used the fact that $|E_0(t)\rangle$ was the adiabatic ground state. We used only the fact that it was separated from all other adiabatic eigenstates by a gap $\gg \hbar\omega$.

A simple example of the adiabatic theorem, of remarkably general utility, is a two state system with Hamiltonian

$$H(t) = \sum B_a(\omega t)\sigma_a, \tag{15.9}$$

---

[1] We use the fact that $P(t)$ commutes with $H(t) - E_0(t)$ to show that the operator appearing in the equation for the time derivative of $|E_0(t)\rangle$ is anti-Hermitian.

with $\hbar\omega \ll |B|$. The adiabatic eigenvalues are $\pm|B|(t)$. The adiabatic eigenstates are solutions of

$$\sum B_a(\omega t)\sigma_a|E_\pm(t)\rangle = \pm|B|(t)|E_\pm(t)\rangle. \tag{15.10}$$

In the basis where σ_3 is diagonal, these equations read

$$\begin{pmatrix} \sqrt{1-BB*} & B* \\ B & -\sqrt{1-BB*} \end{pmatrix} \begin{pmatrix} a_1^\pm \\ a_2^\pm \end{pmatrix} = \pm \begin{pmatrix} a_1^\pm \\ a_2^\pm \end{pmatrix}, \tag{15.11}$$

where $B = \frac{B_1(t)+iB_2(t)}{|B(t)|}$. The solution of these equations is

$$a_1^\pm = -\frac{B*}{\sqrt{1-B*B}\pm 1}a_2^\pm, \tag{15.12}$$

which determines both coefficients up to an overall phase, when combined with the normalization condition $|a_1^\pm|^2 + |a_2^\pm|^2 = 1$.

If we think of this Hamiltonian as a description of a spin in a time-dependent magnetic field, then the simplest way to describe the adiabatic approximation is to say that the state of the spin lines up with the magnetic field (we have incorporated the dipole moment into the vector B_a, so in the case of a real magnetic field, we could have antialignment as well). The operator that controls the corrections to the adiabatic approximation is

$$\delta H = \omega P \frac{B_a'\sigma_a}{-|B|-B_a\sigma_a}, \tag{15.13}$$

where $P = 1 - |E_-(t)\rangle\langle E_-(t)|$. It is an operator of norm 1 and its eigenvalue is bounded by $\omega\frac{|B'|}{|B|}$, which is much smaller than one, when the conditions for validity of the adiabatic theorem are valid.

15.3 THE BERRY PHASE

The phase factor ϕ decomposes into a *dynamic phase*, which depends on the adiabatic energy level of the system, and a term $\gamma_0(t) = \int_0^t ds\, i\langle E_0(t)|\partial_t|E_0(t)\rangle$, called the *geometric phase*. The geometric phase suffers from a certain degree of arbitrariness, because one is always free to redefine the eigenstate $|E_0(t)\rangle$ by multiplying it by an arbitrary phase factor $|E_0(t)\rangle \to e^{i\theta(t)}|E_0(t)\rangle$. This changes the integrand of the geometric phase by $\partial_t\theta$. No measurement at a single time t can be sensitive to such a change. Furthermore, the geometric phase difference at two different times depends only on the wave functions at those times and not on the intervening history. For 60 years, the geometric phase was considered unobservable.

It was the genius of Michael Berry [69] to recognize that the ambiguity in the geometric phase was a kind of $U(1)$ gauge invariance.[2] Let us assume that the time-dependent change in the Hamiltonian is a change in the coefficients of k different operators in an expansion $H = \sum Y^i H_i$ in some canonical basis of operators. In our two state system, the Y^k would be the three components of the magnetic field B_a. Then

$$d|E_0\rangle/dt = \dot{Y}^k \partial Y_k |E_0\rangle \tag{15.14}$$

and

$$\gamma = \int_0^s \dot{Y}^i(s) A_i(Y(s)), \tag{15.15}$$

where the *Berry Connection* or *Berry Vector Potential*[3] is defined by $A_i = i\langle E_0 | \nabla_i | E_0 \rangle$. The phase can therefore be thought of as the line integral of a generalized vector potential. If the parameter space of the Y^k has the topology of flat k dimensional space,[4] then the line integral of A_i around a closed path is equal to the integral of $\partial_i A_j - \partial_j A_i$, the *Berry Magnetic Field* or *Berry Curvature*, over any two-dimensional surface whose boundary is that path. This is a generalization of Stokes' theorem from electromagnetism.

A simple way to generate a nonzero Berry phase is to split a beam of particles initially prepared in the same quantum state, and guide the two halves of the beam around two different paths in space, which intersect at some point in the future. Subject one of the beams to an adiabatic change in its Hamiltonian, by changing some external fields localized around that beam's path and turned off before the beams cross. Then the particles in one beam have wave function ψ, while those in the other have a wave function $\psi e^{i\Gamma} e^{-i\int_0^t \frac{E(s)}{\hbar}}$, where Γ is the geometric phase. The interference of these two beams can measure Γ, if one can separate out the contribution from the dynamic phase. It turns out that this is possible experimentally, and the Berry phase has been measured. We will discuss a particular example of this in the next section.

In the case of a two state system, the Hamiltonian depends on three "magnetic field" components $B_a(t)$. The Berry potential is thus a function $A_a(B_a)$ of three variables, just like a vector potential in electrodynamics. Thus, given any Berry potential with nonzero curl, there will be closed curves $B_a(t)$ in the space of couplings in the two state Hamiltonian, such

---

[2] Berry's gauge invariance should not be confused with the $U(1)$ gauge invariance of electromagnetism, although the Aharonov–Bohm effect, which we will explore in the next section, does lead to a conflation of the two. They are, however, logically distinct concepts, which become intertwined in that particular example.

[3] The choice between these two names depends on one's attachment to physical or mathematical terminology. Mathematicians would call this the connection in the line bundle over the parameter space Y^k defined by the state vector. Physicists tend to think of it as a k dimensional generalization of a vector potential.

[4] If the parameter space has noncontractible closed loops, like a torus, then the line integral can be nonvanishing even when the Berry Curvature is zero.

that the Berry phase is nonzero. In particular, if the magnitude $|B|$ of $B_a(t)$ is fixed, the closed curve will lie on the surface of a sphere of radius $|B|$ and the Berry potential will only have two components, and its curl only one component.

15.4   THE AHARONOV–BOHM EFFECT [70]

Let us return to the problem of a charged particle in a magnetic field, this time the field of an infinite solenoid. The **B** field of such a solenoid vanishes outside of a cylinder whose cross section is a small circle of radius r_0 in the $x_1 - x_2$ plane. We will be interested in the wave function of charged particles in regions separated from that circle by a distance $R \gg r_0$. For example, we can put the electrons in a square well potential ($V_0 > 0$)

$$V(r) = -V_0[\theta(R + a - r) - \theta(R - r)].$$

r is the two-dimensional radius and $\theta(x)$ is the Heaviside step function, equal to 1 when its argument is positive and 0 otherwise. We will consider electrons with energy $E + V_0 \ll V_0$.

For any value of r, the Schrödinger equation is that of a free electron in the solenoid field, with $E \to E \pm V_0$. For r outside the negative region of the potential, the wave function is small and we will neglect it (make V_0 very large). Just as in the constant field case, the motion in the x_3 direction decouples, and we will ignore that as well. It is convenient to work in cylindrical coordinates.

Stokes' theorem from classical electrodynamics tells us that the line integral of **A** around any curve encircling the solenoid once is equal to the magnetic flux F through the solenoid. These line integrals around closed curves are completely gauge invariant. In cylindrical coordinates, a potential with this property is

$$(A_z, A_r, A_\phi) = \frac{F}{2\pi r}(0, 0, 1). \tag{15.16}$$

The Schrödinger equation with energy E in this background field, plus an external potential $V(z, r, \phi)$, has the form

$$-\frac{\hbar^2}{2m}[\partial_z^2 + \partial_r^2 + \frac{1}{r^2}(\partial_\phi - i\frac{qF}{2\pi r})^2]\psi = (E - V)\psi. \tag{15.17}$$

The solutions of this equation have the form

$$\psi_F = \psi_0 e^{i\frac{qF}{2\pi\hbar}\phi}, \tag{15.18}$$

where ψ_0 is a solution with vanishing flux.

Now begin with a beam of electrons approaching the origin along the trajectory $\phi = 0$, in the plane $z = 0$, and consider a potential, which includes a term describing a device that can split a beam of electrons at some point ($z, r \neq 0, \phi = 0$). On average, half the electrons are given velocities $\mathbf{v} = (0, \pm v_1, 0)$ in Cartesian coordinates. The potential away from the beam splitting point attracts the particles back toward the origin. Each electron in the beam is in the same initial spatial wave packet, but with time delays, so that we can neglect electron interactions. However, the potential is asymmetric between left and right, so that there is a finite probability that an electron which moves to the right after the beam splitting, will come back to $\phi = \pi$ at the same time as one which is diverted to the left. The scattering amplitude between the two electrons will include a phase

$$S_{AB} = e^{iq\frac{F}{2\pi\hbar}\int_0^{2\pi}}, \tag{15.19}$$

which comes from the phases accumulated by the two electron wave functions as each traverses half the circle. Note that the logarithm of this phase is proportional to $\int d\theta A_\theta$. Since the magnetic field vanishes everywhere except at $r = 0$, this phase is, by Stokes' theorem from electrodynamics, independent of the detailed paths followed by the two electrons. Only the topology of the closed curve formed by the trajectories of the two electrons is relevant to the phase in scattering, which is also completely gauge invariant. If the closed curve circles the line $r = 0$ in three-dimensional space, along which the magnetic flux is concentrated, then we pick up a flux-dependent phase, otherwise we do not.

The Aharonov–Bohm (AB) effect has been verified experimentally, first in the 1960 experiment of Chambers [71], and in many subsequent experiments, but astonished physicists when it was first proposed, because it is an electromagnetic effect on the electron in regions where all electric and magnetic fields vanish. Only the vector potential is nonvanishing and only the nonlocal topological integral of the vector potential appears in the AB scattering phase. Aharonov told me that his Ph.D. thesis committee brought in R.F. Peierls, an outside expert, to debunk the claims of Aharonov and his advisor Bohm. After grilling Aharonov for hours, Peierls said to him "Young man, I know that you're wrong, but you've defended your claim so well that you deserve the Ph.D." Nowadays, the AB effect is considered obviously correct and lies at the root of our understanding of the generalizations of electrodynamics that describe the Standard Model of particle physics.

As Berry first pointed out in his seminal paper, the AB effect is an example of a Berry phase. To see this, we look at a different setup than the original AB experiment. The potential is now taken to be an infinite square well, confining a single electron to a box sitting at some position $\mathbf{Z}$ in three-dimensional space, whose r coordinate is nonvanishing. The size of the box is much smaller than the r coordinate of $\mathbf{Z}$. As before, we can solve the Schrödinger equation in the presence of the flux by

$$\psi = e^{iq\frac{F}{2\pi\hbar}\phi}\psi_0. \tag{15.20}$$

Since ψ_0 vanishes outside the box, all that matters in this formula is the value of ϕ in the box. Now consider an adiabatic variation $\mathbf{Z(t)}$ of the position of the box, which starts and ends at the same point. The Berry phase is

$$\gamma = i \int \langle \psi | \nabla_{\mathbf{Z}} \psi \rangle \cdot d\mathbf{Z}. \tag{15.21}$$

This has two terms, the Berry phase from ψ_0 and that coming from the variation of ϕ in the flux term, as we vary $\mathbf{Z}$. Since ϕ comes back to the same point, the latter is just the AB flux.

The wave function of any stationary state in the square well potential is a function of $\mathbf{x} - \mathbf{Z}$, so the gradient w.r.t. $\mathbf{Z}$ that appears in Berry's formula may be replaced by a gradient w.r.t. $\mathbf{x}$. The latter is proportional to the momentum operator so the ψ_0 part of the Berry phase is proportional to the expectation value of the momentum in an eigenstate of the square well. The square well eigenstates all obey Dirichlet (vanishing ψ_0) boundary conditions at the walls of the box. That is, they are *standing waves* and have zero momentum expectation value. Thus, the AB phase is just the Berry phase for this special case.

15.5 ANYONS, FERMIONS, AND THE SPIN STATISTICS THEOREM

The Aharonov–Bohm effect gives us insight into the meaning of Fermi statistics, but to understand this insight, we have to take a detour into a world with one less spatial dimension than our own. This is called $2 + 1$ dimensional physics: a three-dimensional space-time with one time dimension.

In $2 + 1$ dimensions, an infinitely thin solenoid is just a point object, so we can imagine point particles, which carry "magnetic" flux, as well as charge. We put magnetic in quotation marks, because the gauge field in question is not really electromagnetic. It will be called the "statistical gauge field," because, as we will see, it determines both the particle's statistics and its spin. More precisely, we will see that coupling a collection of identical particles, with Bose statistics, to such a statistical gauge field, leads to a correlated change of both spin and statistics. In $2 + 1$ dimensions, we can have noninteger spin and forms of statistics different from either Bose or Fermi, but we will see that this is incompatible with rotational symmetry in higher dimensions. Only Bose and Fermi statistics are possible in higher dimensions, and a change of statistics leads to a change of spin by $1/2$.

In the real world, the connection between spin and statistics is very tight. Fermions all have half integer spin and bosons all have integer spin. We can explain this by postulating a set of bosons with integer spin, some of which are coupled to the statistical gauge field, giving them half integer spin and Fermi statistics. However, there is nothing in the formalism of nonrelativistic quantum mechanics which prevents us from postulating the existence of bosons with half integral spin.

The $2 + 1$ dimensional "electric field" is a two-vector $\mathcal{E}_i$ and the "magnetic" field is an antisymmetric tensor $B\epsilon_{ij}$. They can be combined together into a three-dimensional antisymmetric tensor. Let us introduce the speed of light, simply to convert time units into space units: $ct \equiv x_0$. This does not imply that we are writing a relativistic theory. For application of these equations to some condensed matter system, we might choose instead to use some characteristic propagation speed c in the material under study. Then the field strength tensor is

$$F_{\mu\nu} = \partial_\mu S_\nu - \partial_\nu S_\mu, \tag{15.22}$$

where S_μ is a three-dimensional vector potential, which we call the statistical gauge potential. We have $F_{0i} = \partial_0 S_i - \partial_i S_0 \equiv E_i$, and $F_{ij} = \partial_i S_j - \partial_j S_i \equiv \epsilon_{ij} B$.

We say that a particle has statistical charge q if there is a term in its classical action[5] of the form

$$\delta S = q \int ds \, \frac{dx^\mu}{ds} S_\mu(x(s)). \tag{15.23}$$

In this equation, we have introduced an arbitrary parameter s to describe the path of the particle $x_\mu(s)$ in space-time. The term δS in the action does not depend on the choice of this parameter, as you will demonstrate in Exercise 15.1.

In nonrelativistic physics, there is a natural choice of time parameter, $s = t$, the *absolute time* of Newton's *Principia*. For this choice, $x^0 = ct$. The full action for a statistically charged particle is

$$\int dt \, \frac{m\dot{\mathbf{x}}^2}{2} + S_0(\mathbf{x}(t)) + \frac{dx^i}{dt} S_i(\mathbf{x}(t)). \tag{15.24}$$

The equation of motion following from the condition of stationarity of this action is

$$m\ddot{\mathbf{x}} + \frac{dS_i(\mathbf{x}(t))}{dt} = \nabla_i S_0 + \frac{dx^j}{dt} \nabla_i S_j, \tag{15.25}$$

which is the same as the Lorentz force equation.

To define carefully what we mean by a particle with both charge and flux, we model such an entity as a charged particle stuck to the wall of a circle surrounding a pointlike flux Φ, and then shrink the circle to zero size [72]. It can be shown [73] that the results we will obtain are valid for a much wider range of models of what a particle with both charge and flux looks like. If we rotate our model charge–flux composite by 2π, the charge picks up an Aharonov–Bohm phase $e^{i\frac{q\Phi}{\hbar}}$. On the other hand, the response to this rotation is described by the action of the rotation operator $e^{-2\pi i J}$ on the quantum state of the composite system. We conclude that

$$J = m - \frac{q\Phi}{2\pi\hbar}, \tag{15.26}$$

where m is an integer. As usual, the fractional part of this is considered to be internal spin. The simplest composite would have $J = -\frac{q\Phi}{2\pi\hbar}$.

---

[5] See Chapter 4 for the definition and properties of the classical action.

Now consider two such charge–flux composites at positions $\pm x$ in the x, y plane. We can exchange them with each other by doing a rotation of π around the origin. This is an adiabatic motion of the particles. Its only effect comes through the statistical Aharonov–Bohm phases particles pick up by moving charges around fluxes. The particles remain stationary with respect to their own fluxes during this motion, so the phases come from the motion around the other particle's flux. Assuming that all particles/fluxes are identical, these add up to exactly the phase experienced when a single particle traverses around its own flux in a 2π rotation. *This shows that there is a relation between the change in statistics of a particle due to interaction with S_μ and its change in spin.* In $2 + 1$ dimensions, no principle prevents $q\Phi$ from being an arbitrary, even irrational, number in units of $2\pi\hbar$. Particles with general spin and statistics are called *anyons* [74].

For general values of the angular momentum $J = -\frac{q\Phi}{2\pi\hbar}$, the multibody constraints on the wave function become very complicated if we also insist that the wave function satisfy the Schrödinger equation for a free particle. They have only been solved in the limit of infinite mass, where the particle's kinetic energy is dropped from the Hamiltonian. This limit turns out to be relevant to the collective excitations of matter in a certain kind of insulating phase called the fractional Quantum Hall regime [75]. The Hall Effect is a well known phenomenon in classical electrodynamics, in which an electric field applied to a material in one direction, leads to a flow of current in a perpendicular direction. For a planar sample, this is summarized by an equation

$$J^i \propto \epsilon^{ij}\mathcal{E}_j. \tag{15.27}$$

This equation is related by boosts to an equation relating charge density J_0 to magnetic flux density $B = \partial_1 A_2 - \partial_2 A_1$

$$J_0 \propto \epsilon^{0ij}\partial_i A_j. \tag{15.28}$$

Indeed, if we introduce a time coordinate $x^0 = vt$, where v is some characteristic velocity of the material under study, then we can write the two equations as

$$J_\mu \propto \epsilon^{\mu\nu\lambda}\partial_\nu A_\lambda, \tag{15.29}$$

where we have written the scalar potential of electrodynamics as A_0. The Greek indices run over $0, 1, 2$.

This equation resembles the equations for the *statistical* gauge field, but involves the electromagnetic potentials instead of the statistical one. The theory of the fractional quantum Hall effect (FQHE) is based on localized collective excitations of effectively planar materials, which have negligible kinetic energy. They behave like anyons with a variety of fractional values of the statistics parameter qF and also carry fractional electric charge. We do not have space here for a full description of the FQHE, but students should be aware that it is one of our most beautiful illustrations of quantum effects, and of the importance of Aharonov–Bohm–Berry phases in particular.

Returning to $3 + 1$ dimensions, we can of course obtain the same effects by attaching infinite statistical flux tubes to particles. If these were real electromagnetic fluxes, they would contribute an infinite energy to the particles, but the statistical gauge field equations do not have Maxwell like terms in them. They define a local relation between the flux and the particle currents.[6] So there is no infinite energy associated with these flux tubes. Still, they appear to introduce a violation of rotation invariance into the description of particles, by picking out a line in space associated with the tube. This intuitive observation is made mathematical by noting that the spin-statistics connection we have derived is valid in four dimensions for rotations in the plane perpendicular to the flux tube. Thus, the only way that a particle with flux attached can have the same rotation properties as a normal particle is if its Aharonov–Bohm statistical spin takes on one of the values allows by the full rotation group $SO(3)$. We have learned that the eigenvalues of J for rotations in any plane must be integer multiples of $1/2$, so the only allowed fractional spin/statistics is fermionic.

To summarize: we can change the statistics of particles in d space-time dimensions, or more properly of the quantum field which changes the number of particles, by coupling it to a statistical gauge field which attaches a flux tube of dimension $d - 2$ to each point-localized excitation of the field. If $d \geq 4$, this is consistent with the rotational properties of particles, only if the statistical phase is either trivial or fermionic. The argument that quantizes the statistical phase comes from quantization of angular momentum. This way of viewing fermions, as bosons coupled to a field that generates only Aharonov–Bohm phases of ± 1,[7] and has no other physical effect, seems quite intuitive, and allows for an explanation of the spin-statistics connection which is transparent and independent of special relativity or the existence of antiparticles.

However, nothing in our argument so far prevents us from starting the process of statistical flux attachment with bosons of spin one half. We would then obtain integer spin fermions, by adding an interaction to a statistical gauge field. Thus, although the idea of a statistical gauge interaction is an attractive way to understand the origin of Fermi statistics, it is not enough to prove the tight connection between spin and statistics, which we observe in the world. To *derive* the spin statistics connection, we observe in the real world, we have to make the additional assumption that bosons with half integer spin cannot exist. In fact, this assumption follows from combining quantum mechanics, special relativity and the principle that Einstein–Podolsky–Rosen (EPR) correlations cannot be used to send messages faster than the speed of light.[8]

---

[6] In dimensions higher than $2+1$, the flux is an extended object of dimension $d-2$, where d is the space-time dimension. The corresponding "electric and magnetic fields" form an antisymmetric tensor of rank $d - 1$ in space-time and the flux-current relation is $J^\mu \propto \epsilon^{\mu\nu_1\ldots\nu_{d-1}} F_{\nu_1\ldots\nu_{d-1}}$.

[7] Such a gauge interaction is called a *gauge theory with* Z_2 *gauge group*, or Z_2 gauge theory for short.

[8] See Chapter 22.

The Z_2 statistical gauge field interpretation of Fermi–Dirac statistics implies that fermions can only be created locally in pairs. Notice that this also follows from the fact that, in a theory invariant under rotations, half integer spin particles can only be created in pairs, because the only operators carrying half integer spin are creation/annihilation operators of such particles. Relativity ties together these two very similar restrictions on particle interactions. We note that the discussion of anyons here has been very condensed. The reader is urged to consult the excellent lecture notes of J. Preskill [72] for more details.

15.6 EXERCISES

15.1 Prove that $\int ds \, S_\mu(x(s)) \frac{dx^\mu}{ds}$ is independent of the choice of the parameter s.

15.2 Let us consider the Hamiltonian $H = -\mathbf{x} \cdot \sigma$, for a two state quantum system in more detail. The three parameters $\mathbf{x}$ can be slowly varying functions of time, so that we can apply the adiabatic approximation. The adiabatic eigenstates are denoted $|n\mathbf{x}\rangle$, where $n = 1, 2$. The two states have different eigenvalues generically, but at $\mathbf{x} = 0$ they are degenerate. We will see that this degeneracy shows up as an interesting structure in the Berry potential. The Berry potential for the state n is

$$\mathbf{A}^{(\mathbf{n})} = i\hbar \langle n(\mathbf{x})|\nabla|n(\mathbf{x})\rangle.$$

Use the fact that for *all* $\mathbf{x}$

$$\langle n|m \rangle = 0 = \langle n|H|m \rangle,$$

to show that

$$\langle n|\nabla|m \rangle = \frac{\langle n|\nabla H|m \rangle}{E_m - E_n}.$$

This result is true for any finite dimensional Hilbert space, not just the two state system.

15.3 Compute the "magnetic field" of the Berry vector potential using the result of Exercise 15.2. It will be convenient to insert a complete set of states in this calculation and use the fact that there are only two states and therefore, one energy difference.

15.4 Consider a Hamiltonian $H(\alpha)$, for any number of particles, depending on parameters α_i. Assume the wave function in some basis is real for all values of α_i. Show that the Berry phase vanishes.

15.5 Consider a system with νN electrons in a very strong magnetic field. We know from Chapter 9 that the lowest Landau level for noninteracting electrons in this field is highly degenerate. N is the maximum number of electrons in the lowest Landau level and ν is the fraction of that number in our sample. The lowest Landau level is parameterized by any analytic function $F(z_1, \ldots, z_{\nu N})$, which is antisymmetric in the coordinates. The full wave function is

$$Fe^{-\frac{B}{4\hbar c}\sum_i z_i^* z_i}.$$

Now reintroduce the electron interactions. Since the gap to the next Landau level goes to infinity with the strength of the external field, and the strength of the interactions is independent of the field, the ground state of the system is given by some particular choice of F (assuming the interactions lift the degeneracy). Note that the electron spins are polarized by the field so every electron has a single preferred spin state in the Landau level. Laughlin guessed that when $\nu = \frac{1}{2k+1}$ the ground state wave function was

$$F_{Laughlin} = \prod_{i=1}^{\nu N}\prod_{j<i}(z_i - z_j)^{2k+1}.$$

Note that when a pair of electrons are close, the wave function vanishes rapidly, thus minimizing the Coulomb repulsion. Numerical studies and spectacular agreement with the qualitative properties of real systems in strong magnetic fields have convinced everyone that Laughlin's guess was right, and he was awarded the Nobel Prize. Laughlin also proposed that localized excitations, called quasiholes, of the ground state had the form

$$F_{quasihole} = \prod_{i=1}^{\nu N}(z_i - z_0)F_{Laughlin}.$$

A state with two quasiholes, one at the origin, and one at z is

$$\prod_{i=1}^{\nu N}(z_i - z)\prod_{i=1}^{\nu N}(z_i)F_{laughlin}.$$

Show that if we adiabatically move z in a circle around the origin then the wave function picks up a Berry phase $e^{i\nu\pi}$. Thus, Laughlin's quasiholes are *anyons*. The existence of these excitations (which also carry fractional electric charge) has been verified by experiment [75].

CHAPTER 16

Scattering Theory

16.1 INTRODUCTION

The phrase *scattering theory* refers to a set of general results about systems of particles with a Hamiltonian $H_0 + V$, where H_0 describes the free motion of the particles, and V falls off rapidly in the regions where particle coordinate differences $r_{ij} = |\mathbf{r_i} - \mathbf{r_j}|$ are all large. Typically, "falls off rapidly" means at least as fast as r_{ij}^{-3}. We have seen in Chapter 8 that some aspects of scattering theory apply even to the Coulomb potential. We always imagine that our system is translation invariant, but that we have diagonalized the total momentum of the system. If the system is also Galilean invariant, we can go to the rest frame of the center of mass.

Scattering theory can be thought of in two different ways, both coming from an analysis of the Schrödinger equation in the asymptotic region. The eigenstates of H_0 are free particle wave functions of all possible momenta. In particular, we can divide them into momenta that are pointing further in the asymptotic direction (outgoing) and those which point toward smaller values of r_{ij} (incoming). In a given part of the asymptotic region, the incoming and outgoing states look independent, but they are not. If there is no interaction, we know that the exact eigenfunctions are plane waves, and in incoming plane wave at e.g., $x_3 \to -\infty$ is the same as an outgoing plane wave at $x_3 \to \infty$. So also in the interacting case, the incoming and outgoing states are simply two different bases of the Hilbert space.

The physical description of scattering is to start at time $t_0 \to -\infty$ with localized incoming wave packets for each particle, whose Fourier transforms are smooth functions of momentum $\psi_{in}^i(\mathbf{k_i})$. One then imagines acting on this state with the evolution operator of the interaction picture

$$S = T e^{-i \int_{-\infty}^{\infty} ds \, V(s)}. \tag{16.1}$$

This scattering operator or S-matrix tells us what any incoming free particle state evolves into. The hypothesis of scattering theory, which can be proven rigorously for some class of interactions V, is that the asymptotic states are complete (see the discussion of bound states below). Then we can compute an S-matrix, from the matrix elements of S in the incoming basis. It is a nontrivial fact that this same matrix can be computed as the unitary matrix describing the transformation between the incoming and outgoing bases of the Hilbert space

$$\langle i \ (in)|S|j \ (in)\rangle = \langle i \ (out)|j \ (in)\rangle. \tag{16.2}$$

In terms of this second formula for the S-matrix, we will formulate the Lippman–Schwinger equations for scattering. Our explicit discussion will apply only to single particles scattering from a potential, which is equivalent to two particles interacting via a translation invariant potential. We will introduce a perturbative series for the S-matrix, called the Born series, and show that for smooth potentials, the first term dominates at high energy. For spherically symmetric potentials, we can solve the scattering problem for each fixed value of angular momentum. The S-matrix is then diagonal and written in terms of *phase shifts* $e^{i\delta_l(E)}$. We will show that at low energy, the $l = 0$ phase shift dominates.

16.2 GENERAL FORMALISM

For two body systems, the bound state and continuous spectrum are disjoint from each other, and we cannot have transitions between scattering states and bound states, but for systems of three or more particles, we can form bound states of some subset of particles even when the whole system has positive energy. This is sometimes called the slingshot effect in the classical mechanics of astronomical systems. A satellite falls into a negative energy bound orbit around a planet by dumping energy into a third system of particles, which escapes to infinity. In scattering theory, we consider such true bound states as new particles and include their free asymptotic motion in what we call H_0. Thus, in general, the number of each type of particle does not have to match between initial and final states.

Let us consider the subspace of the Hilbert space with a fixed positive energy E, and let $|\psi_0, k\rangle$ be a basis for the subspace of the Hilbert space with that same eigenvalue of H_0. The label k stands for all of the momenta of any collection of free particles (including freely moving bound states) with that energy, as well as the spins and other quantum numbers of those particles. We expect there are eigenstates of the full Hamiltonian H, which can be labelled by those same quantum numbers. That expectation is based on the fact that we can solve the Schrödinger equation approximately in the region where all r_{ij} are large, by simply dropping V from the equation.[1] We call $|\psi, k\pm\rangle$ the basis of H eigenstates, which asymptotically approach $|\psi_0, k\rangle$ in the limit $r_{ij} \to \infty$. We will see the reason for the extra $\pm$ label in a moment.

---

[1] It is important to note that we are implicitly including effects of V when we include bound states in the list of free asymptotic particles described by H_0.

The Schrödinger equation is

$$(H_0 + V - E)|\psi, k\pm\rangle = 0. \tag{16.3}$$

We write

$$|\psi, k\pm\rangle = |\psi_0, k\rangle + |\phi, k\pm\rangle, \tag{16.4}$$

and rewrite the equation as

$$(H_0 - E)|\phi, k\pm\rangle = -[V|\psi_0, k\rangle + V|\phi, k\pm\rangle]. \tag{16.5}$$

The operator $H_0 - E$ is not invertible, because it has zero eigenvalues, but the operators $H_0 - E \pm i\epsilon$ are invertible for arbitrarily small positive ϵ. These operators are not Hermitian, but they are normal. Consider for a moment using them as time evolution operators. They give rise to evolution of the form

$$\mathcal{E}_\pm = e^{-\frac{it}{\hbar}[(H_0 - E)\pm i\epsilon]}. \tag{16.6}$$

We have avoided the usual notation $U(t)$ because these are not unitary operators. As $t \to -\infty$, $\mathcal{E}_+$ sends all states to zero, while $\mathcal{E}_-$ annihilates all states at $t \to \infty$. Now define

$$|\phi, k\pm\rangle = \frac{1}{E - H_0 \pm i\epsilon}[V|\psi_0, k\rangle + V|\phi, k\pm\rangle]. \tag{16.7}$$

That is to say, $|\phi, k\pm\rangle$ are the solutions of these equations, which are called the *Lippmann Schwinger equations*. In the limit $\epsilon \to 0$ $|\psi, k\pm\rangle$, solve the Schrödinger equation with energy E and define two different bases of the eigenspace with energy E. $|\psi, k+\rangle$ is called the *out* basis and $|\psi, k-\rangle$ is called the *in* basis. The S-matrix or scattering matrix is the matrix of overlaps between these two bases.

$$S_{k'\ k} = \langle -\psi, k'|\psi, k+\rangle. \tag{16.8}$$

This is the matrix of the unitary S-operator, which transforms in states to out states

$$S|\psi, k, -\rangle = |\psi, k+\rangle. \tag{16.9}$$

16.3 POTENTIAL SCATTERING

The details of multiparticle scattering are quite intricate, and we will content ourselves with a description of scattering theory of a single particle from a potential. Equivalently, this formalism defines two body scattering in the rest frame of the center of mass, when the interaction is translation invariant. With that interpretation, the mass parameter appearing in the equations is the reduced mass of the two body system.

Let us write the Lippman–Schwinger equation for this system:

$$\psi(\mathbf{x}) = \phi(\mathbf{x}) + \int d^3y \; \langle\mathbf{x}| \frac{1}{E - H_0} |\mathbf{y}\rangle V(\mathbf{y})\psi(\mathbf{y}). \tag{16.10}$$

It is convenient to write the potential as $V = \frac{\hbar^2}{2m} U$, the energy as $E = \frac{\hbar^2 k^2}{2m}$, and scale a similar factor out of ϕ which is a solution of the free Schrödinger equation. Then we get

$$\psi(\mathbf{x}) = \phi(\mathbf{x}) - \int d^3y \; G_0(\mathbf{x} - \mathbf{y})U(\mathbf{y})\psi(\mathbf{y}), \tag{16.11}$$

where

$$G_0(\mathbf{x} - \mathbf{y}) = -\frac{1}{4\pi} \frac{e^{ik|\mathbf{x}-\mathbf{y}|}}{|\mathbf{x} - \mathbf{y}|} \tag{16.12}$$

is the Green function of the Helmholtz operator $\mathcal{H} = \nabla^2 + k^2$.

Since U falls off rapidly at infinity, we can write an approximate solution at large $r = |\mathbf{x}|$, using

$$G_0(\mathbf{x} - \mathbf{y}) \approx \frac{e^{ikr}}{r} e^{-ik\hat{\mathbf{x}}\cdot y}, \tag{16.13}$$

where $\hat{\mathbf{x}}$ is the unit vector in the $\mathbf{x}$ direction. We choose the normalization of ϕ to be simply $\phi = e^{i\mathbf{k}\cdot\mathbf{r}}$ and get

$$\psi - \phi = -\frac{1}{4\pi} \frac{e^{ikr}}{r} \int d^3y e^{-ik\hat{\mathbf{x}}\cdot y} U(\mathbf{y})\psi(\mathbf{y}) \equiv \frac{e^{ikr}}{r} f(\hat{\mathbf{x}}, k). \tag{16.14}$$

f is called the scattering amplitude. It is the matrix element of the operator $\frac{1}{i}(S - 1) \equiv T$, which is called the *transition operator*.

16.4 THE BORN APPROXIMATION

There are two methods of approximating this formula, which have proven useful. The first, called the Born approximation or Born series, is useful when the potential is weak, but also when k is very large. In this limit, if U is a smooth function of position (has finite derivatives of all orders), then the integral falls off rapidly at large k, so even if the potential is large somewhere, it is a small perturbation at large k. Then, we can solve the equation by iteration, as we did for the equations in the interaction picture. This gives us a series in powers of U, the first term of which is

$$f_{Born}(k\hat{\mathbf{x}}) = -\frac{1}{4\pi} \int d^3y \; e^{ik\hat{\mathbf{x}}\cdot y} U(\mathbf{y}). \tag{16.15}$$

$\mathbf{q} = k\hat{\mathbf{x}}$ is the momentum transferred to the particle by the scattering event. It is worth remembering that the first Born approximation to the scattering amplitude is the Fourier transform of the potential. The higher order terms in the Born series depict the particle scattering from the potential multiple times, with free propagation between encounters.

16.5 PHASE SHIFT ANALYSIS

For spherically symmetric potentials, there is another general method of analysis, which is best suited to low-energy scattering. This is called phase shift analysis, and we have already encountered it for the Coulomb potential in Chapter 8. For any spherically symmetric potential, which falls off sufficiently rapidly, the asymptotic Schrödinger equation is just

$$[\partial_r^2 + \frac{2}{r}\partial_r - \frac{l(l+1)}{r^2} + k^2]\psi = 0. \tag{16.16}$$

The large r behavior of the solutions is captured by the JWKB approximation (see Chapter 17): $\frac{e^{\pm ikr}}{r}$. The minus sign corresponds to incoming and the plus sign to outgoing spherical waves. On the other hand, near $r = 0$, the equation is singular, and only one of the two linearly independent solutions to the equation belongs to a Hilbert space where the Hamiltonian is Hermitian.[2] Thus, the unique solution is a fixed linear combination of the incoming and outgoing wave. The equation is real, so the real and imaginary parts of the wave function are independent solutions, but both obviously obey the required regularity condition at the origin. Thus, the solution that is part of the Hilbert space behaves at infinity like

$$\psi \to a_l \frac{\sin(kr - \frac{l\pi}{2} + \delta_l(k))}{kr}. \tag{16.17}$$

The *phase shift* $\delta_l(k)$ is defined so that it vanishes when the potential is zero. The shift by $-\frac{l\pi}{2}$ comes from the fact that for free motion an incoming plane wave exits the sphere at an antipodal point. Spherical harmonics of odd (even) l pick up a minus (plus) sign under antipodal reflection.

Since the space of solutions at fixed l, m is one dimensional, the incoming and outgoing wave solutions of the Lippman–Schwinger equation coincide up to an overall constant. The S-matrix is a one-dimensional unitary operator, with logarithm equal to the phase difference between the incoming and outgoing parts of $\frac{\sin(kr - \frac{l\pi}{2} + \delta_l(k))}{r}$ minus the phase difference that would have been there for free motion. In other words, $S_l(k) = e^{2i\delta_l(k)}$. The scattering amplitude f is a function only of k and θ, the polar angle measured from the incoming direction. It is obtained by expressing the T-matrix, $S = 1 + iT$, in the angle "basis," summing over all spherical harmonics. Thus,

$$f(\theta) = \frac{1}{2ik}\sum_{l=0}^{\infty}(2l+1)P_l(\cos\theta)[e^{2i\delta_l(k)} - 1]. \tag{16.18}$$

Referring back to the general formula for the scattering amplitude, for the case of spherically symmetric potentials, we see that the expansion of the amplitude in powers of k corresponds to integrating the wave function against powers of $\hat{y}$.

---

[2] The proof of Hermiticity requires integration by parts and the surface term at $r = 0$ must vanish.

$$\psi - \phi = -\frac{1}{4\pi} \frac{e^{ikr}}{r} \int d^3y \, e^{-ik\hat{\mathbf{x}}\cdot\mathbf{y}} U(\mathbf{y})\psi(\mathbf{y}) \equiv \frac{e^{ikr}}{r} f(\hat{\mathbf{x}}, k). \tag{16.19}$$

But the l-th power of $\hat{y}$ contains only spherical harmonics up to order l, so the low-energy expansion of the scattering amplitude keeps only low order terms in the angular momentum expansion. Scattering amplitudes at low energy are dominated by the s-wave, or $l = 0$, phase shift.

16.5.1   The Effective Range Approximation

Let us consider the low-energy s-wave phase shift for a potential that falls off rapidly at infinity. If $R(k, r)$ is the radial wave function, then $u(k, r) \equiv rR(k, r)$ satisfies a one-dimensional Schrödinger equation. Let $A(k, r)$ be the asymptotic form of the solution $A(k, r) = \frac{\sin(kr + \delta(k))}{\sin(\delta(k))}$. It satisfies the free Schrödinger equation. The derivative of the Wronskian of two solutions satisfies

$$\partial_r(f(k, r)\partial_r f(0, r) - f(0, r)\partial_r f(k, r)) = k^2 f(k, r)f(0, r). \tag{16.20}$$

This is true for $f = u$ and for $f = A$. If we subtract the relation for u from that for A and integrate from 0 to R, where R is much greater than the range of the potential, then three simplifications occur. On the left-hand side, the value of the difference of Wronskians at R is practically 0, because the exact solution u has approached its asymptotic form A since the potential is almost zero. Secondly, at $r = 0$ the Wronskian of u solutions must vanish in order for the radial Schrödinger operator to be Hermitian. This is manifestly not true for the A functions, and it need not be, since these only represent the asymptotic large r behavior. Indeed, the nonzero phase shift shows us that this is not the solution of the free Schrödinger equation that is regular at the origin. Thus, the left-hand side is evaluated purely in terms of A. On the right-hand side, we can take the upper limit of integration to infinity since $A(k, r)A(0, r) - u(k, r)u(0, r) \sim 0$ for $r > R$. The result is

$$k \cot(\delta(k)) - \lim_{p\to 0} p \cot(\delta(p)) = k^2 \int_0^\infty dr \, [A(k, r)A(0, r) - u(k, r)u(0, r)]. \tag{16.21}$$

We have had to be careful about the zero energy contribution because $\delta(0)$ vanishes, so the term involving the zero energy phase shift approaches a finite limit. This was emphasized by Fermi and Marshall and the limit is called $-1/a$ where a is the *scattering length*.

The coefficient of k^2 on the right-hand side approaches a finite limit as k goes to zero. Thus, we can write

$$ka \cot(\delta(k)) = -1 + \frac{1}{2}k^2 a r_0,$$

where r_0 is a new parameter with the dimensions of length, called the *effective range*. Given a measurement of the phase shifts at low energy, we can fit this formula with any short range potential having two parameters. In particular, a spherical well or barrier with a variable depth and width, can fit the low-energy data for scattering from *any* short range spherically symmetric potential.

The effective range approximation has been used in a wide variety of physical situations. It was invented in early studies of nuclear physics and has recently been utilized to study the interactions of cold atoms. In the latter context, it has been found to fail, even at quite low energies, near narrow resonances in the scattering cross sections. Nonetheless, the approximation has been so ubiquitous and useful that researchers often parameterize the phase shifts in terms of an "energy dependent scattering length and effective range."

16.6 RESONANCES

We have already included an extensive discussion of scattering theory in Chapter 8, for the special case of the Coulomb potential. Rather than repeating that discussion, the reader should probably to reread Chapter 8 at this point, before proceeding.

The differential scattering cross section $\frac{d\sigma}{d\Omega}$, when multiplied by the flux of particles, given as a number per unit area per unit time, tells us the probability that a particle will be scattered into a solid angle Ω with respect to its incoming direction. From Chapter 8, we recall the formulae

$$\frac{d\sigma}{d\Omega} = |f(\theta)|^2. \tag{16.22}$$

$$f(\theta) = \frac{1}{ik} \sum_{l=0}^{\infty} (2l+1)[e^{2i\delta_l(k)} - 1]P_l(\cos\theta). \tag{16.23}$$

The total cross section $\sigma(k)$ for scattering is the integral of the differential cross section over the sphere. Using the orthogonality of the Legendre polynomials on the sphere, we can write

$$\sigma(k) = \sum_l \sigma_l(k)(2l+1)^2, \tag{16.24}$$

where

$$\sigma_l(k) = |a_l|^2 = |e^{2i\delta_l(k)} - 1|^2. \tag{16.25}$$

This takes on a maximum if the phase shift goes through $\pi/2$, which is called a resonance. Writing the amplitude as $1/(\cot\delta_l(E) - i)$, then near the resonance we have

$$\cot(\delta_l(E) - i) \sim 2\frac{(M-E)}{\Gamma}, \tag{16.26}$$

where M is the position of the maximum and Γ, which is called the *width* of the resonance, has dimensions of energy. The amplitude near the peak can thus be written

$$a_l(E) = \frac{\Gamma/2}{E - M - i\Gamma/2}, \tag{16.27}$$

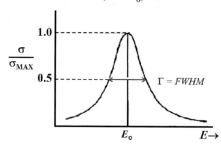

Breit-Wigner Resonance Curve

$$\sigma(E) = \sigma_{max} \frac{\Gamma^2/4}{(E-E_0)^2 + \Gamma^2/4}$$

Figure 16.1 The Breit–Wigner cross section.

which is called the Breit–Wigner form. In mathematics, $|a_l|^2$ is called the Cauchy distribution (when normalized so that its integral is one), though physicists tend to call it the Lorentzian distribution. It has the form shown in Figure 16.1.

Its Fourier transform, which is a function of time (when $\hbar$ is inserted) has the form $e^{-i\frac{M}{\hbar}t}e^{-\frac{\Gamma}{\hbar}t}$, which suggests that a resonance should be thought of as an unstable state, whose energy would have been M if the width were zero. If $\Gamma \ll M$, the Breit–Wigner–Cauchy–Lorentz shape will produce a pronounced bump in the partial cross section σ_l.

16.7   A PARTIALLY WORKED EXERCISE: THE δ SHELL POTENTIAL

The δ shell potential has the form

$$V(r) = -\frac{\hbar^2 k_0}{2m}\delta(r-a) = \frac{\hbar^2 k^2}{2m}U(r). \tag{16.28}$$

The radial equation is

$$\left(-\partial_r^2 - \frac{2}{r}\partial r + \frac{l(l+1)}{r^2} - k^2\right)\psi_l(r) = k_0\delta(r-a)\psi(r). \tag{16.29}$$

Defining $z = kr$, this is

$$\left(-\partial_z^2 - \frac{2}{z}\partial z + \frac{l(l+1)}{z^2} - 1\right)\psi_l(z) = b(k)\delta(z-ka)\psi_l(ka). \tag{16.30}$$

$b(k) = \frac{k_0}{k}$.

For $z < ka$, the only allowed solution is $Aj_l(z)$, where j_l is the solution of the spherical Bessel equation regular at the origin. For $z > ka$, it can be a linear combination of $h_l(z)$ the spherical Bessel function that behaves like $i^{-(l+1)}\frac{e^{iz}}{z}$ at infinity, and its complex conjugate. This is a normalized outgoing spherical wave. Thus, we have

$$\psi_l = \theta(ka - z)Aj_l(z) + \theta(z - ka)[h_l(z) + Bh_l^*(z)]. \tag{16.31}$$

Continuity at $z = ka$ implies that

$$Aj_l(ka) = B[h_l(ka) + h_l^*(ka)]. \tag{16.32}$$

Note that $2j_l(z) = h_l(z) + h_l^*(z)$. The correct discontinuity at ka is obtained if

$$(A - 2)(h_l'(ka) + \frac{h_l(ka)}{ka}) + (A - 2B)(h_l^{*'}(ka) + \frac{h_l^*(ka)}{ka}) = b(h_l + h_l^*)/2. \tag{16.33}$$

An equivalent approach to this problem is to solve the Lippman–Schwinger equation. The Lippman–Schwinger equation for the radial wave function for angular momentum l is

$$\psi_l(r) = j_l(kr) + \int_0^\infty ds\, G_k^l(r, s)U(s)\psi(s)\psi_l(s), \tag{16.34}$$

where G_k^l is Green's function of the radial Schrödinger operator that is nonsingular when either r or s is at the origin, and is such that the incoming radial wave at infinity is proportional to $-\frac{1}{2ikr}[e^{-i(kr-l\pi/2)}]$. This is

$$G_k^l(r, s) = -ik[j_l(kr)h_l(ks)\theta(s - r) + j_l(ks)h_l(kr)\theta(r - s)], \tag{16.35}$$

where $h_l(z)$ is the solution of the spherical Bessel equation which behaves like (recall that $j_l(kr)$ is the solution that is regular at zero and that it behaves like a cosine at infinity). In Exercise 16.6, you will verify all assertions made here in more detail.

For the case of the delta shell potential, the integral in the Lippmann Schwinger equation collapses to a point and the equation becomes a simple algebraic consistency condition for $\psi_l(ka, k)$, whose solution (Exercise 16.6) is

$$\psi_l(ka, k) = \frac{j_l(ka)}{1 - ikk_0 a^2 j_l(ka)h_l(ka)}. \tag{16.36}$$

$$\psi_l(k, r) = j_l(kr) + ik^2 a^2 \psi_l(ka, k) \times [\theta(a - r)j_l(kr)h_l(ka) + \theta(r - a)j_l(ka)h_l(kr)]. \tag{16.37}$$

Taking the large r limit of this, we get

$$\psi_l(k, r) \to j_l(kr) + ik^2 a^2 j_l(ka)h_l(kr). \tag{16.38}$$

We can then identify the phase shift as the coefficient of the outgoing spherical wave, and obtain the partial wave scattering amplitude (Exercise 16.6)

$$f_l(k) = e^{i\delta_l(k)}\sin(\delta_l(k)) = \frac{kk_0a^2 j_l(ka)^2}{1 - ikk_0a^2 j_l(ka)h_l(ka)}. \tag{16.39}$$

Recalling that bound states are given by poles of the scattering amplitude at positive imaginary k, and that Bessel functions are entire functions, we find that the condition for a bound state is

$$1 + |k|k_0a^2 j_l(i|k|a)h_l(i|k|a) = 0. \tag{16.40}$$

If we ask for the minimum coupling strength that will bind a given angular momentum, we can expand this complicated equation around $k = 0$ because the bound state energy will be just below zero. We find the criterion to be

$$k_0a = 2l + 1. \tag{16.41}$$

Note that this increases with l as might be expected because of the repulsive angular momentum barrier. Thus, we expect all lower angular momenta to have bound states for this value of k_0.

For $l = 0$, the spherical Bessel functions are simple and the exact bound state equation becomes

$$\frac{1}{k_0a} = \frac{e^{-|k|}}{2|k|}. \tag{16.42}$$

As k_0 gets larger, the binding energy grows as well. Note that for a repulsive potential $k_0 < 0$ there are no solutions, and by our remarks above, there will be no bound states for any l.

Now let us examine scattering states in the limit $k_0a \to \infty$. In this limit the phase shift, for either sign of k_0 approaches $\arctan(\frac{j_l(ka)}{h_l(ka)})$ which is the phase shift of an impenetrable hard sphere (Exercise 16.5). On the other hand, we know that our wave functions for k_0 positive are nonzero inside the sphere. As part of Exercise 16.6, you will show that this is only true for quantized values of k such that $j_l(ka) = 0$. Thus, in the limit of infinitely strong attractive coupling, the scattering states decouple from a set of states bound inside the well. As the final part of Exercise 16.6, you will investigate what happens to leading order when k_0a is large and finite. Do this for $l = 0$ only. You should find zeroes of the denominator of the scattering amplitude in the complex plane, with *negative imaginary part of k*. These are *resonances*. Compute the scattering amplitudes when the energy is close to one of these complex resonance poles.

16.8 EXERCISES

16.1 Develop a set of diagrammatic rules (Figure 16.2) for the Born series: the diagrams should have k vertices at each order, each one involving one factor of the potential, followed by free propagation of the particle between vertices. Write rules, associating algebraic/integral expressions translating each such diagram into the corresponding contribution to the scattering amplitude.

16.2 The nonsingular solutions of the free Schrödinger equation with angular momentum l are called spherical Bessel functions. We studied them in Chapter 8. Show that the spherical Bessel equations for different k but the same l all have the same solution when expressed in terms of the dimensionless variable kr.

16.3 Consider a spherically symmetric potential, $V(r)$, which vanishes identically for $r > a$ and is constant inside that sphere. In the region $r > a$, one has to solve the spherical Bessel equation, but since we are not near $r = 0$, we can include the solution $y_l(z)$, which is singular at the origin, usually called the Neuman function. Show that the correct linear combination of solutions is

$$R_l(z) = e^{i\delta_l}[\cos(\delta_l)j_l(z) - \sin(\delta_l)y_l(z)].$$

To solve this problem, you need to remember the definition of phase shift in terms of the large z behavior of the wave function, and use a table or mathematical search engine to find the large z behavior of the spherical Bessel functions. Express the function h_l that we used in the delta shell exercise, in terms of j_l and y_l.

16.4 Use the result of Exercise 16.3 to find an expression for the phase shift in terms of the logarithmic derivative $L_l = \frac{R_l'(z)}{R_l(z)}$ of the radial wave function, evaluated at $z = ka$. The expression you should find is

$$\tan(\delta_l) = \frac{j_l'(ka) - L_l(ka)j_l(ka)}{y_l'(ka) - L_l(ka)y_l(ka)}.$$

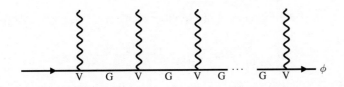

Figure 16.2 Graphical representation of the solution of the Lippman–Schwinger equation.

16.5 Argue that the logarithmic derivative of the wave function at ka is continuous, and so the phase shift can be calculated by solving the Schrödinger equation for $r < a$, applying a nonsingular boundary condition at $r = 0$, and then calculating the logarithmic derivative by evaluating that solution at $z = ka$. The argument that the logarithmic derivative is continuous is similar to the continuity arguments we used to solve the one-dimensional square well potential. Calculate all of the phase shifts for an infinitely repulsive barrier inside the sphere.

16.6 Solve the worked exercise on the delta shell potential more explicitly than we did in the text. Verify all the assertions we made there.

16.7 In the case $l = 0$, the spherical Bessel functions are simple. Show that in that case the phase shifts for the spherical well satisfy

$$\tan(\delta_0(k)) = \frac{k \tan(k_0 a) + k_0 \tan(k r_0)}{k_0 + k \tan(k r_0) \tan(k_0 r_0)}.$$

Here the potential is written as $V = -\frac{\hbar^2 k_0^2}{2m} \theta(a - r)$.

16.8 Find the approximate s-wave scattering amplitude when $ka \ll 1$ and compute the partial wave cross section.

16.9 Note than when $k_0 r_0 = n\pi$ the cross section vanishes, and the well becomes invisible. This is the Ramsauer–Townsend effect, which we explored in the exercises of Chapter 4.

16.10 The l-th partial wave cross section is defined by

$$\sigma_l = \frac{4\pi}{k^2}(2l + 1)\frac{1}{1 + \cot^2(\delta_l(k))}.$$

It has a maximum when

$$\cot(\delta_l(k)) = 0.$$

If this is happening by an increase of δ_l through a half integer multiple of π as the energy is varied, then we have what is called a *resonance*. Near the resonance

$$\cot(\delta_l(k)) = \frac{2(E_R - E)}{\Gamma(E)}.$$

The function Γ is slowly varying. Show that the partial cross section is given by

$$\frac{4\pi}{k^2}(2l + 1)\frac{\Gamma^2(E_R)}{4(E - E_R)^2 + \Gamma^2(E_R)}.$$

This is called the Breit–Wigner resonance formula. Draw a graph of the Breit–Wigner function and explain why the number $\Gamma(E_R)$ is called the *width* of the resonance.

CHAPTER 17

The JWKB Approximation

17.1 INTRODUCTION

How does quantum mechanics give rise to classical mechanics? We have given a detailed discussion of this in Chapter 10. The Jeffreys–Wenzel–Kramers–Brillouin (JWKB) approximation gives us another angle on the problem, for systems described by Hamiltonians of the form $g^2 \sum P_i^2 + \frac{1}{g^2} V(X_i)$. For such systems, the leading term of the small g^2 expansion of the Schrödinger equation is the classical Hamilton–Jacobi equation for the same system. The solutions of the classical equations determine the quantum mechanical wave functions. Remarkably, the expansion is also valid in regions where classical motion is forbidden by energy conservation. In those regions, one uses classical solutions in imaginary time. Wave behavior turns into exponential damping. The JWKB approximation always breaks down in the transition region between classically allowed and forbidden motion, but often one can solve the Schrödinger equation exactly in the transition region.

We will discuss applications of the JWKB approximation to the derivation of the Bohr–Sommerfeld rules for determining energies, and to the phenomenon of quantum mechanical tunneling and the decay of metastable states.

The validity of the JWKB approximation is not enough to guarantee real classical behavior. JWKB wave functions corresponding to different classical motions can have large interference. It is only for collective coordinates describing the average behavior of large complex systems, that decoherence, and therefore real classical behavior occurs.

17.2 THE JWKB EXPANSION

Consider a classical system with action

$$S = g^{-2} \int dt \, [\frac{(\dot{q}^i)^2}{2} - V(q)].$$

(17.1)

The dimensionless parameter g does not affect the equations of motion, but it will appear in the quantum mechanical Schrödinger equation.

$$i\hbar\partial_t\psi = [-\frac{\hbar^2 g^2}{2}\partial_i^2 + g^{-2}V(q)]\psi. \tag{17.2}$$

The Jeffreys–Wenzel–Kramers–Brillouin (JWKB) approximation [76] is a way of obtaining a systematic small g approximation to the solution of the Schrödinger equation.

We write the solution as

$$\psi = e^{-\frac{i}{g^2\hbar}S} \tag{17.3}$$

and obtain

$$\partial_t S = \frac{(\partial_i S)^2}{2} + V - \frac{i\hbar}{2}g^2\partial_i^2 S. \tag{17.4}$$

It is now clear that S has an expansion

$$S = S_0 + g^2 S_1 + \cdots .$$

$$\partial_t S_0 = \frac{(\partial_i S_0)^2}{2} + V, \tag{17.5}$$

$$\partial_t S_1 = \partial_i S_0 \partial_i S_1 - i\hbar\partial_i^2 S_0. \tag{17.6}$$

The first equation is called the *Hamilton–Jacobi* equation. You will recognize it if you have had an advanced class in classical mechanics.

No matter if you have not. To solve it, note that the equation tells us how to construct $S_0(t, q^i)$ starting from $S_0(0, q^i)$. Its value at any point q^i at time t can be constructed from the value of $p_i \equiv \partial_i S$ at that point and a slightly earlier time, and the value of $V(q^i)$. The value of p_i is equivalent to knowing the value of $\frac{dq^i(r)}{dr}\partial_i S$ along any trajectory $q^i(r)$ in the space of q^i (the configuration space of the system). r is just a parameter specifying the trajectory. We can choose it to be anything, and we will choose it to be the time coordinate. Thus, we can write

$$S_0(t, q^i) = -s(t) + S_0(0, q^i(t)), \tag{17.7}$$

where $s(0) = 0$ and $q^i(t)$ is some trajectory beginning at $q^i(0) = q^i$. The equation now looks like

$$\dot{s}(t) = p_i\dot{q}^i(t) - \frac{1}{2}p_i(t)p^i(t) - V(q(t)), \tag{17.8}$$

which is solved by

$$s(t) = \int_0^t ds[-q_i(s)\dot{p}^i(s) - \frac{1}{2}p_i(s)p^i(s) - V(q(s))], \tag{17.9}$$

where we have done an integration by parts. The left-hand side (LHS) of this equation is independent of $q^i(0)$, so we must insist that the same is true of the right-hand side (RHS). When differentiating the RHS w.r.t. $q^i(0)$, we encounter terms with $\partial[p_i(s)/\partial q^j(s)]\partial q^j(s)/\partial q_k(0) = [\partial^2 S_0/\partial q_i \partial q_j]\partial q^j(s)/\partial q_k(0)$, as well as terms with only $\partial q^j(s)/\partial q_k(0)$. The first kind of term depends on the initial condition $S_0(0, q^k)$, while the second does not, so they must vanish independently. In Exercise 17.1, you will verify that this leads to the two equations

$$p_i = \dot{q}^i, \tag{17.10}$$

and

$$\dot{p}_i = -\frac{\partial V}{\partial q^i}. \tag{17.11}$$

These are the classical equations of motion for the system, so the JWKB approximation is often called the *semiclassical* approximation. The quantity S that appears in the logarithm of the wave function is, in leading approximation, just the classical action.

17.3   THE JWKB APPROXIMATION IN ONE DIMENSION

In one dimension, we can work all of this out pretty explicitly. The Hamilton–Jacobi equation is

$$\partial_t S_0 = \frac{1}{2}(\partial_x S_0)^2 + V(x). \tag{17.12}$$

Define $S_0(t, x) = Et + S_0^E(t, x)$. Then

$$E + \partial_t S_0^E = \frac{1}{2}(\partial_x S_0^E)^2 + V(x). \tag{17.13}$$

This is solved by taking S_0^E to be time independent, and to be a solution of

$$E = \frac{1}{2}(\partial_x S_0^E)^2 + V(x). \tag{17.14}$$

These solutions to the H–J equation are analogous to the solutions of the Schrödinger equation with initial conditions equal to one of the eigenfunctions of the Hamiltonian. The Schrödinger equation is linear, and we can get any solution as a superposition of these special solutions.

To understand the general solution of the H–J equation start with an arbitrary function of x, $S_0(0, x)$. The equation tells you how to find the value of $S_0(t, x)$ for infinitesimally small t, and we get the full solution by iterating this procedure from time slice to time slice of the (t, x) plane. Given a solution $S(t, x)$ of the H–J equation, consider a path in a three-dimensional space

$$(x(u), p(u), p_t(u)), \tag{17.15}$$

where $p(u) \equiv \partial_x S_0(t, x)$ and $p_t(u) \equiv \partial_t S_0(t, x)$. u is a parameter along the path, which we initially take to be independent of t. The statement that S_0 satisfies the H–J equation at every point along the path is equivalent to

$$p_t(u) = \frac{1}{2} p(u)^2 + V(x(u)) = 0, \qquad (17.16)$$

all along the path. Differentiating w.r.t. u, we get

$$\partial_u p_t = p(u) \partial_u p(u) + \partial_{x(u)} V(x(u)) \dot{x}(u), \qquad (17.17)$$

where the dot denotes derivative w.r.t. u. This will be satisfied automatically if

$$\partial_u p = -\partial_{x(u)} V(x(u)), \qquad (17.18)$$

$$\partial_u x = p, \qquad (17.19)$$

$$\partial_u p_t = 0. \qquad (17.20)$$

We recognize the first two of these as the equations of motion of a classical particle with energy $E = \frac{1}{2} p^2 + V$, which is conserved as a consequence of the equations of motion. The third equation is just a statement of this conservation of energy, at fixed x. One can now identify the path parameter u with time, but for the third equation this must be done with care.

Going back to the original equation, we can calculate

$$\partial_t^2 S_0 = \partial_x S_0 \partial_{tx}^2 S_0 = \partial_x (E(t, x)) \neq 0, \qquad (17.21)$$

because we are free to set $S_0(0, x)$ equal to an arbitrary function. If we view this in terms of the classical particle dynamics, we have a trajectory emerging from each initial point on the $t = 0$ surface, but they are not all required to start with the same energy. The equation with $\partial_u \partial_t S$ is always true, but we cannot identify t and u in this equation unless we start from initial conditions where the value of $p(0)$ at each initial $x(0)$ is fixed by demanding that the energy is x independent. These solutions are the ones that correspond to the time-independent H–J equation.

17.4   COMMENTS ON THE SOLUTIONS

Note that the wave function we have constructed depends on many different classical trajectories, with different initial conditions and nothing about it picks out a particular trajectory. Indeed, in leading order approximation, the wave function is a pure phase if $S_0(0, q^i)$ is real and the probability of being at any particular point in configuration space is the same. The first contribution to the imaginary part of the logarithm of the wave function comes, in this case, from the first-order term S_1, as we will see in a moment. We can, if we wish, choose $S_0(0, q^i)$ to be complex, so that the initial wave function can be chosen peaked around some particular position. We are, however, familiar with the uncertainty principle, which leads us to suspect that we still cannot insist on a *particular* classical history. The equation

$$p_i = \partial_{q^i} S_0, \tag{17.22}$$

is particularly disturbing if we want to describe a narrowly peaked wave function. First of all, it has an imaginary part, in the region where the imaginary part of S_0 is suppressing the probability. Secondly, a narrowly peaked e^{-iS_0} implies large gradients of S_0 so there is no sense in which the initial value of p_i behaves like a unique classical velocity for a particle which will remain on a fixed trajectory. Therefore, the validity of the JWKB approximation alone is insufficient to explain the emergence of classical physics from quantum mechanics.

17.5   THE JWKB APPROXIMATION FOR THE PROPAGATOR

Indeed, it is best not to think of applying the semiclassical approximation to the initial wave function. This frees us to take the limit of singular wave functions that are concentrated at a point, i.e.,

$$\psi_0(q) = \delta^N(q^i - q^i(t_i)), \tag{17.23}$$

and ask for the probability amplitude that at time t_f, one finds the system at $q^i(t_f)$. From the point of view of the Schrödinger equation, this amplitude, called, *the propagator* is a Green's function, which finds the influence of a delta function source at the initial time. In operator language, it can be written

$$G(q^i(t_f), q^i(t_i)) = \langle q(t_f)|e^{-i\frac{H}{\hbar}(t_f - t_i)}|q(t_i)\rangle. \tag{17.24}$$

If there is a classical solution which gets to $q(t_f)$ at t_f starting from $q(t_i)$ at t_i, then the leading semiclassical approximation for the propagator is

$$G(q^i(t_f), q^i(t_i)) = e^{\frac{i}{g^2\hbar}S[q(t_f), q(t_i)]}. \tag{17.25}$$

In classical physics, action is like virtue: a quantity more often talked about than seen. So it is worth our while to compute it for the harmonic oscillator. The action is

$$\int_{t_i}^{t_f} dt \, \frac{1}{2g^2}[m\dot{q}^2 - m\omega^2 q^2]. \tag{17.26}$$

The complex variable $z = \frac{1}{\sqrt{2\hbar g^2 m\omega}}(m\omega q - ip)$ evolves as $z(t_f) = e^{-i\omega(t_f - t_i)}z(t_i)$. We insert the factor of g^2 here so that the quantum version of z has the commutation relations of creation and annihilation operators. No classical motion can change the value of z^*z, which is proportional to the energy, but the boundary conditions on the propagator fix only Re z. For any initial value of q, we can reach any final value of q at any specified time t_f by choosing the energy appropriately. We have

$$q(t) = \cos[\omega(t - t_i)]q(t_i) + \sin[\omega(t - t_i)]\frac{p(t_i)}{m}, \qquad (17.27)$$

and

$$\dot{q}(t) = \omega\left(-\sin[\omega(t - t_i)]q(t_i) + \cos[\omega(t - t_i)]\frac{p(t_i)}{m}\right), \qquad (17.28)$$

at all times. We solve the first equation at $t = t_f$ to determine $v_i \equiv \frac{p(t_i)}{m}$. The Lagrangian is

$$L = \frac{m}{2g^2}[\dot{q}^2 - \omega^2 q^2] = \frac{m\omega^2}{2g^2}[v_i^2 - q_i^2]\cos(2\omega t) - v_i q_i \sin(2\omega t).$$

Doing the integral and simplifying terms using the solution for v_i, we obtain the action

$$S = \frac{m\omega}{2g^2\hbar\sin(\omega\Delta t)}[(q(t_i)^2 + q(t_f)^2)\cos(\omega\Delta t) - 2q(t_i)q(t_f)], \qquad (17.29)$$

where $\Delta t = t_f - t_i$. Note that when $\Delta t \to 0$ all trace of ω disappears from the formula. We can understand this physically, and see that it is a general result, by noting that for very short times, one can only get between two positions with a finite separation if one has a very high velocity. This means that the kinetic term dominates the potential term and the system can be treated as if it were moving freely. This observation will be important later, in our discussion of the Feynman path integral.

In summary, the JWKB approximation gives us an explicit solution of the Schrödinger equation, written in terms of the collection of all solutions to the classical mechanics problem which gives rise to that Schrödinger equation. The position space propagator $G(q(t_f), q(t_i)) = \langle q_f|e^{-\frac{i}{\hbar}H(t_f - t_i)}|q_i\rangle$ is described, to leading order in g^2 in terms of a single classical solution. From this, we can get the general solution of the Schrödinger equation via

$$\psi(q, t) - \int d^N q(0)\, G(q(t), q(0))\psi(q(0), 0). \qquad (17.30)$$

17.6   THE JWKB APPROXIMATION FOR ENERGY LEVELS

To study energy levels, we first rewrite the problem of finding energy levels in terms of the propagator, and then use the JWKB approximation for the propagator. The *resolvent operator*

$$R(z) \equiv \frac{1}{z - H} \qquad (17.31)$$

exists for all complex values of z that are not in the spectrum of the Hamiltonian H. Formally, the trace of the resolvent is given by

$$T(z) = \text{Tr } R(z) = \int dE \frac{\rho(E)}{z - E}. \tag{17.32}$$

$\int_A^B dE \rho(E)$ is the number of states in the energy interval from A to B, if the eigenstates are discrete. $\rho(E)$ is called the density of states. It is the function that appears in the spectral decomposition of the Hamiltonian

$$H = \int dE \ E \ \rho(E)|E\rangle\langle E|.$$

The trace formula might diverge in regions of E where there are an infinite number of states, typically at very high energies. In that case, we insert projection operators to make it finite and avoid those regions of integration. If the divergence is only power law, then a sufficiently high derivative of $T(z)$ is finite, without any such cutoff. $T(z)$ can be constructed in terms of this derivative, up to a polynomial ambiguity, which will not affect the considerations below. In the application of the JWKB approximation, we will be interested in the vicinity of a particular finite energy level, and none of these technical problems will be important.

We can see from the (Cauchy) integral formula that $T(z)$ is analytic away from the spectrum of H, with poles at the discrete spectrum and cuts along the continuous spectrum. Our goal will be to find the position of those poles, in the JWKB approximation, for Hamiltonians of the form $H = \sum \frac{g^2 p_i^2}{2m} + \frac{1}{g^2} V(q^i)$. To do this, we write

$$\frac{1}{z - H} = i \int_0^\infty dt \ e^{i \frac{z-H}{\hbar} t}. \tag{17.33}$$

The integral converges, as long as z has a positive imaginary part.

We can now take the trace in the position basis, and use the JWKB approximation for the propagator to write

$$T(z) = i \int_0^\infty dt \ e^{izt} \int d^N q \ G(q(t) = q, q(0) = q) =_{JWKB} i \int_0^\infty dt \ e^{izt} \int d^N q \ e^{iS(q(t) = q(0) = q)}. \tag{17.34}$$

The classical solutions which contribute to $T(z)$ are all periodic solutions, with any period t. Writing $L = p_i \dot{q}^i - H$, we write the answer as

$$\sum_{periodic \ solutions \ S} e^{it_S(z - E_S)} e^{i \int p_i(q) dq^i}. \tag{17.35}$$

The integrals over t and q have been subsumed into the sum over periodic solutions, so the divergences we worried about above appear, if at all, in that sum, and do not affect the contribution from any particular solution. $t_\mathcal{S}$ and $E_\mathcal{S}$ are the period and energy of the periodic solution. The integral in the last exponential is taken over the closed path in configuration space traced out by the solution.

Give such a closed path, there are an infinite number of periodic solutions, which have the same energy, and trace out that path n times for any nonnegative integer n. The n-th solution has period $nt_\mathcal{S}$. Summing over all these solutions, we get a contribution

$$T_\mathcal{S}(z) = \frac{1}{1 - e^{it_\mathcal{S}(z - E_\mathcal{S})} e^{i \int p_i(q) dq^i}}. \tag{17.36}$$

If

$$\int p_i(q) dq^i = 2\pi K, \tag{17.37}$$

where K is an integer, then this contribution to $T(z)$ has a pole at $z = E_\mathcal{S}$. We have derived the *Bohr–Sommerfeld* quantization rule! We will work out some examples below, and you will do more in the problem sets.

17.7 THE JWKB APPROXIMATION TO THE WAVE FUNCTIONS OF EIGENSTATES

It is easy to see that the insertion of the ansatz $\psi = e^{i\frac{S}{g^2\hbar}}$ into the time-independent Schrödinger equation leads to the equation

$$\frac{1}{2m}(\nabla S)^2 + ig^2\hbar\nabla^2 S + V(q) = g^2 E. \tag{17.38}$$

Assuming $E_0 \sim \frac{1}{g^2}$, and expanding S in powers of g^2, the leading order approximation to this is what is known as the time-independent Hamilton–Jacobi equation. In simple words, it is just the energy equation with the substitution $p_i \to \partial_i S_0$.

We see immediately that ∇S_0 is real or imaginary, according to whether the energy is larger or smaller than V/g^2. These inequalities divide the configuration space into regions where classical motion is either allowed or forbidden. But what exactly do we mean by forbidden? If we examine the equation

$$p_i = \frac{dq^i}{dt},$$

we see that we can get imaginary momentum, if we analytically continue the time to imaginary values. If we do this at the boundary of the forbidden and allowed regions, where $V = g^2 E$, and $p_i = 0$, we can expect a smooth continuation of the real-time classical solution, which "bounces off the wall" and an imaginary time solution which "propagates" in the forbidden region. The semiclassical approximation is valid in the classically forbidden region!

Despite the smoothness of the classical solution, one can see that the JWKB approximation breaks down in the vicinity of the zero momentum wall $V = g^2 E$. The point is that all of the big terms in the equation for S are going to zero on this wall, according to the leading order solution. Thus, when we get close enough to the wall that the leading order terms are of order g^2, it is no longer a valid approximation to drop the order $g^2 \hbar \nabla^2 S$ term. A better approximation in this region is to expand the potential around its value at the wall.

$$V(z, y^i) = E + z u(y^i). \tag{17.39}$$

Here the y^i are a set of $n - 1$ coordinates parameterizing the wall $V(q^i) = E$, and z is a coordinate locally perpendicular to the wall. In general, the wall is a complicated curved surface in the configuration space, so these are curvilinear coordinates and we have to rewrite the gradient terms in this curvilinear system. For this reason, we will restrict our attention to one-dimensional problems when we write explicit formulae.

Before doing that, let us explore the behavior of the wave function in the classically forbidden region. As in the allowed region, the value of the logarithm of the wave function at any point q in the forbidden region can be obtained from the value on the $V = E$ surface by integrating $\pm \int_0^t ds L_E(\dot{q}(s), q(s))$. Here $L_E = \frac{m}{2} \dot{q}^2 + V(q)$, and $q(s)$ is the solution of the imaginary time equations of motion, which interpolates between a point $q(0)$ on the $V = E$ surface and the point $q(t) = q$. We are exploring a regime where L_E is positive so that the two solutions either grow or fall off exponentially as the parameter t increases. The quantity

$$\pm \int_0^t ds L_E(\dot{q}(s), q(s))$$

is called the *Euclidean Action* of the solution.[1] Physical intuition suggests that the wave function is small in this region of configuration space. We are working in the semiclassical approximation and classical physics tells us that a particle of energy E cannot penetrate into this region. If the small g expansion is a good guide to the correct behavior, then it cannot predict an exponentially large probability to be in this region. This implies that the coefficient in front of the exponentially growing solution, must be at least exponentially small, so that the exponentially growing term never gets larger than the exponentially falling one.

This intuition is indeed borne out when one solves the Schrödinger equation with the correct boundary conditions. The details depend somewhat on the actual problem at hand. One general class of problems deals with bound states in infinite space. In this case, for a potential with a single minimum, the semiclassical bound state energies are determined by the Bohr–Sommerfeld conditions applied to periodic real-time classical solutions. The behavior of the bound state wave functions in the regime outside the large r turning point of the periodic

---

[1] The terminology comes from relativistic quantum field theory (QFT), where the analytic continuation of time turns Minkowski space into a four-dimensional Euclidean space.

solution is determined by the imaginary time solution obtained by analytic continuation of the periodic solution. In this case, the boundary condition implied by normalizability of the wave function implies that we must keep only the exponentially falling solution. The matching of the two wave functions in the region near the turning point, where the JWKB approximation breaks down, leads to a small correction to the Bohr–Sommerfeld conditions. You will explore this in Exercises 17.2–17.5.

17.7.1 Examples

Let us use the JWKB approximation to calculate the bound state energies of the harmonic oscillator and the hydrogen atom, for which we know the exact answers. For the oscillator, the general classical solution is

$$x(t) = A\cos(\omega(t - t_0)). \tag{17.40}$$

and

$$p(t) = -Am\omega\sin(\omega(t - t_0))). \tag{17.41}$$

The solutions are all periodic with period $\frac{2\pi}{\omega}$. Here $\omega = \sqrt{\frac{k}{m}}$ and $V = \frac{1}{2}kx^2$. The energy is

$$E = \frac{p^2}{2m} + \frac{1}{2}kx^2, \tag{17.42}$$

so that

$$p = \sqrt{2m(E - \frac{1}{2}kx^2)}. \tag{17.43}$$

The period solutions go between $x = \pm\frac{2E}{k}$ and a full period traverses this interval twice. Defining $x = \sqrt{\frac{2E}{k}}y$, the action for a period is

$$S = \int p\,dx = 4\frac{E}{\omega}\int_{-1}^{1} dy\sqrt{1 - y^2} = 2\pi\frac{E}{\omega} = 2\pi n\hbar, \tag{17.44}$$

where the last equality is the Bohr–Sommerfeld quantization rule. Thus, we obtain the correct spectrum of the exact quantum oscillator except for the $\hbar\omega/2$ shift of the ground state. This is actually picked up by the first correction to the JWKB approximation, and all higher order corrections vanish. It turns out that this is true for all Hamiltonians quadratic in the canonical variables, a fact which we will understand when we learn the Feynman path integral formula.

It is somewhat more surprising that the first correction to the JWKB approximation for the hydrogen atom also gives the exact result. This is connected to the fact that the system has a complete set of conservation laws, which are smooth functions of the canonical variables: the full equations of motion are equivalent to the conservation laws. Such systems are called *completely integrable*. There is a vast literature on completely integrable systems. An elementary introduction to the topic can be found in [18].

The conserved energy for the hydrogen atom is given by

$$E = \frac{(\mathbf{p})^2}{2m} - \frac{a}{r}. \tag{17.45}$$

Here $a = \frac{e^2}{4\pi\epsilon_0}$. Angular momentum conservation implies that the motion takes place in a plane, and for a bound orbit, conservation of the Eccentricity vector tells us that the orbit is an ellipse. The equation for an ellipse is

$$r^2 = \frac{R^2}{1 + \kappa \sin^2(\phi)}. \tag{17.46}$$

Here r is the distance from one of the foci and ϕ is the angle with that focus as center. The action integral we have to compute in order to evaluate the Bohr–Sommerfeld equation is

$$S = \int p_r dr + p_\phi d\phi = 2\pi l + 2 \int_{r_-}^{r_+} dr \sqrt{2m(E - \frac{l^2}{2mr^2} + \frac{a}{r})}. \tag{17.47}$$

We have used the fact that the angular momentum $p_\phi = l$ is a conserved constant. The radial integral runs between the two zeros of the integrand and the factor of 2 is there because a full cycle covers this radial path twice. Equating this to $2\pi k\hbar$, we get the quantization condition. For $a = 0$ (circular orbits), where $r_+ = r_-$, this reduces to Bohr's condition that the angular momentum is an integer multiple of $\hbar$. You will do the integral in the exercises and verify that the quantization condition gives

$$E = -\frac{1}{(k+l)^2} \; Rydberg. \tag{17.48}$$

The derivation of the Bohr–Sommerfeld condition from the quantum theory in the JWKB approximation allows us to conclude that l is an integer even for noncircular orbits, because the wave function is periodic in ϕ. This formula looks exact, and we can think of $k = j_{max} + 1$, where j_{max} is the highest power in the relevant Laguerre polynomial, but there is some illegal trickery being pulled in making that statement. In the exact formula, l is the integer defined by the square of the total quantum angular momentum operator, $l(l + 1)\hbar^2$. In the Bohr–Sommerfeld rule, $l\hbar$ is the angular momentum in the plane of the orbit, which one would want to identify with the quantum number m. For the classical orbit, these are the same thing, and both are quantized as integers, but they are not the same. These problems with the leading order result are resolved by the first correction to the JWKB approximation, but we do not have space to go into that here.

For general one-dimensional potentials, one can write the Bohr–Sommerfeld condition as

$$\int_{x_-}^{x_+} dx \ \sqrt{2m(E-V)} = \pi k\hbar, \qquad (17.49)$$

where the integral is taken between two consecutive turning points of the classical motion. The factor of 2 difference between this formula and our previous statement of the Bohr–Sommerfeld condition comes from the fact that the classical periodic solution goes back to x_- after being reflected at x_+. Notice that this condition does not refer to the region of space beyond the turning points, and so cannot distinguish between two potentials with different behavior outside the turning points. In particular, the classical motion may be in a local but not global minimum of the potential, but the B–S condition only recognizes that when k gets large enough that there is no longer a periodic motion in that local well. In the quantum theory, such states are not true eigenstates of the Hamiltonian, but rather are long-lived, metastable states. In the next section, we will use the JWKB approximation to compute the decay rate of such a state.

17.8   THE DECAY OF METASTABLE STATES

Another general class of problems, to which the JWKB approximation is often applied, has a barrier interpolating between two classically allowed regimes. One is interested in the fate of a wave function initially concentrated in a well near what we can choose as the origin of coordinates. The potential either goes to a constant at infinity, or has other wells which are deeper and/or wider. It is then very often the case that a normalized wave function concentrated in the well near the origin is an unstable situation. The probability density propagates outward, ending up concentrated near a different well, or flowing out to infinity. In the extreme semiclassical approximation $g \ll 1$, the fraction of probability concentrated in the original well is either exponentially small, or vanishes, as time goes to infinity.

On the other hand, let us suppose that there is a semiclassical bound state near the origin. That is, a periodic solution of energy E, which stays trapped near the origin, even though there are other regions of configuration space, separated from the origin by a finite barrier which it could in principle explore. Then, in the limit $g \ll 1$ we expect to find a *metastable state*, with a life-time that goes to infinity as $g \to 0$. The semiclassical approximation allows us to compute the lifetime of the state, as a systematic expansion around $g = 0$. We call this the *decay* of the metastable state, and our aim is to calculate the *decay rate* of this state; the probability per unit time that it will decay. The mechanism of decay, in the semiclassical approximation, is called *quantum mechanical tunneling through a barrier*, or *quantum tunneling* for short.

To get a feel for the intuition behind this nomenclature, it is best to watch the following video.

www.youtube.com/watch?v=cV2fkDscwvY

It shows the modulus of the wave function, for a particular solution of the time-dependent Schrödinger equation, hitting a barrier higher than the energy of the incoming particle. One can see how the wave function "tunnels through the barrier."

It is important to understand that there is no violation of energy conservation in tunneling processes. In the decay of a metastable state, the initial wave function, related to the periodic classical solution whose energy is below the barrier height, is *not* an eigenstate of the Hamiltonian. The true eigenstate of the Hamiltonian at that value of the energy is either part of a scattering continuum (if the potential simply asymptotes to a constant on the other side of the barrier), or is concentrated in some other well of the potential.

We have seen that in the JWKB approximation, the behavior of the wave function in the tunneling region (the region "under the barrier") is determined by the solution of the imaginary time classical equations of motion. The surface $V = E$ in configuration space has two disconnected components, which are distinguished by their distance from the origin. In the simplest case, they have the topology[2] of spheres in configuration space. The imaginary time solution, which interpolates between two real-time classical motions on either side of the barrier, has $\dot{q}^i = 0$ when it hits either of the surfaces $V = E$. If it did not, we could not match the imaginary derivative under the barrier to the real derivative inside it. There is a solution for each choice of a pair of points on the two disconnected components $V = E$. Each of these solutions has a different Euclidean action. Since all the Euclidean actions are $\propto \frac{1}{g^2}$, all these contributions are exponentially suppressed compared to that of minimal action. Therefore, to leading exponential order as $g \to 0$, only the minimal action path contributes to the tunneling amplitude. This *most probable escape path* [77] has been called an *instanton*[3] following the work of [78].

We can avoid the search over all pairs of points $(q^i_{inner}, q^i_{outer})$ on the $V = E$ surface by noting that since $\dot{q} = 0$ on this surface, we can find another solution of the imaginary time equations which starts at q^i_{inner} bounces off q^i_{outer} and returns to q^i_{inner}, by simply retracing the original solution. So all we have to do is find the minimal action bounce solution, which starts and ends at q^i_{inner} and then minimize over the choice of q^i_{inner}. The action of this bounce solution is twice that of the original, so the square of the wave function, which gives the tunneling probability, is

$$P_{tunneling} \sim e^{-S_{bounce}}, \tag{17.50}$$

---

[2] But not generally the geometry, unless the potential is spherically symmetric in the full configuration space.

[3] Again, the terminology comes from QFT, where the relevant solutions are localized in space as well as time. Static classical solutions, which are localized in space, can be interpreted as heavy particles and were dubbed *solitons*. 't Hooft therefore invented the term instanton to refer to an imaginary time solution localized in all dimensions.

for the minimal action bounce. Things become even simpler, if we are talking about the metastable ground state, which corresponds to the static classical solution localized at the origin. In this case, the inner $V = E$ surface collapses to a point. Since $q^i_{inner} = 0$, we just have to find a single bounce solution. For the decay of states other than the ground state, we have to find the bounce for all values of q^{inner} and find the minimum action.

17.9 MORE EXAMPLES

A very general situation in which JWKB formulae are useful is the determination of the asymptotic behavior of the wave function of a one-dimensional system with a confining potential. This is a potential that grows at infinity, so that the system only has bound states. The equation for the logarithm of the wave function ($\psi = e^{-S}$) of an eigenstate of energy E is

$$-\frac{\hbar^2}{2m}\left[(\frac{dS}{dx})^2 - \frac{d^2S}{dx^2}\right] = E - V. \tag{17.51}$$

In the asymptotic region $V \gg E$. We make the ansatz that the second derivative of S is negligible and find

$$\frac{dS}{dx} = \pm\sqrt{\frac{2m}{\hbar^2}V}. \tag{17.52}$$

The second derivative is then smaller than the first. The ratio to the first derivative term is $V'/V^{3/2}$. Unless V has rapid and growing oscillations (e.g., $V \sim x^a \sin(x^{b+1})$ with $b > 3a/2$), this term is indeed negligible. In all cases of practical interest then, the JWKB approximation is valid in the asymptotic regime. It gives rise to two linearly independent solutions, behaving as $e^{\pm\int\sqrt{V}}$. Only the falling solution is normalizable.

Another very common situation is the decay of a metastable minimum of the potential. Generically, the minimum will be a quadratic potential, centered around a point that we can take as the origin of coordinates. The Euclidean (imaginary time) equations of motion are

$$\ddot{x} = V'(x). \tag{17.53}$$

These are the ordinary classical equations of motion in the upside down potential $-V(x)$. The nature of the solutions is completely determined by "energy" conservation in this upside down potential. The metastable minimum at $x = 0$ is a maximum of the upside down potential. The bounce solution for the metastable ground state must begin at this maximum at time $t_E = -\infty$, and return to it at $t_E = \infty$. This implies that the velocity at this point is zero (i.e., asymptotes to zero as $t_E \to \pm\infty$), so the Euclidean "energy" of the solution is equal to $V(0)$. The conservation law implies that the solution passes through the minimum of $-V$ (the maximum separating the metastable minimum from the true minimum of the potential) with nonzero velocity, and climbs the opposite side until it reaches a point where $V(x_m) = V(0)$ (see Figure 17.1).

We then have $dx/dt_E(x_m) = 0$, by "energy" conservation, so the solution can turn around and return to the origin. An additional consequence of the vanishing velocity at x_m is that the Euclidean solution can be analytically continued into a real-time, real-valued solution at this point. That solution gives the JWKB phase of the wave function in the classically allowed region.

The bounce solution satisfies

$$\frac{1}{2m}\dot{x}^2 - V(x) = -V(0), \tag{17.54}$$

so that the Euclidean action is

$$S = \int_{-\infty}^{\infty} dt_E \left[\frac{1}{2m}\dot{x}^2 + V(x)\right] = \int_{-\infty}^{\infty} dt_E \left[2(V - V(0))\right]. \tag{17.55}$$

We have dropped a term $\int V(0)$ which would have been present also for the classical solution where $x(t) = 0$. This has to do with the proper normalization of the wave function. The integral over the constant term diverges and would lead to a vanishing wave function. It must be subtracted out. The general rule, when computing decay amplitudes in the JWKB approximation is to compute the difference between the Euclidean actions of the bounce solution, and the static solution sitting at the metastable minimum.

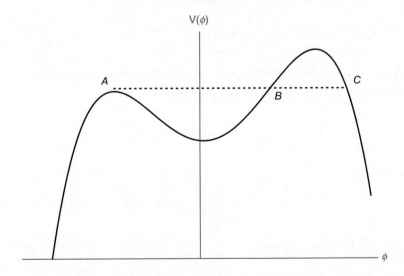

Figure 17.1 The upside down potential for an instanton calculation.

The equation $\frac{1}{2m}\dot{x}^2 - V(x) = -V(0)$ allows us to convert the t_E integral to an integral over x. The presence of a point where $\dot{x} = 0$ means that the t_E integral covers the x-axis twice. This is a consequence of computing the action of the bounce solution. We end up with a formula for the decay probability

$$P_{decay} = e^{-\frac{2}{\hbar} \int_0^{x_m} \sqrt{2m(V(0)-V(x))}dx}.$$

(17.56)

In principle, the interpretation of this quantity as a decay probability is valid for cases in which the potential to the right of x_m is such that the system has no discrete eigenstate localized at finite x. In that case, a Gaussian wave function, initially localized near the origin, eventually has a probability distribution that flows out to infinity. P_{decay} gives the approximate probability per unit time that the Gaussian will decay to a state-localized near infinity.

If, on the other hand, the potential has a discrete eigenstate whose eigenfunction is localized at some $x_0 > x_m$, then there is a finite probability that a particle in that eigenstate will actually be found at the origin. In this case, P_{decay} corresponds roughly to *that probability*. An interpretation, which is always valid, is that P_{decay} is the probability per unit time that a Gaussian wave function, initially localized near the origin, with a width given by the harmonic oscillator ground state of the potential $\frac{1}{2}V''(0)x^2$, will be found to the right of x_m.

17.10   THE JWKB APPROXIMATION FOR PHASE SHIFTS

For large angular momentum, the effective potential

$$U(r) = V(r) + \frac{\hbar^2 l(l+1)}{2mr^2},$$

(17.57)

in the radial Schrödinger equation for $u_l(r) = R_l(r)/r$, satisfies the criteria for validity of the JWKB approximation. Furthermore, in the expression

$$f(\theta) = \frac{1}{2ik} \sum_{l=0}^{\infty} (2l+1)P_l(\cos\theta)[e^{2i\delta_l(k)} - 1],$$

(17.58)

we can substitute the large l expansion of the Legendre polynomials

$$P_l(\cos\theta) \to \frac{2i}{\sqrt{2\pi l \sin(\theta)}} \cos((l+1/2)\theta + \pi/4),$$

(17.59)

to write the large l part of the sum as

$$f(\theta) = \frac{1}{k\sqrt{2\pi \sin(\theta)}} \sum_{l=l*}^{\infty} [e^{i(2\delta_l - (l+1/2)\theta - \pi/4)} - e^{i(2\delta_l + (l+1/2)\theta + \pi/4)}],$$

(17.60)

where $l*$ is large enough for the expansion of the Legendre polynomial to be accurate. In the JWKB approximation, the phases of the wave function are large, so all the terms in this sum are rapidly oscillating and the sum is dominated by the stationary phase point

$$2\frac{d\delta_l}{dl} = \pm\theta. \tag{17.61}$$

The JWKB approximation to the wave function has the phase

$$\phi_{JWKB} = \pi/4 + \frac{1}{\hbar}\int_{r*}^{r} dy \sqrt{\hbar^2 k^2 - 2mV(y) - \frac{\hbar^2 l(l+1)}{y^2}}. \tag{17.62}$$

The JWKB approximation to the phase shift is the limit as $r \to \infty$ of the difference between ϕ_{JWKB} and the phase $kr - l\pi/2$ of the free wave function. For potentials satisfying the postulates of scattering theory, the difference is finite:

$$\delta_l^{JWKB} = \frac{1}{\hbar}\int_{r*}^{\infty} dy \left[\sqrt{\hbar^2 k^2 - 2mV(y) - \frac{\hbar^2 l(l+1)}{y^2}} - k\right] + \frac{1}{2}\pi(l+1/2) - kr*. \tag{17.63}$$

$r*$, the turning point, depends on l, but that dependence cancels against the last term in the formula. Thus, the stationary phase equation becomes (to leading order in l, with $L \equiv \hbar l$ the "classical angular momentum")

$$L\int_{r*}^{\infty} \frac{dy}{y^2\sqrt{2m(E-V) - L^2/y^2}} = \frac{1}{2}(\pi \pm \theta). \tag{17.64}$$

We have written $\hbar^2 k^2 = 2mE$ so that all terms in this formula have a classical interpretation. This equation is the same as that we could have obtained in classical mechanics, by computing the scattering angle in terms of the angular momentum. These two variables are quantum mechanically complementary, but the stationary phase approximation allows them to have fixed values simultaneously. The stationary phase equation says that $\delta_l \sim l\theta$ and our derivation assumed that δ_l was large, so the criterion for the validity of this approximation is that $l\theta$ be large.

17.11 EXERCISES

17.1 Show that the condition that

$$\int_0^t ds[-q_i(s)\dot{p}^i(s) - \frac{1}{2}p_i(s)p^i(s) - V(q(s))]$$

is independent of the initial value $q^i(0)$ leads to the classical Hamilton equations for the system.

17.2 The JWKB approximation for eigenfunctions breaks down when one is near a turning point of the classical motion, a place where $V(x) = E$. Near a generic turning point we have $V(x) - E \sim ax$. Show that in this region you can solve the Schrödinger equation, which is called the Airy equation, exactly. Show that the Fourier transform turns the equation into a first order equation. This solution is called the Airy function $Ai(x)$. Show that the second solution has the form $Bi(x) = b(x)Ai(x)$, where

$$\frac{b''}{b'} = -2\frac{Ai'}{Ai}.$$

17.3 Find the behavior of the two solutions to the Airy equation when $x \to \pm\infty$. In this limit, you can do the Fourier transform integral by stationary phase approximation.

17.4 Consider a periodic solution of the classical equations and the eigenfunction of the Schrödinger equation for each of the Bohr–Sommerfeld levels. For simplicity, consider a potential symmetric around $x = 0$, with bound states in the region near the origin. The bound state wave functions are either even or odd, so we can concentrate on turning points of the classical motion to the right of the origin. To the right of the turning point, the solution must fall off exponentially. Show that only one linear combination of solutions to the Airy equation has this property. Match that solution to the JWKB wave function to the left of the turning point and show that the match can only be accomplished by a shift in the energy level, compared to that computed by Bohr–Sommerfeld.

17.5 Consider the Schrödinger equation in the complex plane and show that the JWKB approximation can be used to connect directly between the two sides of a turning point without ever solving the Airy equation. Show how to obtain the energy level shift described in the previous exercise.

17.6 Finish the computation of the Bohr–Sommerfeld quantization condition for the hydrogen atom.

17.7 In this exercise, we will use units in which $\hbar = c = 1$. Mass, energy, and momentum then have the same dimensions. The natural energy scale of the strong interactions is ~ 100–200 MeV. The modern theory of the strong interactions is based on a relativistic QFT known as QCD (short for quantum chromodynamics). The theory contains a variety of particles known as quarks, which have different masses. They interact via forces generated by other fields, called gluons. If a very heavy quark–antiquark pair is separated by a large distance, the theory is supposed to generate a confining potential, which varies linearly, $V = kr$, with $k \sim \frac{(100 MeV)^2}{\hbar c}$. At short distances compared to

$\frac{\hbar c}{100 \text{ MeV}}$ the potential is supposed to be approximately Coulomb-like $V \alpha_c \frac{\hbar c}{r}$. α_c is a dimensionless constant.

Low lying bound states of quarks and antiquarks that are much heavier than this scale should be treatable in a nonrelativistic approximation, because the kinetic energy and the binding energy are of the same order of magnitude, and much less than the mass. There are three types of quarks, charm, bottom and top, to which this analysis might apply. However, top quarks decay (via the weak interactions) into bottom quarks too rapidly to form bound states. Assuming the nonrelativistic approximation, one can try to compute the spectrum of quark antiquark bound states as a function of the quark mass, and the parameters α_c and k. The JWKB approximation gives a handy analytic tool for making a first estimate of the bound state spectrum. We will make an even more drastic approximation, and replace the Coulomb term by a constant V_0, of order 100 MeV. Write the Bohr–Sommerfeld quantization rule for zero angular momentum in this potential. Show that knowledge of the ground state, the first radial excitation, and the quark mass enable you to calculate the masses of higher excitations. The mass of a bound state is twice the quark mass plus the (negative) binding energy. For the charmed quark system, the quark mass is $1.5 \frac{\text{Gev}}{c^2}$ and the rest energies of the ground state and first radial excitation (these are called the $\psi(1S)$ and $\psi(2S)$ particles) are 3.1 and 3.7 GeV. The bottom quark rest energy is 4.18 GeV and the $v(1S)$ and $v(2S)$ particles have rest energies 9.46 and 10.02 GeV. Give the JWKB estimate for the energy of the next radial excitation in each case. Compare to the values you can find in the Particle Data Group compilation [79].

17.8 If a one-dimensional potential rises at infinity (e.g., like a power $|x|^p$), then there are an infinite number of solutions of the Bohr–Sommerfeld condition

$$\int dx \ \sqrt{2m(E - V(x))} = 2\pi n\hbar. \tag{17.65}$$

Show that as $n \to \infty$, these JWKB energies are closer and closer to the exact eigenvalues.

17.9 Model the strong and electromagnetic forces between an alpha particle (Helium nucleus) and a nucleus by

$$V(r) = \theta(r - a)\frac{2Ze^2}{4\pi\epsilon_0 r}. \tag{17.66}$$

a is of order 10 times the pion Compton wavelength $a \sim 10^{-12}$ cm, or $\frac{1}{14}$ inverse MeV in natural units. This is the potential for an $l = 0$ or s-wave bound state. For more general l there is an additional repulsive centrifugal potential. The idea behind this potential is that strong attractive nuclear forces dominate the Coulomb potential for $r < a$ and that they are of such short range that we can just model them by a constant potential. Argue that this problem has no bound states.

17.10 Consider the scattering states in the potential above. For small enough positive energy, the well near the origin in the potential would create a metastable *resonance* in which the particle was trapped for some time in the region $r < a$. Classically this resonance would be stable, but there is a quantum tunneling amplitude for the particle to escape to infinity for positive energy, and you argued in the previous exercise that there were no true negative energy-bound states. Calculate the tunneling probability in this potential as a function of the energy E.

17.11 Calculate S_2, the third term in the JWKB expansion of the logarithm of the wave function of an energy eigenstate.

17.12 Show that the formula

$$P_l(x) = \frac{1}{\pi} \int_0^\pi d\phi \, (x + i\sqrt{1 - x^2} \cos \phi)^l$$

satisfies the Legendre equation and that it is square integrable on the interval $-1 \leq x \leq 1$. Argue that this means it is a representation of the Legendre polynomials.

17.13 Use the formula of Exercise 17.12 to derive the large l limit of the Legendre polynomials, which we used in our semiclassical analysis of phase shifts.

CHAPTER 18

The Variational Principle

18.1 INTRODUCTION

We have already used the variational principle in Chapters 11 and 13, but here we will give a more careful exposition and describe some more general examples. The basic idea is simple: the expectation value of the Hamiltonian in any normalized state is given by

$$\sum |c_k|^2 E_k \geq E_0 \sum |c_k|^2 = E_0, \qquad (18.1)$$

where E_0 is the ground state energy. Thus, if we start from any state, and make changes that reduce the expectation value of the energy, we get a better approximation for the ground state. The art in employing the variational principle lies in using one's intuition to find an ansatz for the ground state wave function, which captures important features of the physics of the system, while giving rise to calculations that are relatively simple. A lot of the words that are used to describe complicated physical systems are actually derived from clever variational approximations. The most famous example is the concept of single electron orbitals in complicated systems with interacting electrons.

In this chapter, we expose some general properties of the variational method and do a number of examples. The variety of variational approximations in the literature illustrates the creativity of physicists, and cannot be captured in a single book.

18.2 GENERAL PROPERTIES

We will begin by making some general remarks about the variational principle, and then give some examples.

Given a Hamiltonian H, we can get a variational bound on its ground state energy by choosing *any* exactly soluble Hamiltonian H_0 and noting that

$$G(H) \leq \langle \psi_0 | H | \psi_0 \rangle = G(H_0) + \langle \psi_0 | (H - H_0) | \psi_0 \rangle. \qquad (18.2)$$

We have introduced, *for this chapter only*, the notation $G(H)$ to denote the ground state energy of a Hamiltonian. The second form of the variational estimate suggests the possibility of finding corrections to the estimate by treating $H - H_0$ as a small perturbation of H_0. This can sometimes be useful, even though there is no apparent small parameter in the problem.

   This consideration can be generalized to statistical mechanics. Write the partition function as

$$Z \equiv \text{Tr } e^{-\beta H} = \frac{\text{Tr } e^{-\beta(H-H_0)}e^{-\beta H_0}}{\text{Tr } e^{-\beta H_0}} \text{Tr } e^{-\beta H_0}. \tag{18.3}$$

Now it is a very general fact about functions that $\langle e^X \rangle \geq e^{\langle X \rangle}$ for any probability distribution.[1] To prove this, consider the expectation value of e^{sX} as a function of s. We have

$$\partial_s^2 \ln \langle e^{sX} \rangle = \frac{\langle X^2 e^{sX} \rangle}{\langle e^{sX} \rangle} - \left(\frac{\langle X e^{sX} \rangle}{\langle e^{sX} \rangle} \right)^2 = \frac{\langle ((X-a)^2 e^{sX} \rangle}{\langle e^{sX} \rangle}, \tag{18.4}$$

where $a = \frac{\langle X e^{sX} \rangle}{\langle e^{sX} \rangle}$. That is, the log of the expectation value is a convex function of x. A convex function is always larger than the linear approximation to that function around any point, so the inequality is true. Since the free energy is defined by $Z = e^{-\beta F}$, we have

$$F \leq F_0 + \langle (H - H_0) \rangle, \tag{18.5}$$

where the expectation value is taken in the thermal density matrix $Z_0^{-1} e^{-\beta H_0}$.

18.3   EXAMPLES

With these general ideas behind us, let us try some particular problems. As a first example, let us consider a one-dimensional problem with a monomial potential $V = g^2 x^{2q}$. Work in units where $\frac{\hbar^2}{2m} = 1$. Let us try a Gaussian ansatz for the ground state wave function

$$\psi_0 = e^{-\frac{a}{2}x^2} \frac{a}{\sqrt{\pi}}. \tag{18.6}$$

This satisfies

$$-\nabla^2 \psi_0 = 2a^2 - 4a^4 x^2 \psi_0, \tag{18.7}$$

so the expectation value of the Hamiltonian is

$$\langle \psi_0 | -\nabla^2 + g^2 x^{2q} | \psi_0 \rangle = 2a^2 - 4a^4 \langle x^2 \rangle + g^2 \langle x^{2q} \rangle. \tag{18.8}$$

The integral

$$\int x^{2q} e^{-a^2 x^2} = (-\frac{d}{db})^q [\int e^{-bx^2}]_{b=a^2} = (-\frac{d}{db})^q 2\sqrt{\frac{\pi}{b}} = \frac{1}{2} \cdots \frac{2q-1}{2} \frac{2\sqrt{\pi}}{a^{2q+1}}. \tag{18.9}$$

---

[1] The notation $\langle A \rangle$ just means the expectation value of A, not necessarily a QM expectation value.

Thus,

$$\langle x^{2q} \rangle = \frac{1}{2} \cdots \frac{2q+1}{2} a^{-2q} \tag{18.10}$$

and

$$\langle H \rangle = \frac{1}{2} a^2 + g^2 \frac{1}{2} \cdots \frac{2q+1}{2} a^{-2q}. \tag{18.11}$$

This is minimized at

$$\frac{1}{2} = q \frac{g^2}{a_0^{2(q+1)}} \frac{1}{2} \cdots \frac{2q+1}{2} \tag{18.12}$$

and the bound on the ground state energy is of order $a_0^2 \sim g^{\frac{2}{q+1}}$.

By scaling $x = g^{\frac{1}{q+1}} y$, we can see that the variational estimate has exactly the same dependence on g as the exact answer (Exercise 18.1). There are a number of other general remarks about variational estimates, which are illustrated nicely by this example. The quantum ground state energy of a bound state of some collection of particles is determined by a competition between the tendency of the potential to collect all of the particles in one place, and the kinetic energy which the confined particles have by virtue of the uncertainty principle. A Gaussian ansatz for the bound state wave function incorporates both of these features, with the width of the Gaussian determined by the relative balance between potential and uncertainty principle kinetic energy. In our example, we see that as g is decreased, the width gets larger and the energy lower. Similarly, if we increase q for fixed g, the width gets larger because the potential gets flatter and flatter for $x < 1$ and then rises abruptly when $x > 1$.

Note that the shape of the Gaussian wave function at infinity is completely wrong. Indeed, as $x \to \infty$ we are in a regime where the JWKB approximation of the previous section is valid and the logarithm of the exact wave function behaves like

$$\ln \psi_0 \sim g x^{q+1}. \tag{18.13}$$

Thus, although the variational ansatz gives a good approximation to the ground state energy, it does not give accurate answers for questions about the probability of finding a particle at large distances from the origin, given that the system is in its ground state. The reason for this discrepancy in accuracy is that in the ground state, the distant regions of x do not give much of a contribution to the integral that computes the total energy. More generally, it is true that variational estimates give us a much better approximation to the ground state *energy* than they do to the ground state *wave function*.

A second example to which we can apply the variational technique is the ground state of the helium atom. In the Bohr–Rydberg units, which we used in our discussion of atomic and molecular physics, the Hamiltonian is

$$H = \frac{\mathbf{p_1^2} + \mathbf{p_2^2}}{2} - \frac{2}{r_1} - \frac{2}{r_2} + \frac{1}{|\mathbf{r_1} - \mathbf{r_2}|}. \tag{18.14}$$

We have neglected the differences between the electron mass and the actual reduced mass of the electrons. There are two simple variational estimates of the ground state energy. The first uses our general observation above relating the ground state energies of two different Hamiltonians

$$G(H) \le G(H_0) + \langle (H - H_0) \rangle, \tag{18.15}$$

where the expectation value is taken in the ground state of H_0. For helium, we take

$$H_0 = \frac{\mathbf{p}_1^2 + \mathbf{p}_2^2}{2} - \frac{2}{r_1} - \frac{2}{r_2}. \tag{18.16}$$

The ground state energy of H_0 is -8 Rydberg, since H_0 is just two copies of the Hamiltonian of the doubly charged hydrogenic ion. The ground state wave function is the spin singlet, multiplied by the product of two hydrogen ground states with $a_B \to a_B/2$:

$$\psi_{s_1/s_2}(\mathbf{r_1}, \mathbf{r_2}) = \epsilon_{s_1/s_2} \frac{8}{\sqrt{2\pi}} e^{-2(r_1 + r_2)}. \tag{18.17}$$

The expectation value of $H - H_0$ is

$$\langle (H - H_0) \rangle = 4\pi \frac{32}{\pi^2} \int_0^\infty dr_1 \, dr_2 (r_1 r_2)^2 e^{-4(r_1 + r_2)} d\phi d\theta \, \sin(\theta) \frac{1}{\sqrt{r_1^2 + r_2^2 - 2r_1 r_2 \cos(\theta)}}. \tag{18.18}$$

The first factor of 4π comes from the overall angular integration. θ is the angle between the vectors $\mathbf{r_1}$ and $\mathbf{r_1}$. To do the angular integration, it is convenient to write

$$\frac{1}{\sqrt{r_1^2 + r_2^2 - 2r_1 r_2 \cos(\theta)}} = \frac{1}{\sqrt{\pi}} \int_0^\infty ds s^{-1/2} e^{-s[r_1^2 + r_2^2] - 2r_1 r_2 \cos(\theta)]}, \tag{18.19}$$

where we have used the by now familiar identity

$$\Gamma(1/2) = \int ds s^{-1/2} e^{-s} = \sqrt{\pi}.$$

Now we use the change of variables

$$\int_0^\pi \sin(\theta) d\theta \, f(\cos(\theta)) = \int_{-1}^1 dx \, f(x)$$

to perform the angular integral. The result is

$$\langle (H - H_0) \rangle = \frac{256}{\sqrt{\pi}} \int_0^\infty dr_1 \, dr_2 (r_1 r_2)^2 \int_0^\infty ds \, s^{-1/2} e^{-4(r_1 + r_2)} \frac{1}{2 s r_1 r_2} (e^{-s(r_1 - r_2)^2} + e^{-s(r_1 + r_2)^2}). \tag{18.20}$$

The remaining integrals are all elementary, or reduce to Euler Gamma functions. The end result of the computation is that

$$E_0^{helium} \leq -75\text{eV}. \tag{18.21}$$

The experimental result is -78.975 eV. Note that the original approximation, which neglected the electron repulsion, was *not* a variational approximation, and so undershot the right answer. In fact, we can prove that it undershot, because it is obtained by neglecting a positive term in the Hamiltonian. Thus, these simple calculations show that the helium ground state is between -109 eV (8 Rydbergs) and -75 eV. More elaborate calculations have been done, with up to 18 variational parameters and they reproduce the correct answer within experimental error. The simplest improvement is to use a variational ansatz for the wave function that is the product of two single electron ion wave functions but with the "effective nuclear charge," Z treated as a variational parameter. In Exercise 18.2, you will show that this puts an upper bound on the ground state energy of helium of -77.5 eV.

18.4   THE HARTREE AND HARTREE–FOCK APPROXIMATIONS

Let us begin by recalling the quantum field theoretic (also called *second quantized*) treatment of multiparticle states of bosons and fermions. We start with a single particle Hilbert space with some orthonormal basis $|i\rangle$, and introduce a set of operators $a_i^\dagger$, one for each element of the basis. These satisfy

$$[a_i, a_j^\dagger]_\pm = \delta_{ij}, \tag{18.22}$$

$$[a_i, a_j]_\pm = 0. \tag{18.23}$$

The plus sign is for fermions and the minus sign for bosons. The Hermitian conjugate of the second equation is also valid, and tells us that the states

$$|i_1 \ldots i_n\rangle = \frac{1}{\mathcal{N}} a_{i_1}^\dagger \ldots a_{i_n}^\dagger |0\rangle, \tag{18.24}$$

where

$$a_i|0\rangle = 0, \tag{18.25}$$

for all i, are a basis for the symmetrized (bosons) or antisymmetrized (fermions) n-fold tensor product of the single particle Hilbert space with itself. The state $|0\rangle$ is called the *no-particle* state or *vacuum* state. These tensor product Hilbert spaces are the Hilbert spaces of n independent bosons or fermions. The normalization factor $\mathcal{N}$ differs from 1 only for bosons. It is the product of $\sqrt{n_k!}$, where n_k is the number of bosons in the state $|k\rangle$ of the single particle Hilbert space.

The most general Hamiltonian describing noninteracting particles has the form

$$H_0 = a_i^\dagger h_{ij} a_j, \tag{18.26}$$

where we use the summation convention. These are called *one body operators* because they are a sum of terms, each of which acts on only a single particle. The Hilbert space on which the creation and annihilation operators act, contains any number of particles. An operator which does not change the total number of particles will be a sum of terms, each of which is a monomial containing equal numbers of creation and annihilation operators. We can always write operators in *normal-ordered form*, with all the creation operators to the left of all the annihilation operators.

Operators of the form

$$a_i^\dagger a_j^\dagger V_{ij,kl} a_k a_l \tag{18.27}$$

are called two body operators. Written in terms of the states $|i_1 \ldots i_n\rangle$ in the tensor product Hilbert space, they have matrix elements which act on only two particles at a time. Many model Hamiltonians for interacting particles, like the Coulomb interaction, can be written in terms of such two body operators. The Hartree and Hartree Fock approximations, which we are about to introduce, can be applied to Hamiltonians that contain operators affecting k particles at a time (k annihilation and creation operators), but are simplest for simple two body interactions.

For a one body Hamiltonian H_0, the ground state is simple. For bosons it takes the form

$$|\Psi_0\rangle = \frac{1}{\sqrt{n!}} (a_i^\dagger \phi_1^i)^n |0\rangle, \tag{18.28}$$

where

$$h_{ij} \phi_1^j = \epsilon_1 \phi_1^i. \tag{18.29}$$

In other words, the state $\sum \psi_1^i |i\rangle$ is the ground state of the single particle Hamiltonian whose matrix is h_{ij}. The ground state energy of the n particle system is just $n\epsilon_1$.

For fermions, because of the Pauli principle, which is implemented by the anticommutation relations between the creation operators, we cannot put all of the particles in the same single particle eigenstate, but instead must choose the n lowest states of h_{ij} and put one fermion in each of them. The multifermion ground state is called the *filled Fermi sea* and has the form

$$|\Psi_0\rangle = \sum a_{i_n}^\dagger \phi_n^{i_n} \ldots \sum a_{i_1}^\dagger \phi_1^{i_1} |0\rangle. \tag{18.30}$$

The wave function (coefficients) of this state in the tensor product basis $|i_1 \ldots i_n\rangle$ is det $[\phi_i^{i,j}]$ and such states are called *Slater determinants*.

Note that in the boson ground state, the individual single boson states are *unentangled*, while in the fermion ground state there is a very simple pattern of entanglement, total antisymmetrization, required by Fermi Dirac statistics. The basic idea of the Hartree and Hartree–Fock approximations is to use multiparticle states with these simplest patterns of multiparticle entanglement, as trial variational states for more complicated Hamiltonians, which continue k-body operators with $k \geq 2$. Equivalently, given a Hamiltonian like

$$H = a_i^\dagger k_{ij} a_j + a_i^\dagger a_j^\dagger V_{ij,kl} a_k a_l, \tag{18.31}$$

we search for the one body Hamiltonian H_0 whose ground state gives the lowest expectation value of H.

For bosons, the expectation value is simply computed. Annihilation operators operating to the right, and creation operators operating to the left on $|\Psi_0\rangle$ vanish unless they have a projection on the single particle ground state ϕ_1^i:

$$a_j \frac{1}{\sqrt{n!}} (a_i^\dagger \phi_1^i)^n |0\rangle = \frac{n}{\sqrt{n!}} (a_i^\dagger \phi_1^i)^{n-1} \phi_1^j |0\rangle. \tag{18.32}$$

$$a_j a_k \frac{1}{\sqrt{n!}} (a_i^\dagger \phi_1^i)^n |0\rangle = \frac{n(n-1)}{\sqrt{n!}} (a_i^\dagger \phi_1^i)^{n-2} \phi_1^j \phi_1^k |0\rangle. \tag{18.33}$$

This gives

$$\langle \Psi_0 | H | \Psi_0 \rangle = n \phi_1^{*\,i} k_{ij} \phi_1^j + n(n-1) V_{ij,kl} \phi_1^{*\,i} \phi_1^{*\,j} \phi_1^k \phi_1^l. \tag{18.34}$$

We are instructed to minimize this w.r.t. the coefficients ϕ_1^i, subject to the constraint $\phi_1^{*\,i} \phi_1^i = 1$. This leads to a nonlinear equation for the coefficients. It is the general form of the Gross–Pitaevski equation for dilute Bose gases, which we discussed in Chapter 12. These are the Hartree equations.

To evaluate the expectation value of the Hamiltonian in a ground state of independent fermions, we note that such a state is determined by an n-dimensional subspace of the single particle Hilbert space, defined by the first n levels of the Hamiltonian H_0. The fermion ground state is invariant under unitary transformations in the single particle space which leave this subspace invariant. Under such a transformation, the creation and annihilation operators go into

$$a_i \rightarrow U_{ij} a_j, \tag{18.35}$$

$$a_i^\dagger \rightarrow U_{ij}^* a_j^\dagger. \tag{18.36}$$

We want to evaluate

$$\langle \Psi_0 | a_i^\dagger a_j | \Psi_0 \rangle, \tag{18.37}$$

and

$$\langle \Psi_0 | a_i^\dagger a_j^\dagger a_k a_l | \Psi_0 \rangle. \tag{18.38}$$

These expectation values are collections of pure numbers. The second set of numbers is antisymmetric under interchange of either of the pairs i, j or k, l. They also have to vanish whenever, considered as a vector in any of the indices i, j, k, l, that vector is orthogonal to the special subspace. The only collections of numbers invariant under all the unitary transformations and having these orthogonality and symmetry properties are

$$\langle \Psi_0 | a_i^\dagger a_j | \Psi_0 \rangle = A P_{ij}, \qquad (18.39)$$

and

$$\langle \Psi_0 | a_i^\dagger a_j^\dagger a_k a_l | \Psi_0 \rangle = B(P_{ik} P_{jl} - P_{il} P_{jk}). \qquad (18.40)$$

A, B are numbers and

$$P_{ij} = \sum_n \phi_n^{*\ i} \phi_n^j \qquad (18.41)$$

is the projection matrix on the single particle subspace corresponding to the ground state. We can determine the coefficients A, B by noting that if we contract the first equation with δ_{ij} and the second with δ_{jk} we get (using $\delta_{ij} P_{ij} = n$ and the fact that P is a projection)

$$\langle \Psi_0 | N | \Psi_0 \rangle = An, \qquad (18.42)$$

and

$$\langle \Psi_0 | a_i^\dagger N a_l | \Psi_0 \rangle = B(1 - n) P_{il}. \qquad (18.43)$$

In these equations, N is the number operator, which simply counts how many particles are in the state. The first equation thus gives $An = n$ while the second implies that

$$(n - 1)A = (1 - n)B,$$

so that

$$A = -B = 1.$$

Let us write these equations explicitly for the case of electrons interacting with nuclei in the Born–Oppenheimer approximation. The index i labeling the single particle Hilbert space is replaced by the electron spin index a and its position $\mathbf{x}$. The Hamiltonian in Bohr–Rydberg units is

$$H = \int d^3x \; [\psi_a^\dagger(\mathbf{x})[-\frac{\nabla^2}{2} + V(\mathbf{x})]\psi_a(\mathbf{x}) + \int d^3x \; d^3y \; \rho(\mathbf{x})\frac{1}{2|\mathbf{x} - \mathbf{y}|}\rho(\mathbf{y}). \qquad (18.44)$$

The density operator is defined by

$$\rho(\mathbf{x}) = \psi_a^\dagger(\mathbf{x})\psi_a(\mathbf{x}). \qquad (18.45)$$

This Hamiltonian is not in normal ordered form, but we can put it in that form by moving $\psi_a(x)$ to the right of $\psi_a^\dagger(y)$ using the relation

$$[\psi_a(x), \psi_b^\dagger(y)]_+ = \delta_{ab}\delta^3(\mathbf{x} - \mathbf{y}). \qquad (18.46)$$

The singular term obtained by this maneuver represents the self-interaction of the electrons and is usually omitted in the first quantized representation of the Coulomb interaction. The justification for this omission comes from the theory of Quantum Electrodynamics. There it is shown that the effect of these self-interactions is felt only through a shift in the electron mass. We take them into account, by using the experimental value of the mass.

The expectation value of the normal ordered Hamiltonian in a Slater determinant state is

$$\langle \Psi_0 | H | \Psi_0 \rangle = \int d^3x \sum_n \phi_n^{*\ a}(\mathbf{x}) [\frac{-\nabla^2}{2} + V(\mathbf{x})] \tag{18.47}$$

$$+\phi_n^a(\mathbf{x}) \int d^3x\, d^3y \sum_{m,n} [\phi_n^{*\ a}(\mathbf{x})\phi_n^a(\mathbf{x})\phi_m^{*\ b}(\mathbf{y})\phi_m^b(\mathbf{y}) - \phi_n^{*\ b}(\mathbf{y})\phi_n^a(\mathbf{x})\phi_m^{*\ a}(\mathbf{x})\phi_m^b(\mathbf{y})]\frac{1}{2|\mathbf{x}-\mathbf{y}|}. \tag{18.48}$$

We must vary this w.r.t. $\phi_k^{*\ c}(\mathbf{z})$, subject to the constraints

$$\int \phi_n^{*\ a}\phi_m^a = \delta_{mn}.$$

Imposing those constraints with a Lagrange multiplier λ_{mn}, we get

$$0 = [\frac{-\nabla_z^2}{2} + V(\mathbf{z})]\phi_k^c(\mathbf{z}) + \lambda_{kn}\phi_n^c(z) + \int d^3x \frac{\rho(x)\delta_{km} - \rho_{km}(x)}{|\mathbf{x}-\mathbf{z}|}\phi_m^c(\mathbf{z}), \tag{18.49}$$

where

$$\rho_{km}(x) = \phi_m^{*a}(x)\phi_k^a(x), \tag{18.50}$$

and

$$\rho(x) = \sum_n \rho_{nn}(x). \tag{18.51}$$

These are the Hartree–Fock equations for atomic, molecular, and condensed matter physics. The term involving the matrix ρ_{km} but not its trace is called the *exchange term*. It would be absent if the particles satisfied only the Pauli exclusion principle (no two particles in the same state), rather than the full requirement of antisymmetrization. If we drop it, we get Hartree's *self consistent field approximation*. That is, the equations look like an ordinary Schrödinger equation, with a potential

$$V_{Hartree} = V + \int d^3x \frac{\rho(x)}{|\mathbf{x}-\mathbf{z}|}. \tag{18.52}$$

The intuitive idea behind this approximation is that each electron feels a potential equal to the sum of the nuclear potential plus the Coulomb repulsion of all the other electrons, taken into account in a sort of classical probabilistic average, with the charge density derived from the quantum probability of finding each electron at a given position. The full Hartree–Fock equations have no such intuitive picture associated with them, but they are the correct expression of Fermi statistics in the variational approximation of the ground state by that of a Hamiltonian with only single particle terms.

It is interesting to compare the Hartree–Fock approach to the Density Functional approach. In the Hartree approximation to Density Functional Theory, the Kohn–Sham orbitals satisfy the Hartree equations without the exchange term. It is not really correct to think of the antisymmetrized Kohn–Sham orbital wave function as the actual electronic wave function of the problem, though many authors do so. Exchange effects and more complicated corrections to the Hartree approximation are taken into account by the corrections to the Hartree approximation to the density functional. In principle, if one had the exact density functional, one would still solve for the minimum energy in terms of "Exact Kohn–Sham orbitals," which satisfy single particle Schrödinger equations, but the energy is not computed as the expectation value of the Hamiltonian in an antisymmetrized state made from those orbitals.

The standard way to solve either the Hartee or Hartree–Fock equations is to start with some choice for the $\phi_n(x)$ (e.g., hydrogen wave functions for an atomic physics problem), compute the matrix $\rho_{mn}(x)$ and then solve for the eigenstates of the new linear operator defined by this choice. Then iterate. It turns out that the iteration is numerically stable, because one is seeking a minimum of the energy.

18.5   THE LANCZOS METHOD

There is a general approach to improving variational approximations to the ground state energy, which is related to a method invented by Lanczos[80], for finding *all* of the eigenvalues and eigenstates of a Hermitian matrix by iteration. Imagine that we have found a decent variational approximation to the ground state $|\psi_0\rangle$, which is not an actual eigenstate. It has an eigenstate expansion

$$|\psi_0\rangle = \sum c_n|E_n\rangle. \tag{18.53}$$

Generally, apart from constraints due to exact symmetries, all of the c_n will be nonzero. Consider the quantities

$$\mathcal{E}(k) \equiv \langle\psi_0|H^k|\psi_0\rangle = \sum |c_n|^2 E_n^k. \tag{18.54}$$

The vectors $|v_k\rangle = H^k|\psi_0\rangle$ for $0 \leq k \leq N - 1$ are not orthonormal, but will generally be a basis of an N-dimensional subspace of the Hilbert space. Indeed the scalar products between those basis elements are given by

$$\langle v_p|v_q\rangle = \mathcal{E}(p + q). \tag{18.55}$$

Now let us express a vector $|\psi_N\rangle$ in the N-dimensional subspace as a linear combination of the $|v_k\rangle$.

$$|\psi\rangle = \sum_{n=0}^{N-1} a_n |v_n\rangle. \tag{18.56}$$

The coefficients are arbitrary complex numbers, so they are independent of their complex conjugates. The expectation value of the Hamiltonian in this state is

$$\frac{\sum a_n^* a_m [\mathcal{E}(n+m+1)]}{\sum a_n^* a_m \mathcal{E}(n+m)}. \tag{18.57}$$

Varying this w.r.t. a_n^* (remember that a complex variable and its conjugate are independent)

$$\sum_{m=0}^{N-1} [\mathcal{E}(n+m+1) - \langle H \rangle_0 \mathcal{E}(m+n)] a_m = 0. \tag{18.58}$$

The second term in brackets comes from varying the denominator, which accounts for the minus sign and the fact that it is proportional to $\langle H \rangle$. Let us call $\langle H \rangle_0 \equiv E_0$ and introduce a collection of real N-dimensional vectors

$$\mathcal{E}_k = \begin{pmatrix} \mathcal{E}(k) \\ \mathcal{E}(k+1) \\ \vdots \\ \mathcal{E}(k+N-1) \end{pmatrix}. \tag{18.59}$$

Thinking of the coefficient a_n as a complex N vector $\mathbf{a}$, we can write the variational equations as

$$P_{m+1} \equiv \mathcal{E}_{m+1} \cdot \mathbf{a} = E_0 \mathcal{E}_m \cdot \mathbf{a}, \tag{18.60}$$

so that

$$\mathcal{E}_m \cdot \mathbf{a} = E_0^m \mathcal{E}_0 \cdot \mathbf{a}. \tag{18.61}$$

The vectors $\mathcal{E}_k$ for $1 \le k \le N$ will generally be a basis for the space of all N vectors so that

$$\mathcal{E}_0 = \sum_{k=1}^{N} c_k \mathcal{E}_k. \tag{18.62}$$

We should emphasize that given a choice of Hamiltonian and $|\psi_0\rangle$, the c_k are relatively easily determined in terms of $\mathcal{E}(m)$ for $1 \le m \le 2N+1$. The variational equations now read

$$P_m = E_0^m \sum_{n=1}^{N} c_n P_n. \tag{18.63}$$

The consistency condition for this equation to have solutions is

$$\sum_{n=1}^{N} E_0^m c_m = 1, \tag{18.64}$$

and the estimate for the ground state energy is the lowest solution of this polynomial equation. This approach, and the more elaborate Lanczos method for finding all the eigenvalues in the subspace generated by acting with the Hamiltonian $N-1$ times on $|\psi_0\rangle$, have been used extensively in a variety of problems. An entry to this literature can be found in[80]. The particular approach to the ground state described here does not require one to orthonormalize the vectors $|v_k\rangle$, and is likely to be computationally simpler than the general Lanczos technique. One general problem with the Lanczos method is that for systems with a ground state energy proportional to a large volume V, the estimates do not automatically scale linearly with V. One must find a regime in V with stable linear scaling, for each N, and establish that that regime goes to infinite volume as N goes to infinity.

18.6 EXERCISES

18.1 Show that the exact eigenvalue for the potential gx^{2q} scales with g just like the Gaussian variational answer.

18.2 Calculate the ground state energy of helium using the variational ansatz that the wave function is the product of wave functions for a single electron ion with a "shielded" charge Z. Use Z as the variational parameter.

18.3 Consider a particle moving in two dimensions with Hamiltonian

$$H = p_x^2 + p_y^2 + V_0\{\theta(|x| - 2a)[1 - \theta(a - |y|)] + \theta(|y| - 2a)[1 - \theta(a - |x|)],$$

in the limit $V_0 \to \infty$. The particle is forbidden to enter the regions where the potential is nonzero. Classically it can run out to infinity in the x direction if its y position satisfies $|y| < a$ and its y velocity vanishes. Use a variational argument to show that, quantum mechanically, this can only occur above some threshold energy. That is, there are positive energy bound states in the system.

18.4 Obtain a variational bound for the ground state energy of the Yukawa potential $V(r) - g\frac{e^{-kr}}{r}$. Try a variational wave function of exponential form, like the ground state of hydrogen.

18.5 Define the "harmonic atom" by replacing the Coulomb potential by harmonic oscillator potentials. Show that this problem is exactly soluble. For simplicity, do this problem in one dimension. You will find it easiest to do this problem without using quantum field theory. The Hamiltonian (in units where $\hbar$ and the mass are set equal to 1 is $H = \sum P_i^2/2 + 1/2\Omega^2 X_i^2 - 1/2\omega^2 \sum_{ij} X_i X_j$, which is a collection of coupled oscillators. You solve it by finding the normal modes, but then you have to impose the constraints of Fermi statistics. Work in the limit $N \gg 1$, $\Omega \gg \Omega - N\omega > 0$.

18.6 Define the Harmonic Hartree approximation to the harmonic atom by looking for the single particle harmonic potential $\frac{\nu^2}{2} \sum X_i^2$ centered at the origin, whose ground state for N electrons is the best variational approximation to the Hamiltonian of Exercise 18.5. The ground state is the Slater determinant made from the N lowest eigenstates of this oscillator. Compare the result to the exact answer.

18.7 Consider a Hamiltonian of the form $\sum_{i<j} K_{ij}\sigma^a(i)\sigma^a(j)$. i and j are some finite set of "sites," whose total number is N. Consider the ground state of a Hamiltonian $\sum_i h^a(i)\sigma^a(i)$ as a variational approximation to this problem. Find the equation determining the best values for the "local magnetic fields" $h^a(i)$.

18.8 Let us apply the Lanczos method to the Hamiltonian $P^2/2m + m\omega_1^2 X^2/2$, starting from the ground state wave function of the oscillator with frequency ω_0. Compute the expectation values of the first two nontrivial powers of the Hamiltonian H^2 and H^3. Use these to evaluate the first three moments $\langle 0|H^k|0\rangle$. In doing this, you will be helped by noticing that if P^2 acts to the left or right on the oscillator ground state, it can be replaced by $\hbar\omega - \omega^2 X^2$. Also use the fact that X^2 acting on the ground state is a linear combination of that state and the second excited state. This enables one to reduce the calculation to expectation values of powers of X. Evaluate the first Lanczos approximation to the ground state energy and compare to the exact answer.

CHAPTER 19

The Feynman Path Integral

19.1 INTRODUCTION

Feynman's path integral approach to quantum mechanics (QM) is the most useful way to approach most QM problems. It is based on the Lagrangian approach to mechanics, and thus manifests symmetries much more clearly than the Hamiltonian formalism. This is particularly true for symmetries like Lorentz boosts, which do not leave the Hamiltonian invariant.

The path integral is a formal solution to the equations of QM in terms of an infinite dimensional integral. It avoids many subtle issues having to do with the domains of unbounded operators, and directly computes observable quantities. It is also relatively straightforward to discretize path integrals and use sophisticated numerical integration routines to evaluate them.

Indeed, the virtues of the path integral formalism are so numerous that one is tempted to rewrite all quantum textbooks in path integral language. The one drawback is that it is much harder to explain the probability interpretation of the formulae in the path integral formula. For example, if one adds a term $\dot{x}^4$ to the Lagrangian of a free particle, the (imaginary time) path integral is perfectly well defined as long as the coefficient of this term is positive. However, the quantities computed from the new path integral cannot be interpreted as expectation values of operators in a Hilbert space. For this and other reasons, we have hidden the path integral chapter at the back of the book. Perhaps one can think of it as saving the best for last.

19.2 TWO DERIVATIONS OF THE PATH INTEGRAL FORMULA

We are all used to the fact that the equations of physics become simpler over very short time intervals. We can write closed form differential equations for many problems whose finite time evolution is extremely difficult to figure out. The insight behind the Feynman Path Integral formulation is that, in QM at least, the simplification of very short time intervals leads to a "reduction to quadratures" of the finite time solutions. That is, one can write a very explicit solution of the problem in terms of an integral formula. The catch is that the integral is always infinite dimensional, if we want the exact answer. Most infinite dimensional integrals are very hard to compute. On the other hand, if one is willing to live with finite precision, one can evaluate the integrals with powerful numerical integration techniques. The path integral (also called functional integral) technique has led to many important insights into QM problems and is virtually the universal tool of choice in the study of many particle problems.

The cleanest derivation of the path integral is obtained by analytically continuing to imaginary time. The fact that quantum amplitudes are analytic is a rigorous theorem for finite dimensional Hamiltonians. For Hamiltonians that are bounded from below, $e^{-\tau H}$ is a very well behaved operator. τ is the analytic continuation of $\frac{it}{\hbar}$ to positive imaginary values. From this point on, in this Chapter, we will set $\hbar = 1$. The infinite dimension of the Hilbert space almost always comes from very high-energy states.[1] If the high energy density of states grows less rapidly than an exponential,[2] the exponential damping of imaginary time makes all infinite sums over states convergent.

Now let us write

$$\langle \psi(T)|e^{-\tau H}|\psi(0)\rangle = \langle \psi(T)|(e^{-\frac{\tau}{N}H})^N|\psi(0)\rangle. \tag{19.1}$$

At this point, we insert a complete set of intermediate states between every pair of operators, and write the initial and final states in terms of their wave function in the chosen basis. Different choices of basis lead to different forms of the path integral. If we actually use a countable basis of normalizable states, we get a *path sum* rather than a path integral. This is useful for fundamental variables like spin operators, which operate in finite dimensional Hilbert spaces. However, we will begin by studying a Hamiltonian $H = \frac{P^2}{2m} + V(X)$ and inserting position eigenstates. We are then led to the computation of

$$\langle x(t_i)|e^{-\frac{\tau}{N}H}|x(t_{i-1})\rangle, \tag{19.2}$$

with an accuracy of order τ/N in the exponent.

Recall the Zassenhaus formula from Exercise 6.3

$$e^{t(W+Y)} = e^{tW}e^{tY}e^{-\frac{t^2}{2}[W,Y]}e^{\frac{t^3}{6}(2[Y,[W,Y]]+[W,[W,Y]])}$$

$$\times e^{-\frac{t^4}{24}([[[W,Y],W],W]+3[[[W,Y],W],Y]+3[[[W,Y],Y],Y])} \dots \tag{19.3}$$

---

[1] Most infinite degeneracies associated with low energy can be regulated by putting the system in finite volume.

[2] This is the case for all known systems, which do not involve gravity or string theory.

When $W = \frac{P^2}{2m}$ and $Y = V(X)$, and $t = -\tau/N$, the logarithms of the terms with powers of t higher than 1 are $o(1/N^2)$ or smaller, and therefore should not contribute to the final answer as $N \to \infty$. On the other hand,

$$\langle x(t_i)|e^{-\frac{\tau}{2mN}P^2}e^{-\frac{\tau}{N}V(X)}|x(t_{i-1})\rangle = \sqrt{\frac{2mN}{\hbar^2\tau}}e^{-\frac{mN(x(t_i)-x(t_{i-1}))^2}{2\tau}}e^{-\frac{\tau}{N}V(x(t_{i-1}))}. \tag{19.4}$$

You can derive this easily (Exercise 19.1) by inserting a complete set of momentum eigenstates. Remember that the $x(t)$ variables are integration variables. As $N \to \infty$, the first exponential factor suppresses integration regions for which the x variables at neighboring times are not close to each other. So we can think of $x(t)$ as a continuous function of time. For differentiable functions, the product of exponentials from all the infinitesimal time intervals converges to the classical action (for imaginary time). Dirac [81] was the first to notice this, but Feynman was the person who exploited this observation to do new physical calculations.

Formally, ignoring the prefactors and the fact that not all continuous functions are differentiable, we can write the answer as

$$\langle x(T)|e^{-\tau H}|x(0)\rangle = \int [dx(t)]e^{-S[x(t)]}, \tag{19.5}$$

where the functional integral is over all continuous paths, which have fixed endpoints. We can deal with the nasty prefactors by noting that they depend only on N and τ and so are the same for a free particle as they are for any potential, and are independent of the endpoints. So we can write the ratio of amplitudes for any potential and the free particle, as a ratio of two path integrals and not worry about overall position-independent factors in the individual path integrals.

What is really going on here is that in the limit of small times, the system is completely dominated by free motion, as long as the potential is not wildly varying with position. For continuous paths, the contribution of the potential to the short time motion is more or less constant. This must be rethought for some singular potentials, but it is a rule of very wide general applicability.

The foregoing was Feynman's derivation of the path integral formula. What follows is a derivation, of a similar formula, which follows an argument due to Schwinger. Consider the Hamiltonian $\frac{P^2}{2m} + V(X)$ again, but now subject it to a time-dependent perturbation

$$\delta H = -J(t)X. \tag{19.6}$$

Following our discussion of time-dependent perturbation theory, the evolution operator of the perturbed system in the interaction picture can be written in terms of the Heisenberg operators $X(t)$ of the unperturbed system:

$$U_I(t, t_0) = T \, e^{i \int_{t_0}^{t} J(s)X(s)}. \tag{19.7}$$

Taking any matrix element of this operator, we get a complex valued *functional* $Z[J]$ of the source $J(t)$. The physical meaning of that functional is clear: it is the amplitude that the source $J(t)$ induces a transition between a pair of Heisenberg states of the original system. We have not indicated the dependence on the initial and final state. In most applications, they are identical and we take the thermal average over them at some temperature T. As $T \to 0$, the thermal density matrix becomes the projection on the (possibly degenerate) ground state. Z is then called the ground state persistence amplitude.

The time-ordered exponential is defined by its power series expansion in powers of J, so if we make a small change in the source $J \to J + \delta J$, we find that

$$\delta Z \equiv \int \frac{\delta Z}{\delta J(s)} \delta J(s), \tag{19.8}$$

which defines the *functional derivative* of Z. The functional derivative of Z is *equal to*

$$\frac{1}{i} \frac{\delta Z}{\delta J(t)} = \langle final \, |T[X(t)e^{i \int J(s)X(s)}]| \, initial \rangle. \tag{19.9}$$

Using the Heisenberg equations of motion, we have

$$\langle final \, |T[P(t)e^{-i \int J(s)X(s)}]| \, initial \rangle = m \frac{d}{dt} \frac{1}{i} \frac{\delta Z}{\delta J(t)}. \tag{19.10}$$

This equation is correct, but we have to be careful of its derivation, because there is implicit t dependence in the definition of the time ordering symbol, through factors like

$$\theta(t - s)X(t)X(s) + \theta(s - t)X(s)X(t).$$

Fortunately, when we take the derivative of the Heaviside step functions, we get two terms, which combine to give the commutator of $X(t)$ with itself at equal times. This is a result of the identity

$$\frac{d}{dt}\theta(t - s) = \delta(t - s) = -\frac{d}{dt}\theta(s - t). \tag{19.11}$$

We now repeat this exercise for the second time derivative, this time picking up a term from the commutation of X and P at equal times. Since this commutator is just a number, it comes out of the time-ordered product. In Exercise 19.2, you will show that

$$m \frac{d^2}{dt^2} \frac{1}{i} \frac{\delta Z}{\delta J(t)} + \langle final \, |T[V'(X(t))e^{-i \int J(s)X(s)}]| \, initial \rangle = -J(t)Z. \tag{19.12}$$

Exercise 19.2 Give a careful derivation of the equation directly above.

We can rewrite this as

$$0 = \left[m \frac{d^2}{dt^2} \frac{1}{i} \frac{\delta Z}{\delta J(t)} + V'\left(\frac{1}{i} \frac{\delta}{\delta J(t)} \right) + J(t)Z \right]. \tag{19.13}$$

This equation is both familiar looking and peculiar. It looks like the classical equation of motion, except that the classical position $x(t)$ is replaced by the functional differential operator $\frac{1}{i} \frac{\delta}{\delta J(t)}$, inserted into the classical equations of motion, and allowed to act on the functional Z.

If we expand the functional Z out into a *functional power series*

$$Z[J] = \sum_{n=0}^{\infty} \frac{1}{n!} \int ds_1 \dots ds_n \ G_n(s_1 \dots s_n) J(s_1) \dots J(s_n), \tag{19.14}$$

the functional differential equation breaks up into an infinite set of ordinary differential equations for the coefficients. These are called the *Schwinger–Dyson (SD) equations* [82].

Exercise 19.3 Derive the SD equations from the functional equation for $Z[J]$.

How do we think about, much less solve, this scary set of equations. Everything infinite is a limit of something finite. We can make these equations look more familiar by breaking the time interval $t - t_0$ up into N intervals of equal length and replacing the time derivatives by finite time differences. When we do that, the number of source variables $J(t_n) \equiv J_n$ (evaluated for example at the midpoint of each interval) becomes finite and the scary looking functional equations become ordinary partial differential equations via $\frac{\delta}{\delta J(t)} \to \frac{\partial}{\partial J_n}$. Now we can think about them.

Our first observation is that despite their obvious relation to classical equations, these equations are *linear* equations for Z.[3] Secondly, although they may contain high powers of the partial derivative operator $\frac{\partial}{\partial J_n}$, they are linear in the source J_n itself. This suggests a strategy: the Fourier transform turns derivatives into multiplication by the Fourier conjugate variable. If we Fourier transform the SD equations w.r.t. all of the variables J_n, we will get a *first*-order linear partial differential equation (PDE) in the Fourier conjugate variables. We will call the latter x_n. The Fourier transform will involve an integral over all of these variables x_n, which will become a *functional integral* in the limit of continuous time.

Now that we have got the idea, let us do all of this directly in the continuum limit. We write

$$Z[J] = \int [dx(t)] e^{i(S[x] + \int J(s)x(s)ds)}. \tag{19.15}$$

---

[3] This is connected to the fact that the Schrödinger equation is a linear equation for the time evolution operator.

The SD equations then become

$$0 = \int [dx(t)] e^{i(S[x] + \int J(s)x(s)ds)} [m\frac{d^2x}{dt^2} + V'(x) + J(t)]. \qquad (19.16)$$

The term in square brackets is of course equal to $\frac{\delta(S_c[x] + \int Jx)}{\delta x(t)}$, where $S_c[x]$ is just the classical action for the unperturbed system. If we choose the Fourier transform functional $S[x]$ equal to S_c, then the integrand is just

$$\frac{\delta}{i\delta x(t)} e^{i(S_c[x] + \int J(s)x(s)ds)}. \qquad (19.17)$$

This integral of a total derivative will vanish if the quantity being differentiated vanishes at infinite values of $x(t)$. We again see the virtue of giving time a positive imaginary part and defining the real time theory as a limit of this analytic continuation. For purely imaginary time, the functional integral formula looks like

$$Z[J] = \int [d[x(t)]] e^{-(S_E[x] + \int J(s)x(s)ds)}. \qquad (19.18)$$

The Euclidean action S_E is just $-S_c$ with $t \to i\tau$.

In this way of approaching the path integral formalism, we did not have to try to define what the path integral formula was. It was *any* linear function on functionals, with the property that it gives the same result if we replace $x(t)$ by $x(t) + y(t)$. For a finite number of variables, this defines the integral up to an overall multiplicative constant. That cannot be right here, because we have implicitly defined Z for *any* choice of initial and final state. We also have to deal with the fact that the overall constant is likely to be infinite. It is best to deal with these issues in the context of an explicit example, so we now turn to:

19.3   THE PATH INTEGRAL FOR A HARMONIC OSCILLATOR

The simplest way to do the Euclidean path integral for the harmonic oscillator is to write $x(t) = x_c(t) + \sqrt{\frac{2}{T}} \sum c_n \sin(n\pi t/T)$, where $x_c(t)$ is the solution of

$$\ddot{x}_c(t) = \omega^2 x_c(t), \qquad (19.19)$$

with $\omega = \sqrt{k/m}$, which satisfies the boundary conditions imposed by the initial and final position eigenstates. By Fourier's theorem, c_n parameterizes the space of functions satisfying $\delta x(0) = \delta x(T) = 0$, so this decomposition is a way of describing all functions which satisfy the boundary conditions. Since $x_c(t)$ minimizes the Euclidean action, the result for the path integral is

$$\int [dx(t)]e^{-S_E[x(t)]} = e^{-S_E[x_c(t)]} \int \prod_n [dc_n]e^{-\frac{m}{T}\sum_{n,k}c_n c_k \int_0^T dt\, [\sin(n\pi t/T)(-\frac{d^2}{dt^2}+\omega^2)\sin(k\pi t/T)]}.$$

(19.20)

The functions $\sqrt{\frac{2}{T}}\sin(n\pi t/T)$ are orthonormal, so

$$\int [dx(t)]e^{-S_E[x(t)]} = e^{-S_E[x_c(t)]} \int \prod_n [dc_n]e^{-\frac{m}{2}\sum_n c_n^2[(\frac{n\pi}{T})^2+\omega^2]}.$$

(19.21)

The result is just an infinite product of independent Gaussian integrals. Note that all of the dependence on the initial and final positions comes from the classical action.

The infinite product is not convergent. However, if we divide the path integral by that for the free particle ($\omega = 0$), we will see that the result is finite. So we can write

$$\langle x(T)|e^{-tH_{osc}}|x(0)\rangle = e^{-S_E[x_c(t)]}\frac{\langle 0|e^{-tH_{osc}}|0\rangle}{\langle 0|e^{-tH_{free}}|0\rangle}\sqrt{\frac{m}{2\pi t\hbar^2}}.$$

(19.22)

The careful reader will want to remember that we have defined Euclidean time with a factor of $\hbar$, so that it has dimensions of inverse energy, when comparing this equation to previous formulae for the free particle propagator.

The infinite product represented by the ratio is

$$\prod_{n=1}^{\infty}\left(1+\frac{\omega^2 T^2}{n^2\pi^2}\right) = e^{\sum_{n=1}^{\infty}\ln(1+\frac{\omega^2 T^2}{n^2\pi^2})},$$

(19.23)

which converges. You will evaluate this in Exercise 19.4. In Exercise 19.5, you are asked to repeat the computation using Feynman's original derivation of the path integral.

To finish the path integral evaluation of the transition amplitude, we have to compute the action. The general solution of the classical equations is

$$x_c(\tau) = Ae^{\omega\tau} + Be^{-\omega\tau}.$$

(19.24)

The boundary conditions imply that

$$A + B = x(0),$$

(19.25)

$$Ae^{\omega T} + Be^{-\omega T} = x(T).$$

(19.26)

The action is given by

$$S_E = \frac{m\omega^2}{2}\int_0^T [(Ae^{\omega T}-Be^{-\omega T})^2+(Ae^{\omega T}+Be^{-\omega T})^2] = \frac{m\omega[A^2(e^{2\omega T}-1)-B^2(e^{-2\omega T}-1)]}{4}.$$

(19.27)

Working this out is a little exercise in 2×2 matrix multiplication, which gives

$$S_E = \frac{m\omega}{4\hbar} \left[\frac{\sinh(2\omega T)}{\sinh^2(\omega T)} (x(T)^2 + x(0)^2) - \frac{4x(0)x(T)}{\sinh(\omega T)} \right]. \tag{19.28}$$

The full propagator is thus

$$\sqrt{\frac{m\omega}{2\pi \sinh(\omega T)}} e^{-\frac{m\omega}{2\sinh(\omega T)} [(x(0)^2 + x(T)^2)\cosh(\omega T) - 2x(0)x(T)]}. \tag{19.29}$$

When analytically continued back to real time, this gives us

$$\sqrt{\frac{m\omega}{2\pi i\hbar \sin(\omega t)}} e^{i\frac{m\omega}{2\sin(\omega t)} [(x(0)^2 + x(t)^2)\cos(\omega t) - 2x(0)x(t)]}. \tag{19.30}$$

This formula gives us yet another way to compute the normalized eigenfunctions and the eigenvalues of the harmonic oscillator. Indeed,

$$\langle x(T)|e^{-\frac{H\tau}{\hbar}}|x(0)\rangle = \sum_{n=0}^{\infty} \psi_n(x(T))\psi_n^*(x(0))e^{-\frac{E_n\tau}{\hbar}}. \tag{19.31}$$

In Exercise 19.6, you will be asked to compute a few of the Hermite functions by this method.

Using this formula, we can compute the transition amplitude between any initial and final state by integrating the position eigenstate result against $\psi^*(x(T))\psi(x(0))$. However, if both the initial and final states are the ground state, there is a simpler formula. Let us look at this first from the Hamiltonian point of view, in imaginary time. It is clear that if we take the imaginary time to infinity, the ground state contribution dominates, so

$$\lim_{\tau \to \infty} \langle X(T)|e^{-\frac{H\tau}{\hbar}}|x(0)\rangle \to \psi_0(X(T))\psi_0^*(x(0))e^{-\frac{E_0\tau}{\hbar}}. \tag{19.32}$$

On the other hand, the path integral computation gives, in this limit

$$\lim_{\tau \to \infty} \langle X(T)|e^{-\frac{H\tau}{\hbar}}|x(0)\rangle \to = e^{-\frac{\omega T}{2}} \sqrt{\frac{m}{2\pi}} e^{-\frac{m\omega}{4\hbar}(x(T)^2 + x(0)^2)}. \tag{19.33}$$

Comparing the two expressions, we see that we can read off both the ground state energy and the ground state wave function (up to the usual overall constant phase) from the path integral computation.

Now let us generalize these computations to the case where the oscillator is subjected to an external force, constant in space but with an arbitrary time dependence. This changes the action by adding a term $\int J(\tau)x(\tau)$. By reviewing our derivation of the path integral formula, you will see that the only change is that we have to solve the classical equations with an extra source term. The resulting classical action differs from that with $J = 0$ by a term quadratic in J.[4] It has the form

$$\delta S = \frac{1}{2} \int d\tau d\sigma \, [J(\tau)G(\tau,\sigma)J(\sigma)], \tag{19.34}$$

where

$$[-\frac{d^2}{d\tau^2} + \omega^2]G(\tau,\sigma) = \delta(\tau - \sigma). \tag{19.35}$$

This is a Green's function equation and has many solutions. The one appropriate to the transition amplitude between position eigenstates is fixed by imposing the boundary conditions

$$x(T) = \int_0^T d\sigma G(T,\sigma)J(\sigma); \quad x(0) = \int_0^T d\sigma G(T,\sigma)J(\sigma). \tag{19.36}$$

What is new and exciting about the external force problem is that we can also get ground state to ground state transition amplitudes by imposing boundary conditions on the Green's function equation.

Our Hamiltonian discussion of the $J = 0$ problem suggests that we should consider the $T \to \infty$ limit. In the time-dependent problem, this should be a sensible thing to do if J is turned off asymptotically in time. To be more precise, we consider the interval $[T, -T]$ and consider a force $J(\tau)$ such that $J(\pm T) \to 0$ as $T \to \infty$. For large imaginary times, the time-independent Hamiltonian analysis shows that we should be projecting out the ground state. So, in the limit of large imaginary time, we are studying the amplitude for the ground state of the $J = 0$ problem, to remain the ground state after being subjected to a time-dependent source, which is turned on only for a finite time. This was the sort of problem we studied in Chapter 13 on time-dependent perturbation theory. To be more precise, we are studying such a problem if we take the source function $J(\tau)$ and analytically continue it to a real function of real time.

The Dirac picture analysis, in imaginary time, tells us that the expression for this amplitude is

$$Z[J] \equiv \langle \Psi_0 | T[e^{i \int_{-\infty}^{\infty} J(\tau)x(\tau)} | 0 \rangle. \tag{19.37}$$

On the other hand, the path integral analysis tells us that the same expression is computed (modulo a prefactor, which depends only on time, and which we have already computed at finite time) by solving the classical equations of motion with some boundary conditions. The obvious boundary conditions are that $x(\tau)$ go to the classical minimum of the potential at $T \to \pm\infty$. The expression for the classical action as a functional of $J(s)$ is

---

[4] Without even solving the equations, you can see that the possible linear term in J is absent, because the action is invariant under the simultaneous reflection $J \to -J$, $x \to -x$.

$$S_{E\ cl} = \frac{1}{2} \int_{-\infty}^{\infty} d\tau d\sigma \, [J(\tau)G(\tau - \sigma)J(\sigma)]. \tag{19.38}$$

$$[-\frac{d^2}{d\tau^2} + \omega^2]G(\tau - \sigma) = \delta(\tau - \sigma). \tag{19.39}$$

We can solve the Green's function equation by Fourier transformation

$$G(t - s) = \int \frac{dz}{2\pi} \frac{e^{iz(t-s)}}{z^2 + \omega^2}. \tag{19.40}$$

Since the integrand has two poles at $z = \pm i\omega$, different choices of contour in the z plane will give different answers. In Exercise 19.7, you will show that these differences correspond, as they must, to different homogeneous solutions that can be added to any particular solution of the Green's function equation.

Choosing the contour along the real z axis, we can evaluate the integral by closing the contour in the upper (lower) half z-plane if $\tau - \sigma > 0$ ($\tau - \sigma < 0$). We get

$$G(\tau - \sigma) = \frac{1}{\omega} e^{-\omega|\tau - \sigma|}, \tag{19.41}$$

which falls to zero at infinity, as it should. If we now try to analytically continue $\tau - \sigma$ to negative imaginary values, $-i(t - s)$ to obtain the real ground state persistence amplitude, we can do this explicitly, or by rotating the contour of the z integration to $z = iE$. The resulting real time Green's function has the expression

$$G(t - s) = \int \frac{dE}{2\pi} \frac{ie^{-iE(t-s)}}{E^2 - \omega^2 + i\epsilon}. \tag{19.42}$$

ϵ is an infinitesimal positive number. Inserting ϵ is equivalent to deforming the contour of the real E integration so that the positive ω pole is below the contour, and the negative frequency pole above it. An equivalent statement of this contour choice is that no matter what the sign of $t - s$, only positive energies propagate forward in time.

Comparing the Hamiltonian and path integral expressions for the ground state persistence amplitude, and expanding to second order in J, we find the equation

$$\langle \Psi_0 | Tx(t)x(s) | \Psi_0 \rangle = G(t - s). \tag{19.43}$$

If we pick a time order and evaluate the left-hand side by putting in a complete set of intermediate states, then we see that indeed only positive energy propagates forward in time. Higher order time-ordered products of the $x(t)$ operators are evaluated in terms of sums of products of $G(t_i - t_j)$ functions. This result is called *Wick's Theorem*, and you will explore it in Exercise 19.8.

19.4 MORE GENERAL POTENTIALS

One can develop a perturbation expansion around the harmonic oscillator problem by expanding a general potential around a local minimum and treating the corrections to the quadratic term as small. Wick's theorem allows us to evaluate each term in this series, and there is a nice pictorial algorithm for computing them, known as Feynman diagrams. You will explore simple examples in the exercises. The series generated in this way give only asymptotic expansions, rather than a convergent one. It is worth understanding the origin of this, even though we do not have space in a course like this one, to go into the details.

Given a function $f(z)$, an asymptotic series $\sum_{n=0}^{\infty} f_n z^n$ is a formal expression, such that if one takes the polynomial $P_N(z)$ formed by the first N terms, then

$$|P_N(z) - f(z)| \sim |z|^N, \tag{19.44}$$

as $|z| \to 0$ within some wedge of finite opening angle in the z plane. If az_0 with a real is a line in this wedge, then $f(az_0)$ is infinitely differentiable at 0 w.r.t. the variable a and the k-th term in the series is $k! f^{(k)}(0)$. Now think about the function $e^{-1/z}$. As long as the real part of $1/z$ is positive, the derivatives of this function at the origin all vanish. So, anywhere in the wedge with Re $\frac{1}{z} > 0$, we can add this function to any $f(z)$ that has an asymptotic series in any smaller wedge, and get a new function with the same asymptotic series. If the series is really convergent, then among the many functions with the same asymptotic series, there is a unique one, which is analytic in an entire disk surrounding the origin.

Here is an example of a function with an asymptotic series, which is not analytic.

$$f(z) = \int_{-\infty}^{\infty} dx \, e^{-x^2 - zx^4}. \tag{19.45}$$

It obviously has an asymptotic series for real positive z, but the coefficient of z^n is

$$\frac{(-1)^n}{n!} \int e^{-x^2} x^{4n}.$$

For large n the integral grows like $(2n)!$ so the series diverges. The reason is sort of obvious. The integral diverges when z is negative, so the function cannot be analytic in a full disk. The functional integral formulation of QM makes it clear that similar issues will be encountered in QM perturbation series.

For problems with a small number of variables, there are other methods of computing the coefficients in the perturbation expansion of eigenvalues and eigenfunctions. In particular, for single variable problems with potentials well approximated by polynomials, there are powerful difference equation techniques [83] for computing both the eigenvalues and the eigenfunctions, which are much more efficient than either path integral or Rayleigh–Schrödinger computations.

Path integral methods really come into their own in systems with many variables, basically because one can compute Gaussian integrals in any number of dimensions. Feynman's diagrammatic series are quite universal in many body problems.

19.5   PATH INTEGRALS AT FINITE TEMPERATURE

We have seen that the thermal density matrix for a quantum system is $\rho_\beta = \frac{e^{-\beta H}}{Z}$, where the partition function is given by $Z = \text{tr } e^{-\beta H}$. Given a system at thermal equilibrium, we can consider two different kinds of physical processes. First, in order to understand the thermodynamics of the system, we want to compute thermal expectation values of various operators. We can do this by perturbing the Hamiltonian $H \to H + \lambda_a(t)O_a$, computing the partition function $Z(\lambda)$ and taking derivatives of its logarithm w.r.t. the λ_a. t is the Euclidean time parameter.

On the other hand, perturbing the system kicks it out of equilibrium, and we might be interested in studying the time dependence of various quantities as they decay back to an equilibrium state. The first kind of computation is a relatively simple variation on things we have already done, while the time-dependent calculations are intricate and messy. For that reason, we will treat only equilibrium expectation values.

Our task then is to compute $\text{tr}e^{-\beta H}$, and for a system described by a number of canonical coordinates Q^i, we do this by taking the trace in the basis where these coordinates are diagonal. Thus,

$$Z = \int d^N q \langle q|e^{-\beta H}|q\rangle. \tag{19.46}$$

The integrand is just the Euclidean time continuation of the amplitude to start at a particular point in configuration space and return to that point in a fixed time.

The matrix element of the time evolution operator between two points has, as we have seen, a path integral representation over all paths that start and end at those points in the required time. In this case, the points are the same, so the paths are periodic with period β, and we integrate over the end point, so we get *all periodic paths*. Thus,

$$Z_\beta = \int [dq(t)]e^{-S_E[q(t)]}, \tag{19.47}$$

where we integrate over all paths in imaginary time, which are periodic, with period $\hbar\beta$. Thermal expectation values are computed by taking logarithmic derivatives of this formula with respect to perturbing parameters, and then setting those parameters equal to zero.

For the harmonic oscillator, we have already computed the path integral for

$$\langle y|e^{-\beta H}|x\rangle = \sqrt{\frac{m\omega}{2\pi\hbar\sinh(\hbar\omega\beta)}}\exp[-\frac{m\omega}{2\pi\hbar\sinh(\hbar\omega\beta)}([x^2 + y^2]\cosh(\hbar\omega\beta) - 2xy)]. \tag{19.48}$$

To evaluate the thermal partition function, we have to set $x = y$ and integrate over x. The integral is Gaussian

$$\int dx\, e^{-Ax^2} = (\frac{\pi}{A})^{1/2},$$

(19.49)

with

$$A = 2\frac{m\omega}{2\pi\hbar \sinh(\hbar\omega\beta)}(\cosh(\hbar\omega\beta) - 1).$$

Thus,

$$Z = \sqrt{\frac{1}{2[\cosh(\hbar\omega\beta) - 1]}}.$$

(19.50)

On the other hand, straightforward evaluation of

$$Z = \sum_{n=0}^{\infty} e^{-\hbar\omega\beta(n+\frac{1}{2})}$$

(19.51)

yields

$$Z = e^{-\frac{\hbar\omega\beta}{2}}\frac{1}{1 - e^{-\beta\hbar\omega}}.$$

(19.52)

Simple algebra shows that these two expressions are equal.

One can also do the imaginary time path integral with an external source added to the Lagrangian via

$$\delta L = \int ds\, j(s)x(s).$$

(19.53)

The source must be periodic in imaginary time. Dividing by the partition function Z, we get the generating functional $Z[j]$ for thermal expectation values

$$\mathrm{tr}\, e^{-\beta H}x(s_1)\ldots x(s_2).$$

(19.54)

The answer follows from calculations that we have already done and we obtain

$$Z[j] = e^{\frac{1}{2}\int dsdt\, j(s)j(t)G(t,s)},$$

(19.55)

where

$$[-\frac{d^2}{dt^2} + \omega^2]G(t,s) = \frac{1}{m}\delta(t - s).$$

(19.56)

The solution with periodic boundary conditions is

$$G(t,s) = \sum_{n=-\infty}^{\infty} \frac{e^{2\pi i \frac{n(t-s)}{\beta\hbar}}}{m\omega^2 + \frac{2\pi mn^2}{\beta^2\hbar^2}}.$$

(19.57)

Below, we will formulate path integrals for anticommuting variables (fermions) and we will see that the finite temperature results require antiperiodicity rather than periodicity in Euclidean time.

19.6 PATH INTEGRALS AND THE JWKB APPROXIMATION

The path integral method provides an instantaneous derivation of the leading order JWKB approximation to a variety of quantum amplitudes. As an example, let us consider $\langle x|e^{-\frac{iHt}{\hbar}}|y\rangle$, where x, y are a short hand notation for any configuration space with a Lagrangian whose kinetic term takes the form

$$L_{kin} = \frac{1}{2}\dot{x}^i M_{ij}\dot{x}^j, \tag{19.58}$$

with M_{ij} a constant symmetric matrix. In Exercise 19.9, you will see what happens when M depends on x. The path integral formula for this amplitude is just

$$\langle x|e^{-\frac{iHt}{\hbar}}|y\rangle = N^{-1}\int [dx(s)]e^{i\frac{S_{cl}[x(s)]}{\hbar}}, \tag{19.59}$$

where the integral is over all paths $x(s)$ which go from $x(0) = y$ to $x(t) = x$. The normalization factor N is determined by the requirement that

$$\lim_{t\to 0}\langle x|e^{-\frac{iHt}{\hbar}}|y\rangle = \delta(x - y). \tag{19.60}$$

In the semiclassical approximation, this functional integral is evaluated as

$$\langle x|e^{-\frac{iHt}{\hbar}}|y\rangle = e^{i\frac{S_{cl}[x_{cl}(s)]}{\hbar}}\det^{-1/2}(D/D_0), \tag{19.61}$$

where D is the matrix differential operator

$$D_{ij} = M_{ij}\frac{-d^2}{ds^2} + \frac{\partial^2 V}{\partial x^i \partial x^j}(x_c(s)), \tag{19.62}$$

and D_0 is a constant times $M_{ij}\frac{-d^2}{ds^2}$, with the constant determined by the normalization condition. $x_c(s)$ is the classical solution of the equations of motion satisfying the boundary conditions. The determinants come from doing the Gaussian integral for fluctuations around the classical solution, which have boundary conditions $\Delta x(t) = \Delta x(0) = 0$. There is a rather general and beautiful theorem for calculating these functional determinants, due to Gelfand and Yaglom [84].

Gelfand and Yaglom showed that in the case of a single x variable, the functional determinant could be calculated by solving the initial value problem

$$[\frac{-d^2}{ds^2} + \frac{d^2V}{dx^2}(x_c(s))]v(s) = 0. \tag{19.63}$$

with boundary conditions $v(0) = 0$, and $\frac{dv}{dx}(0) = 1$. Then

$$\frac{\det[\frac{-d^2}{ds^2} + \frac{d^2V}{dx^2}(x_c(s))]}{\det[\frac{-d^2}{ds^2}]} = \frac{v(t)}{v_0(t)}. \tag{19.64}$$

Here $v_0(s)$ is the solution of the same boundary value problem for the action with vanishing potential. You will prove this result in Exercise 19.10. Generalizations of it can be found in the review by Dunne [84].

19.7 PATH INTEGRALS FOR SPIN AND OTHER DISCRETE VARIABLES

In his famous book on the Path Integral formulation of QM [85], Feynman lamented the fact that he had not found a path integral formulation for single nonrelativistic particles with spin. Thinking about this problem leads one in a number of interesting directions. A first approach is just to note that we can use the path integral strategy of dividing time up into small intervals for Hamiltonians that depend on spin. When we do that, we end up inserting complete sets of intermediate states of both spin and position, at each time. Suppose the spin-dependent Hamiltonian takes the form of a magnetic dipole moment in an external magnetic field $H_{spin} = \mu B_a(\mathbf{X})J_a$. J_a are the spin matrices of the particle. For short times, we can write

$$e^{-\frac{\tau}{N}H} = e^{-\frac{\tau P^2}{2mN}}e^{-\frac{\tau}{N}V(X)}e^{-\frac{\tau}{N}B_a(\mathbf{X})J_a}. \tag{19.65}$$

When we sandwich this between X eigenstates, we get the discrete approximation to the imaginary time path integral over $x(\tau)$ from the first two factors, while the second gives the discrete approximation to

$$Z[B] \equiv T\exp[-\int_0^\tau B_a(\mathbf{x}(\tau))J_a(\mathbf{x}(\tau))].$$

This derivation illustrates one of the most useful features of the path integral formalism, the decomposition of a quantum problem involving two interacting systems into individual problems, in which each of the systems evolves in a time-dependent background determined by the path integration variable of the other. Feynman called this the *influence functional* method. In the case of spin, it tells us that if we can find a path integral formula for the problem of a spin in a time-dependent background field $B_a(\tau)$, then we have a path integral that will apply to the spinning particle (or any other problem in which the spin J_a interacts with another quantum system).

The key question is thus how to write path integral formulae for time-ordered products of operators in a finite dimensional Hilbert space. If we follow Feynman's procedure of evaluating matrix elements of the short time evolution operator in some particular basis, then we get a "path sum" formula, in which there are no obvious simplifications in the short time limit. On the other hand, if the evolution operator $U(t, t_0)$ is a continuous function of t, then the vector in Hilbert space does not change very much when we make a small change in t. This suggests that we use an overcomplete basis, consisting of *all* the vectors in Hilbert space to write the path integral for a finite dimensional system. The quantum state of such a system is a set of N complex numbers z_i satisfying $\sum_i |z_i|^2 = 1$. The overall phase rotation $z_i \to e^{i\theta}z_i$

does not change the state. Equivalently, we can think of the space of states as unnormalized complex N vectors, with the identification $z_i \sim \lambda z_i$, where λ is a general nonzero complex number. The space of quantum states is just a complex projective space. The conventional mathematical name for this space is CP^{N-1}, because it has $N - 1$ complex dimensions.

One way to think about the space of states of a quantum system is that it is the degenerate eigenspace of a Hamiltonian which is equal to zero on every state. We can write a classical Lagrangian for N complex variables, which gives vanishing Hamiltonian, as follows

$$L = \frac{i}{2}(z_i^* \partial_t z_i - z_i \partial_t z_i^*). \tag{19.66}$$

The canonical commutation relations following from this Lagrangian are

$$[z_i, z_j^*] = \delta_{ij}, \tag{19.67}$$

which are the relations for N creation–annihilation operator pairs. If we impose the constraint

$$\sum z_i^* z_i = 1, \tag{19.68}$$

then there are only N states which satisfy it, namely the states where one of the harmonic oscillators built from the creation and annihilation operators is excited to its first level, while all the others are in their ground state.

This constrained subspace of states can be viewed as the result of doing the path integral, with the above Lagrangian, with the variables restricted to live on the compact manifold CP^{N-1}. That manifold is obtained from complex N-dimensional space by imposing the "gauge equivalence" $z_i \equiv \lambda z_i$, with λ an arbitrary complex number. This just means that the parameterization of CP^{N-1} by N complex numbers is redundant. We partially fix the ambiguity by imposing the constraint $\sum z_i^* z_i = 1$, which shows that the manifold is compact. What remains is a phase ambiguity $z_i \equiv e^{i\theta} z_i$. In a path integral, we can multiply $z_i(t)$ by a different phase $\theta(t)$ at each time. If our integral is really over variables defined on CP^{N-1}, the action should be invariant under such a change. Indeed, since $\sum z_i^* z_i = 1$, the action changes to

$$\int L \to \int (L + \partial_t \theta) = \theta(t_2) - \theta(t_1). \tag{19.69}$$

The propagator between the two points $z_{1,2}^i$ on the manifold of complex unit N-vectors will thus change under a gauge transformation by $G(z_2, t_2; z_1, t_1) \to e^{i(\theta(t_1) - \theta(t_2))} G(z_2, t_2; z_1, t_1)$. The solution of the time-dependent Schrödinger equation is

$$\psi(z_2, t_2) = \int d^N z_1 G(z_2, t_2; z_1, t_1) \psi(z_1, t_1). \tag{19.70}$$

We see that the phase ambiguity in the path integral is equivalent to the statement that the phase of the wave function is unphysical.

We can now add a general Hamiltonian for the system by adding a Hamiltonian function via the standard prescription $L \to L_0 - H(z_i^*, z_i, t)$. This will be a well-defined Lagrangian on the manifold CP^{N-1}, if H is invariant under the phase transformation $z_i \to e^{i\theta} z_i$. Note that we have allowed for explicit time dependence in the Hamiltonian to accommodate interactions between our N-dimensional system and other variables, as explained above.

19.8 FERMIONS AND GRASSMANN INTEGRATION

When $N = 2^n$, there is another, often more convenient, way of representing finite dimensional systems in terms of *fermionic* path integrals. The Hilbert space of dimension 2^n can be presented in terms of n operators satisfying

$$[\psi_k, \psi_l]_+ = 0, \tag{19.71}$$

$$[\psi_k, \psi_l^\dagger]_+ = \delta_{kl}. \tag{19.72}$$

Start from a state satisfying

$$\psi_k|0\rangle = 0, \tag{19.73}$$

and construct

$$\psi_{k_1}^\dagger \dots \psi_{k_p}^\dagger |0\rangle, \tag{19.74}$$

for $1 \leq p \leq n$. The total number of states is 2^n. As you will prove in Exercise 19.11, we can write these operators in terms of $2n$ Hermitian operators γ_a satisfying

$$[\gamma_a, \gamma_b]_+ = \delta_{ab}. \tag{19.75}$$

Given a time-dependent Hamiltonian built out of these variables, we can construct matrix elements of time-ordered products of $\gamma_a(t)$. We define the time-ordered products with minus signs, so that they are totally antisymmetric under interchange. For example,

$$T\gamma_a(t)\gamma_b(s) = \theta(t-s)\gamma_a(t)\gamma_b(s) - \theta(s-t)\gamma_b(s)\gamma_a(t). \tag{19.76}$$

A generating functional for such totally antisymmetric products can be constructed with the help of Grassmann numbers. Grassmann numbers are complex linear combinations of a finite number of generators η_i, which satisfy the multiplication rule

$$[\eta_i, \eta_j]_+ = 0.$$

We will need to talk about Grassmann valued functions. These are simply defined by letting the complex parameters be functions of time. Since there are an infinite number of linearly independent functions, we will need an infinite number of Grassmann generators.

Grassmann introduced Grassmann variables as part of the mathematical theory of *differential forms*. They have properties attributed to *infinitesimals* in less rigorous presentations of that subject. They square to zero, and a product of k of them requires k different "dimensions" and can represent an infinitesimal k-plane. We will be interested in them because we can define notions of derivative and integral for Grassmann algebras, just as we can for ordinary functions.

Functions of a finite number of Grassmann generators (i.e., general elements of the Grassmann algebra) are finite order polynomials. The coefficients are antisymmetric tensors. So, we can define the derivative, $\frac{\partial}{\partial \eta_i}$, by a simple algebraic rule: it is zero if η_i does not appear in the monomial one is differentiating. If it does appear, it appears linearly, so we can define the derivative by simply dropping that variable, and multiplying by $(-1)^P$ where P is the number of other Grassmann variables you have to move the derivative through in order to "get to" η_i. For example,

$$\frac{\partial}{\partial \eta_1}(\eta_1 \eta_2 \eta_3) = \eta_2 \eta_3. \tag{19.77}$$

$$\frac{\partial}{\partial \eta_2}(\eta_1 \eta_2 \eta_3) = -\eta_1 \eta_3. \tag{19.78}$$

$$\frac{\partial}{\partial \eta_3}(\eta_1 \eta_2 \eta_3) = \eta_1 \eta_2. \tag{19.79}$$

Of course the minus sign depends on whether you start differentiating from the left or the right and some people define both left and right Grassmann derivatives. We will stick with the left derivative. Once we have defined the derivative for monomials, we extend it to arbitrary polynomials by insisting that it act linearly.

Linearity is also the key to defining an integral over Grassmann variables. For a single Grassmann variable, the most general function is $a + b\eta$. By linearity, we must have

$$\int d\eta(a + b\eta) = a \int d\eta 1 + b \int d\eta \eta. \tag{19.80}$$

There is no analog of an indefinite integral for Grassmann numbers. For the definite integral, we want the integration by parts rule

$$\int d\eta \frac{\partial}{\partial \eta} f = 0. \tag{19.81}$$

Since $1 = \frac{\partial \eta}{\partial \eta}$, we must have $\int d\eta 1 = 0$. We normalize the only nonzero result by

$$\int d\eta \eta = 1. \tag{19.82}$$

So the integral is just the derivative! For multiple Grassmann variables, we define the integral by iteration. The N-dimensional integral is defined by doing each one-dimensional integral in turn, taking care to pick up minus signs as we move $\int d\eta_i$ through η_j. Obviously the order in which we do the integrations matters to the overall sign. It is also obvious that the only monomial in a general Grassmann function, which survives integration, is the last term proportional to $\eta_1 \ldots \eta_N$.

We define the order by

$$\int d^N \eta (\eta_{a_1} \ldots \eta_{a_N}) = \epsilon_{a_1 \ldots a_N}. \tag{19.83}$$

The most important Grassmann integral for quantum mechanical applications is the Gaussian

$$I[\chi] = \int d^N \eta \; e^{\frac{1}{2}\eta^a A_{ab}\eta^b} e^{\chi_a \eta^a}, \tag{19.84}$$

where χ_a are an independent set of Grassmann variables, and $A = -A^T$ is an antisymmetric matrix. Using the fact that Grassmann integrals are invariant under shifts of the variables, and assuming N is even and $\det A \neq 0$, we can write

$$I[\chi] = e^{\frac{1}{2}\chi_a \chi_b (A^{-1})^{ab}} I[0], \tag{19.85}$$

$$I[0] = (\frac{1}{2})^{N/2} \int d^N \eta \; [\eta^{a_1}\eta^{b_1} A_{a_1 b_1} \ldots \eta^{a_{N/2}}\eta^{b_{N/2}} A_{a_{N/2} b_{N/2}}]. \tag{19.86}$$

Doing the integral we get

$$I[0] = (\frac{1}{2})^{N/2} \epsilon^{a_1 b_1 \ldots a_{N/2} b_{N/2}} A_{a_1 b_1} \ldots A_{a_{N/2} b_{N/2}} \equiv \mathrm{Pf}(A). \tag{19.87}$$

The right-hand side is the definition of the *Pfaffian* of an even dimensional antisymmetric matrix.

Now let us double the number of Grassmann variables by inserting an index η_i^a with $i = 1, 2$ and write

$$I^2[0] = \int d^{2N}\eta \; e^{\frac{1}{2}\eta_i^a A_{ab}\eta_i^b} = \mathrm{Pf}^2(A). \tag{19.88}$$

Define the complex Grassmann numbers

$$\psi^a = \frac{1}{\sqrt{2}}(\eta_1^a + i\eta_2^a). \tag{19.89}$$

$$\bar{\psi}^a = \frac{1}{\sqrt{2}}(\eta_1^a - i\eta_2^a). \tag{19.90}$$

The Jacobian of the transformation is one so

$$I^2[0] = \int d^N\psi \; d^N\bar{\psi} e^{\bar{\psi}^a A_{ab}\psi^b} = \det(A). \tag{19.91}$$

The evaluation of the complex Grassmann integral is the same as that for two independent real Grassmann integrals, as one can verify by elementary algebra. We have proven a celebrated theorem, namely that the square root of the determinant of an antisymmetric matrix is a polynomial in its matrix elements.

We have also gotten the formula for complex Grassmann integration

$$I[\chi] = \int d^N\psi \; d^N\bar{\psi} e^{\bar{\psi}^a M_{ab}\psi^b} e^{\bar{\chi}_a\psi^a + \bar{\psi}_a\chi^a} = e^{\bar{\chi}_a\chi_b(M^{-1})^{ab}} \det(M). \qquad (19.92)$$

This is just like the ordinary complex Gaussian integral except that we have the determinant rather than its inverse. Note that we have written this formula for a general matrix M, because antisymmetry is no longer necessary.

To apply these formulae to QM, we have to have complex Grassmann functions of time $\psi(t)$, which requires an infinite dimensional Grassmann algebra.

$$\psi(t) = \sum_{n=0}^{\infty} \psi_n f_n(t). \qquad (19.93)$$

$f_n(t)$ are some complete set of functions. The functional integral is

$$I[\chi(t)] = \int [d\psi(t)][d\bar{\psi}(t)] e^{iS + \int ds \; (\bar{\chi}(s)\psi^a(s) + \bar{\psi}_a(s)\chi^a(s)}. \qquad (19.94)$$

where

$$S = \int ds[\bar{\psi}i\partial_s\psi - h(\psi(s), \bar{\psi}(s))]. \qquad (19.95)$$

To see the relationship between this formula and a quantum system, we consider a single fermion creation–annihilation operator pair

$$[\Psi, \Psi^\dagger]_+ = 1; \qquad \Psi^2 = \Psi^{\dagger\,2} = 0. \qquad (19.96)$$

We can realize this operator algebra on a Hilbert space whose kets are functions of a single complex Grassmann variable ψ but with the rule that the corresponding bra is the complex conjugate function of $\bar{\psi}$. The Hilbert space is two-dimensional, since the most general function is $a + b\psi$. The operator Ψ acts as multiplication by ψ (by analogy with ordinary position coordinates)

$$\Psi[a + b\psi] = a\psi, \qquad (19.97)$$

while $\Psi^\dagger$ is $\frac{d}{d\psi}$, that is

$$\Psi^\dagger[a + b\psi] = b. \qquad (19.98)$$

Grassmann differentiation is easy, since the most general function of a finite number of Grassmann variables is a polynomial. One only has to be careful about order, because, for consistency we have to have

$$[\partial_{\psi^a}, \psi^b]_+ = \delta_{ab}, \qquad (19.99)$$

if there are multiple Grassmann variables. So, for example, taking a derivative from the left might give a different answer than taking it from the right. We always think about differentiating from the left.

The scalar product in this Hilbert space has the form

$$\langle f|g\rangle = \int d\psi d\bar{\psi}\mu(\psi,\bar{\psi})\bar{f}(\bar{\psi})g(\psi). \tag{19.100}$$

We choose the weight function μ so that the two functions 1 and ψ form an orthonormal basis. Thus,

$$\int d\psi d\bar{\psi}\mu(\psi,\bar{\psi}) = 1. \tag{19.101}$$

$$\int d\psi d\bar{\psi}\mu(\psi,\bar{\psi})\bar{psi}\psi = 1. \tag{19.102}$$

$$\int d\psi d\bar{\psi}\mu(\psi,\bar{\psi})\psi = 0. \tag{19.103}$$

$$\int d\psi d\bar{\psi}\mu(\psi,\bar{\psi})\bar{\psi} = 0. \tag{19.104}$$

Thus,

$$\mu = 1 + \bar{\psi}\psi = e^{\bar{\psi}\psi}. \tag{19.105}$$

Evaluating the time evolution operator over a sequence of infinitesimal intervals, for this simple quantum system with Hamiltonian $h(\Psi, \Psi^\dagger)$, reproduces the path integral formula above. The procedure generalizes easily to multiple fermions, Ψ^a.

In principle, we can use fermions to describe *any* system with a finite dimensional Hilbert space. Simple embed the N-dimensional space into the smallest fermion Hilbert space with $2^k > N$. The N-dimensional subspace satisfies some linear constraints

$$L_p|s\rangle = 0, \tag{19.106}$$

if and only if $|s\rangle$ is in the subspace. Consider the full fermion system, but with a term

$$\delta H = \sum \alpha_p L_p^\dagger L_p, \tag{19.107}$$

added to the Hamiltonian, with positive coefficients α_p. When $\alpha_p \to \infty$ only states in the N-dimensional subspace survive as finite energy states.

19.9 FURTHER EXERCISES

19.4 Evaluate the infinite product

$$\prod_{n=1}^{\infty} \left(1 + \frac{\omega^2 T^2}{n^2 \pi^2}\right) = e^{\sum_{n=1}^{\infty} \ln\left(1 + \frac{\omega^2 T^2}{n^2 \pi^2}\right)}.$$

19.5 Evaluate the path integral for the harmonic oscillator using Feynman's original derivation.

19.6 Use our evaluation of $\langle y|e^{-iHt/\hbar}|x\rangle$ for the harmonic oscillator, to compute the first three Hermite polynomials.

19.7 Prove that the differences between different choices of contour for the Fourier transform representation of the Green function obey the homogeneous equation.

19.8 Use the evaluation of the path integral for the harmonic oscillator coupled to a source, to compute the higher time-ordered products of any number of $X(t)$ operators.

19.9 Develop the path integral formalism for a Lagrangian of the form $\dot{q}_i M_{ij}(q)\dot{q}_j$. Note that you must choose an operator ordering prescription for the Hamiltonian, to define the quantum theory.

19.10 Prove the Gelfand–Yaglom theorem:

$$\frac{\det\left[\frac{-d^2}{dx^2} + V(x)\right]}{\det\left[\frac{-d^2}{dx^2}\right]} = \frac{u(1)}{u_0(1)}.$$

$u(x)$ is the solution of

$$\left[\frac{-d^2}{dx^2} + V(x)\right] v(s) = 0.$$

with boundary conditions $v(0) = 0$, and $\frac{dv}{dx}(0) = 1$. $v_0(s)$ is the solution of the same boundary value problem for the Hamiltonian with vanishing potential. In this problem $x \in [0, 1]$ and the eigenfunctions have Dirichlet boundary conditions $\psi(0) = \psi(1) = 0$.

a. The Riemann ζ function is defined by

$$\zeta(s) = \sum_{n=1}^{\infty} n^{-s},$$

for values of s for which the sum converges, and then by analytic continuation to the complex s plane. For a Hermitian operator with discrete spectrum, we define

$$\zeta_M(s) = \text{Tr}M^{-s}.$$

Show, formally, that if the sums converge for some values of s, then

$$\det M = e^{-\zeta'_M(0)}.$$

b. Suppose that the eigenvalues of M are bounded from below. We can always add a constant such that the bound is 0. Let $f(\lambda)$ be a function with simple zeroes at the eigenvalues $\lambda = \lambda_n$ and nowhere else. Suppose further that f is analytic in the λ plane with a cut along the negative real axis. Then the function $L(\lambda) = \frac{d\ln f}{d\lambda}$ has simple poles at the eigenvalues, with residue 1. By Cauchy's theorem, we can write

$$\zeta_M(s) = \frac{1}{2\pi i} \int_C d\lambda \lambda^{-s} L(\lambda),$$

where the contour C starts at $\infty - i\epsilon$, encircles the origin and returns to $\infty + i\epsilon$. Show that if one can neglect the contribution from a circle at infinity, one can move the contour to encircle the cut of λ^{-s} at $s = 0$ and then

$$\zeta_M(s) = \frac{\sin(\pi s)}{\pi} \int_0^{-\infty} \lambda^{-s} L(\lambda),$$

so that

$$-\zeta'_M(0) = \ln\frac{f(0)}{f(-\infty)}.$$

c. Let $M = -\frac{d^2}{dx^2} + V(x)$, with boundary conditions $\psi(0) = \psi(1) = 0$. Define the function $u(\lambda, x)$ to be the solution of

$$[-\frac{d^2}{dx^2} + V(x)]u = \lambda u,$$

with boundary condition $u(\lambda, 0) = 0$, $u'(\lambda, 0) = 1$. This is like a classical mechanics problem in a time-dependent harmonic oscillator and always has a solution. On the other hand $u(\lambda, 1)$ vanishes precisely at the eigenvalues $\lambda = \lambda_n$. For $\lambda \to -\infty$, the potential becomes negligible, but u does not have a limit. This annoying problem goes away if we divide by the determinant of the operator M_0, with $V = 0$. Thus

$$\det\frac{M}{M_0} = \frac{u(0, 1)}{u_0(0, 1)} = u(0, 1).$$

This remarkable result can be generalized in a number of ways.

d. Verify the Gelfand–Yaglom theorem by direct computation of the spectrum in the case where the potential is a constant.

19.11 Given $2n$ Hermitian operators satisfying

$$[\gamma_a, \gamma_b] = \delta_{ab},$$

show that the complex linear combinations $a_1 = \gamma_1 + i\gamma_2$, $a_2 = \gamma_3 + i\gamma_4$, etc. satisfy the algebra of fermionic creation and annihilation operators.

CHAPTER 20

Quantum Computation?

20.1   INTRODUCTION

This chapter is meant as a very brief introduction to the ideas of *theoretical* quantum computation. It is supposed to be enough to allow you to begin reading serious literature on the subject [86]. The most important part of the subject of quantum computation is experimental and practical. If it proves impossible to build a machine that can perform large quantum computations efficiently and reliably, then much of the theoretical work will prove useless. On the other hand, theoretical quantum computer science is extremely important, both because it has already shown that a quantum computer can, in principle, solve certain problems much more efficiently than a classical computer, and because it provides clues (topological quantum computing [87], etc.) to how one might build a real quantum computing device. In addition, the theory of quantum computing sheds light on the peculiar properties of quantum entanglement and has led to important developments (the theory of tensor networks [88]) in the attempt to construct approximate ground state wave functions for complicated condensed matter systems.

A modern digital computer stores data in physical systems that can reliably encode a binary number with k digits. There are 2^k such numbers, and k is called the number of *bits* in the computer. A *byte* is a subsystem with $k = 10$ and the largest computers ever built have of order 10^{15} bytes of Random Access Memory (the part of the machine on which active computation takes place) and a total memory that is about 10^{18} bytes. So we are talking about $k \sim 10^{18}$ at current technological limits.

Computation consists of transformations among these 2^k numbers. Conceptually, the simplest example is a program, which computes the values of some function $i \to f(i)$. The art of writing computer code consists of figuring out efficient ways to convert the problem you want to solve into a sequence of operations on binary numbers. Among those operations, there are reversible ones, namely operations that take the list of numbers in some canonical order into a permutation of that order.

In the first chapter of this book, we learned that any finite list of data can be viewed as a list of the ortho-normal basis vectors of a finite dimensional Hilbert space. It should be clear to someone who has read this book up to this point, that the Hilbert space of states of a classical computer is a k-fold tensor product of two-dimensional Hilbert spaces. Choose the classical computational basis to be the basis defined by the eigenvalues of $\sigma_3^{(p)}$ in the p-th factor of the tensor product, given some choice of ordering for the factors. The binary number representation is reproduced by writing the p-th digit of the number as the eigenvalue of $\frac{1+\sigma_3^{(p)}}{2}$. Examples of reversible classical computational operations, permutations, are the operators $\sigma_1^{(p)}$, which flip the p-th bit and leave the others alone. Computer scientists call this operation NOT, and denote it by $\mathbf{X}$. We will stick to the standard physics notation, but if you want to read the quantum computation literature you will have to get used to their notation and terminology.

20.2 QUANTUM INFORMATION

We will begin by discussing *quantum information science*, which is a collection of general results about quantum mechanics (QM) that are useful in quantum computation, but have more general interest. Recall that a general quantum state is given by a density matrix ρ, which is a positive Hermitian operator whose eigenvalues sum to one. We will stick to finite dimensional Hilbert spaces, so we do not have to worry about technicalities in defining concepts like traces, tensor products, etc.

A density matrix is *pure* if $\rho^2 = \rho$, which means that ρ is the projector on some specific quantum state vector $|s\rangle$. In such a state, in a Hilbert space of dimension D, there are D independent operators, whose value is predicted with certainty. These are ρ itself, and $D-1$ other one-dimensional projectors, on vectors orthogonal to $|s\rangle$. If the system is in an impure state, then less is predicted with certainty. There is an interesting and extremely deep connection between this notion of maximal information, and the entanglement of two independent quantum systems.

Recall that we can think of the Hilbert space of a pair of independent systems as a tensor product of the Hilbert spaces of the individual systems

$$\mathcal{H} = \mathcal{H}_A \otimes \mathcal{H}_B. \tag{20.1}$$

In quantum information theory, we call a Hilbert space with such a preferred factorization a *bipartite system*. Without loss of generality, we take the dimension of $\mathcal{H}_A$ to be less than or equal to that of $\mathcal{H}_B$. There is a special subgroup of unitary transformations on a bipartite system, which has the form $U = U_A \otimes U_B$. If the initial state of the system has the form $\rho = \rho_A \otimes \rho_B$, this will be preserved under conjugation by that subgroup. Such states are said to be *unentangled*. Any vector in $\mathcal{H}$ can be written in terms of ortho-normal bases of the individual spaces as

$$= \sum_{iJ} C_{iJ} |a_i\rangle |b_J\rangle. \tag{20.2}$$

The coefficient matrix C is generally rectangular, and satisfies

$$\sum_{iJ} |C_{iJ}|^2 = 1. \tag{20.3}$$

This can be read as the statement that the two positive Hermitian matrices $\rho_A = CC^\dagger$ and $\rho_B = C^\dagger C$, both have trace 1 and therefore define states on the individual systems. You will prove in Exercise 20.1 that these states are pure only if

$$C_{iJ} = c_i d_J, \tag{20.4}$$

which is to say that the systems are unentangled and the state on $\mathcal{H}$ is a tensor product. These matrices are the matrices of density operators $\rho_{A,B}$ in the bases which we used to define the coefficients C_{iJ}.

Entanglement thus implies that the probability of finding the A system in some particular state, which can be determined experimentally by doing measurements of operators of the form $O_A \otimes 1$, is correlated with the probability that the second system is in one of its particular states J. The most famous such correlation is the one explored in the paper of Einstein Rosen and Podolsky [30]. Entangled states can appear bizarre and nonlocal, because they can occur even when the two systems are very far apart in space. If, however, we admit that the entanglement occurred because of some interaction in the past, there is nothing that violates causality in the existence of entanglement between distant objects. Indeed, the factorization of unitary operators, $U = U_A \otimes U_B$ which preserves *lack of entanglement* is precisely what we would expect from a Hamiltonian evolution of a system whose Hamiltonian had the form

$$H = H_A \otimes 1 + 1 \otimes H_B, \tag{20.5}$$

characteristic of two noninteracting systems. More general time evolution will create entanglement for an initially unentangled state. Of course, we cannot rule out the possibility that someone started the universe off in a state which had entanglement between distant degrees of freedom but one usually assumes that any entanglement revealed by an experiment has a causal explanation.

In a bipartite system, we can always predict the probabilities of single system measurements in terms of the reduced density operator ρ_A of that system. This is unlikely to be pure, and we can define a measure of its purity called the Shannon–von Neumann entropy

$$S \equiv -\text{Tr}[\rho_A \ln \rho_A] = \lim_{n \to 1} = \frac{\text{Tr}[\rho_A^n]}{1 - n}. \tag{20.6}$$

For $n \neq 1$ and integer, the quantities on the right-hand side are called Renyi entropies. The Shannon–Von Neuman entropy is always nonnegative, and vanishes only when the state is pure. It depends only on the spectrum of the density operator p_i. It is also bounded from above by the entropy of the maximally mixed (also called maximally uncertain) density operator $\rho_{max} \equiv \frac{1}{D_A} 1$, which is the logarithm of the dimension of the Hilbert space $\mathcal{H}_A$. Finally, the entropy is a convex function of the density operator. Given a finite set of density operators ρ_i, one can form a new one via $\rho = \sum t_i \rho_i$ where t_i are nonnegative numbers summing to one. Then $S(\rho) \geq \sum t_i S(\rho_i)$, as you will prove in Exercise 20.2. You will also prove (Exercise 20.3) that the entropies of the two reduced density matrices derived from a pure state in a bipartite system are equal. These two equal entropies of a pure state in a bipartite system are called the *entanglement entropy* of one part of the system with the other.

Given a density operator ρ_A on $\mathcal{H}_A$, we can ask the question: is it possible to find a pure state ρ in a larger bipartite system, such that ρ_A is the reduced density matrix. Consider any Hilbert space $\mathcal{H}_B$ with dimension larger than or equal to that of $\mathcal{H}_A$. Write the spectral decomposition

$$\rho_A = \sum p_i P_i, \tag{20.7}$$

where P_i is a complete set of commuting one-dimensional projectors on states $|a_i\rangle$. Now let $|b_i\rangle$ be a basis of a subspace of $\mathcal{H}_B$ whose dimension is equal to D_A. Consider the pure state

$$|S\rangle = \sum_i \alpha_i |a_i\rangle \otimes |b_i\rangle. \tag{20.8}$$

Then the reduced density operator in $\mathcal{H}_A$ is

$$\rho_A = \sum_i |\alpha_i|^2 P_i, \tag{20.9}$$

so we need to only choose α_i to have absolute value $p_i^{1/2}$. This procedure is called the Schmidt decomposition, and the resulting pure state is called a *purification* of the original state. It is clear that the purification is highly nonunique. We can perform any unitary transformation in $\mathcal{H}_B$ on the states $|b_i\rangle$ and get a new purification, or choose any other subspace of the larger space with the same dimension.

20.3 PAGE'S THEOREM REDUX, MONOGAMY, AND CLONING

We have already encountered entanglement in both our discussions of measurement theory and of statistical equilibrium. Let us first recall Page's theorem, from the latter discussion. The context of the theorem is a bipartite system with $D_B \gg D_A$. The reduced density matrix $\rho_A = C_{iJ} C_{Ji}^*$ can then be analyzed statistically. Let us choose the basis in which ρ_A is diagonal. The D_A diagonal matrix elements p_i of ρ_A are each written as a sum of a squares of D_B complex numbers subject to the constraint $\sum_{iJ} |C_{iJ}|^2 = 1$. This constraint is invariant under unitary transformations in the group $U(D_A) \otimes U(D_B)$ and in particular under permutations in the indices i. Therefore, if we choose the complex coefficients randomly, then on average, we will have all p_i equal to $1/D_A$. The root mean square deviation of each p_i from the average will be $1/\sqrt{D_B}$, so if $D_A \ll D_B$ we get the maximally uncertain density matrix with very high probability.

In general, any purification of the maximally mixed density operator will have the form

$$|Pur\rangle = \frac{1}{\sqrt{D_A}} \left(\sum_i |a_i\rangle \otimes |b_i\rangle \right), \tag{20.10}$$

where we have chosen some orthonormal basis of a D_A-dimensional subspace of $\mathcal{H}_B$. For any such state, we say that system A is *maximally entangled* with system B. The significance of this phrase is that a measurement of $|b_i\rangle\langle b_i|$ will automatically tell us what state system A is in, even though system B and system A might be separated by a million light years. As we have emphasized repeatedly, no real information transfer is involved here. We have simply prepared a quantum state in which the probability of system A being in any of the states $|a_i\rangle$ is equal, but is correlated exactly with the probability that system B is in $|b_i\rangle$. There is nothing spooky or nonlocal about this, if we can explain the entanglement by some earlier causal contact between the two systems.

Maximal entanglement is *monogamous*. Suppose we ask that the system is *simultaneously* maximally entangled with two different bases of $\mathcal{H}_B$. Then

$$\sum_i |a_i\rangle \otimes |b_i\rangle = \sum_i |a_i\rangle \otimes |c_i\rangle. \tag{20.11}$$

Applying the projection operator $|a_k\rangle\langle a_k| \otimes 1$ to this equation we get

$$|a_k\rangle \otimes |b_k\rangle = |a_k\rangle \otimes |c_k\rangle, \tag{20.12}$$

for each k, which means that $|b_k\rangle = |c_k\rangle$. The monogamy of maximal entanglement is very important to the problem of quantum encryption of data, to be discussed below.

Another important application of these ideas is the so-called no cloning theorem. The question is the following. Given a system A in state $|a\rangle$ can we allow it to interact with another system in such a way that system, B carries away with it a clone of $|a\rangle$? Translated into mathematics, this is the question of whether there exists a unitary transformation in a bipartite system, which will transform the state $|a\rangle \otimes |b\rangle$ into $|a\rangle \otimes |a\rangle$. Since either of these two states is a normalized vector in the full bipartite Hilbert space, we can obviously do this for any *given* state $|a\rangle$. The no-cloning theorem refers to the fact that it is impossible to find a single unitary transformation, which will perform this operation *for every* state $|a\rangle$. Thus, we ask whether

$$|a_i\rangle \otimes |a_i\rangle = U(|a_i\rangle \otimes |e\rangle), \tag{20.13}$$

for all $|a_i\rangle$. Consider the scalar product of two different states on the right-hand side of these equations, before the action of U. Since $|e\rangle$ is normalized, this is just $\langle a_1|a_2\rangle$ in the Hilbert space $\mathcal{H}_A$. Since U is unitary, this is the same as the scalar product on the left-hand side, so that

$$\langle a_1|a_2\rangle^2 = \langle a_1|a_2\rangle. \tag{20.14}$$

Thus, the claim that one can clone two states with the same unitary is possibly valid only if the two states are the same, or if they are orthogonal. Thus, one could clone all states in a given orthonormal basis, but not superpositions of them. This last sentence is the statement that in a *classical theory*, where one forbids superpositions of the preferred basis, one *can clone* arbitrary states with the same "machine." The nonclassical principle of superposition defeats this attempt.

20.4 QUANTUM KEY DISTRIBUTION

While we have generally tried to steer away from discussing human intervention in QM, the subject of quantum encryption is all about human agents trying to send messages to each other, and trying to protect them from being read by an enemy. We will therefore follow the practice in the field and assume that Alice wants to send a secret message to Bob. A standard practice in cryptography is the *one time pad*. Secret messages are encoded by constructing a map between the clear version of the message, written in some well-known human or computer language, and an encoded version. The map is constructed with the use of a *key*, some string of bits which, in the simplest version of encryption tells us which symbol to substitute for each symbol in the clear message. If Alice uses the same key multiple times, her encryption is vulnerable to analysis of letter frequencies, etc., so if multiple messages are to be sent, she wants the ability to make frequent switches of the key. However, in attempting to send the multiple keys to the recipient, Alice runs the risk of interception of her key distribution messages by an eavesdropper, Eve.

Alice could encode her key distribution message as a set of bits in a classical computer, which we represent by introducing N spins. A message consists of a choice of the state in the 2^N-dimensional Hilbert space of the system. N will be taken fairly large, in order to make statistical reasoning about the results of quantum measurements a reliable tool.

The problem we are trying to solve is to discern the possibility that, in the process of transmitting the bit string from Alice to Bob, it was intercepted by the enemy, Eve. With the usual procedures of cryptography, there is really no way to tell if a message has been intercepted. If, however, we could transmit quantum states, the situation is completely different. The point is that a quantum state gives only a probability for a given outcome, unless one knows in advance which complete set of commuting operators is diagonal in that state. For purposes of encryption, it is sufficient, at least to illustrate the point, to consider only the operators $\sigma_{1,3}(i)$ where i denotes the position along the bit string.

The first step of the encryption procedure is for Alice to choose at random, for each (i) whether she will encode the bit value 1 as $\sigma_3(i) = 1$ or $\sigma_1(i) = 1$. Her message is then encoded in the state of a one-dimensional quantum spin chain. The next step is the part for which no technology yet exists, except for very small numbers of Q-bits. Alice sends a physical system prepared in the quantum state of the spin chain she has chosen, to Bob, without losing any quantum information. One can imagine a *quantum wire*, which sends N electrons to Bob, with each electron in a spin state along either the 1 or 3 axes, according to the random choice made by Alice, with the sign of the spin encoding the message. Bob does not yet have a message. If he chooses his own random sequence of $\sigma_3(i)$ *or* $\sigma_1(i)$ measuring devices, he has equal probability of reading the wrong message as the right one, every time his random choice fails to coincide with that made by Alice.

Now Alice gets on the telephone and reveals (perhaps to Eve as well, if Eve has Alice's phone tapped) to Bob what her choice of axes was. It is important to realize that at this point, the Q-bits have already been sent. If Eve has not already intercepted them, she can no longer do so. They are sitting safely in Bob's ultrasecure laboratory. Bob discards those bits of the message where his axis differed from Alice's, because there is a 50% probability that he read that part of the message incorrectly, since the quantity he measured was actually uncertain in the state sent by Alice. At this point, Alice and Bob publicly compare half of the bits where their axes coincided.

If Eve *did not* intercept the message, they should find perfect agreement. If Eve *did* intercept the message, she did so by choosing her own random sequence of axes and measuring either $\sigma_3(i)$ or $\sigma_1(i)$ for each bit. In doing so, she changed the quantum state of the system. To say this more precisely, she entangled the system with her spin measuring devices, which are macroscopic. The quantum state, conditioned on the fact that those devices read *anything at all*, is different from the state sent by Alice. This is the quantum state that Bob actually queried with his spin measuring devices. Consider those bits where Bob and Alice made the same choice of axes. For each of those bits, Eve had 50% probability of making the same choice. When Alice and Bob compare the values they assigned to the bits where their axes agreed, they will get the wrong answer about 50% of the time, and they will know that Eve has intercepted the message. Presumably, in this case, Alice would just try again. None of the textbooks on the subject tells us what to do if Eve intercepts every message. Quantum encryption can only alert us to the fact that our messages are being read, not get them through untainted.

However, once Alice and Bob know that Eve has *not* intercepted the message, Alice can share the bits of the message that Bob did not get yet, through normal classical channels. He has learned the value of roughly 1/4 of the bits in the key, randomly distributed in the message, and Alice and Bob know that Eve can have no clue what those bits were. Even if she Eve-sdrops on the classical transmission of the remaining data, she is unlikely to learn Alice and Bob's secret key. The probability that she does guess it is $2^{-\frac{N}{4}}$, even assuming that she has monitored Alice's classical communication of $N/4$ bits to Bob. The key, which is $2^{N/2}$ bits long, can now be used to send secure messages.

The bottleneck in the actual implementation of this algorithm is sending the quantum information from Alice to Bob. The simplest practical implementation of such a *quantum channel* is sending free photons through space, or a fiber-optic cable. Coherent quantum states of photon polarization can be sent over distances of order 10^2 km. The most commonly discussed technique for sending quantum information is the quantum teleportation protocol, to which we now turn.

20.5   QUANTUM TELEPORTATION

The subject of this section has an unfortunate name, since it does not really involve transportation of localized objects from one place to another, and certainly does not involve faster than light communication. Quantum teleportation is a method for transferring information about the quantum state of a system, to a remote location. Let us imagine that it is the state of some two state system, and label a particular basis in its Hilbert space by the binary numbers $|0\rangle$ and $|1\rangle$. We assume the state of the system is

$$|s\rangle = \alpha|0\rangle + \beta|1\rangle. \tag{20.15}$$

Assume this system is in Alice's laboratory.

The first step in the quantum teleportation communication protocol has to occur before the actual teleportation begins. One prepares a pair of spin one half particles, or photons, in for example the state

$$|\mathcal{B}_1\rangle_{AB} = \frac{1}{\sqrt{2}}(|+\rangle_A \otimes |+\rangle_B + (|-\rangle_A \otimes |-\rangle_B). \tag{20.16}$$

The $\pm$ signs represent polarization along the three direction, while the subscripts refer to Alice and Bob, because this part of the protocol also sends one of the entangled particles to Alice and the other to Bob. In Exercise 20.4, you will verify that this state is part of an orthonormal basis given by

$$|\mathcal{B}_2\rangle_{AB} = \frac{1}{\sqrt{2}}(|+\rangle_A \otimes |+\rangle_B - (|-\rangle_A \otimes |-\rangle_B). \tag{20.17}$$

$$|\mathcal{B}_3\rangle_{AB} = \frac{1}{\sqrt{2}}(|+\rangle_A \otimes |-\rangle_B + (|-\rangle_A \otimes |+\rangle_B). \tag{20.18}$$

$$|\mathcal{B}_4\rangle_{AB} = \frac{1}{\sqrt{2}}(|+\rangle_A \otimes |-\rangle_B - (|-\rangle_A \otimes |+\rangle_B). \tag{20.19}$$

This is called the Bell basis. Alice and Bob have received the information about the preferred axis, 3, and the choice of the Bell vector $|\mathcal{B}_1\rangle_{AB}$, from the agent who created and sent the entangled pair. They each carefully store their precious entangled particle in a way that prevents it from interacting with anything else.

Now the state of the system in Alice's laboratory

$$|s\rangle \otimes |\mathcal{B}_1\rangle_{AB}, \tag{20.20}$$

which consists of the two state system one hopes to communicate, and one of the particles in the Bell pair, is entangled with the system in Bob's laboratory (the other particle in the Bell pair). The actual teleportation consists of two steps. First Alice uses an experimental apparatus, whose pointer positions are designed to become entangled with the states in the Bell basis *in the tensor product of the state of her spin one half particle and that of the two state system.*

To understand what happens, we should simply expand the premeasurement state in that basis

$$|s\rangle \otimes |\mathcal{B}_1\rangle_{AB} = \frac{1}{2}[|\mathcal{B}_1\rangle_{AC} \otimes (\alpha|+\rangle_B + \beta|-\rangle_B) + |\mathcal{B}_2\rangle_{AC} \otimes (\alpha|+\rangle_B - \beta|-\rangle_B)+ \tag{20.21}$$

$$|\mathcal{B}_3\rangle_{AC} \otimes (\beta|+\rangle_B + \alpha|-\rangle_B) + |\mathcal{B}_4\rangle_{AC} \otimes (\beta|+\rangle_B - \alpha|-\rangle_B). \tag{20.22}$$

This is the same state we had before, just written in another basis. What is interesting about it is that *if Alice can construct a macroscopic device, which is entangled with the Bell basis, then for each choice of Bell state, the state of Bob's particle has the same quantum information as the quantum state* $|s\rangle$ *of the original system.* Indeed, for $|\mathcal{B}_1\rangle_{AC}$ the coefficients of $|\pm\rangle_B$ are the same as those of $|0/1\rangle$ in $|s\rangle$. For other members of the Bell basis, the state of Bob's particle is a given unitary transformation on $\alpha|+\rangle_B + \beta|-\rangle_B$.

Recall what it means for Alice to do a measurement. The quantum state of the three body system is just a probability distribution for all normal operators on the full Hilbert space. The measurement entangles each component of Alice's two body subsystem, in the Bell basis, with the state of one of the macroscopic pointers on Alice's machine. As always, we can now apply Bayes' conditional probability rule to collapse the quantum state to the one corresponding to what actually happened to the macroscopic needles in a particular run of the experiment. One can then communicate information about the result of this measurement to Bob, using macroscopic apparatus, which follows almost deterministic laws starting from the initial condition of what Alice's needle registered. This violates unitarity, since Bayes' rule always requires us to renormalize probabilities to eliminate those predictions that "did not happen." This is why the quantum teleportation scheme can appear to violate the no-cloning theorem.

Note that independently of which Bell state Alice's measurement finds, Bob can clone the quantum state $|s\rangle$ in his lab, once he receives Alice's classical communication. If Alice finds

$|\mathcal{B}_1\rangle_{AC}$, Bob just has $|s\rangle$, whereas if she finds one of the other Bell states, he knows precisely which unitary operation he must perform on the state of his particle, in order to retrieve $|s\rangle$. Actually, it is more proper to say that Bob has cloned only the quantum information in $|s\rangle$ since we have taken pains to say that $|s\rangle$ was the state of *some* two state system, not necessarily that of a spin one half particle identical to the one in Bob's laboratory. One of the most successful recent demonstrations of quantum teleportation is [89]. Interested readers should note the way in which this experiment transfers quantum information back and forth between nuclear spins, electron spins, and photons, as well as the delicate control needed to keep the quantum coherence of the system.

20.6   GATES FOR QUANTUM COMPUTERS

Computer scientists use the term *gates* for operations which change the state of a computer. For quantum computer scientists, gates are unitary transformations. We have already discussed the NOT gate which changes the state of the i-th bit by applying the operator $\sigma_1(i)$. Our next example is the *controlled* NOT, or c-NOT operation. This is $C_{ij} = \frac{1+\sigma_3^{(i)}}{2}\sigma_1^{(j)} + \frac{1-\sigma_3^{(i)}}{2}$. In writing this formula, we have used a compressed notation for operators in a tensor product space, which is the notation used in lattice spin systems. We write only the parts of the operator which act nontrivially on some of the bits, leaving implicit the tensor product with the unit matrix acting on the rest of the bits. An operator is called p−local if it acts in a nontrivial way on exactly p bits. Note that $C^2 = 1$, and $C = C^\dagger$, so that c-NOT is a unitary and therefore reversible operation. A more compact notation for C_{ij} is

$$C_{ij} = P_+^{(i)}\sigma_1^{(j)} + P_-^{(i)}, \qquad (20.23)$$

where $P_\pm^{(i)}$ are the projectors on the subspaces where $\sigma_3(i)$ is ± 1.

An extremely important operation in computer science is the swap operation S_{ij}, which exchanges the contents of the i-th and j-th bits. That is, it takes $|\ldots 0 \ldots 1 \ldots\rangle$ to $|\ldots 1 \ldots 0 \ldots\rangle$, for any values of the bits represented by $\ldots$. In Exercise 20.5, you will verify by operator multiplication that

$$S_{ij} = C_{ij}C_{ji}C_{ij}. \qquad (20.24)$$

In words, suppose the j-bit is 0, then after the operation of C_{ij} the two-bit system is in the state $|11\rangle$ or $|00\rangle$ depending on the initial state of the i-th bit. Let us choose the case $|11\rangle$. Now the operation C_{ji} changes this to $|01\rangle$ and the second application of C_{ij} leaves this state intact. Similarly, if the i-th bit were 0, then all three c-NOT operations leave the state invariant, so we again get the effect of the swap. The same thing happens if the j-th bit is 1. Note that the operators $S_{i,i+1}$ are just transpositions, and the theory of the permutation group shows that the transpositions generate all permutations, so the S_{ij} operations form an overcomplete set of generators.

Another important operator in quantum computation is the Walsh–Hadamard operator, for which QC theorists have unfortunately chosen the symbol $H^{(i)}$. This clash with the standard QM notation for Hamiltonian is mitigated by the fact that in QC one always deals with discrete unitary evolution, so there are no Hamiltonians. As its symbol indicates, $H^{(i)}$ acts on the Hilbert space of a single bit, as follows

$$H^{(i)} = \frac{1}{\sqrt{2}}(\sigma_1^{(i)} + \sigma_3^{(i)}). \tag{20.25}$$

From the point of view of spin, H is just the Pauli matrix in the direction $\pi/4$ in the 1–3 plane. This means that it takes the state of a single bit in the computational (i.e., σ_3 diagonal) basis into a superposition of states where the bit is 1 and 0. In fact, those superpositions are just the eigenstates of $\sigma_1(i)$, and $H^{(i)}$ is just the unitary transformation between the eigenbases of $\sigma_3(i)$ and $\sigma_1(i)$.

At this point, we should introduce some conventional QC language. A single bit, which can only be in the 1 or 0 states is called a C-bit, while if we allow superpositions, it is called a Q-bit. Obviously, from the point of view of Hilbert space, we are talking about the same two-dimensional space. Insisting on using only C-bits is analogous to our classical theory of the ammonia molecule. The difference between the classical and quantum theories of ammonia is whether we allow time evolution to change a C-bit into a more general pair of orthonormal vectors in Hilbert space. The difference between a quantum and classical computer program is similarly whether we force the discrete transformation between different states of the computer to be in the permutation subgroup S_{2^k} or allow more general unitary transformations. $H^{(i)}$ is not allowed in a classical program, but note that

$$H^{(i)}H^{(j)}C_{ij}H^{(i)}H^{(j)} = \frac{1}{2}\begin{pmatrix} H^{(i)} & H^{(i)} \\ H^{(i)} & -H^{(i)} \end{pmatrix}\frac{1}{2}\begin{pmatrix} P_-^{(i)} & P_+^{(i)} \\ P_+^{(i)} & P_-^{(i)} \end{pmatrix}\frac{1}{2}\begin{pmatrix} H^{(i)} & H^{(i)} \\ H^{(i)} & -H^{(i)} \end{pmatrix}. \tag{20.26}$$

Carrying out the rightmost matrix multiplication, we get

$$H^{(i)}H^{(j)}C_{ij}H^{(i)}H^{(j)} = \frac{1}{2\sqrt{2}}\begin{pmatrix} H^{(i)} & H^{(i)} \\ H^{(i)} & -H^{(i)} \end{pmatrix}\begin{pmatrix} \sigma_-^{(i)} - 1 & \sigma_+^{(i)} + 1 \\ \sigma_+^{(i)} + 1 & 1 - \sigma_-^{(i)} \end{pmatrix}, \tag{20.27}$$

where $\sigma_\pm$ are the spin raising and lowering operators. It is easy to see that

$$H_i\sigma_\pm^{(i)} = P_\mp^{(i)} \pm \sigma_\pm^{(i)}. \tag{20.28}$$

Using these results, you will verify in Exercise 20.6 that

$$H^{(i)}H^{(j)}C_{ij}H^{(i)}H^{(j)} = C_{ji}. \tag{20.29}$$

Switching the control and target bits of a c-NOT operation is an important device in computing. Using only classical operations one can accomplish the same goal by conjugating with the swap operator S_{ij}. However this is a 2 Q-bit operation which is not a product of two single Q-bit operations. In hypothetical physical realizations of quantum computers, just as in classical computers, operations which act on a larger number of bits are more difficult to construct. The reason for this is locality. The bits, whether classical or quantum, are arranged on a lattice in some solid. Local Hamiltonians will typically do controlled things only to one bit, or two nearest neighbor bits.

QC scientists use the term N–Q–bit $gate$ to describe a unitary transformation on N Q-bits, which is to say an element of $U(2^N)$. Technologically feasible quantum computers will have to have only one and two Q-bit gates, and programs should minimize the number of two Q-bit gates they use, in order to be efficient. A general one cubit gate has the form $e^{i\alpha}(n_0 + \mathbf{n} \cdot \sigma)$, where $(n_0, \mathbf{n})$ is a real unit four vector. It has four real parameters. A general 2-Q-bit gate contains 16 real parameters. Now consider the three classes of gates

$$U_i, \ U_j, \ V_i C_{ij} V_j. \tag{20.30}$$

Clearly the first two classes are independent of each other, and of the third. Suppose that for two different choices of V_i and V_j the corresponding operators in the third class are equal.

$$V_i C_{ij} V_j = W_i C_{ij} W_j. \tag{20.31}$$

Then there is some pair of single Q-bit unitaries $U_i = W_i^{-1} V_i$ and $T_j = V_j W_j^{-1}$, such that

$$U_i C_{ij} U_j = C_{ij}. \tag{20.32}$$

We can write this in 2×2 block form as

$$\begin{pmatrix} u_{11} & u_{12} \\ u_{21} & u_{22} \end{pmatrix} \begin{pmatrix} \sigma_1 T & 0 \\ 0 & T \end{pmatrix} = \begin{pmatrix} \sigma_1 & 0 \\ 0 & 1 \end{pmatrix}. \tag{20.33}$$

We have omitted the j label on σ_1 and T. From this equation, we immediately see that $u_{12} = u_{21} = 0$, and unitarity then implies that $u_{ii} = e^{i\alpha_i}$. But then we also have $e^{\alpha_1}\sigma_1 T = \sigma_1$ and $e^{\alpha_2}T = 1$. Thus, T is proportional to the unit matrix, $\alpha_1 = \alpha_2$, and the phase of T is the opposite of this one. So the only duplication is the transformation that multiplies one Q-bit by a phase and the other by the opposite phase. We conclude that any two Q-bit gate can be obtained by concatenating the c-NOT operation with one Q-bit gates.

20.7 COMPUTATIONAL COMPLEXITY

Turing's mathematical model of a classical computer consists of a transformation in the Hilbert space of N Q-bits, which takes an initial state into a desired final state, using only elements of an S_{2^N} subgroup of $U(2^N)$. Without loss of generality, we can work in the basis in which the initial state is a tensor product of N single Q-bit states, each of which is an eigenstate of $\sigma_3(i)$. In quantum computing language, the set of all such states is called *the computational basis*. The S_{2^N} subgroup transforms each computational basis state into another, permuting the elements of the basis. For practical reasons, one wants to have each computational step involve only a small number of Q-bits. Turing showed that any element of S_{2^N} could be built by concatenating the single Q-bit operations $\sigma_1(i)$ and the two Q-bit operations S_{ij}, as we have discussed above.

We can think of a state in the computational basis as a binary number, with the value of the projection operator on $\sigma_3(i) = 1$ giving the i-th digit of the number. A classical computation consists of computing the value of some function $f(i)$ for each i. If the computer has N bits, we are restricted to calculating functions whose values are less than or equal to the binary number $1\ldots1 = 2^N + 2^{N-1} + \cdots + 1$. If the function is invertible, then it corresponds to some S_{2^N} transformation on the N Q-bit Hilbert space. Each permutation has a characteristic *size*, *viz.* the minimal number of 1 or 2 Q-bit gates that it takes to build that transformation. The running time for a classical computer, which performs its operations serially, to perform the computation $i \to f(i)$ is proportional to the size of the corresponding S_{2^N} transformation.

A *parallel computer* is a device which can perform some number k of 1 or 2 Q-bit transformations simultaneously on k disjoint pairs of Q-bits. In this case, we define the *depth* of a computation as the number of such simultaneous transformations it takes to perform the computation. The *computational complexity* of a problem is the depth of the computation required to solve it. For any given problem, the computational complexity is a fixed number. However, many problems have a natural scale to them, which can be increased either without limit, or up to some huge value. For example, we can ask, given some integer M, what its prime factors are. We would like to know how the computational complexity of this problem scales with M. Problems whose complexity scales like M^p for some fixed p and $M \to \infty$ are called Polynomial or said to belong to class P, whereas those which depend on M exponentially belong to class NP. Obviously, if one hopes to solve a problem for some very large value of M, one hopes that it is in class P. Conversely, if one is trying to encrypt data with a code that is hard to crack, one encrypts it using an algorithm that is in NP. It turns out that for a classical computer, the factorization problem is in class NP. Much of the interest in quantum computing comes from the fact that some problems, the factorization problem among them, which are in class NP for a classical computer, are in class P for a quantum computer.

The complexity of a quantum computational problem is defined in a manner completely analogous to the classical case. A quantum computation is a unitary transformation in $U(2^N)$. Since unitary transformations depend on continuous parameters, we are unlikely to be able to construct one exactly by performing any finite number of one and two Q-bit gates. However, all we really need to do is to get close to the required unitary. Closeness can be defined in terms of the quantities

$$||U - V||_p \equiv \left(\text{Tr} \left[(U - V)^\dagger (U - V) \right]^{\frac{p}{2}} \right)^{\frac{1}{p}}, \qquad (20.34)$$

with $p = 1$ being the usual choice. We insist that $||(U - U^{\mathcal{C}}_{gate})||_1$ be less than ϵ.

One can show that for almost all choices of U, which is to say, with probability one for a unitary picked from the uniform probability distribution on the space of all unitary transformations $U(2^M)$, the computational complexity is bounded by

$$2^{2M} \ln \left(\frac{1}{\epsilon} \right) < \mathcal{C} < 2^{2M} [\ln \left(\frac{1}{\epsilon} \right)]^c, \qquad (20.35)$$

where $1 < c < 2$ (the exact value of c is unknown). So most quantum computations are exponentially hard. However, there are some exponentially hard classical computations, which can be performed in polynomial time on a quantum computer. Since a full exposition of quantum computer science would require another book the size of this one, and since a number of excellent books on the subject exist already, we will just illustrate this with a simple example, known as the Deutsch–Josza (DJ) problem [90].

The DJ problem is one of a class of *black box* problems, in which one is given a function f defined on M Q-bits, with certain properties, and asked to verify whether it satisfies some other property Z. The function is implemented by some program, called *the oracle* and an evaluation of it on some particular Q-bit is called a *query*. The object of the game is to minimize the number of queries of the oracle one needs, in order to determine whether Z is true or false. For the DJ problem, the function f is known to map every input state in the computational basis of the M Q-bit Hilbert space to either 0 or 1, and we are told further that *either* it is constant *or* it takes the value 0 on half the states in the computational basis and 1 on the other half. In the second case, we are not told the specific states on which it is zero.

Classically, we would have to call the function $2^{M-1}+1$ times to determine whether or not it was constant. This is the only way to be sure that the function is or is not constant. If we make only k calls to the function, then we will fail to produce the right answer with probability $\leq \frac{1}{2^{k-1}}$. DJ produced a quantum algorithm, which determines the answer definitively with order M queries.

We use a Hilbert space with $M + 1$ Q-bits, the last one serving to encode the action of the function f. We can think of the function f as an operator, P_f, on the Hilbert space of M Q-bits, diagonal in the computational basis. It is a projection operator on that subspace where the function f has the value 1. Define an operator in the $M + 1$ Q-bit space by

$$U_f = P_f \otimes \sigma_1(M + 1) + (1 - P_f) \otimes 1. \qquad (20.36)$$

This operator is unitary.

Now consider an input state $|s_0\rangle = |+, +, \ldots +, -\rangle_3$, in the Hilbert space where the first M Q-bits have $\sigma_3 = +1$, and the last one has $\sigma_3(M + 1) = -1$. Act on $|s_0\rangle$ with the tensor product of all the Hadamard operators $\frac{1}{\sqrt{2}}(\sigma_1(i) + \sigma_3(i))$. When the Hadamard operator acts on the $\sigma_3 = \pm 1$ state it gives the eigenstate $\sigma_1 = \pm 1$, so the result of multiplying by all the $H(i)$ is

$$|s_1\rangle = |+, +, \ldots +, -\rangle_1. \tag{20.37}$$

Since $\sigma_1(M + 1)|s_1\rangle = -|s_1\rangle$, we have

$$U_f|s_1\rangle = (1 - 2P_f)|s_1\rangle. \tag{20.38}$$

Note that the operator $(1 - 2P)$ acts like the unit operator on the last Q-bit.

The state $|s_1\rangle$ is a superposition of all possible eigenstates of $\sigma_3(i)$, $1 \le i \le M$, with equal amplitude. If f is a constant function, P_f is just the unit operator. If we now act again with the product of all Hadamard operators, the state we get is just $-|s_0\rangle$, so if we measure the projector on $|s_0\rangle$, we will get 1. However, if P vanishes on half the joint eigenstates of all the $\sigma_3(i)$, this is no longer true. Now, we want to evaluate

$$\langle s_0| \otimes_i H(i)(1 - 2P_f)|s_1\rangle = \langle s_1|(1 - 2P_f)|s_1\rangle = 0. \tag{20.39}$$

Indeed, since $|s_1\rangle$ is an equal amplitude superposition of all the joint eigenstates, and $(1 - 2P_f) = -1$ on half of these eigenstates and $+1$ on the other half, we get "destructive interference."

In carrying out the computation, we have used $M + 1$ gates to construct the state $|s_1\rangle$, and "queried the oracle" U_f one time. This represents an exponential speedup of the computation, relative to a classical computer. Note also that despite the fact that we are using QM, the final answer is definite. The states $U_f|s_1\rangle$ and $|s_1\rangle$ coincide for the constant function and are orthogonal for any projector of rank 2^{M-1}. The Deutsch–Josza algorithm does not in itself have any practical utility, but it provided Shor [91] with vital clues for constructing an algorithm for factoring large numbers in polynomial time. Modern methods of encryption of financial data and military secrets depend on the exponential difficulty of the factorization problem for classical computers. If a practical quantum computer is ever developed, all data encrypted by these methods is vulnerable. Quantum computer scientists assure us that if this ever happens, they will be ready with quantum encryption algorithms, which are impervious to quantum computation.

20.8 CAN WE BUILD A QUANTUM COMPUTER?

A useful quantum computer must be able to store large amounts of information. We noted in the beginning that the best current supercomputers have 10^{18} Q-bits, so a quantum computer with the same storage capacity would be like a macroscopic system. In fact, all current approaches to real quantum computers involve very large devices, much larger than comparable classical computers. The point is that robust *transistors*, the bits of a classical computer, can be made from the collective coordinates of collections of $\sim 10^6$ atoms. However, precisely because these coordinates are robust, they cannot encode quantum information.

Among the variables that have been proposed for the Q-bits of a quantum computer are the spins of ions in an ion trap, electron spins, energies or positions in a quantum dot, and magnetic fluxes in superconductors. As of the date of this paragraph, the number of Q-bits that have been successfully used for a computation is less than 10^2, using any of these technologies. By comparison, a single computer chip has $> 10^9$ transistors. The problem is the fragility of quantum information. The microscopic states of real substances are typically changing very rapidly and are entangled with their environment, so that quantum information is lost. Elaborate theoretical protocols, called *Quantum Error Correcting Codes*, have shown that it is theoretically possible to make robust quantum states of a set of Q-bits, by entangling them with a much larger system, in such a way that most local changes of the state of the large system will not affect the state of the small *Code Subspace*. Practical implementations of these protocols have, so far, not succeeded in getting anywhere close to the number of Q-bits one would need for a genuine computation.

Quantum computation is thus, as of this writing, a purely theoretical subject. Some references about the state of the art of building a real quantum computer can be found here [92].

20.9 EXERCISES

20.1 Show that $S_{ij} = C_{ij}C_{ji}C_{ij}$, by using the rules for multiplying Pauli matrices. Remember that $[\sigma_a^{(i)}, \sigma_b^{(j)}] = 0$ and explain why this is so.

20.2 Show that $S_{ij} = \frac{1}{2}(1 + \sigma_a^{(i)}\sigma_a^{(j)})$. This relation was originally discovered by Pauli, studying the spin states of two electrons. S_{ij} is called the Pauli Spin Exchange operator in that context.

20.3 Given a pure state in a bipartite system, prove that the von Neumann entropies of both reduced density matrices ρ_A and ρ_B are equal.

20.4 Prove that the state

$$|\mathcal{B}_1\rangle_{AB} = \frac{1}{\sqrt{2}}(|+\rangle_A \otimes |+\rangle_B + (|-\rangle_A \otimes |-\rangle_B)$$

has norm 1 and is orthogonal to all of the other vectors in the Bell Basis. Prove that all of those vectors are orthonormal.

20.5 Prove that

$$H^{(i)} H^{(j)} C_{ij} H^{(i)} H^{(j)} = C_{ji}.$$

20.6 The problem of determining whether a large integer N is factorizable into primes is extremely difficult. Any classical algorithm for solving it requires a number of operations growing like e^{cN}. This is why the problem is used in RSA encryption. Shor's demonstration that this problem can be solved in polynomial time on a hypothetical quantum computer is one of the most important results in the field. Explaining Shor's algorithm would require us to learn too much number theory. The number theory can be found in many places, among them Mermin's book on quantum computing. The main point of it is that the problem can be reduced to finding the *periods* of a function defined on binary numbers that are N bits long. If we write the binary number as $(x_1, \ldots, x_N)$ where each x_i can be either zero or one, then a function is a rule that maps each N-vector $\mathbf{x}$ to another N-vector $\mathbf{f}(\mathbf{x})$.[1] We can add two such vectors by adding their components, using mod 2 arithmetic, and define their scalar product by $\mathbf{x} \cdot \mathbf{y} = \sum x_i y_i$, again mod 2. A period of a function $\mathbf{f}$ is a vector $\mathbf{a}$ such that $\mathbf{f}(\mathbf{x} + \mathbf{a}) = \mathbf{f}(\mathbf{x})$ for every vector $\mathbf{a}$. *Simon's Algorithm* is the task of finding what period of a function $\mathbf{f}$ is, given that one knows such a period exists. Classically, this problem is exponentially hard and is the core of the factoring problem. Simon showed that there is a quantum algorithm for solving it in a time linear in N, with probability that goes to 1 as N gets large. To start our exploration of Simon's algorithm, introduce a Hilbert space of dimension 2^{2N}. Show that a pair of binary vectors $\mathbf{x}, \mathbf{y}$ corresponds to an orthonormal basis vector $|\mathbf{x}, \mathbf{y}\rangle$ in this Hilbert space, according to the rule

$$\sigma_3(i)|\mathbf{x}, \mathbf{y}\rangle = (-1)^{x^i}; \quad 1 \le i \le N,$$

$$\sigma_3(i + N)|\mathbf{x}, \mathbf{y}\rangle = (-1)^{y^i}; \quad 1 \le i \le N.$$

We will always call unit vectors in the Hilbert space states, in order not to confuse them with vectors of binary numbers. The notation x^i denotes the i component of a vector of binary numbers. Show that the mapping

$$U(\mathbf{f})|\mathbf{x}, \mathbf{y}\rangle = |\mathbf{x}, \mathbf{f}(\mathbf{x}) + \mathbf{y}\rangle$$

is a unitary operator on the Hilbert space.

20.7 Now, start with the state $|\mathbf{0}, \mathbf{0}\rangle$ for which all of the $\sigma_3(i)$ operators have eigenvalue 1. Act on this state with the unitary

---

[1] Mathematically sophisticated readers will have noted that the space of bit strings is really a module over the ring Z_2 rather than a vector space. This does not change the argument. The fact that the sum of two period vectors is zero is an indication of that fact.

$$W = \prod_{i=1}^{N}[\frac{1}{\sqrt{2}}(\sigma_3(i) + \sigma_1(i))]U(\mathbf{f})\prod_{i=1}^{N}[\frac{1}{\sqrt{2}}(\sigma_3(i) + \sigma_1(i))].$$

To do this, show that

$$\prod_{i=1}^{N}[\frac{1}{\sqrt{2}}(\sigma_3(i) + \sigma_1(i))]|\mathbf{x}, \mathbf{y}\rangle = 2^{-N/2}\sum_{\mathbf{z}}(-1)^{\mathbf{x}\dot{\mathbf{z}}}|\mathbf{z}, \mathbf{y}\rangle,$$

where the sum is over all possible N bit strings $\mathbf{z}$. Applying W we then get

$$W|\mathbf{0}, \mathbf{0}\rangle = \prod_{i=1}^{N}[\frac{1}{\sqrt{2}}(\sigma_3(i) + \sigma_1(i))]U(\mathbf{f})\sum_{\mathbf{z}}(-1)^{\mathbf{0}\dot{\mathbf{z}}}|\mathbf{z}, \mathbf{0}\rangle$$

$$= \prod_{i=1}^{N}[\frac{1}{\sqrt{2}}(\sigma_3(i) + \sigma_1(i))]\sum_{\mathbf{z}}|\mathbf{z}, \mathbf{f}(\mathbf{z})\rangle$$

$$= \sum_{\mathbf{z}}\sum_{\mathbf{y}}(-1)^{\mathbf{z}\dot{\mathbf{y}}}|\mathbf{y}, \mathbf{f}(\mathbf{z})\rangle.$$

20.8 Given that we have prepared our system in the quantum state $W|\mathbf{0}, \mathbf{0}\rangle$, we now imagine entangling the subsystem in the first tensor factor, with macroscopic collective coordinates, so that a fixed value of the collective coordinates corresponds to a fixed value of $\mathbf{y}$. The remaining subsystem, the second tensor factor, is now in the quantum state

$$\sum_{\mathbf{z}}(-1)^{\mathbf{z}\cdot\mathbf{y}}|\mathbf{y}, \mathbf{f}(\mathbf{z})\rangle.$$

Now use the fact that $\mathbf{f}$ is periodic and that the period bit string $\mathbf{a}$ satisfies $\mathbf{a} + \mathbf{a} = \mathbf{0}$ because of mod 2 arithmetic. Show that this implies that this quantum state is zero (which means that it had zero probability in the premeasurement state) unless $\mathbf{y} \cdot \mathbf{a} = 0$. That is to say, *any* measurement in the computational basis will return a value that is orthogonal to the period vector.

20.9 Now count the number of steps in the above procedure. Argue that implementation of W takes a number of elementary operations that scales like N. It is important that we assume the function is given to us as a table of values. It takes no computational work to evaluate it. The *key* observation is that acting with N individual $\sigma_3(i) + \sigma_1(i)$ automatically gives us a sum over all the bit strings. So we get a sum over 2^N things, by doing N operations. Now imagine we have done of order kN measurements, with $k > 1$ but independent of N. In each measurement we have found a vector in the N-dimensional space of bit strings that is orthogonal to the period vector. Argue that this means it is very unlikely that we have not found the period vector.

L'Envoi: Relativistic Quantum Field Theory

This book has been about quantum mechanical models compatible with the Galilean Principle of Relativity (though we usually worked in a fixed inertial frame), and we know that this must be replaced by Einstein's Special Theory of Relativity if we want results accurate to better than v/c where v is the speed of the fastest particle participating in the process under study. c is of course the speed of light, $c = 3 \times 10^8$ m/s. In this short chapter, we want to indicate the remarkable complications that result from insisting that quantum mechanics (QM) be invariant under Lorentz transformations.

Since Special Relativity introduces a fundamental constant with dimensions of velocity, it is reasonable to define a time coordinate $X_0 = ct$ with dimensions of space, and a four vector X_μ, which indicates the position of a particle in *space-time*. The most concise statement of special relativity is that physical properties involved in traveling between two space-time points should depend only on the interval

$$I = (X_0 - Y_0)^2 - (\mathbf{X} - \mathbf{Y})^2 \equiv (X - Y)^2. \tag{21.1}$$

A fixed interval is the locus of points on a hyperboloid in space-time, and Lorentz transformations are just the hyperbolic rotations that keep the shape of hyperboloids fixed.

If $(X - Y)^2 > 0$, the hyperboloid has two sheets, with opposite signs for $X_0 - Y_0$. Such an interval is called time-like and Lorentz transformations cannot change which point is in the future. This behavior persists as $I \to 0$ through positive values. However, if I is negative then time order is not Lorentz invariant, and detectors traveling at different velocities will not agree on the causal order of X and Y, when they are space-like separated. Classically this seems OK. No system traveling at speeds below the speed of light can travel between space-like separated points.

Now let us repeat our quantum mechanical calculation of the probability amplitude for a free particle to travel between X and Y. Recall that the key to it is that the Hamiltonian is just a function of momentum. The relativistic relation between energy and momentum is

$$P_0^2 = \mathbf{P}^2 + m^2, \tag{21.2}$$

where we have set the speed of light equal to one so that energy, mass and momentum all have the same dimensions. As long as m is real, this is the equation for a time-like hyperboloid (this is the reason we called the energy P_0), so the sign of P_0 can be taken positive. Systems with energy that is not bounded from below tend to be unstable to the slightest perturbation.

We solve for the time evolution of quantum amplitudes by inserting a complete set of momentum eigenstates. The only new thing is that we have to integrate over spatial wave numbers with the Lorentz invariant volume element

$$d^4k \ \theta(k_0)\delta\left(k^2 - \frac{m^2}{\hbar^2}\right) = \frac{d^3k}{2k_0}. \tag{21.3}$$

$\frac{\hbar}{m} = \frac{\hbar}{mc} = l_c(m)$ is called the Compton wave-length of the particle. $k_0(m)$ is a frequency, given by $k_0(m) = \sqrt{\mathbf{k}^2 + l_c(m)^{-2}}$. The propagation amplitude is

$$A(x, y) = \int \frac{d^3k}{2k_0(m)(2\pi)^3} \ e^{-ik(y-x)}. \tag{21.4}$$

The frequency k_0 is positive, so this looks like a properly causal amplitude if y is in the future of x. It is also Lorentz invariant. The only problem is that it is nonzero when $y - x$ is a space-like vector.

To see this note that for a space-like vector, we can always find a Lorentz frame where the two points have the same time coordinate. In this frame,

$$A(x, y) = \int \frac{d^3k}{2k_0(m)(2\pi)^3} \ e^{-i\mathbf{k}\cdot(\mathbf{y}-\mathbf{x})}. \tag{21.5}$$

This is a Bessel function. It vanishes exponentially when the spatial separation is larger than l_c but is never exactly zero. To avoid violating causality, we have to introduce a new principle: any system that can emit a particle at a point is also capable of absorbing the particle, and vice versa. We can then write a causally sensible amplitude

$$A(x, y) = \theta(y_0 - x_0) \int \frac{d^3k}{2k_0(m)(2\pi)^3} \ e^{-ik(y-x)} + \theta(x_0 - y_0) \int \frac{d^3k}{2k_0(m)(2\pi)^3} \ e^{ik(y-x)}. \tag{21.6}$$

The step functions in time are Lorentz invariant if $x - y$ is time-like or null, and when it is space-like the two integrals are identical so the step functions sum to one. Our new principle enables us to write an amplitude that is both Lorentz invariant and causal.

Clever readers will call a halt: Wait a minute! What if the particle carried a Lorentz invariant quantum number like electric charge? Then we could tell where the particle was emitted or absorbed by examining the charge balance. This is not a bug, but a feature of our hypothesis: it is the prediction of *antiparticles*! An antiparticle has exactly the same mass but opposite values of all conserved Lorentz invariant charges, as the particle. Needless to say, this prediction has been spectacularly confirmed by experiment, and the equality of masses of particle and antiparticle (and lifetimes for decaying particles) has been verified to many significant digits.

The same sort of argument shows that any mechanism that can scatter particles can create particle antiparticle pairs. Figure 21.1 shows a particle being scattered from an external field. The amplitude is nonzero when the causal order of the points has the field operating in a region to the past of both particles. This is creation of a pair. The arrow follows the charge flow and one can see that the second particle is an antiparticle.

Finally, though we will not be able to prove it here, Lorentz invariance, QM and causality imply the spin statistics connection $(-1)^F = e^{2\pi i J_3}$. We can see a hint of this by thinking of the amplitude $A(x, y)$ on a plane of simultaneity of two space-like separated points, as the amplitude for two particles to be present at the two points. It is symmetric under interchange. We assumed that the particle state was characterized only by its momentum, which means that its spin was zero. It is straightforward, though a bit formal and tedious, to generalize this argument to spinning particles, and one finds that the spin statistics connection just falls out.

One of the most important consequences of these observations is that one *must* use quantum field theory to describe relativistic QM. From the point of view of this book, this remark is not such a big deal. We have advocated the field point of view because it explains the reason behind particle statistics (which is otherwise an additional postulate tacked on to the rules of QM), and illuminates the meaning of wave particle duality and of the fact that the single particle Schrödinger wave function obeys the same equations as a classical

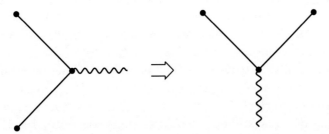

Figure 21.1 Particle anti-particle production goes into particle scattering under a Lorentz transformation, when points are at space-like separation.

field. However, if the total number of particles is conserved, we could always revert to a multiparticle Schrödinger wave function. It is occasionally useful to do this, but it is simply impossible to do it in the relativistic theory, because the number of particles can change once the energy is larger than the mass.

For example, in the nonrelativistic theory, we can make the initial probability of finding a particle at position $\mathbf{x}$ as narrow a distribution as we wish. If the particle is free, this distribution will spread, but we can start it as narrowly peaked as we like. In relativistic quantum field theory, this is not possible. We can make the wave packet narrow by allowing it to have very high probability at high momentum. But momentum carries energy. Roughly speaking, if we try to localize a particle in a region smaller than its Compton wave length, by doing position measurements, then we have a probability of inserting enough energy into that region to create particle antiparticle pairs. The whole concept of a single particle localized at a point in space-time no longer has meaning. Any sufficiently localized excitation has some probability to be a multiparticle state, and the particle number becomes uncertain.

Relativistic quantum field theory has given rise to a model of the world, the Standard Model of Particle Physics, which explains every experimental result we have obtained, sometimes with shocking precision. For example, the agreement between theory and experiment for the magnetic momentum of the electron is accurate to one part in 10^9. The only gross phenomena that are not explained by this model are the nature of the gravitational force, the nature of the dark matter we have observed to permeate the cosmos, and the reason why the universe contains more baryons than antibaryons.[1] There are plausible extensions of the standard model, still based on quantum field theory, which explain the last two items as well. Indeed, there are so many of them that we do not know which is correct, if any.

The relationship between Einstein's theory of gravity and the quantum theory has not yet been elucidated. Einstein's field equations resemble the field equations of the standard model, so one's initial temptation was to view it as just another quantum field theory. There are many reasons why this point of view is incorrect. We will mention only the one that is least well understood, but perhaps holds the clue to the quantum theory of gravity: the connection between geometry and entropy. In quantum field theory, the dimension of the Hilbert space associated with a finite region of space-time is always infinite. However, if we treat the expectation values of the energy and momentum densities of a typical state in this Hilbert space as sources for the gravitational field, then most of the states lead to a distortion of space-time geometry called a black hole. The boundary of the black hole is a null surface, a surface along which light waves can propagate (Figure 21.2). In ordinary flat space-time, if we follow such a surface into the future, then the cross sectional area grows at later and later times. For the surface of a black hole, the area remains constant; $A = 16\pi(GM)^2$, where G is Newton's gravitational constant and M the mass of the black hole.

---

[1] If this were not true, we would not be here to speculate about why it is true.

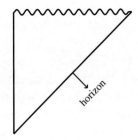

Figure 21.2 The Penrose space-time diagram of a black hole. The tangents to particle trajectories must make and an angle $< \pi/4$ with the vertical axis.

When Planck introduced his famous constant $\hbar$, he pointed out that when combined with the speed of light and Newton's gravitational constant G, it led to a complete system of natural units. In particular, there is a natural length scale, *the Planck length*, given by

$$L_P^2 = \frac{G\hbar}{c^3} \sim 10^{-66} cm.^2 \tag{21.7}$$

Bekenstein and Hawking [93] put forward persuasive arguments that a black hole is a thermal system, with an entropy $S = \frac{A}{4L_P^2}$, where A is the area defined above.

A causal diamond in space-time is defined by a trajectory that is always pointed toward the future and two points along that trajectory. The diamond is the region of space-time to which one can send a signal traveling at speed $\leq c$, and receive a reflection back, in the time interval between the two points. Each such diamond has an area associated with it: the maximal area on its null boundary. In 1995, Jacobson [94] showed that if one applies the first law of thermodynamics, $dE = TdS$, locally to every causal diamond in space-time, with the entropy expressed as $1/4$ of the area in Planck units, then one derives Einstein's equations relating the geometry of space-time to the energy and momentum densities of matter.[2]

There is one more clue to the quantum theory of gravity that can be gleaned from rather general considerations. If we consider two causal diamonds in a space-time, then it might be the case that there are no trajectories traveling at a speed $\leq c$, which connect any points in one diamond, to any point in another. Two such diamonds are said to be at space-like separation. Any operator localized in one diamond, had better commute with any operator localized in the other. If this were not the case, then an experiment done in one diamond, could affect the outcome of an experiment done in the other, even though nothing physical could connect them. The "spooky nonlocality" of QM would not be just the

---

[2] Jacobson did not phrase his argument in terms of causal diamonds, but this formulation is equivalent to his original one. It should also be noted that Jacobson's derivation does not capture the cosmological term in Einstein's equations. This should be viewed as telling us that the cosmological term is not a local energy density.

result of mistakenly interpreting probability distributions as physical fields, but would lead to measureable consequences.

String Theory [95] and the Anti-de Sitter Space Conformal Field Theory Correspondence [96], derived from String Theory, are the only mathematically well formulated models of quantum gravity, which exist at the present time. These models apply to space-times with a cosmological constant $\Lambda \leq 0$, which have asymptotic boundaries containing causal diamonds of infinite area. None of these models appears to apply to the world we inhabit. Furthermore, the operators that one is allowed to consider measureable in these models are localized on the boundaries at infinity and do not tell us directly how to interpret the measurements we actually do within finite area diamonds. Thus, the challenge of constructing models of QM, which are compatible with the cosmology we inhabit is still before us. Perhaps some of the readers of this book will be able to contribute to the task of meeting that challenge.

In the meantime, there is a wealth of QM to be learned, which applies directly to the earth we live on and the substances inhabiting it. Hopefully, this book has helped to give you a step in the right direction toward learning that material and applying it.

Interpretations of Quantum Mechanics

The interpretation of quantum mechanics (QM) championed in this book is an intrinsically statistical one. Einstein, the most famous critic of QM, was completely satisfied with such an interpretation, but as an aficionado of classical statistical mechanics, he felt that there had to be a nonstatistical "reality" underlying QM, which evolved in a deterministic manner, such that the quantum statistics was obtained by averaging over some unmeasured variables.

The inventors of the Copenhagen Interpretation disagreed with Einstein, but IMHO made several mistakes in at least their presentation of the disagreement.[1] The Copenhagen crowd insisted vehemently, and correctly, that the wave function (really the density operator derived from it) was a complete description of a quantum system—there were no hidden variables. Whatever they meant by this, it is been taken by many modern authors to mean that the wave function is a "real thing" and that we can take it to be a description of what is "actually going on" in any given run of an experiment. This is true both for authors who accept the conventional (Copenhagen plus decoherence) interpretation of QM and those who propose alternatives.

There is nothing in the actual practice of QM that requires or justifies such an interpretation. The wave function is used as a probability distribution and checks of the predictions it makes are performed by doing multiple experiments and using the frequentist rule. No one has ever proposed an experiment or a set of simultaneous experiments, which can measure the wave function without doing such repetition. Indeed, the EPR argument and those of

---

[1] I am not an historian of science, and have not studied the old literature, so I am not in a position to defend statements about who said what when. My knowledge of the positions of Heisenberg, Bohr, Born, Jordan, etc. is mostly limited to the presentations of them in textbooks and modern popular accounts.

Bell show that if one could make such a measurement, it would imply that signals could transfer information faster than the speed of light.

Let us recall Feynman's discussion of the double slit experiment in order to emphasize this fact. There are two important parts of that discussion, which we should recall. The first is that every actual experimental result consists of a click, when some particular detector at some point of the screen is struck by a particle. The second occurs somewhat later on, when a picture of an interference pattern is shown, and we a see a phenomenon we have come to associate with actual physical waves. However, there is no experiment, which actually detects that wave pattern. Rather, one collects many clicks, and uses the relative frequency of clicks at different places, via the frequentist rule, to compare the *probabilistic* predictions of QM to the statistics of many experiments. The only time we actually see physical waves corresponding to solutions of the Schrödinger equation is when we look at coherent states of bosons. We explained all of that in Chapter 5.

Thus, the first challenge one would issue to anyone who interprets the wave function as some kind of physical phenomenon, like a water wave or an electromagnetic wave, is to propose a single experiment, done one time, which measures all aspects of the wave function. No mathematical formula in QM tells us how to perform, even in principle, the measurement of the wave function at a given time. By contrast, QM does tell us how to measure, in principle, any set of commuting normal operators at a single time. In order to do so one must couple the system of interest to a large macroscopic system in such a way that distinct eigenstates of the set, get entangled with states of the macrosystem labeled by distinct values of the macrosystem's collective coordinates. Because the quantum probabilities for the coupled system now satisfy Bayes' rule with accuracy exponential in the entropy of the macrosystem, we can state that the value of the commuting set of operators at the time of the measurement becomes part of a decoherent history. As long as the macrosystem continues to exist[2] then the predictions of QM for *the future of this particular experiment* are identical to those of a theory which attributed the unpredictable behavior of the macrosystem's detectables to our failure to measure some classical hidden variables. In that second kind of theory we would discard part of the probability distribution which predicted other outcomes, using Bayes' rule to "update the probability distribution based on new data about what is really going on." We do this in QM by "collapsing the wave function" since that is a procedure consistent with linearity of the Schrödinger equation. As we have emphasized in Chapter 10, this is a convenient approximation for following the evolution of the system, conditioned on the assumption that a series of macroscopic systems carry imprints of the original experiment. Once all macroscopic records of that particular experiment have disappeared, one must, if one wishes to make correct statistical predictions about the future, go back to the wave function

---

[2] or, if it imprints its history on another macrosystem before it disappears,

that is a superposition of all possible outcomes of that experiment, because all traces of the decoherent history have disappeared.

Before summarizing some of the attempts to come up with alternative interpretations of QM, or alternatives to QM that give rise to "objective wave function collapse," it is appropriate to comment on an attitude found all too common among working physicists. That is, the belief that one can simultaneously adhere to the "Copenhagen interpretation" of QM and believe that the wave function is a physical thing, rather than a probability distribution. Presumably the reason for this is that one can find such statements in writings of some of the inventors of the Copenhagen interpretation. This is an inconsistent point of view. The only consistent way to interpret the Copenhagen view of QM is the statistical viewpoint advocated in this book: QM is a new kind of probability theory in which not all variables can have definite values at the same time. If the equations of motion relate such mutually indefinite variables, then there are in general no probabilities for histories. Probabilities for histories arise as approximate descriptions of macrosystems.

A.1 MODAL INTERPRETATION OF QUANTUM MECHANICS

Let us begin our discussion of realist interpretations of QM with a class of interpretations that appear to do the least damage, both to the formalism, and to our desire to have the history of the universe be describable by quantities that have definite values at all times. This vast list of ideas, comes under the name of *modal* interpretations of QM. We will not attempt to study all of those proposals.

A simple modal interpretation is based on the assumption that there is some principle that restricts the initial state of the universe to be some element of a particular privileged orthonormal basis $|r_i\rangle$ (where the letter r for *reality* is chosen intentionally). If R_i are the projection operators on these states, then the Heisenberg operators $e^{i\frac{Ht}{\hbar}} R_i e^{-i\frac{Ht}{\hbar}}$ have a time dependent joint probability distribution, which is determined by the initial probability that the universe was in one of the states $|r_i\rangle$. In a modal interpretation these are considered probabilities for histories of variables that are diagonal in the $|r_i\rangle$ basis. The probability of being in the state $|i\rangle$ at time t, given that one was in state j at time t_0 is

$$P(i,j;t-t_0) = |\langle r_i|e^{-i\frac{H}{\hbar}(t-t_0)}|r_j\rangle|^2 = |U(r_i,r_j,t-t_0)|^2. \qquad (A.1)$$

Note that the quantum amplitudes U for these transitions have phases $e^{i\theta(i,j,t-t_0)}$, which are not determined by these probabilities.

This theory of probabilities for histories of the "real" quantities R_i is not in general deterministic. That is, even if we begin in a pure state, with only one of the R_j nonzero, the probability distribution $P(i,j;t-t_0)$ will be nonvanishing for more than one value of i. Proponents of a realist interpretation of QM are not bothered by this, and there is no reason why they should be. The results of experiments on microscopic systems show without a doubt that such systems display randomness. One might want to attribute that randomness to unmeasured deterministic "hidden variables," but there is nothing apart from the philosophical stance of 18th century mathematical physicists that makes determinism a sacred principle.

Similarly, the insistence that the initial conditions of the universe be restricted to elements of a particular orthonormal basis, and not superpositions of them, is not a source for worry. If we are talking about the dynamics of a small subsystem of the universe, experiment shows us that we can prepare it in any superposition of its quantum states. The initial conditions for the entire universe are not under our control, and we do not yet have a complete theory of the earliest moments of cosmic history. It might be that such a theory will explain the restriction on initial conditions

What *is* disturbing about an interpretation that takes the probabilities for histories of the specified operators R_i as the underlying reality of quantum theory is that these probabilities do not determine all other predictions of the quantum theory, because of the undetermined phases. Furthermore, the probabilities for histories do not obey an intuitively obvious rule, which one might have thought was required of any realist description. This is the rule

$$P(i, j; t - t_0) = \sum_{histories} P(history), \qquad (A.2)$$

where we sum over all histories consistent with the initial and final conditions

The idea behind a realist description is that what actually happens as the universe evolves is some particular history of the real variables. That is, they have values, even if nothing intervenes to measure those values. Statistics should enter into the theory only as a restriction on our ability to predict what those values will be, given their initial values. In our modal model, we have a probability distribution for such histories. In a realist interpretation, one would think that one should be able to say that the probability that the universe was in state $|r_i\rangle$ at time t was the sum of the probabilities of all possible histories that had this property. After all, as realists we must believe that what is actually going on in the world we observe is some particular history of the real variables r_i. If we had not motivated the choice of basis in our model by an initial condition at the beginning of the universe, we could derive this sum rule from the frequentist definition of probability, plus the assumption that every re-run of the world was just a particular history of the r_i.

In standard quantum theory, one can compute the probability amplitude for some history in which at times t_k the system was in the state $|r_k\rangle$, given that it was in state $|r_j\rangle$ at $t = t_0$. It is

$$A(r_k, t_k; r_j, t_0) = U(r_N, r_{N-1}; t_N - t_{N-1}) \ldots U(r_1, r_j; t_1 - t_0). \qquad (A.3)$$

In the standard interpretation of QM, we consider the absolute square of this number to be the probability that the system was in state $|r_k\rangle$ at time t_k, *i.e.* the probability of a given history.

Anyone who's gotten to this point in the book, will know that in general, the probabilities for histories defined in this way will *not* (in general) satisfy the realist sum rule. Rather, by the linearity of the Schrödinger equation, we must sum up the *amplitudes* for the different histories and take the absolute square of the resulting sum. We have gone to great lengths to argue that if we take variables which are collective coordinates of macroscopic objects, then the probabilities for their histories *do* satisfy the sum rule with accuracy exponential in the size of the object. Part of the demonstration involved acknowledging that these collective variables gave only a coarse grained description of the objects. They are not candidates for the variables r_k, which are supposed to be a complete basis for the Hilbert space at a microscopic level.

It appears that there is only one way out of the paradox that our set of probabilities for histories do not satisfy the sum rule. Namely, we should restrict attention to the special class of quantum systems where this problem does not arise. For a system with a finite number of states, this is discrete time dynamics with unitary transformations restricted to the permutation subgroup S_M which transforms every element in the $|r_k\rangle$ basis into $|r_{P(k)}\rangle$, where the overused letter P is here being made to stand for permutation. For such systems, the values of the operators R_i are definite for all times and the dynamics is deterministic. The system is a *cellular automaton*.

't Hooft has made the remarkable conjecture [97] that the quantum world we observe might be described by a model like this. 't Hooft's formalism accepts the general quantum framework in which there are mutually incompatible observables. His conjecture is that the variables we measure, which show all of the randomness and uncertainty associated with QM, are simply operators in the Hilbert space of the cellular automaton, which do not commute with the fundamental operators R_i. The success of the conventional QM description of matter does not contradict this conjecture if one can show that the action of conventional Hamiltonians can be realized as an approximation to an evolution under the symmetric group, perhaps in a system larger than that we associate with conventional theories. For example, one might attempt to show that some kind of cutoff on quantum field theory could be embedded in a model with S_N dynamics. 't Hooft has shown [98] that this is indeed possible for fields satisfying the equation

$$(\partial_t \pm \partial_x)\phi = 0, \tag{A.4}$$

and has tried to connect this result to superstring theory [95], which is a proposed unified theory of particle interactions and gravitation. Obviously, a thorough investigation of these claims is far beyond the scope of the present book.

't Hooft's approach raises fascinating mathematical questions about which quantum systems can be well approximated by S_N dynamics (perhaps of a larger system). Even if these questions can be answered in a way that is favorable to 't Hooft's program, it is clear that the proposal does not resolve a number of issues that one might have expected a realist interpretation of QM to resolve. First of all, like other modal interpretations of QM, the dynamics of the real variables R_i does not determine the evolution of all quantum operators in the system. If V_D is a unitary operator diagonal in the $|r_i\rangle$ basis, then we can replace the permutation evolution operator U_P by

$$U_P \rightarrow\rightarrow V_D U_P. \tag{A.5}$$

The Heisenberg operators

$$R_i(t) = (V_D U_P)^{\dagger \frac{t}{\tau}} R_i (V_D U_P)^{\frac{t}{\tau}}, \tag{A.6}$$

where τ is the underlying discrete time interval, and t is an integer multiple of it, are equal to

$$R_i(t) = U_P^{\dagger \frac{t}{\tau}} R_i U_P^{\frac{t}{\tau}}. \tag{A.7}$$

On the other hand, V_D does effect the time development of operators which are not diagonal in the $|r_i\rangle$ basis. This ambiguity is essentially the same as the phase ambiguity that we pointed out above.

The consequence of this ambiguity, which is quite general to modal interpretations, is that the histories of the "real" variables, do not determine the dynamics of the variables we actually measure. This observation lends a hollow ring to the claim that the "real" variables are the underlying reality behind QM randomness. Recall our discussion of the Koopman model of classical mechanics as QM in Chapter 2. There we pointed out that the quantum rules reduced to those of classical statistical mechanics for systems with a "Liouville" Hamiltonian, *if we restrict attention to operators that commute with the commuting phase space variables p, q*. A fancy mathematical way of imposing this restriction is precisely to insist that the transformations

$$\psi(p, q) \rightarrow e^{i\theta(p,q)} \psi(p, q), \tag{A.8}$$

which commute with the Liouville Hamiltonian, are *gauge transformations (i.e. redundancies)*, so that only operators that commute with p and q are physical. A corresponding gauge principle for 't Hooft's model of QM would defeat its purpose. We need the noncommuting variables to reproduce the experimental successes of the standard interpretation of QM.

Most modal interpretations of QM do not accept the quantum predictions for the probabilities of histories of the r_i as part of their definition of reality. Rather, they invent a stochastic dynamics for the r_i, which is designed to reproduce only the probability distributions $\text{Tr}\,(\rho(t) R_i)$, where $\rho(t)$ is the quantum density matrix. Modal interpreters of QM accept that the usual quantum predictions for probabilities of histories *do* represent what will happen to macroscopic systems whose collective coordinates become completely entangled with (some subset of) the $|r_i\rangle$[3] and attribute the disagreement with the "more fundamental"

---

[3] This is usually expressed in phrases like "this is what will happen as the result of measurements."

stochastic probabilities to the "complications of the measurement process." People who are comfortable with a theory in which one is unable to probe the most fundamental aspects of a theory with experiments, will be satisfied with such an approach. However, there is no clear agreement on a "correct" modal interpretation of QM, and widespread agreement among advocates that the last word has not yet been said on this subject.

For the present author, the most satisfactory version of modal QM is the one advocated by 't Hooft, because it retains the measurable probabilities for histories of the r_i as the fundamental definition of reality, and insists on dynamics which preserves the probability sum rule for real histories. The 't Hooft interpretation should be supplemented with a convincing principle that determines the unitary evolution operator U_D, and should realize the goal of 't Hooft's program, which is to show that a theory of the world we observe can be derived from deterministic dynamics.

Even if those goals are realized, there are two further challenges, of perhaps a more philosophical nature, that must be faced by any interpretation of the modal sort. The notion of "physical reality" that one tries to recover in modal interpretations of QM is based on our intuitive understanding of the behavior of the collective coordinates of macroscopic objects. It seems a bit at odds with the basic philosophical posture of modal interpretations, to derive that behavior as an approximation to the nonintuitive quantum behavior of operators that do not commute with the R_i. At the very least, one would want to show that the commutators $[R_i(t), C_K(t)]$ were very small for all i and all collective coordinates of all macroscopic objects. Further one would want to show that the experimental absence of interference between different histories of collective coordinates followed not from the kinds of considerations we presented in Chapter 10, but from the fact that a history of the macroscopic world was in one to one correspondence with a history of the $r_i(t)$.

Remarkably, the phenomenon of unhappening, if it could ever be observed experimentally, would provide a sharp test of any interpretation of QM, which claimed that real histories of collective coordinates were in one to one correspondence with real histories of some complete set of microscopic commuting quantities. *In such an interpretation of QM, unhappening should not occur.* Perhaps a clever experimentalist can come up with a testable version of the exploding laboratory thought experiment, and see whether unhappening really occurs in nature, as the standard formulation of QM leads one to expect. If such an experiment were to succeed, one would be, at the very least, forced to conclude that the reality we perceive in the behavior of macroscopic objects cannot be connected directly to the "real" variables in any modal interpretation of QM.

The final challenge to any modal interpretation of QM comes from one of the founders of the subject, J.S. Bell, in his famous generalization of the arguments of Einstein–Podolsky–Rosen and Bohm. Bell emphasized that the disturbing question posed by the success of QM, was whether or not one could attribute values to *all of the* **potentially measurable** *quantities in a quantum system* in histories of the system in which some of them were not measured.

He showed that for simple and experimentally accessible systems, such an assumption contradicted the rules of QM, unless the hidden real variables r_i could influence the behavior of locally measurable quantities, which were at space-like separation. We have reviewed this argument in Chapter 10. While it is clear that the variables $r_i(t)$ of modal interpretations of QM are not locally related to any measurable properties of particles or fields, there have been no demonstrations of how such interpretations resolve Bell's conundrum. Do they provide histories in which *all measurable quantities* have definite values, and if so, how? Do they allow for nonlocal (*i.e.* faster than light) signaling? One doubts that a "realist" interpretation of QM, which did not provide definitive answers to these questions would have satisfied Bell or Einstein.

A.2 BOHMIAN MECHANICS

The earliest attempt to come up with a realist interpretation of QM is the Pilot Wave theory of De Broglie [99] and Bohm [100]. Its modern incarnation is called *Bohmian Mechanics* [101]. Its most basic tenet is that the wave function $\psi(Q^i, t)$ is a "real" field in the multidimensional configuration space of the system. Discrete quantum numbers, like spin, are treated by introducing multiplets of fields,[4] transforming in various representations of symmetry groups. One is supposed to consider a fixed history of the world to be determined by a fixed configuration of this field, developing through time. The field satisfies the Schrödinger equation. For simplicity, we will ignore spin and other indices in the following

Given such a complex field $\psi = Ae^{i\frac{S}{\hbar}}$, with $A \geq 0$, one can always introduce a set of trajectories in the configuration space, via the equation

$$\dot{Q}^i(t) = \nabla^i S, \tag{A.9}$$

although this equation is problematic in the vicinity of points where $\psi = 0$. The $Q^i(t)$ are called the trajectories of *Bohmian particles*. It would be a mistake to identify these with actual localized particles in space. Indeed, in the most straightforward extension of this formalism to quantum field theory [100], the Bohmian particle is really a *Bohmian field configuration*. Despite this, much of the literature on Bohmian mechanics is full of phrases conflating the notion of Bohmian particle with actual particles. Note in particular that for a given history of the wave function, at any time, there are Bohmian particles going through all points of space, where $\psi \neq 0$. Points with $\psi = 0$ are problematic because the phase S is not defined there. Since the standard theory of the Schrödinger equation allows us to contemplate initial

---

[4] When multiparticle systems are considered, the tensor product spin indices are appended in a manner which appears more *ad hoc*.

conditions which are smooth functions of compact support[5] one must make some sort of rule, which prevents Bohmian particles from being in regions with $\psi = 0$.

One way to deal with trajectories that run through all points where $\psi \neq 0$ is to assume that the initial conditions for the variables $Q^i(t)$ are uncertain, with a probability distribution $P_0(Q(0))$ that vanishes wherever $\rho \equiv \psi^*\psi(Q(0))$ does. The two quantities $P(Q,t) \equiv P_0(Q(t))J(Q(t),Q(0))$ and ρ satisfy the same continuity equation:

$$\partial_t P = -\nabla^i[P\nabla_i S]. \tag{A.10}$$

$$\partial_t \rho = -\nabla^i[\rho\nabla_i S]. \tag{A.11}$$

Here $J(Q(t),Q(0))$ is the Jacobian of the mapping from the initial values $Q(0)$ to the values of $Q(t)$ at time t. For ρ, this equation follows from the Schrödinger equation, while for P it follows from simply solving the differential equation for $Q(t)$ and taking account of the fact that P is a *probability density on configuration space* and that the mapping $Q(0) \to Q(t)$ does not usually preserve the Euclidean volume element $d^N Q$.

If we choose an initial probability distribution, which is nonvanishing at points where the $\psi = 0$, then the equations of Bohmian mechanics are problematic; the initial velocities of the Bohmian particles are ill defined. The only apparent way to ensure that this never happens, is to insist that the probability distribution is a function $P(Q) = f(\rho(Q))$, where f is a function which vanishes sufficiently rapidly at $\rho = 0$. Since P and ρ satisfy the same linear partial differential equation, this relation can be preserved in time only if, in fact, $P = c\rho$ where c is a constant. Since the Schrödinger equation is linear, we can always choose a solution that satisfies the probabilistic normalization condition $\int d^N Q \rho = 1$, and conclude that for this solution, $c = 1$.

To complete a closed set of equations, we should also record the equation for S, which follows from the Schrödinger equation:

$$\hbar\partial_t S = \frac{\hbar^2}{2m}[(\nabla^i S)^2 - \frac{\nabla^i \nabla_i \rho}{\rho}] + V(x). \tag{A.12}$$

If we write the dimensionless quantity $S = \frac{I}{\hbar}$ then $\hbar$ appears only in the term involving ρ. If we neglect this term, we find that I satisfies the classical Hamilton–Jacobi equation, and this sheds some light on the "reality" of the wave function. In classical Hamilton–Jacobi theory, we do not consider the classical action I to be a physical field on configuration space. The classical equations of motion have a plethora of solutions, which depend on initial conditions for the canonical variables Q^i and P_i. The solution of the H–J equation is a device for discussing all of these solutions at once. Any given situation, in which something definitely happens is

---

[5] Indeed, smoothness is too strong a constraint. We need only require that $\frac{\nabla^2 \psi}{\psi}$ have the same degree of differentiability as the potential.

described by a single trajectory $Q^i(t)$ and only involves the variation of the H–J function S, along that trajectory. One can use the full H–J solution, over all of configuration space, in a probabilistic situation, where we specify only a distribution for the initial conditions $Q^i(0)$. Then the combination of the H–J equation and the continuity equation for probability tells us how to solve completely for the evolution of the probability distribution.

In the quantum version of these equations, probability distributions which are sharp in the $Q^i(0)$ give singular contributions to the equations of motion for S. In the linear Schrödinger equation, a singular initial condition $\psi \propto \delta^N(Q^i(0) - Q_0^i)$ generates some linear combination of the retarded or advanced Green functions, which have nonnormalizable but spread out probability distributions. Thus, the idea of a nonprobabilistic, purely mechanical interpretation of "Bohmian mechanics," does not make sense.

The fact that the "quantum Hamilton–Jacobi equation" for S depends nonlinearly on the probability distribution $P = \rho$, implies that there is no classical stochastic interpretation of these equations. By a classical stochastic system we mean equations of the form

$$\dot{Q}^i(t) = f^i(t; Q), \tag{A.13}$$

where the initial conditions $Q^i(0)$ and perhaps some coefficients in the forcing functions f^i are random variables chosen from some probability distribution. If the correlations in time of the random elements in the forcing function are delta functions, then one obtains a local linear Fokker–Planck equation for the probability distribution $P(Q, t)$.[6] Probabilities obeying such an equation always satisfy Bayes' conditional probability rule, which means that one can write the probabilities as sums over probabilities of histories. In the nomenclature used by many philosophers of QM, and popularized by J.S. Bell, equations like this obey the principle of "counterfactual definiteness." The probabilities defined by Bohmian mechanics do not satisfy this rule, which is not surprising, because they are equal to the probabilities computed in QM. Mathematically, within the formalism of Bohmian mechanics, the failure of Bayes' rule arises from the terms in the evolution equation for S, which are nonlinear in ρ.

If instead of adopting a nonclassical probability interpretation of the equations, we insist on a stance that the wave function is a "real" field on configuration space, much as the classical electric and magnetic fields are real fields in space, then the interpretation of the Bohmian trajectories must be completely different. They are the analog of the Madelung trajectories of hydrodynamics [102]. If a particle is dropped into a fluid obeying *e.g.* the Navier–Stokes equation, it will "surf the wave" along one of these trajectories, with the initial condition determined by the time and place the particle is dropped in. In this case ρ is interpreted as the fluid density and has nothing to do with the particle. Although Bohm–Madelung trajectories exist throughout the configuration space, only the one picked out by some *deus ex machina* (DEM) has a particle on it. In this interpretation of the equations,

---

[6] If the time correlations are not delta functions, the resulting Fokker–Planck equation is nonlocal in time.

there is nothing probabilistic, and no real connection between the wave function's history, and that of the particle. One cannot, in such an interpretation, associate the motion of the Madelung surfer and the measured positions of particles. The former is a deterministic single trajectory in configuration space, determined by a choice of initial condition, while the latter are random variables, about which the theory only makes statistical predictions.

One can try to make up an *origins story* to restore the agreement between this interpretation of the Bohmian equations and experimental facts. At some initial time, the DEM drops particles at the initial points of Bohmian trajectories, at random. We have seen that the only consistent way of doing this is to force the initial probability distribution chosen by the DEM to be the absolute square of the wave function. Thus, we are led back to the intrinsically probabilistic interpretation of the equations, in which the *soi disant* probabilities for histories of the Bohmian particles do not satisfy Bayes rule.

Bohmian mechanics differs from generic modal interpretations of QM, in that it makes full use of the quantum phase information in defining the histories of its modal variables, so that those histories determine expectation values of *all* operators. However, the price to be paid is that it does not define probabilities for histories which satisfy the realist sum rule. One cannot imagine that every actual run of an experiment corresponds to a particular Bohmian history. More general modal interpretations can try to invent a stochastic dynamics, satisfying Bayes' sum rule, which reproduces the quantum probabilities for their particular choice of a commuting set of variables. They need only explain why that dynamics can never be probed by experiment.

As for Bell's original complaints about the impossibility of assigning historical values to unmeasured spin components of actual particles, it does not appear that Bohmian mechanics says any more about that problem, than a general modal interpretation. Its treatment of spin is entirely formal, and it does not assign histories to the values of all components of the spin.

A.3 THE RELATIVE/STATE—MANY WORLDS INTERPRETATION

Probably the most popular of the interpretations of QM that regards the wave function as an actual physical thing, is the Everett relative state interpretation, which morphed into the DeWitt–Graham *Many Worlds* interpretation [103]. Everett insisted that collapse of the wave function was not real, but just signified a correlation between different subsystems of a global system representing the entire universe. Subsystems are defined by a tensor factorization of the Hilbert space. The Hilbert space state is considered real, and satisfies the deterministic Schrödinger equation for the whole universe. While Everett was aware, at some level, of the phenomenon of decoherence, he regarded it as a mere tool for deciding which tensor factors of the Hilbert space were good models of classical observers. He always insisted that decoherence was not an essential part of his interpretation, and that at some level, the quantum coherence of superpositions of macroscopically different states was testable. This agrees with our discussion of the phenomenon of unhappening. Indeed, much of Everett's discussion sounds very much like the presentation of QM in this book.

The place where Everett parts ways with statistical interpretations of QM is his insistence that the fundamental equations stand on their own, without any probabilistic interpretation. Everett claims that one can somehow derive the Born rule from the Schrödinger equation and the postulate that operators have definite values in their eigenstates. Many authors, including this one, find his arguments for this unconvincing. The basic equation of the relative state interpretation is the von Neumann decomposition of the "state of the universe" in the tensor product basis of "system plus apparatus"

$$|\Psi\rangle = \sum a_{nM} |\psi_n\rangle \otimes |\psi_M\rangle. \tag{A.14}$$

Each term in this decomposition of the normalized state $|\Psi\rangle$ is a relative state of the apparatus relative to the system.

Everett asserts that the natural measure on the set of relative states is simply the squared norm $|a_{nM}|^2$, essentially putting in the Born rule by hand. There have been a variety of attempts to really derive the Born rule, but the fact remains that if one regards the wave function as real, then its coefficients in some tensor product basis are equally real. They should be measurable by some single experiment at a fixed time. That is, they are not probabilities. The suggestion that the probability rule might be derivable from the other postulates of QM is a fascinating one.

In a series of recent papers, Zurek [28] has taken an interesting intermediate stance on this problem. What follows is my own translation of his arguments into the language of this book. There is no guarantee whatsoever that it is equivalent to Zurek's own point of view. In Chapter 2, we saw that classical mechanics is a particular kind of QM. In the context of Hilbert spaces with a finite number of states it is a restriction to unitary transformations that are in the permutation subgroup, which permutes the elements of a particular basis. Zurek proposes to allow the generalization to arbitrary unitary transformations, without mentioning the word probability. The probability interpretation of the wave function is to be derived from the more primitive notion of the decomposition of the system into two subsystems. Zurek takes as a postulate that this decomposition corresponds to a tensor product decomposition of the Hilbert space. I would characterize this postulate as IOTTMCO.[7] Let $\mathcal{H} = \mathcal{H}_S \otimes \mathcal{H}_E$. Given bases in the two subfactors, we have the Schmidt decomposition

$$|\psi\rangle = \sum_{iA} c_{iA} |i\rangle \otimes |A\rangle. \tag{A.15}$$

We will now depart from Zurek's presentation by assuming that the system E, for *environment* is vastly larger and that its dynamics is described by a Hamiltonian with the properties

---

[7] Intuitively obvious to the most casual observer.

of the local lattice Hamiltonians we studied in Chapter 10. In particular, the system E possesses collective coordinates, with the properties outlined in Chapter 10. It is important to emphasize that, although we used the language of probability to describe those properties, they are simply statements about the commutators of operators, and the behavior of "coarse grained histories," defined by the mathematical formula

$$A(C_a(t_L)) \equiv \langle B|P[C_a(t_L)]e^{-iH(t_L-t_{L-1})}\ldots e^{-iHt_1}|A\rangle. \tag{A.16}$$

$$Prob\ (C_a(t_L)) = |A(C_a(t_L))|^2. \tag{A.17}$$

This formula is written in the Hilbert space $\mathcal{H}_E$, and we have set $\hbar = 1$. $P[C_a(t_L)]$ is the projection operator on a range of eigenstates of the collective coordinates with the eigenvalues indicated in square brackets. According to the definition of collective coordinate, we take the range of eigenvalues around the central value to scale like $\frac{1}{\sqrt{N}}$, and the dimension of the projector is of order e^{cN} with c of order one. N is the logarithm of the dimension of $\mathcal{H}_E$. The arguments of Chapter 10 show that the numbers $Prob\ (C_a(t_L)$ have the mathematical properties we would expect of probabilities for histories, including, up to exponentially small corrections the classical probability sum rule, the property which allows us to define Bayes' conditional probability rule. Given the dynamics of the system, these probabilities are completely determined by the initial eigenvalue of the collective coordinates. We postulate in addition that if the system is in an eigenstate of an operator, then the theory predicts that the physical quantity represented by that operator actually has the value given by the mathematical eigenvalue.

Now we take the initial state of the composite system to be $|\psi\rangle$ and choose the basis in $\mathcal{H}_E$ to be eigenstates, with different eigenvalue, of the collective coordinates of B. The postulated size discrepancy between the two Hilbert spaces assures us that we can do this, actually in a huge number of different ways. Once we have chosen the basis in this way we can write $c_{iA} = c_i\delta_{iA}$. We now assume that the initial state of the world was a simple tensor product

$$|\psi_{init}\rangle = \sum_i c_i|i\rangle \otimes |Ready\rangle, \tag{A.18}$$

and that some interaction between the small and large systems allowed this to evolve into the entangled state $|\psi\rangle$. As first noted by von Neumann, this is completely compatible with unitary evolution.

The squares of amplitudes of decoherent histories have all the properties expected in a classical theory of probabilities for those histories. The macrosystem has been entangled with the microsystem in such a way that the value of the initial collective coordinates is correlated with the values of the projection operators $|i\rangle\langle i|$ on the microstate. Observation shows that macroscopic collective coordinates perform individual histories, not weird superpositions of histories, so we have no choice but to interpret the lack of certainty of which history will actually occur to anything else but a probability theory. The reduced density matrix for the values of the collective coordinates is a diagonal matrix with entries $|c_i|^2$ and so these must be interpreted as the initial probabilities predicted by the theory, for which history will occur. Since they are independent of the particular macrosystem, which has been entangled with the system S and were coefficients in the pre-entanglement state of S, they must be identified as the probabilities for the microsystem to be in the state $|i\rangle$ given that the value of the projection operator $|\psi_{init}\rangle\langle\psi_{init}|$ is 1.

So, we have started from the assumption that the state of a system (physical interpretation not assumed known) is a vector in Hilbert space, which evolves under unitary time evolution, and that when the state is an eigenstate of an operator, the physical quantity represented by the operator has value equal to the operator's eigenvalue. We have added the assumption that composite systems are described by tensor products, and invoked mathematical properties of collective coordinates of large systems with local Hamiltonians. Entanglement between a macroscopic and microscopic system then leads to a state which is a superposition of different initial values of macroscopic collective coordinates. The observational fact that the world seems to involve only single histories of collective coordinates, combined with the fact that histories of collective coordinates decohere, leads us to the Born rule for ALL systems, whether microscopic or macroscopic. The quantum state must then be interpreted as a probability distribution, not a real physical object. analogous to a classical field.

Another attempt to "derive" the probability interpretation was given in a famous paper by Hartle [29], which argued that, starting only from the assumption that the value of an operator A in the state $|a_i\rangle$ was a_i, then if one studied the ensemble of N identical copies of the system, each prepared in the state $|\psi\rangle = \sum c_i|a_i\rangle$, then one could define a *frequency operator*, which calculated how many times one would obtain the result a_1 in simultaneous measurements of A in each of the systems. The frequency operator is

$$f(a_1) = \sum_{a_i^1...a_i^N} [N^{-1}\sum_{k=1}^{N}\delta(a_1,a_i^k)]|a_i^1,\ldots a_i^k\rangle\langle a_i^1,\ldots a_i^k|. \tag{A.19}$$

It is obvious that if the ensemble is in a state where A has definite values in each copy of the system, this operator counts the frequencies of getting any particular result.

Hartle shows that in the limit $N \to \infty$, we have

$$f(a_1)[|\psi\rangle \otimes \ldots \otimes |\psi\rangle] = |c_1|^2 f(a_1)[|\psi\rangle \otimes \ldots \otimes |\psi\rangle]. \tag{A.20}$$

Thus, in the limit of an infinite ensemble of identically prepared systems, the frequency operator, which gives us the fraction of systems in the state $|a_1\rangle$ when it acts on eigenstates of $A\otimes\ldots\otimes A$, has vanishing uncertainty on any identical tensor product state of the ensemble and its value is the Born rule probability $|c_i|^2$. This is certainly an interesting result, though many have questioned whether it is in any sense a derivation of Born's rule.

The Many Worlds Interpretation, is a philosophical expansion of Everett's point of view, even though the two are often conflated. It asserts that, at least when the subsystem in the above decomposition, whose states are labeled by capital letters, represents a collective coordinate of a macroscopic object, then the decomposition reflects a physical branching of the world into noncommunicating branches.

The quantum state, according to this point of view, describes not just the universe we observe, but also an infinite number of alternate universes, all of which are real. The whole ensemble, which one might call a multiverse (though that term is more often used in a different context), is a system in which every possibility encoded in a superposition involving macrosystems, is realized. This version of the interpretation has a much harder time coping with the phenomenon of unhappening, and an experimental demonstration of unhappening would rule it out.

The worlds of the MWI are defined by different decoherent histories, which branch from each other every time a quantum system is entangled with the collective coordinates of a macroscopic system. Most macrosystems do not leave permanent traces on the world, and one cannot even decide whether a particular system will leave a permanent trace without understanding the ultimate fate of the universe. Consider again our laboratory in intergalactic space, in which we have set up the "Schrödinger's bomb" experiment, but with a variable time delay between the two clock settings determined by the outcome of the measurement of some quantum spin component. I would defy any adherent of MWI, who had not yet heard about this particular thought experiment, to come up with a description different than "The spin measurement creates a branch into two alternate worlds, which exist objectively, and in each of which the time between the measurement and the explosion is different." Yet we have seen that the standard interpretation of QM implies that a detector that remains out of causal contact with the laboratory until after the second possible explosion time, can see collisions between particles in the debris "from the first explosion" and photons "emitted in the second." It is hard to see how to make this compatible with the idea that the wave function is an element of reality, which is telling us about how a single run of the experiment behaves. Unhappening is, to use a particularly awkward phrase, the *undecoherence* of two temporarily decoherent histories. If we use decoherence as a definition of the coming into being of a new branch of the multiverse, then unhappening signals "the annihilation of two worlds, back to a coherent wave function with no classical interpretation." This has been illustrated in a somewhat obscure movie [34].

Both versions of Everett's interpretation of QM give no explanation of why one cannot describe an experimental procedure, even in principle, which can measure all aspects of the wave function at a single time. This is probably due to the fact that Everett takes as a postulate that the quantities in the mathematical formalism of QM, which are identified with detectable quantities measurable in experiments, are precisely the Hermitian operators on the Hilbert space. This postulate implies, because not all operators commute, and because the values of a complete set of commuting operators do not determine the wave function, that it is impossible to determine the quantum state with experiments done at a single time. This property, while acceptable and expected for a probability distribution, is unusual for a normal physical quantity.

A.4 OBJECTIVE COLLAPSE

In the statistical interpretation of QM advocated in this book, wave function collapse is an application of Bayes' conditional probability rule to situations involving the histories of macroscopic collective coordinates. If a wave function is a superposition of states corresponding to two different values of the collective coordinates of a macrosystem, then the interference terms in the computation of probability from this wave function are exponentially small in the size of the macrosystem. The quantum theory then predicts a classical trajectory for the collective coordinates. There are fluctuation corrections to these predictions, which are inverse power laws in the macrosystem size, and neglecting the exponentially small interference terms we can attribute these corrections to uncertainties in initial measurements or interaction with classical hidden variables. Thus we can pretend that the theory makes predictions about "what will happen in the real world." The concept of something definitely happening, in the sense of having a definite history, is an emergent one, applicable only to collective coordinates. Since Bayes' rule is satisfied with exquisite accuracy, we can use it to throw away the piece of the wave function that does not correspond to a particular macro-observation, renormalize the remainder, and allow it to propagate via the Schrödinger equation.

Objective collapse theories attempt instead to alter the Schrödinger equation so that, even at the microscopic level, one can have evolution of pure states into mixed states, so that the density matrix $\rho_{pure} = |\psi\rangle\langle\psi|$, where $|\psi\rangle = \alpha_1|\psi_1\rangle + \alpha_2|\psi_2\rangle$, can evolve into $\rho_{mixed} = |\alpha_1|^2|\psi_1\rangle\langle\psi_1| + |\alpha_2|^2|\psi_2\rangle\langle\psi_2|$. The mixed density matrix is, in a naive reading of the measurement process, "what is really observed in experiments." In the statistical interpretation, ρ_{mixed} *is* the reduced density for the microsystem after it has become entangled with a macrosystem in such a way that the microstates are correlated with different states of macroscopic collective coordinates.

In objective collapse theories, the evolution from pure to mixed states is supposed to occur as part of the microscopic dynamics. No one has explained why this makes these theories any more "objective" than the statistical interpretation. Both of these approaches admit that Einstein was wrong about God not playing dice: probability is, in either case, a fundamental feature of the equations describing the world, rather than a consequence of our lack of information about parts of the dynamical system. Both describe the apparent collapse of the wave function as an application of Bayes' rule to a mixed density matrix for quantum states of microsystems.

The origin of the unease that objective collapse theorists have with the statistical interpretation stems from the frequent use of the word "measurement" by the founders of the Copenhagen interpretation. This word seems to imply human agency, and thus would be "subjective." On the other hand, anyone who has understood the import of the theory of decoherence, whether from the presentation of it in this book, or the many more lucid presentations in the literature, realizes that one can always substitute the phrase "entanglement with the collective coordinates of a macroscopic object" for the word measurement, in the discussion of the decoherence of the density matrix of a microsystem.

At any rate, the best thing about objective collapse theories is that they are subject to experimental test.[8] Indeed, if we could actually explore the phenomenon of "unhappening" experimentally, it would give us an immediate test of objective collapse theories. In our thought experiment with an exploding laboratory, described in Chapter 10, the statistical interpretation of QM implies the existence of a proton-photon scattering event, coming from a superposition of histories where the laboratory exploded at different times. In an objective collapse theory the actual microscopic density matrix, no longer has matrix elements connecting the two histories.

More feasible experiments, which have actually been done, already put very strong constraints on objective collapse theories. To understand them, we have to decide what mathematical form to allow for such theories. Since we want to use Bayes' rule for the density matrix ρ_{mixed}, it would not make much sense to assume a nonlinear time evolution equation for the density matrix. The time evolution equation should preserve the nonnegativity of the density matrix's eigenvalues, and the fact that they sum to 1. It can be shown that the most general evolution equation consistent with these rules is the Lindblad equation [104]

$$\dot{\rho} = -\frac{i}{\hbar}[H, \rho] + \sum_{a,b} C_{ab}[O_a, [O_b, \rho]]. \tag{A.21}$$

An illuminating example, which gives rise to such an equation, can be constructed [105] by coupling the Hamiltonian H to time dependent external sources $H \to H + j_a(t)O_a$. If the sources are treated as classical Gaussian random variables, with vanishing expectation value $\langle\langle j_a(t) = 0\rangle\rangle$ and

$$\langle\langle j_a(t)j_b(0)\rangle\rangle = C_{ab}\delta(t), \tag{A.22}$$

then the expectation value of the density matrix satisfies this equation. In this context, C_{ab} is positive definite, and the $O_a(t)$ are Hermitian operators, and one can show that, as a consequence, the von Neumann entropy, $-\operatorname{Tr} \rho \ln \rho$ is increasing, so that there is no violation of the second law of thermodynamics.

---

[8] This is in contrast to many worlds or Bohmian interpretations, which do not deviate from the mathematical formulation of QM.

A disturbing property of the Lindblad equations, is that energy is only conserved statistically. The white noise correlation of the sources shows us that there is a finite probability for injecting arbitrarily large amounts of energy into the system described by H. Only the expectation value of the injected energy is vanishing. In the context of quantum field theory with an ultra-violet cutoff at some experimentally acceptable energy Λ, my colleagues and I [105] showed that unless the dimensionless[9] coefficients in C_{ab} were very small, the Lindblad equation fills the universe with a gas of particles with energies of order Λ in a time scale of order Λ^{-1}. We argued that the only way to avoid this conclusion, is to have the external sources couple to very nonlocal operators, and that this would lead to violations of the locality property of quantum field theory. The locality property states that two quantum operators situated at points in space-time, which cannot communicate via signals sent at a speed no larger than the speed of light, must commute with each other. If it is not valid, then Einstein–Podolsky–Rosen experiments can be used to send signals faster than light. EPR experiments have been done, and no such signals detected. Moreover, nonlocal modifications of quantum field theory have a hard time accounting for the 12 decimal place agreement between local quantum field theory and experiment.

Energy conservation has also been tested very stringently by the success of quantum tunneling formulae in accounting for the lifetimes of highly meta-stable states. We are quite familiar with the fact that in thermal equilibrium states, with temperature of order the barrier height, states that are meta-stable at zero temperature decay very rapidly. The most stringent constraints probably come from the extremely long lifetimes of certain radioactive nuclei, which you explored in Exercise 17.7. In [105] we estimated that the dimensionless coefficients in C_{ab} had to be smaller than 10^{-125} in order not to interfere with the agreement between theory and experiment in alpha particle decays of nuclei. It is hard to see how these tiny terms in the Lindblad equation could play the role they are supposed to play in explaining the apparent evolution of pure quantum states into mixed states, which occurs when microscopic and macroscopic systems interact over time scales of order seconds or minutes. This bound assumed that the violations of locality in space induced by the C_{ab} terms took place only on the Planck scale 10^{-33} cm. If instead, we allow violation of locality at scales where experiment assures us that no violation occurs, then the bound is reduced to 10^{-65}. For each decade in energy that particle accelerators gain, without finding evidence of locality violation, the bound becomes stricter by a factor of 10^4.

---

[9] In quantum field theory with a cutoff, one does dimensional analysis by multiplying each term in the Hamiltonian by the appropriate power of $\frac{\Lambda}{4\pi}$. In the text, we are referring to the dimensionless C_{ab} with this scaling taken out. In [105] we also assumed that all positive powers of $\frac{\Lambda}{4\pi}$ were simply absent from the Lindblad equation.

The literature on objective collapse theories contains very few papers, which attempt to address the arguments of [105]. The reason for this is another defect of the Lindblad equation, namely its lack of relativistic covariance. In fact, although [105] uses the formalism of relativistic quantum field theory, its authors were well aware of this problem. They considered that the problematic aspects exposed in their work were not likely to be resolved by a hypothetical version of the Lindblad equation, which was compatible with special relativity.

We do not have space here to explore the literature on this topic, and readers should not go away with the impression that objective collapse theories have been proven wrong. What is clear is that such theories have severe problems with the invariance principles which seem to hold in the real world (both space time translations, and Lorentz transformations), and with the fact that signals cannot be sent faster than the speed of light. All of these properties are subject to experimental checks, and it is certainly worth doing more experiments to check their validity. Doing so could either confirm the kind of violations of these principles that the Lindblad equation leads us to expect, or put stricter bounds on the unitarity violating terms in the equation. Advocates of this sort of interpretation of QM should spend more time trying to make a version of Lindblad dynamics, which is more compatible with symmetries and locality.

Although I like the falsifiability of the dynamical collapse approach, from a philosophical point of view I do not view it as a real advance on the standard statistical interpretation. Given the modern updating of the ideas of Bohr and von Neumann via the theory of decoherence, the statistical interpretation gives us a complete understanding of why the mixed density matrix ρ_{mixed} is equivalent to the true quantum density matrix, with accuracy exponential in the size of the macrosystem, whenever a microsystem's states become entangled with those of the collective coordinates of a macrosystem. The real philosophical objection to this approach is Einstein's "God doesn't play dice," *i.e.* that probability should enter into a physical theory only as an admission of ignorance of the complete physical state of the system. *Objective Collapse theories do not deal with this objection.* They admit that the most refined version of physical theory is in fact statistical, in a way that can't be attributed to an inability to measure all the properties of the system's state. The only difference in philosophical posture is that dynamical collapse theories view the eventual validity of Bayes' rule, which allows us to pretend that there is an underlying predicted trajectory in certain circumstances, as following from Lindblad dynamics, rather than the entanglement with macrosystems. Occam's razor would suggest that we reject such theories for adding a complicated new set of rules, which have obvious problems with Lorentz invariance and causality, and do not really solve the fundamental philosophical problem of QM. QM is incompatible with the idea that the mathematical equations of physics describe an underlying physical reality, "what really happens," to which we have to add probabilistic notions for purely practical reasons. So are dynamical collapse theories. At a time before the theory of decoherence clarified the actual nature of wave function collapse, perhaps such theories held some attraction, but that is no longer the case.

Still, it is preferable to reject physical theories on the grounds that they do not fit experimental data. Dynamical collapse theories, unlike Bohmian mechanics and the Many Worlds interpretation, can be falsified in this way. I would contend that to a large extent they have been, but perhaps more sophisticated models of this type, and more precise experimental tests, will finally put this set of ideas to rest.

A.5 SUMMARY

Attempts to "explain" QM in terms that are compatible with human prejudice about the nature of the world, are extremely diverse, and it is probably foolish to try to force them all into a small number of boxes. Nonetheless, in what follows, we try to outline some common themes, which should be addressed by any such attempt.

- The first question one must ask of a theory is whether it considers quantum amplitudes $\langle e_n | \psi \rangle$, perhaps in some preferred basis, to be *real* in the sense that we think of solutions of Newton's equations or Maxwell's equations as real. The alternative is to think of the amplitudes only as a probability distribution for normal operators on the Hilbert space.

- A theory, which considers the amplitudes to be real, should explain how we can determine them by a collection of measurements at a fixed time. Our notion of reality is based on our experience with being able to check whether something is true or not. Many advocates of realist interpretations of QM accept the conventional wisdom that measurements are complicated and use this as an excuse to *not* address the question of how one actually checks the reality of the wave function. IMHO any notion of "reality" that is not subject to experimental check should be discarded, by Occam's razor.

- Any model which treats the quantum amplitudes as merely a probability distribution, checkable only by doing multiple experiments on identically prepared systems, has to face the conundra of any intrinsically statistical theory. The frequentist definition of probability can never be carried out, so even frequentists have to settle for a certain amount of Bayesian "expectation," with all of its philosophically unpleasant subjectivity. This problem is exacerbated for theories of cosmology, where, in principle, one does not get to repeat the experiment at all.

- Probabilistic models must then give a prescription for dealing with the apparent existence of a macroscopic "reality," in which randomness appears to be consistent with the existence of probabilities for histories of bodies at least as small as Brownian particles.

The practical interpretation of a probability theory always utilizes Bayes' rule. Bayes' rule is a mathematical procedure, which can be implemented for any set of probabilities for histories which obeys the *decoherence sum rule*. The classical rationale for Bayes' rule is that there is an underlying reality with fixed initial conditions, and physical laws, which predict a definite outcome given fixed initial conditions. Probability only appears because of our ignorance of some of the initial conditions. However, we can apply Bayes' rule to any set of probabilities for histories obeying the sum rule, even if the sum rule is only obeyed approximately. This is what we do in the statistical interpretation of QM. We have already commented about the philosophical confusions and experimental embarrassments of objective collapse theories.

- Advocates of realist interpretations of QM often ask the question "Probabilities for what?" when confronted with a purely statistical interpretation. The answer to this question is clear: QM is a theory that describes systems at a microscopic level, but some systems have collective variables, whose QM properties are exponentially close to those predicted by a classical statistical theory. The QM predictions for some microscopic variable are predictions about the frequencies of occurrence of states of collective coordinates of some macrosystem, given that they have become entangled with the states of the microsystem. This entanglement can occur without any experimental intervention, human or otherwise. The predicted probabilities are independent of the nature of the macroscopic object whose collective coordinates are entangled with the microsystem.

A.6 WHAT IS WRONG WITH FUNDAMENTAL PROBABILITY?

Our experience of the world teaches us that it is a place where things happen over the course of time. Physicists try to make that intuitive notion precise by assigning numerical values to each "thing" and "measuring" those values. That is how we know that things happen over the course of time. The measured values change. When we think about a baseball, two of the things we can measure are its center of mass position and the rate of change of that position with time. If we tried to measure them more and more precisely we would find that we could not actually do it with arbitrary precision. Repeated measurements of accuracy the inverse square root of the volume of the baseball in Bohr units, would reveal inevitable random fluctuations, which obeyed the Heisenberg uncertainty relations. However, with an accuracy $e^{-c\frac{V}{a_B^3}}$ we would be able to account for those fluctuations by classical models of random forces. So we can get an accurate predictive theory of the motion of the baseball, and an even more accurate classical statistical theory, which would assert that we could have gotten precise predictions if only we would carefully identified the source of the random forces acting on the baseball. Both of those theories would be wrong, but extremely accurate.

There are two different important concepts in the previous paragraph. One is that things have values at a particular time. If we identify those values with the results of measurements at that time, this is tautologically true, by the only measure of truth a scientist should accept. The second concept is that we can make *predictions* about the future behavior of "things," given enough knowledge at a given time.

Predictive theories of physics do not predict initial conditions. Thus, if we just talk about the values of variables at a fixed time, there is no experimental difference between probabilistic and predictive theories. Each will predict that any given experimental outcome is possible, depending on the initial state of the system. The real difference between probabilistic and deterministic theories has to do with prediction. Probabilistic theories do not predict the outcomes in the future, given a complete set of initial data in the past[10] The set of probabilistic theories divides into two classes, defined by whether or not one can define exact probabilities for histories of all variables, satisfying the probability sum rule, which states that the probability of observing some outcome at time t is the sum of the probabilities for all histories which lead to that outcome. The only mathematically consistent theory we know, which does *not* satisfy this rule is QM.

One can call the vast class of probabilistic theories which do satisfy the probability sum rule exactly, *pseudo-deterministic*. That is to say, they are built on some set of variables satisfying deterministic equations, and then averaging over some of the variables. The theorist might declare that certain of the variables are *hidden; i.e.* in principle impossible to measure, so that we could never verify that the world was following some particular history of the variables, but the system behaves as if such a history existed.

In QM, we can only have pseudo-determinism for the limited subset of variables we have called collective coordinates of macrosystems. The phenomenon of unhappening shows that such pseudo-determinism is likely to be an ephemeral property for all but the most coarse grained aspects of the universe. However, this pseudo-determinism, limited both in the range of variables to which it applies, and the time over which it applies to them, is enough to account for all of the macroscopic phenomena that shaped the evolution of our brains and our intuitive conception of how the world works.

Since pseudo-determinism is not an exact property of QM, we are truly forced to confront the question of what a probabilistic theory *means*. What does it mean, in Einstein's phrase, for *God* to play dice? We are willing to accept our own limitations as data gathering instruments and use probabilistic theories based on *our* ignorance, but we are used to thinking of the laws of nature as an expression of objective rules for describing "what is really going on"

---

[10] The phrase complete set of initial data, means the result of a maximal set of simultaneous measurements on a system. The word measurement means complete entanglement of a basis of states of the system with different values of the collective coordinates of macroscopic objects. It is an approximate notion, but the approximations have an accuracy better than $e^{-10^{20}}$.

in the universe we observe. How can those rules be based on a formalism whose objective frequentist definition is defined by a limiting process, which can never be carried out? Any use of probabilistic concepts inevitably involves subjective language like *expectation* or *confidence*. It appears that all attempts to view the Schrödinger equation and the quantum state as *real things* in the sense that term is used in classical physics, are motivated by the feeling that the subjective notion of probability should not be a fundamental part of physical theory.[11]

I think there are two rejoinders to this. The first is that the whole notion of prediction, on which physical theories are based, is a human enterprise. The universe does what it does, independently of any physical theory. Humans want to know, for both practical and emotional reasons, what it is going to do before it does it. Classical physical theories are based on a stance of philosophical omnipotence: human beings can invent mathematical models, which are a precise map of the mechanism by which the universe evolves from one instant of time to another, limited only by the ability to gather data on all relevant initial conditions. QM denies the possibility of doing that. It says we cannot make predictions, except statistical ones. Experiment is in very broad agreement with that conclusion. Data on microscopic phenomena are definitely random. When we can do "repeated experiments on identically prepared isolated systems" (the scare quotes remind you that every adjective in this phrase refers to an *approximation*), the statistics of those random events matches the mathematical predictions of QM. The mathematical formulae of QM do not admit a pseudo-deterministic interpretation. So what is wrong with that?

All it really means is that we must give up our dreams of omnipotence to a greater degree than some of us are willing to agree to. We cannot even build a mathematical model of the workings of the universe that would have predicted everything if only we would been allowed to know in advance the particular history of the hidden variables that the universe was following. What a tragedy for human ambition! Given the unprecedented degree of agreement between theory and experiment that quantum theories have achieved, one is tempted to say, "Get over it and grow up. You can't always get what you want, but so far we have gotten more than we need."

The second rejoinder is that the quantum theory of probability for noncommutative variables has a degree of mathematical inevitability and beauty that far outstrips that of classical mechanics. Given *any* list of data, supposed to correspond to some measurable aspects of a physical system, the set of all classical detectables of the system can be viewed as a maximal commuting family of normal operators in a Hilbert space and any classical probability distribution is a density operator on the space, which commutes with the family. That density operator automatically provides a probability distribution for *all* normal operators. Furthermore, any other density operator, not necessarily commuting with the original maximal

---

[11] In my opinion, *all* attempts to get around probability have failed. Some of the evidence was presented in our discussion of specific interpretations of QM above.

abelian subalgebra, defines a probability distribution for all normal operators. Quite remarkably, from a point of view about probability that attributes its use to failure to measure all relevant variables, for pure states all of the nonsharp distributions are predicted with mathematical precision. A 19th century mathematician, familiar with Boole's logic and the theory of Hilbert space,[12] and a sufficient amount of *chutzpah*, might have discovered this noncommutative probability theory and proposed that it might account for some of the randomness observed in experiments in the real world.

If she had been sufficiently insightful, our hypothetical mathematical physicist might have noticed the beautiful unification of detectable quantities in a system, and operations on a system, provided by QM, and scooped Emmy Noether by proving the quantum Noether's theorem before the classical version. The quantum theorem is simpler than the classical one, and much more integrated with the fundamental principles of the noncommutative probability theory. Completely apart from its agreement with experiment, the mathematical structure of QM is much more compelling and beautiful than its classical counterpart.

Wigner and others have speculated [106] about the "unreasonable effectiveness of mathematics in the natural sciences." While this point of view has some disturbing quasi-religious overtones, it nonetheless points out a truth, for which we do not yet have a complete explanation. Mathematics is an invention of humans, but mathematical structures have the surprising ability to become useful in contexts far removed from the original purpose for which they were invented. The geometry of Hilbert space, a straightforward mathematical extension of the (approximate!) Euclidean geometry of ordinary three-dimensional space, turns out to be the fundamental principle underlying physics at its most microscopic level. How remarkable!

Perhaps we just have to accept that the fundamental limitations of our brain's software will forever prevent us from having a more intuitive feeling for the mechanism that underlies the evolution of the universe. We should be thankful that we have complete command of the mathematics, which allows us to make exquisitely precise predictions of the behavior of microsystems which are so far beyond our intuitive ken. We have speculated above that sentient beings whose brain architecture was based on the principles of Quantum Computer Science, might have a better intuitive understanding than we do, of the intrinsic nature of the quantum universe. If we ever encounter or build such beings, we could ask them to explain it to us, but we should worry about suffering the fate of Christopher Kingsley, the hero of Fred Hoyle's "Black Cloud."

---

[12] If the mathematician were willing to content herself with finite dimensional spaces, Boole's logic would have been sufficient.

The Dirac Delta Function

Consider functions on the real line, which have one or more of the properties, *continuity*, *smoothness* (continuous derivatives of all orders), or L_p ($\int dx \; |f|^p < \infty$). Each class of functions defines a vector space of infinite dimension over the complex numbers. It is also true, although we will not prove it in general, that each of these is a *topological vector space*. That is, there is a notion of when two functions are close to each other. The dual space of any vector space is the space of *linear functionals*, linear maps of the space into the space of complex numbers. For topological vector spaces, we also require that the linear functionals be continuous. That is, when two functions $f_{1,2}$ are close to each other, the values of the map $L(f_1)$ and $L(f_2)$ are close to each other.

The Dirac delta function can be thought of as a linear functional on many of these spaces. For example, consider the space of square integrable functions. The delta function, $\delta(x - x_0)$, acting on a function, simply evaluates that function at a point x_0. It is obviously linear. It is not, however, continuous. The distance between two functions in this space is $\sqrt{\int |f_1 - f_2|^2}$. Let $f_1 - f_2$ differ from 0 by an amount of order 1, but only in an interval of size ϵ centered at x_0. Then the distance between the functions is $o(\sqrt{\epsilon})$ but the difference between $\delta[f_1]$ and $\delta[f_2]$ is order 1, no matter how small ϵ is. The delta function *is* continuous if we define the distance between two functions to be the maximum of their pointwise difference over the real line.

Similarly, derivatives of the delta function can be defined as functionals, but on somewhat more restrictive spaces of functions. We simply use integration by parts to write

$$\int \frac{d^k}{dx^k}\delta(x - x_0)f(x) = (-1)^k \frac{d^k f}{dx^k}(x_0),$$

which defines a linear functional on the space of functions that is k times differentiable at x_0. Depending on the notion of distance in the space of functions, these functionals might or might not be continuous.

The delta function (and its derivatives) concentrated at a fixed point y map the space of functions on which they act into the complex numbers. If we let y vary, then we have instead a map from a space of functions into another space of functions, which can often be identified with the original function space. Indeed, the delta function itself is clearly just the identity operator in function space. We can think of such distributions as continuous versions of matrices, called integral operators, mapping a function f into another function g via the formula

$$g(x) = \int dy \ K(x,y)f(y).$$

This formula makes sense when $K(x,y)$ is an integrable function (with further smoothness conditions, if we want g and f to have the same differentiability properties), but also when K is a delta function or one of its derivatives. Given two such integral operators, we can define their product by

$$K_{12}(x,y) = \int dz \ K_1(x,z)K_2(z,y),$$

which gives us the operator product in the corresponding linear space.

In quantum mechanics, we are concerned with a space of square integrable functions, which are, generically, not differentiable. There are, however, bases consisting of infinitely differentiable functions, so that any function can be approximated (in the L^2 norm) by such functions. The delta function is a globally defined operator on this space, but its derivatives define unbounded operators, which are only defined on a dense subspace of the Hilbert space.

Noether's Theorem

Consider a Lagrangian $L(q^i, \dot{q}^i)$, whose action integral is invariant under a one parameter continuous group of symmetries $q^i \to q^i + \epsilon f^i(q, \dot{q})$. We have written the infinitesimal form of the transformation and ϵ is the infinitesimal. The variation of the action integral is

$$\delta S = \int \, dt [\frac{\partial L}{\partial q^i} f^i + \frac{\partial L}{\partial \dot{q}^i} \dot{f}^i].$$ (C.1)

We can write this as

$$\delta S = \int \, dt [\frac{\partial L}{\partial q^i} - \frac{d}{dt}(\frac{\partial L}{\partial \dot{q}^i})] f^i + \frac{d}{dt}(\frac{\partial L}{\partial \dot{q}^i} f^i).$$ (C.2)

If the action is invariant under the symmetry, $\delta S = 0$, for every path $q^i(t)$ and every time interval. Invariance of the action does not imply invariance of the Lagrangian. It is enough that $\delta L = \partial_t \Lambda$. Thus, the condition for invariance is

$$0 = [\frac{\partial L}{\partial q^i} - \frac{d}{dt}(\frac{\partial L}{\partial \dot{q}^i})] f^i + \frac{d}{dt}(\frac{\partial L}{\partial \dot{q}^i} f^i - \Lambda).$$ (C.3)

The Lagrange equations of motion are

$$0 = \frac{\partial L}{\partial q^i} - \frac{d}{dt}(\frac{\partial L}{\partial \dot{q}^i}).$$ (C.4)

$$0 = \frac{d}{dt}(\frac{\partial L}{\partial \dot{q}^i} f^i - \Lambda).$$ (C.5)

That is, every continuous one parameter group of symmetries of the action corresponds to a conserved quantity.

Let's try this out for the symmetry of time translation invariance. Then $f^i = \dot{q}^i$ and $\Lambda = L$. Then the conserved quantity is

$$\frac{\partial L}{\partial \dot{q}^i}\dot{q}^i - L. \tag{C.6}$$

Recall that

$$\frac{\partial L}{\partial \dot{q}^i} \equiv p_i, \tag{C.7}$$

the momentum canonically conjugate to q^i. The application of Noether's theorem to time translation invariance gives us the canonical Hamiltonian as the corresponding conserved "Noether charge."

For invariance under a linear transformation of the q^i, like rotations, we have $f^i = \omega^i_j q^j$, and $\Lambda = 0$, where ω^i_j is the infinitesimal matrix of rotation. The conserved charge for rotations in the ij plane is

$$J_{ij} = q_i p_j - q_j p_i, \tag{C.8}$$

where we have used the Euclidean metric δ_{ij} to lower the index on q^i.

There are three ways to think about the quantum mechanical use of Noether's theorem. The first is to view it as a way to derive the canonical commutation relations. In quantum theory (QM), symmetry transformations are unitary operators in Hilbert space, which commute with the Hamiltonian. A one parameter group of symmetries has a Hermitian generator, Q, which commutes with the Hamiltonian. Noether's theorem gives us an expression $Q = p_i f^i(q)$ (we will only discuss the simple case where f^i is independent of p_j) for this generator in terms of the classical canonical variables q^i and p_i. The equations

$$i[Q, q^i] = f^i(q) \tag{C.9}$$

are solved by the canonical commutator

$$[p_i, q^j] = -i\delta^j_i. \tag{C.10}$$

In the path integral formalism, the use of Noether's theorem is even more direct. Restricting again to transformations for which f^i depends only on q, we do a change of variables in the path integral, which has the form of a symmetry transformation with a time dependent infinitesimal parameter $\epsilon(t)$. The measure of integration is invariant under this transformation[1] and the action changes by

$$\delta S = \int dt \, \partial_t \epsilon p_i(t) f^i(q(t)) = \int dt \, \partial_t \epsilon \, Q(t). \tag{C.11}$$

---

[1] Actually, in quantum field theory there are examples where the measure is not invariant. This leads to quantum mechanical breaking of a symmetry of the classical action, also called an *anomaly* because the phenomenon was so confusing when it was first discovered.

Since we just did a change of variables, the actual value of the integral should not change and so the effect of this new term in the action should be zero.

$$\langle F| \int dt\ \partial_t \epsilon\ Q(t)|I\rangle = 0. \tag{C.12}$$

Choosing ϵ to be a differentiable function with support separated from the ends of the interval, so that the boundary values are not changed, we conclude, by integration by parts, that $\partial_t Q$ has vanishing matrix elements between any initial and final state of $t_F > t > t_I$.

   If we apply the same procedure to the evaluation of expectation values of time ordered products of operators, we get an identity

$$\partial_t \langle F|T[Q(t)O_1(t_1)\ldots O_n(t_n)]|I\rangle = \sum_{j=1}^{n} \delta(t-t_j)\langle F|T[O_1(t_1)\ldots \delta O_j(t_j)\ldots O_n(t_n)]|I\rangle. \tag{C.13}$$

Identities like this are called Ward–Takahashi identities and they are very useful in quantum field theory.

   However, it may be that the ultimate point of view about Noether's theorem in QM is that it is really an intrinsic part of the quantum formalism, for which we have no need of a classical derivation. This is, for the most part, the point of view we have followed in the body of the text. In QM, a symmetry is a unitary operator, which commutes with the Hamiltonian. On the other hand, every normal operator has eigenvalues and that makes it a potential candidate for "measurement," if we can construct a device, which will entangle different eigenstates with different values of a macroscopic collective coordinate. So symmetries lead directly to conserved quantities in QM.

   The one place where the classical analysis leads to results which have not yet been reproduced using this more abstract point of view is the construction of local Noether currents and charge densities in quantum field theory. Noether's theorem leads directly to operators representing the amount of conserved quantum number in a local region, and the abstract formalism has not yet reproduced such a construction.

Group Theory

A symmetry in quantum theory (QM) is a unitary transformation, U, which commutes with[1] the Hamiltonian. Given two such transformations $U_{1,2}$, their products $U_1 U_2$ and $U_2 U_1$ are also symmetries. Furthermore, every unitary transformation is invertible.

The set of all symmetries of a given Hamiltonian thus satisfies the mathematical axioms of a *group*, namely a set G equipped with a multiplication rule, and a unit element $e \in G$ such that

$$eg = ge = g,$$

for all $g \in G$. The axioms for a group also require that every element has an inverse such that

$$gg^{-1} = g^{-1}g = e.$$

A group is defined by its abstract multiplication law: a listing or parameterization of all the elements, and a rule for determining what the product of any two elements is. In principle, this rule takes the form of a big table. However, in most cases, one can find a simpler way of describing the multiplication law. For groups whose underlying set is discrete, this is often expressed in terms of generators and relations. For example, the group Z_N of integers k modulo N is generated by a single element $z = e^{2\pi i k/N}$ satisfying the relation $z^N = 1$. The group $SL(2, Z)$ of fractional linear transformations on a complex variable w, with integer coefficients is

$$w \to \frac{aw + b}{cw + d},$$

where a, b, c, d are integers satisfying $ad - bc = 1$. The group is generated by

$$T : w \to w + 1$$

---

[1] More generally, we could demand that the time evolution operator and the operator U form group with a finite number of infinitesimal generators. The Galilean or Lorentz groups are examples of this kind of structure.

and

$$S : w \rightarrow -\frac{1}{w}.$$

These satisfy

$$S^2 = 1$$

and

$$(ST)^3 = 1.$$

The group S_N of permutations of N elements is generated by transpositions, etc.

Given a group multiplication law, there may be many inequivalent ways to realize it in terms of unitary transformations on a Hilbert space. These are called inequivalent *unitary representations* of the group. Note that given a representation $U(g)$, there is another one $V^\dagger U(g) V$ for every unitary transformation in the Hilbert space in which $U(g)$ acts. Two representations related in this fashion are called *unitarily equivalent*.

Some symmetries, like translations and rotations, depend on continuous parameters. When this is the case, there is always a subgroup which can be continuously deformed to the identity. This is called *the connected component of the identity*. Sophus Lie had the insight that any transformation in the connected component of the identity could be built up by a sequence of infinitesimal transformations. As a consequence of his seminal work, continuous groups are now called Lie groups. If ω^a are the real continuous parameters on which the group elements depend (the number of independent parameters is called the dimension of the group), then in any unitary representation, an infinitesimal transformation has the form

$$U(\omega^a) \sim 1 + i\omega^a T_a,$$

when the parameters are very close to zero. To leading order in ω, we have

$$U^\dagger(\omega^a) \sim 1 - i\omega^a T_a^\dagger,$$

so that unitarity implies $T_a = T_a^\dagger$.

To leading order in the deviation of ω^a from zero, we have $U(\omega_1^a)U(\omega_2^a) = U(\omega_2^a)U(\omega_1^a)$. The leading correction to this relation is given by

$$U(\omega_1^a)U(\omega_2^a)U^{-1}(\omega_1^a)U^{-1}(\omega_1^a) = 1 - \omega_1^a \omega_2^a [T_a, T_b].$$

The right-hand side will be an infinitesimal group element, if and only if

$$[T_a, T_b] = i f_{ab}^c T_c,$$

where the coefficients $f_{ab}^c = -f_{ba}^c$ are real.

The set of *infinitesimal generators* T^a is closed under taking real linear combinations and commutation. This is the definition of a *Lie Algebra*.[2] The Baker–Campbell–Hausdorff formula

$$\ln\left(e^X\, e^Y\right) = X + Y + \frac{1}{2}[X,Y] + \cdots,$$

where the ellipses denote higher order nested commutators, shows that

$$e^{i\omega_1^a T_a} e^{i\omega_2^a T_a}.$$

is a group element, and thus the entire multiplication table of the group is determined by its Lie algebra.

If the group has a faithful (no element is mapped into the identity matrix unless it is the identity element) finite dimensional unitary representation, then the parameter space is compact, since the space of *all* unitary matrices is. Consider the abelian subgroup $e^{i\alpha\omega^a T_a}$ for some fixed vector ω^a. The real symmetric matrix $\mathrm{tr}\,(T_a T_b)$ can be brought to the form δ_{ab}, by redefining $T_a \to S_a^b T_b$ with S a real invertible matrix. With this normalization, we see that the group element is periodic in α with periodicity $\alpha \to \alpha + \frac{2\pi}{\sqrt{\omega^a \omega_a}}$.

A maximal set of commuting generators T_a is called a Cartan subalgebra, and the number of commuting generators is called the rank, r, of the group (the total number of generators is called the dimension, d_G, of the group). If k of those generators commute with *all* other generators, we say that $G = U(1)^k \otimes G_{SS}$. G_{SS} has rank $r - k$ and dimension $d_G - k$. We can obviously study the full symmetry group by studying the $U(1)$s and G_{SS}, separately. We can simplify things further by writing

$$G_{SS} = G_1 \otimes \ldots \otimes G_n,$$

where every element in G_i commutes with every element in G_j, and each G_i cannot be further factorized into commuting subgroups. Such nonfactorizable groups are called *simple*, and the SS on G_{SS} stands for *semisimple*.

A famous theorem [107] classifies all compact simple Lie groups. They fall into four infinite families, and there are five *exceptional groups* $G_2, F_4, E_{6,7,8}$. The subscript denotes the rank of the group. The four infinite families are the groups of all special (i.e., determinant 1) unitary matrices, $SU(n)$ in n complex dimensions, the groups $SO(2n)$ and $SO(2n+1)$ of orthogonal transformations in even and odd numbers of real dimensions, and the groups of unitary symplectic transformations $Sp(n)$ in $2n$ real dimensions. Symplectic transformations

---

[2] For a pure mathematician the definition of a Lie Algebra invokes an operation with all the properties of the commutator of two operators, but is not explicitly realized in some linear space. Such a linear space realization is called a *representation* of the Lie algebra. In QM, continuous symmetries are always unitary transformations, so we are always working with a unitary representation.

are like orthogonal transformations except that they preserve a nondegenerate *antisymmetric* bilinear form.

We will not outline the proof of this theorem, but just indicate the main line of argument. The maximal set of mutually commuting generators of the Lie algebra is called its Cartan subalgebra. The number of generators in the Cartan subalgebra is called the rank of the group. Label the Cartan generators by H_i, $1 \leq i \leq r$. Commutation with the Cartan generators is a linear transformation on the set of generators *not* in the Cartan subalgebra. One argues that this linear transformation is diagonalizable, that is: there exist linear combinations of generators E_r such that

$$[H_i, E_r] = r_i E_r.$$

The r dimensional vectors r_i are called the *roots* of the algebra. The Jacobi identity then implies that $[E_r, E_s]$ is a generator with root vector $r_i + s_i$, so that the roots form a lattice in r-dimensional space. The constraints on the geometry of such lattices lead to the abovementioned classification of Lie algebras.

Given a simple compact Lie group, one can ask for all of its inequivalent unitary representations. There is a trivial way to make new representations from old, somewhat analogous to the product construction of semisimple groups from simple ones. Given two sets of generators $T_a^{1,2}$, each of which satisfies the group commutation relations and is realized as Hermitian operators in Hilbert spaces $\mathcal{H}_{1,2}$ then the operators

$$\begin{pmatrix} T_a^1 & 0 \\ 0 & T_a^2, \end{pmatrix}$$

acting in $\mathcal{H}_1 \oplus \mathcal{H}_2$ are Hermitian and satisfy the commutation relations. Note that operators of the form $a_1 P_1 + a_2 P_2$, where P_i are the projection operators on the subspaces $\mathcal{H}_i$, commute with all the generators of this *reducible* representation. The action of the group leaves invariant the individual subspaces.

A representation is said to be *irreducible* if the only subspaces left invariant by the group action are the zero vector and the whole Hilbert space. For finite dimensional irreducible representations, Schur's Lemma states that in this case, the only operators commuting with all the generators are proportional to the unit operator. This means that operators like $T_a T_a$ are just pure numbers in a finite dimensional irreducible representation. The values of these *Casimir invariants* (their general form is $\text{tr}[T_{a_1} \ldots T_{a_n}] T_{a_1} \ldots T_{a_n}$) characterize the representation. Following the model of our discussion of angular momentum, one uses this numerical information to find all the vectors in the representation.

Given two unitary representations of a Lie group G in Hilbert spaces $\mathcal{H}_{1,2}$, the tensor product space

$$\mathcal{H}_1 \otimes \mathcal{H}_2,$$

carries a representation with generators

$$T_a^1 \otimes 1 + 1 \otimes T_a^2.$$

Even if the individual factors are irreducible, the tensor product representation is not. The coefficients expressing the decomposition of a general vector in the tensor product into vectors in its irreducible pieces are called *generalized Clebsch–Gordan coefficients*, and an entire industry is devoted to computing them. The fact that the tensor product representation has such a direct sum decomposition follows from the Peter–Weyl theorem: for compact semisimple Lie algebras: every representation is a direct sum of irreducible representations. This theorem is *not* true for the noncompact groups we discuss below.

A representation of a group is said to be real if all of the generators T_a can be written as imaginary antisymmetric matrices in some basis, so that the group action is by real matrices. For any representation, the matrices $\bar{T}^a = -T_a^*$ form another representation called the complex conjugate representation. Obviously, in a real representation these two representations are equal to each other. A weaker requirement is that the two are unitarily equivalent.

$$-T_a^* = U T_a U^\dagger.$$

An example where the complex conjugate representation is unitarily equivalent to the original one, where $U \neq 1$, is given by the spin $1/2$ representation of $SU(2)$.

$$-\sigma_a^* = \sigma_2 \sigma_a \sigma_2.$$

Such representations are called pseudo-real. The multiplication rule $\sigma_a \sigma_b = \delta_{ab} + i\epsilon_{abc}\sigma_c$ shows that it is impossible to find a representation in which all matrices are imaginary. Representations which are unitarily equivalent to their conjugates, but are not real, are called *pseudo-real*. One interesting property of all real or pseudo-real representations is that

$$\text{tr}\,[T_{a_1} \ldots T_{a_{2k+1}}] = -\text{tr}\,[T_{a_1}^* \ldots T_{a_{2k+1}}^*] = -\text{tr}\,[T_{a_1}^\dagger \ldots T_{a_{2k+1}}^\dagger] = -\text{tr}\,[T_{a_1} \ldots T_{a_{2k+1}}],$$

so that the trace of an odd product of generators vanishes. In the penultimate equality above, we have used the fact that the traces of M and M^T are equal, and that the trace of a product is invariant under cyclic permutation.

The representation space of a complex or pseudo-real representation is a complex Hilbert space. That of a real representation is a real Hilbert space, but we can complexify it by doubling the number of real dimensions. Thus, given a Lie group with real parameters ω^a we can look at the set of all operators $e^{i\omega^a T_a}$ with $\omega^a = R^a + iI^a$. These will form a group (by the Campbell–Baker–Hausdorff formula) called the complexification of the original Lie group. The complexified groups are no longer compact, because the transformations $e^{-I^a T_a}$ are not periodic functions of I^a. They have unbounded eigenvalues for large values of $|I|$.

We can consider these complexified groups to be real Lie groups, with parameters R_a, I_a but in these finite dimensional representations, not all the generators are Hermitian, so the representation is not unitary. Among the groups we can construct in this way are $SO(m, n)$ the set of all transformations satisfying

$$\det O = 1,$$

$$O^T g O = g,$$

with g a symmetric matrix with m positive and n negative eigenvalues. A particular case is the Lorentz group $SO(1, 3)$.

These non-compact groups *do* have unitary representations, but they are all infinite dimensional. The theory of unitary representations of non-compact groups is quite intricate and we will not study it, except to give an example for $SO(m, n)$. Consider the generalized hyperbolae defined by

$$y^i g_{ij} y^j = \pm 1.$$

We can build a Hilbert space of complex valued square integrable functions on one of these hyperbolae. The scalar product is

$$\int d^{m+n} y \delta(y^i g_{ij} y^j \pm 1) f^*(y) g(y).$$

The action of $SO(m, n)$, via $y^i \to O^i_j y^j$ induces an action of the group on this Hilbert space, which preserves the scalar product, and so is given by unitary transformations.

We conclude this appendix with a quick and dirty proof of the Wigner–Eckart theorem. We study a system with conserved angular momentum **J**. The Hilbert space breaks up into a direct sum of spaces, with each term in the sum being the tensor product of an irreducible representation $|j\ m\rangle$ of the rotation group and a Hilbert space $\mathcal{H}_j$ on which the operators **J** do not act. The same is true of the vector space of all operators A on the original Hilbert space. We can write any operator as

$$A = \sum A_k O_{kn},$$

where

$$[\mathbf{J}, O_{kn}] = i\mathcal{J}^{(j)}_{kp} O_{jp},$$

where $\mathcal{J}^{(j)}$ is the $(2j+1) \times (2j+1)$ dimensional matrix representation of angular momentum. The expansion of A omits all of the indices that A carries, by virtue of transforming like a direct sum of angular momentum representations.

To prove the Wigner–Eckart theorem, we insert the above commutation relation between states $|j\ m\rangle$ and $|j'\ m'\rangle$, and allow the **J** operator in the commutator to act on the states. We then equate that to the result of the operator commutator, obtaining

$$0 = [\mathcal{J}^{(j')} - \mathcal{J}^{(j)} - \mathcal{J}^{(k)}]\langle j'\ m'|O_{kn}|j\ m\rangle.$$

The notation here is very compact: each set of angular momentum matrices acts only on the magnetic quantum numbers associated with its label. This equation is the defining equation for Clebsch–Gordon coefficients and it determines them completely up to a normalization. However, the operators O_{kn} also act on the Hilbert spaces $\mathcal{H}_j$, so in general, the matrix element of a fixed O_{kn} between states in $\mathcal{H}_j$ and $\mathcal{H}_{j'}$ will be a number $R_{k,j,j'}$, which depends on the total angular momenta as well as the choices of individual states in those Hilbert spaces. This is called *the reduced matrix element*, and is only defined when the Clebsch–Gordon coefficient is not zero. You can find a table of Clebsch–Gordon coefficients by consulting Google.

Group theory is a vast and intricate branch of mathematics, full of beautiful theorems and a vast array of fascinating formulae. Many mathematicians spend their entire careers opening up new avenues in the teeming metropolis of groups. This appendix barely scratches the surface of what is known about the relatively small part of group theory with direct relevance to quantum mechanics. It is enough for you to understand the uses of group theory in the text, but no more.

Laguerre Polynomials

Recall that the hydrogen wave functions are written in terms of polynomials in the variable $w = 2\rho$, which we called the associated Laguerre polynomials L_k^a where $a = 2l + 1$ and $k = n - l - 1$. These satisfy

$$w\frac{d^2 L_k^a}{dw^2} + (a + 1 - w)\frac{dL_k^a}{dw} + kL_k^a = 0. \tag{E.1}$$

We have followed Griffiths in our normalization of L_k^a, whereas in this appendix we will follow the mathematical convention that the coefficient of the constant term in every polynomial is 1. Thus, $(L_k^a)_{text} = k!(L_k^a)_{appendix}$. For $k = 0$, the obvious polynomial solution of the equation is $L_0^a = 1$, while for $k = 1$ a trial linear solution leads to $L_1^a = 1 + a - x$. Now assume we know all of the polynomials up to some value of k and write

$$(k + 1)L_{k+1}^a \equiv (2k + 1 + a - w)L_k^a - (k + a)L_{k-1}^a. \tag{E.2}$$

Acting with the operator wd_w^2 on both sides and using the equations for L_k^a and L_{k-1}^a, it is easy to verify that this definition gives a solution of the $k + 1$th equation, which is obviously a polynomial of one higher order. The normalization is correct if we remember that in the math convention, all of these polynomials equal 1 at $w = 0$.

Using the fact that the hydrogen wave functions u_{nlm} are orthogonal, we find that

$$\int_0^\infty x^a e^{-x} L_k^a L_j^a = \frac{\Gamma(k + a)}{\Gamma(k + 1)}\delta_{jk}. \tag{E.3}$$

To understand the normalization here one must use the formula for normalized wave functions in Chapter 13, and remember the change of conventions between this appendix and the text.

Now let us write the recursion relation defining the associated Laguerre polynomials as

$$(k+1)L_{k+1}^a - (2k+1+a)L_k^a + (k+a)L_{k-1}^a = wL_k^a. \tag{E.4}$$

This equation should be interpreted as defining the (infinite) matrix representation of the operator w, in the orthogonal basis of Laguerre polynomials. The expectation values of w^p are given in terms of diagonal matrix elements of the p-th power of this matrix.

Summary of Dirac Notation and Linear Algebra

A finite dimensional Hilbert space can be thought of as all complex linear combinations of a set of orthonormal basis vectors $|e_i\rangle$ ($1 \leq i \leq N$) whose scalar products are defined by the formula

$$\langle e_i | e_j \rangle = \delta_{ij}.$$

A general vector $|v\rangle$ is a linear combination

$$|v\rangle = \sum_n v_n |e_n\rangle,$$

and a general dual vector $\langle w|$ is

$$\langle w| = \sum_n w_n^* \langle e_n|.$$

The scalar product is required to be linear in both of these variables so that

$$\langle w|v\rangle = \sum_{m,n} w_n^* v_n.$$

We can think of $|v\rangle$ (w.r.t. the $|e_n\rangle$) basis, as the column of numbers v_n and $\langle w|$ as the row of numbers w_n^*, in which case the scalar product formula is the same as the formula for multiplying rectangular matrices. Mathematicians write vectors as simple letters, v, and the scalar product as $\langle w|v\rangle = (v, w)$. This is "linear in its first argument and conjugate linear in the second." The reason we introduce complex conjugation into the formula for the dual vectors is that the scalar product of a vector with itself is nonnegative and can be thought of as the square of a length. In quantum mechanics (QM), the scalar product of a vector with itself is the probability that if one is definitely in a particular state, then one is actually in that state, which is of course equal to one.

A linear operator A on the Hilbert space is a function from the Hilbert space to itself, $|v\rangle \rightarrow A|v\rangle$, which satisfies the rule

$$A(a|v\rangle + b|w\rangle) = (aA|v\rangle + bA|w\rangle).$$

Then

$$A|v\rangle = A \sum v_n |e_n\rangle = \sum A|e_n\rangle v_n.$$

Taking the scalar product of this equation with $|e_m\rangle$, we obtain the expansion coefficients $(Av)_m$ (the column vector) corresponding to the vector $A|v\rangle$. They are

$$\sum_n \langle e_m|A|e_n\rangle v_n,$$

which is the rule for acting with a square matrix on a column vector.

The scalar product $\langle w|A|v\rangle$ is also the scalar product of the vector $|v\rangle$ with another vector $A^\dagger|w\rangle$, which in Dirac notation has the awkward notational form

$$\langle w|A|v\rangle = \langle A^\dagger w|v\rangle.$$

$A^\dagger$ is the Hermitian conjugate operator. The matrix of the operator $A^\dagger$ is

$$\langle e_m|A^\dagger|e_n\rangle = \langle e_n|A^\dagger|e_m\rangle^*.$$

The most useful rule for Hermitian conjugates is that when an operator acts to the left in a scalar product, it acts like its Hermitian conjugate.

The expression $|w\rangle\langle v|$ defines a linear operator, taking any vector $|s\rangle$ into $|w\rangle\langle v|s\rangle$. The vector is taken into its projection on $|v\rangle$ multiplied into the vector $|w\rangle$. Schwinger called these measurement symbols. The special case of a projection operator on a vector of length 1 $|e\rangle$ just gives us the component vector of $|s\rangle$ in the $|e\rangle$ direction. One of the two or three most important equations in QM is the *resolution of the identity*

$$1 = \sum_n |e_n\rangle\langle e_n|,$$

where the 1 on the left-hand side stands for the unit operator. In terms of matrices, the matrix of $|e_n\rangle\langle e_n|$ has a 1 in the n-th row and column, and zeroes everywhere else, so the identity is obvious, but it is also a valid identity for every other orthonormal basis. The unit operator looks the same in all bases, and nothing we have said specifies which orthonormal basis we are using.

An operator is called *normal* if it has a complete set of orthonormal eigenvectors

$$A|a_i\rangle = a_i|a_i\rangle.$$

$$\langle a_i | a_j \rangle = \delta_{ij}.$$

These need not coincide with the basis $|e_i\rangle$ or some permutation of it. Any basis satisfies the resolution of the identity, so we can write

$$|a_i\rangle = \sum_n |e_n\rangle\langle e_n | a_i\rangle.$$

The matrix $\langle e_n | a_i \rangle$ is the matrix of the operator U, which transforms the $|e_n\rangle$ basis into the $|a_i\rangle$ basis. Its inverse is obviously the matrix $\langle a_i | e_n \rangle = \langle e_n | a_i \rangle^*$, because of the properties of the scalar product. Thus, its inverse is its Hermitian conjugate

$$U^\dagger U = U U^\dagger = 1.$$

Such operators are called *unitary*. Furthermore, every unitary operator transforms one orthonormal basis into another. One can show that the set of all unitaries can be exhausted by thinking of all the transformations that transform a given orthonormal basis into another.

If we transform the diagonal matrix of the normal operator A via $A \to U^\dagger A U$, we will get the matrix of A in the $|e_n\rangle$ basis, which is not diagonal. It will still satisfy $[A, A^\dagger] = 0$, since this equation transforms into $U^\dagger[A, A^\dagger]U = 0$, and we can remove the U's because they are invertible. Commuting with its adjoint is an equivalent characterization of a normal operator. Every operator with this property is diagonalizable. This is called the spectral theorem and a proof is sketched in the text. Every normal operator can be written

$$A = H_1 + i H_2,$$

where $H_i^\dagger = H_i$ and $[H_1, H_2] = 0$. Thus, the study of normal operators reduces to that of Hermitian operators, and most QM texts use only Hermitian operators. Note, however, that unitary operators are also normal, but rarely Hermitian. Unitary operators, which commute with the Hamiltonian of a system, are symmetries of the system. Noether's famous theorem, which relates symmetries to conservation laws, is valid for discrete symmetries only if we consider the generally complex eigenvalues of unitary operators as valid conservation laws.

Infinite dimensional systems are defined by letting the vectors have an infinite number of components, with the restriction $\sum_{n=1}^\infty |v_n|^2 < \infty$. The Cauchy–Schwarz inequality

$$\langle w | v \rangle \le \sqrt{\sum_{n=1}^\infty |v_n|^2 \sum_{n=1}^\infty |w_n|^2}$$

shows that the scalar products of two such infinite vectors is finite.

A simple way of understanding the limit is to think of a finite dimensional system as describing the states of a particle that lives on the second hand positions of a clock with N

seconds. The finite number of states correspond to the positions on the clock, and can be thought of as labeling the possible values of a unitary matrix U, whose eigenvalues are the N-th roots of unity.

$$U|n\rangle = e^{\frac{2\pi i n}{N}}|n\rangle.$$

Define the shift operator V by

$$V|n\rangle = |n + 1 \ (\text{mod} \ N)\rangle.$$

Then

$$V^N = 1; \quad UV = VUe^{\frac{2\pi i}{N}}.$$

Now imagine taking N to infinity through even values. Then $V^{N-k} = V^{\dagger \, k}$ for $1 \leq k \leq (N-1)/2$. Let us work in the basis of eigenstates of V and label them by positive integers for k in this range, negative integers for $k < N$ above this range, and label the state with eigenvalue 1 of V by 0. In the limit, we get a basis of states labeled by all integers, and normalizable vectors must satisfy

$$\sum_{p=-\infty}^{\infty} |v_p|^2 < \infty.$$

Now let us try to think about what happens to the eigenstates and eigenvalues of U in this limit. The eigenstates satisfy

$$U|\theta\rangle = e^{i\theta}|\theta\rangle,$$

and

$$U|p\rangle = |p - 1\rangle,$$

as long as p is not near the boundary $p = -(N-1)/2$. If we could neglect the boundary contributions, then a solution for

$$|\theta\rangle = \sum_p c_p|p\rangle,$$

would be $c_p \propto e^{ip\theta}$ with a p independent constant of proportionality. Such states are not normalizable. Note also that by neglecting the mod N contributions, we no longer have any constraint on θ except the periodicity condition which says that θ and $\theta + 2\pi K$ are the same state for any integer K.

The function of θ,

$$\delta_N(\theta) \equiv \sum_p e^{2\pi i p\theta},$$

is finite for finite N but obviously approaches infinity in the large N limit if $\theta = 0$. To see what happens for other values, note that when N is finite, it vanishes for all values of $e^{i\theta}$ that are N-th roots of unity not equal to 1. This is just the statement that for finite N all of the nontrivial eigenstates of U are orthogonal to that with $U = 1$. The N-th roots of unity become dense on the circle as N goes to infinity, and for finite N these are the only values of θ that are allowed in our system. So the eigenfunction with $\theta = 0$ must, when viewed as a function on the circle, go to zero everywhere except at $\theta = 0$. On the other hand, if $f(e^{i\theta})$ is a continuous function on the circle, then

$$\sum_{q=1}^{N} \delta_N(\theta_q) f(e^{i\theta_q}) = f(1),$$

for all N. So the function δ_N approaches what mathematicians call a *measure with point support* on the circle and physicists call the Dirac delta function $\delta(\alpha)$ on the circle. $e^{i\alpha}$ is the coordinate on the circle. The eigenfunctions for any other value of θ are simply $\delta(\alpha - \theta)$. The orthonormality relation converges to the relation

$$\int_0^{2\pi} d\alpha \, \delta(\alpha - \theta)\delta(\alpha - \theta_0) = \delta(\theta - \theta_0).$$

This is called delta function normalization. All of this is explained in more detail in Chapter 6.

Answers to Selected Problems

Answer to Exercise 1.1: The Let's Make a Deal Problem: The possible distribution of winning and losing doors is LLW, LWL, WLL, where W means the contestant wins and L that she loses. There is one chance in three that she has picked the right door, and two chances in three that she hasn't. Thus, it is twice as probable that she's picked the wrong door. Once shown that one of the other doors is a loser, she knows that it is twice as probable that the remaining door is the winner. The extra information does not change the probability that she's chosen the right door, it just leads to the knowledge of which of the other two doors holds the prize, in the more probable situation in which she's in fact chosen the wrong door. So she should always switch her choice, to maximize the probability of winning. The mistake many people make is to equate this situation to one in which the door with the booby prize is revealed *before* the contestant makes a choice. In that case, the sample space just consists of WL, LW for the two remaining doors and there is a 50/50 probability of making a mistake by switching.

Answer to Exercise 1.2: Listing the possibilities by birth order we have

$$BBB, BBG, BGB, GBB, BGG, GBG, GGB, GGG.$$

Fifty percent of the cases have two girls. Now consider that we know that one of the children is a girl named Florida. The sample space is

$$BBGf, BGfB, GfBB, BGGf, BGfG, GBGf, GfBG, GfBGf,$$

$$BGfGf, GfGfB, GGGf, GGfG, GfGG, GfGfG, GfGGf, GGfGf, GfGfGf.$$

We have allowed for the possibility that parents will give their children the same name. On the other hand, Florida is no longer very popular, so the probability that more than one child is named Florida is very very tiny, and we can neglect it. The sample space now has ten elements, out of which seven have two girls.

Answer to Exercise 1.3: Assuming an equal probability for each direction, the probability is just the fraction of 2π contained in the wedge, which is just $\frac{p}{N}$. It is independent of k. If one chooses a different origin, the only thing that is affected is the starting direction in our choice of drawing lines. If it is taken to be such that one of the lines from the origin goes through the point of the needle closest to the origin, then the answer is unchanged. If not, there is a fractional change by the angle between the line between the origin and the tip of the needle, and the nearest of the radial lines from the origin.

Answer to Exercise 1.5: If N is odd, there are two maxima, at $K = \frac{N\pm1}{2}$. Using Stirling's approximation to the factorial

$$M! \approx \sqrt{2\pi M}(M/e)^M,$$

we find the distribution

$$[\frac{N}{2}!]^{-2}[(1 - \frac{2x}{N})^{-N/2+x}(1 + \frac{2x}{N})^{-N/2-x}] \approx [\frac{N}{2}!]^{-2}e^{-\frac{4x^2}{N}}.$$

Answer to Exercise 1.7: There are $\frac{N!}{n!(N-n)!}$ different ways of choosing n voters out of a total population of N. We are interested in those ways in which k of the n are chosen from the population of pN who will vote for Jefferson and $n - k$ are chosen from the population of $(1 - p)N$ who will vote for his opponent. This number is

$$\frac{(pN)!}{k!(pN - k)!}\frac{[(1 - p)N]!}{(n - k)![(1 - p)N - n + k]!},$$

and the normalized distribution of the fraction $r = k/n$ is obtained by dividing this by $\frac{N!}{n!(N-n)!}$. In the regime of interest, $N \gg n \gg 1$, we can use Stirling's formula to approximate the factorial $(aN + b)! \approx \sqrt{2\pi aN}e^{-aN}(aN)^{aN+b}$. It is then easy to verify that N drops out of the ratio. The k dependent part of the distribution comes entirely from the denominator of the ratio of factorials and has the form

$$e^{-n[r\ln (r/p)+(1-r)\ln (\frac{1-r}{1-p})}.$$

Since $n \gg 1$, this is dominated by the maximum, which is $r = p$, and the distribution is approximately the Gaussian expansion around this point, whose width scales like $n^{-1/2}$. The confidence intervals similarly shrink with n, and polling experts believe that $n \sim 10^3$ is large enough to give reliable results. What is undoubtedly true is that the systematic errors involved in trying to poll a truly random sample of voters are so large that the additional precision of larger n (but always $\ll N$) would be somewhat illusory.

Answer to Exercise 2.1: According to the general algebraic formula, $v^i \rightarrow \sum_j M^i_j v^j$, the first element of the vector gotten by acting with

$$\begin{pmatrix} 0 & a \\ b & 0 \end{pmatrix} \tag{G.1}$$

on

$$\begin{pmatrix} c \\ d \end{pmatrix} \tag{G.2}$$

is $0c + bd = bd$, while the second is $ac + 0d = ac$, so

$$\begin{pmatrix} 0 & a \\ b & 0 \end{pmatrix} \begin{pmatrix} c \\ d \end{pmatrix} = \begin{pmatrix} bd \\ ac \end{pmatrix} \tag{G.3}$$

Answer to Exercise 2.2: The equation for the inverse is

$$MM^{-1} = \begin{pmatrix} a & b \\ c & d \end{pmatrix} \begin{pmatrix} e & f \\ g & h \end{pmatrix} = \begin{pmatrix} 1 & 0 \\ 0 & 1 \end{pmatrix} . \tag{G.4}$$

The product of the two matrices is

$$\begin{pmatrix} ae + bg & af + bh \\ ce + dg & cf + dh \end{pmatrix} . \tag{G.5}$$

The equations for the inverse are thus

$$ae + bg = 1 = cf + dh, \quad af + bh = 0 = ce + dg. \tag{G.6}$$

So, $g = -(c/d)e$, $h = -(a/b)f$ and $1 = (a - \frac{bc}{d})e = (c - \frac{ad}{b})f$. The last two equations have solutions if and only if $ac - bd \neq 0$. This combination of matrix elements is called the *determinant* of the matrix M and denoted $\det(M)$. It measures whether the rows and columns of the matrix are linearly independent vectors. The inverse matrix is

$$M^{-1} = (ad - bc)^{-1} \begin{pmatrix} d & -b \\ -c & a \end{pmatrix} , \tag{G.7}$$

and it is easy to verify that $M^{-1}M = 1$.

Answer to Exercise 2.3: To show that

$$\sigma_1 |\pm\rangle = |\mp\rangle,$$

write the equations as

$$\begin{pmatrix} 0 & 1 \\ 1 & 0 \end{pmatrix} \begin{pmatrix} 1 \\ 0 \end{pmatrix} = \begin{pmatrix} 0 \\ 1 \end{pmatrix} ; \tag{G.8}$$

$$\begin{pmatrix} 0 & 1 \\ 1 & 0 \end{pmatrix} \begin{pmatrix} 0 \\ 1 \end{pmatrix} = \begin{pmatrix} 1 \\ 0 \end{pmatrix} , \tag{G.9}$$

and use the result of Exercise 2.1. Now write

$$\sigma_1^2 = \begin{pmatrix} 0 & 1 \\ 1 & 0 \end{pmatrix} \begin{pmatrix} 0 & 1 \\ 1 & 0 \end{pmatrix} = \begin{pmatrix} 0 \times 0 + 1 \times 1 & 0 \times 1 + 1 \times 0 \\ 1 \times 0 + 0 \times 1 & 1 \times 1 + 0 \times 0 \end{pmatrix} = \begin{pmatrix} 1 & 0 \\ 0 & 1 \end{pmatrix}.$$

Answer to Exercise 2.4:

$$\sigma_2 = i \begin{pmatrix} 0 & 1 \\ 0 & 1 \end{pmatrix} \begin{pmatrix} 1 & 0 \\ 0 & -1 \end{pmatrix} = i \begin{pmatrix} 0 \times 1 + 1 \times 0 & 0 \times 0 + 1 \times -1 \\ 1 \times 1 + 0 \times 0 & 1 \times 0 + 0 \times -1 \end{pmatrix} = \begin{pmatrix} 0 & -i \\ i & 0 \end{pmatrix}.$$

$$\sigma_2^2 = i \times i \times \sigma_1 \sigma_3 \sigma_1 \sigma_3 = \sigma_1^2 \sigma_3^2 = 1,$$

where we have used $\sigma_1 \sigma_3 = -\sigma_3 \sigma_1$. It is straightforward to show that the same result follows from explicit matrix multiplication.

Answer to Exercise 2.5: Using standard probability theory, evaluate the expectation value of a general polynomial $P(E, D)$ of the energy and dipole moment of the ammonia molecule. Show that this is equivalent to the formula $\mathrm{Tr} P(E, D) \rho$. According to standard probability theory, the expectation value of $P(E, D)$ is just

$$\sum_{i=1}^{2} p_i P(\mathcal{E}_i, D_i).$$

Since $\mathcal{E}_i = E$ and $D_i = \pm d$, the most general polynomial reduces to $A + BD_i$, so the expectation value is

$$\langle P(E, D) \rangle = A + Bdp_1 - Bdp_2.$$

To do the same computation in terms of matrices, note that for a given polynomial $P(E, D) = A + BD$, where A and B are the same coefficients we got for that polynomial in the classical computation. Then

$$\mathrm{Tr}\, (\rho[A + BD]) = A + Bdp_1 - BdP_2.$$

Answer to Exercise 2.6: To show that

$$\epsilon_{abc}\epsilon_{cde} = \delta_{ad}\delta_{be} - \delta_{ae}\delta_{bd},$$

note that all three indices must be different in order for ϵ_{abc} to have a nonzero value. For a given value of c, this means that either $a = d$ or $a = e$, with b being the other value $\neq c$. The two possibilities are mutually exclusive, so for given values of a, b, d, e satisfying this constraint, only one value of c contributes to the sum implicit in the Einstein convention. Consequently the answer is always ± 1 and the $a = d$ and $a = e$ terms have opposite sign. For $a = d$ we have $\epsilon_{abc}\epsilon_{cab} = 1$, so the formula is proven.

Using this equation to evaluate $\nabla \times (\nabla \times \mathbf{V})$, for a vector function $\mathbf{V}(\mathbf{x})$, we have $[\nabla \times (\nabla \times \mathbf{V})]_a = \epsilon_{abc}\nabla_b\epsilon_{cde}\nabla_d V_e = \nabla_a\nabla_b V_b - \nabla^2 V_a$.

Answer to Exercise 2.7: The diagonal matrix element of A in some orthonormal basis is the coefficient of $|\mathcal{E}_i\rangle$ in the expansion

$$A|\mathcal{E}_i\rangle = A_{ij}|\mathcal{E}_j\rangle.$$

Taking the scalar product with $|\mathcal{E}_i\rangle$, this gives $A_{ii} = \langle\mathcal{E}_i|A|\mathcal{E}_i\rangle$. Now insert $1 = \sum_j |f_j\rangle\langle f_j|$ for any choice of basis. Then

$$A_{ii} = \langle\mathcal{E}_i||f_k\rangle\langle f_k|A|f_j\rangle\langle f_j||\mathcal{E}_i\rangle.$$

Performing the sum over i first means we are computing $(UU^\dagger)_{jk} = \delta_{jk}$, where U is the unitary transformation between the two bases. This proves that $A_{ii} = \langle f_i|A|f_i\rangle$.

Now note that $A_{ij}B_{ji}$ can be interpreted as Tr AB if we sum over j first or Tr BA if we sum over i first. The trace of a product of two operators is thus independent of the operator order.

Answer to Exercise 2.8: We evaluate Tr $A\rho$ by summing its diagonal matrix elements in the $|a_k\rangle$ basis, and inserting the identity $\rho = \sum_j |p_j\rangle p_j\langle p_j|$. We obtain

$$\text{Tr}\,(A\rho) = \sum_{jk} a_k\langle a_k|p_j\rangle p_j\langle p_j|a_k\rangle.$$

This has the form

$$\sum_k a_k P(a_k),$$

where

$$P(a_k) = \sum_j p_j|\langle p_j|a_k\rangle|^2.$$

The *soi-disant* probabilities are obviously positive. They are less than one because $\sum_j p_j = 1$ and the quantity multiplying each p_j is the norm squared of the projection on $|p_j\rangle$ of $|a_k\rangle$.

Furthermore, the sum over k just gives the norm of $|p_j\rangle$, which is 1, and the probabilities $P(a_k)$ then sum to $\sum_j p_j = 1$.

Answer to Exercise 2.9: The easiest way to do this problem is by explicit matrix multiplication. Let us take $n_a n_a = 1$, and introduce the complex number $z = n_1 - in_2$ and $n_3 = \pm w = \pm\sqrt{1 - zz^*}$. We will choose the plus sign. The matrix eigenvalue equation is

$$\begin{pmatrix} w & z \\ z^* & -w \end{pmatrix} \begin{pmatrix} a \\ \sqrt{1 - a^*a} \end{pmatrix} = \pm \begin{pmatrix} a \\ \sqrt{1 - a^*a} \end{pmatrix}.$$

We have used the freedom to multiply eigenstates by a phase to make the lower component zero. The absolute square of both sides of the upper component of the matrix equation is

$$|z|^2(1 - |a|^2) = |a|^2(w \mp 1)(w^* \mp 1).$$

Using the relation between z and w, we can write this as

$$|a|^2 = \frac{1}{2}[|z|^2 \pm (w + w^*)].$$

These are the probability to be in one of the two eigenstates of $n_a\sigma_a$, if the system is in $|+\rangle_3$.

Answer to Exercise 2.10: If $M = UDU^\dagger$, where D is diagonal and $UU^\dagger = U^\dagger U = 1$, then

$$[M, M^\dagger] = U[D, D^\dagger]U^\dagger = 0.$$

The most general 2×2 matrix has the form

$$M = \begin{pmatrix} a & b \\ c & d \end{pmatrix}.$$

The off diagonal part decomposes uniquely into a symmetric and antisymmetric piece, and these are proportional to σ_1 and σ_2, respectively. The diagonal part has the form $A + Bn_3\sigma_3$, where $a = A + Bn_3$ and $d = A - Bn_3$. Now we can compute

$$[M, M^\dagger] = [A + Bn_a\sigma_a, A^* + B^*n_b^*\sigma_b] = iBB^*\epsilon_{abc}n_a n_b^*\sigma_c.$$

This is zero only if $n_a = \beta n_a^*$ with β a complex number, which can be absorbed into B. In other words, by redefining B we can make n_a real. We have seen how to diagonalize $n_a\sigma_a$ for a real 3-vector, in the previous exercise.

Answer to Exercise 2.11: Write $n_a = \beta m_a$, where m_a is a unit vector so that $(m_a\sigma_a)^2 = 1$. Then

$$V = e^{in_a\sigma_a} = \cos(\beta) + i\sin(\beta)m_a\sigma_a,$$

which is diagonalizable according to the previous exercise. Using the algebra of the Pauli matrices and the fact that their traces are all zero its easy to see that

$$\text{Tr } V = 2\cos(\beta)$$

and

$$\text{Tr } V^2 = 2[\cos^2(\beta) - \sin^2(\beta)] = 2\cos(2\beta).$$

The eigenvalues are thus $e^{\pm i\beta}$, which means that V is unitary of determinant one. The extra factor $e^{i\alpha}$ gives a unitary with general determinant.

Answer to Exercise 2.12:

$$[A, B]^\dagger = [B^\dagger, A^\dagger] = [B, A] = -[A, B],$$

for any two Hermitian matrices. The matrix $G_{ab} \equiv \text{Tr } (\lambda_a\lambda_b)$ is real and symmetric. If v^a is a real n^2 dimensional vector then

$$v^a v^b G_{ab} = \text{Tr } (v_a\lambda_a)^2.$$

The square of a hermitian matrix only has positive eigenvalues, so G_{ab} is a positive definite matrix. This means that by replacing the matrices λ_a by linear combinations of themselves, we can make $G_{ab} = \delta_{ab}$. Now multiply the commutator

$$[\lambda_a, \lambda_b] = if_{abc}\lambda_c.$$

by λ_d and take the trace. We get

$$if_{abd} = \text{Tr } ([\lambda_a, \lambda_b]\lambda_d).$$

But the trace of a product of three matrices is independent of a cyclic permutation of their order. So we also have

$$if_{abd} = \text{Tr } ([\lambda_d, \lambda_a]\lambda_b) = \text{Tr } ([\lambda_b, \lambda_d]\lambda_a).$$

Thus, f_{abd} is antisymmetric under any transposition of indices.

Any matrix can be written $M = A + v^I\lambda^I$, where A and v^I are complex and I runs over only $n-1$ indices. We have simply identified λ_n as proportional to the unit matrix. The commutator of this matrix with its adjoint is

$$[M, M^\dagger] = f_{IJL}v^I v^{*J}\lambda_L.$$

Thus, the condition for a matrix to be diagonalizable is

$$f_{IJL}v^{I}v^{*J} = 0.$$

Answer to Exercise 2.13: Let $H = UDU^{\dagger}$, where D is diagonal. It follows that $H^2 = UD^2U^{\dagger}$, and more generally that $H^k = UD^kU^{\dagger}$. Using the power series expansion as the definition of the exponential, we have

$$e^{iH} = Ue^{iD}U^{\dagger}.$$

The matrix e^{iD} is a diagonal matrix of phases, and so is obviously unitary. Thus, we have written the exponential as a product of three unitary operators, which is unitary.

Answer to Exercise 2.14: We need the result that, for any two matrices, det $AB =$ det Adet B. As a consequence det Udet $U^{\dagger} = 1$ for any unitary matrix, and det $UDU^{\dagger} =$ det D. The relation

$$\det D = e^{\mathrm{tr}\ln D}$$

is obvious for diagonal matrices, and the remarks above prove that this is a basis independent statement.

Answer to Exercise 2.15:

$$P(E, D) \equiv \sum p_{nm} E_0^n d^m \sigma_3^m.$$

But $\sigma_3^m = \sigma_3$ if m is odd and 1 if m is even. Therefore,

$$P(E, D) = p_1(E_0, d) + p_3(E_0, d)\sigma_3.$$

p_1 is the polynomial gotten by taking only even m, i.e., with coefficients $p_{n,2k}$ and p_3 uses only the odd m coefficients.

Answer to Exercise 3.1: As noted in the hint, the fact that the operators $U(a)$ all commute and are unitary, means that there is a basis where they are simultaneously diagonalizable, with eigenvalues $e^{ik_i(a)}$, with real k_i. We have written this in a manner appropriate for a discrete spectrum. The generalization to continuous spectrum should be obvious. The multiplication law of the $U(a)$ shows that $k_i(a) = k_i a$. The operator K is the Hermitian operator with spectrum k_i.

Answer to Exercise 3.2: The statement that the Heisenberg equations of motion remain invariant under a Galilean boost with infinitesimal parameter δv is

$$\delta \dot{X} = \frac{i}{\hbar}([\delta H, X] + [H, \delta X]).$$

$$\delta \dot{P} = \frac{i}{\hbar}([\delta H, P] + [H, \delta P]).$$

Since δX and δP are both proportional to the unit operator, the second term on each right-hand side vanishes. We get

$$\hbar \delta v = i[\delta H, X].$$

$$0 = i[\delta H, P].$$

The first equation tells us that δH is nonzero, and the second that it commutes with P, and so is a function of it. The first equation tells us that it is in fact a linear function.

$$\delta H = \delta v P.$$

However, we also know that

$$\delta H = -i[N, H]\delta v = \delta v P.$$

This is

$$P = -i[\frac{m}{\hbar}(X - \frac{P}{m}t), H] = -i\frac{m}{\hbar}[X, H].$$

This equation implies that $H = \frac{P^2}{2m}$. We have also proven that

$$[N, H] = iP = \hbar \partial_t N/\partial t,$$

which means that the Heisenberg operator $N(t)$ is time independent.

Answer to Exercise 3.3: Start by computing

$$\frac{i}{\hbar}[\pm i\hbar(\frac{\partial E}{\partial P}\frac{\partial}{\partial X} - \frac{\partial E}{\partial X}\frac{\partial}{\partial P}), X] = \mp\frac{P}{m}. \tag{G.10}$$

$$\frac{i}{\hbar}[\pm i\hbar(\frac{\partial E}{\partial P}\frac{\partial}{\partial X} - \frac{\partial E}{\partial X}\frac{\partial}{\partial P}), P] = \pm\frac{\partial V}{\partial X}. \tag{G.11}$$

The Heisenberg equations of motion for H_{cl} are thus the same as Newton's equations if we choose the minus sign in the definition of H_{cl}. After dividing by $i\hbar$, the Schrödinger equation is then

$$\partial_t \psi(P, X) = (\frac{dV}{dX}\partial_X\psi - \frac{P}{m}\partial_P\psi).$$

The probability distribution $\rho = \psi^*\psi$ satisfies the same equation because the equation is first order in all derivatives.

For this energy function, the classical equations of motion for a function $f(X, P)$ are

$$\partial_t f = \partial_X f \frac{P}{m} - \partial_P f \frac{dV}{dx}.$$

Given an initial probability distribution $\rho(0, X, P)$ for the variables, we compute time-dependent expectation values via

$$\langle f \rangle = \int dX dP f(X(t), P(t)) \rho(0, X, P).$$

Alternatively, in a manner analogous to the transition between Heisenberg and Schrödinger pictures, we can compute the expectation value as

$$\langle f \rangle = \int dX dP f(X, P) \rho(t, X, P).$$

In order to capture the motion, the probability distribution must evolve *back* to the point where $X(t) = X$ and $P(t) = P$, so its equation of motion has the opposite sign from that expected from just following the forward time evolution of a function. Since quantum expectation values of functions of X and P are insensitive to the phase of the wave function, this quantum theory makes no use of the phase unless we want to discuss operators involving ∂_X or ∂_P. If we declare those operators to be "unphysical," its predictions are entirely equivalent to those of classical physics.

Answer to Exercise 3.5: The expection value is

$$\langle X^n \rangle = \frac{\int dx \, x^n e^{-\frac{(x-x_0)^2}{4\Delta^2}}}{\int dx \, e^{-\frac{(x-x_0)^2}{4\Delta^2}}}.$$

Shifting variables to $y = x - x_0$, this is:

$$\langle X^n \rangle = \frac{\int dy \, (y + y_0)^n e^{-\frac{y^2}{4\Delta^2}}}{\int dy \, e^{-\frac{y^2}{4\Delta^2}}}.$$

If we expand the integral in powers of y_0, then the coefficient of odd powers of y_0^k is proportional to the integral of $y^{n-k} e^{-ay^2}$ and vanishes if $n - k$ is odd because of the symmetry under $y \to -y$. The coefficient of y_0^k for $n - k = 2l$ is

$$\frac{n!}{(k)!(2l)!} \frac{\int dy \, y^{2l} e^{-\frac{y^2}{4\Delta^2}}}{\int dy \, e^{-\frac{y^2}{4\Delta^2}}}.$$

The numerator is $(-1)^l$ times the l-th derivative of the denominator with respect to $w \equiv \frac{1}{4\Delta^2}$. The denominator integral is

$$\int dy \, e^{-\frac{y^2}{4\Delta^2}} = \frac{4}{\Delta} \int_0^\infty du u^{-1/2} e^{-u} = \frac{4\sqrt{\pi}}{\Delta}.$$

This is $8\sqrt{\pi w}$, and it is easy to take the l-th derivative.

Answer to Exercise 3.7: Let

$$U_n|k\rangle = e^{i\frac{2\pi k}{N}}|k\rangle,$$

where k is an integer modulo n. The shift operator is

$$V_n|k\rangle|k+1\rangle,$$

where the addition is mod n, i.e., $k + n \equiv k$. Then

$$U_n V_n|k\rangle = e^{i\frac{2\pi(k+1)}{N}}|k+1\rangle,$$

and

$$U_n V_n|k\rangle = e^{i\frac{2\pi k}{N}}|k+1\rangle,$$

so the required operator algebra is satisfied.

Answer to Exercise 3.9: By expanding the exponential, it is easy to see that

$$U^k V_\alpha U^{-k} = e^{i\alpha \frac{U^k P_\theta U^{-k}}{\hbar}}.$$

Now let us evaluate

$$P_\theta U^{-k} f(\theta) = \frac{\hbar}{i} \frac{\partial}{\partial \theta} [e^{-ik\theta} f(\theta)] = -i\hbar k e^{-ik\theta} f(\theta) + e^{-ik\theta} P_\theta[f].$$

It follows that

$$U^k P_\theta U^{-k} = P_\theta - \hbar k,$$

so that

$$U^k V_\alpha U^{-k} V_{-\alpha} = e^{-ik\alpha}.$$

Answer to Exercise 4.1: Consider the unitarity equation $U^\dagger U = 1$. The rule of matrix multiplication is that the ij matrix element of the product matrix AB is the dot product (no complex conjugation) of the i-th row of A with the j-th column of B. The i-th row of $U^\dagger$ is the complex conjugate transpose of the i-th column of U, so the ij matrix element of $U^\dagger U$

is the scalar product between the i and j columns of U. Unitarity is just the statement that these are an orthonormal set. Since their number is equal to the dimension of the Hilbert space, they are an orthonormal basis.

Answer to Exercise 4.2: If

$$U = \sum_k |e_k\rangle\langle f_k|,$$

so that U maps the f basis into the e basis, then $U^\dagger$, its conjugate transpose is

$$U^\dagger = \sum_k |f_k\rangle\langle e_k|.$$

Then

$$U^\dagger U = \sum_k |f_k\rangle\langle e_k| \sum_l |e_l\rangle\langle f_l| = \sum_k |f_k\rangle\langle f_k| = 1.$$

The last step is the resolution of the identity for the basis f. The opposite order $UU^\dagger = 1$. To show that any unitary transformation can be written in this way, start with *any* orthonormal basis $|f_k\rangle$. The column vectors representing $|e_k\rangle = U|f_k\rangle$ in the $|f_k\rangle$ basis are just the columns of the matrix of U in that basis. The unitarity condition is the statement that those vectors form a new orthonormal basis.

Answer to Exercise 4.3:

$$\psi(x) = e^{-ax^2 + bx}.$$

$$\tilde\psi(k) = \int \frac{dx}{\sqrt{2\pi}} e^{-ikx} e^{-ax^2 + bx} = \int \frac{dx}{\sqrt{2\pi}} e^{-a(x - \frac{(b-ik)}{2a})^2} e^{\frac{1}{4a}(b-ik)^2}.$$

Now shift and rescale the variable of integration

$$y = \frac{x - \frac{(b-ik)}{2a}}{\sqrt{a}},$$

and obtain

$$\sqrt{a} \int \frac{dy}{\sqrt{2\pi}} e^{-y^2} e^{\frac{1}{4a}(b-ik)^2}.$$

Now

$$\int dy\, e^{-y^2} = [\int dy_1\, dy_2\, e^{-(y_1^2 + y_2^2)}]^{1/2},$$

and the latter integral can be done in radial/angular coordinates $d^2y = d\phi r dr$. The integrand is independent of π, so we get

$$\int dy\, e^{-y^2} = [\pi \int_0^\infty d(r^2) e^{-r^2}]^{1/2} = \sqrt{\pi}.$$

The end result is

$$\tilde{\psi}(k) = e^{\frac{1}{4a}(b-ik)^2}\sqrt{a/2}.$$

$$\psi(x) = e^{-a|x|}.$$

$$\tilde{\psi}(k) = \int_0^\infty \frac{dx}{\sqrt{2\pi}} [e^{-ikx} + e^{ikx}] e^{-ka} = \frac{1}{\sqrt{2\pi}}[\frac{1}{a+ik} + \frac{1}{a-ik}] = \sqrt{\frac{2}{\pi}}\frac{a}{k^2+a^2}.$$

$$\psi(x) = s\theta(a-x)\theta(x-b).$$

$$\tilde{\psi}(k) = s\int_0^\infty \frac{dx}{\sqrt{2\pi}} \int_b^a e^{-ikx} = \frac{is}{\sqrt{2\pi}}(e^{-ika} - e^{-ikb}).$$

Answer to Exercise 4.4: A smooth function has an infinite number of continuous derivatives. If we use $F^2 = R$ to write $f(-x)$ as the Fourier transform of the Fourier transform of $f(x)$, we have

$$\frac{d^n}{dx^n}f(x) = \int \frac{dk}{2\pi} e^{-ikx} \tilde{f}(k)(-ik)^n. \tag{G.12}$$

The existence of all of these derivatives implies that these integrals all converge, so that the Fourier transform falls off faster than any power. The Fourier transform of $(x^2 + a^2)^{-b}$ is

$$\tilde{f}(k) = \int dx e^{ikx} \int_0^\infty ds\ e^{-s(x^2+a^2)} \frac{s^{b-1}}{\Gamma(b)}.$$

We can do the Gaussian integral over x, obtaining,

$$\tilde{f}(k) = \sqrt{\pi} \int_0^\infty ds\ e^{-sa^2} e^{-\frac{k^2}{4s}} \frac{s^{b-3/2}}{\Gamma(b)}.$$

Rescale $s = kt$ so that the integral becomes

$$\tilde{f}(k) = \sqrt{\pi}k^{b-1/2} \int_0^\infty dt\ e^{-kta^2} e^{-\frac{k}{4t}} \frac{t^{b-3/2}}{\Gamma(b)}.$$

For large k we can do this integral by the steepest descent method. Its behavior is dominated by the stationary point of the function in the exponential $t_* = 1/4a^2$. $\tilde{f}(k)$ falls off exponentially, but the exponential is multiplied by a larger power of k for larger b.

Answer to Exercise 4.5 The Fourier transform is just

$$\tilde{\theta}(k) = \int_0^a \frac{dx}{\sqrt{2\pi}} e^{-ikx} = \frac{1}{\sqrt{2\pi}}\frac{e^{ika} - 1}{ik}. \tag{G.13}$$

This vanishes for $ka \ll 1$. The function is complex and oscillates in the regime $ka \sim 1$, while in the regime $ka \gg 1$, the first term oscillates so rapidly that it vanishes when averaged over small intervals in k and so in this regime, we have an almost featureless $1/k$ falloff. If this is the wave function of a particle at time $t = 0$, it should be multiplied by $1/\sqrt{a}$ in order to have proper normalization. It evolves into

$$\psi(x,t) = \int \frac{dk}{\sqrt{2\pi}} e^{-ikx - i\frac{\hbar k^2 t}{2m}} \frac{e^{ika} - 1}{ik\sqrt{a}}.$$

Factor out $\frac{1}{\sqrt{2\pi a}}$ and take the derivative with respect to a to get

$$\int \frac{dk}{\sqrt{2\pi}} e^{-ik(x-a) - i\frac{\hbar k t}{2m}}.$$

If it is positive this is a Gaussian integral. It is defined by analytic continuation from this regime to the one where t is positive. So the result is

$$e^{-\frac{im(x-a)^2}{2\hbar t}}.$$

Our wave function is the integral of this expression between 0 and a. If $|x| \gg a$, the integrand is very rapidly oscillating and the integral is very small, until $\hbar t \sim mxa$. After that time, the integral is order 1. Thus, the wave function falls off very rapidly outside the original interval but it spreads to cover the region out to x at a velocity $\frac{\hbar}{am}$. This is the velocity corresponding to the momentum uncertainty in the original wave function.

Answers to Exercises 4.6–4.8: The solution to the Schrödinger equation with energy E is

$$\psi_E = e^{\pm\sqrt{\frac{2m(V(x)-E)}{\hbar^2}}}, \tag{G.14}$$

where $V(x) = V_0\theta(a-x)\theta(x+a)$. Integrating the Schrödinger equation near the discontinuities in $V(x)$, we find that the wave function and its first derivative must be continuous there. If $E > 0$ we have delta function normalizable solutions, while if $E < 0$ we can get normalizable bound state solutions by picking only the falling exponential as $x \to \pm\infty$. The problem is invariant under reflections, so for the bound states we can choose to look at even and odd solutions. We can impose the continuity conditions only at $x = a$ and they will automatically be satisfied for negative x. In the region near the origin, the wave function depends on the crucial quantity

$$ik_0 = \sqrt{\frac{2m(V_0 - E)}{\hbar^2}}, \tag{G.15}$$

which is imaginary. The imaginary part of ak_0 is equal to the classical action in units of $\hbar$ of a particle traveling from $-a$ to a, if $E - V_0 > 0$. When the opposite inequality holds, k_0 is imaginary. In either case, we define k_0 with the positive (imaginary) square root. We define

$$ik = \sqrt{-E}\sqrt{\frac{2m(-E)}{\hbar^2}}. \tag{G.16}$$

In addition, define $r \equiv \frac{k_0}{k}$. In the region $|x| > a$, the wave function grows or falls exponentially if $E < 0$. We must choose the falling solution at both positive and negative values of x, to obtain a normalizable wave function. The continuity conditions for even and odd wave functions at $x = a$ are

$$A\cosh(|k_0|a) = Be^{-|k|a}, \quad Ar\sinh(|k_0|a) = -Be^{-|k|a}, \tag{G.17}$$

$$A\sinh(|k_0|a) = Be^{-|k|a}, \quad Ar\cosh(|k_0|a) = -Be^{-|k|a}, \tag{G.18}$$

when k_0 is imaginary. The hyperbolic sine and cosine are both positive, and these equations have no solution. Thus, we must have the intuitively obvious condition $E - V_0 > 0$, which implies a potential well rather than a barrier and asymptotic energy less than the depth of the well, in order to have a bound state. For k_0 real, the matching conditions become

$$A\cos(|k_0|a) = Be^{-|k|a}, \quad Ar\sin(|k_0|a) = -Be^{-|k|a}, \tag{G.19}$$

$$A\sin(|k_0|a) = Be^{-|k|a}, \quad Ar\cos(|k_0|a) = -Be^{-|k|a}. \tag{G.20}$$

Motivated by the bound on E, we define $E = yV_0$, with $0 \leq y \leq 1$. The conditions become

$$\sqrt{\frac{y}{1-y}} = -\tan(s_0\sqrt{1-y}). \tag{G.21}$$

$$\sqrt{\frac{y}{1-y}} = -\cot(s_0\sqrt{1-y}). \tag{G.22}$$

These depend only on a single parameter $s_0 \equiv a\sqrt{\frac{-2mV_0}{\hbar^2}}$. When s_0 is small, we can expand the trigonometric functions. The equation for odd solutions becomes $y = 1/s_0$, which is inconsistent with $0 \leq y \leq 1$. The even solution is $y = s_0$ and is consistent. There is a single bound state. When s_0 is large, we can satisfy the equation with $s_0\sqrt{1-y} \sim \frac{n\pi}{2}$, where n is odd for even solutions and vice versa, as long as $\frac{n\pi}{2s_0} < 1$. The point is that the tangent and cotangent take on any large value in the vicinity of their poles, so that as long as y is sufficiently close to 1 we can match. These explicit formulae are only valid for large n, but there are bound states for every value of n satisfying the inequality. As s_0 goes to infinity, we get an infinite number of states and the explicit formula becomes more exact. In this limit, after adding a constant to make all the energies positive, the spectrum approaches that of the infinite square well. Note, however, that s_0 can be large for a shallow, but very broad, well also. In that situation, we would still have continuum eigenstates.

Turning now to $E > 0$, we define

$$\psi^+ = A^+_{out}e^{ikx} + A^+_{in}e^{-ikx}, \quad x > a, \tag{G.23}$$

$$\psi^+ = A^+_{out}e^{-ikx} + A^+_{in}e^{-ikx}, \quad x < -a. \tag{G.24}$$

The subscripts *in* and *out* refer to the fact that, when multiplied by $e^{-i\frac{\hbar^2 k^2}{2m}t}$, the relevant part of the solution becomes a traveling incoming or outgoing wave. To write the continuity conditions at $x = \pm a$ compactly, it is convenient to define $z = e^{ik_0 a}$ and $\alpha^\pm_{in} = A^\pm_{in}e^{-ika}$ $\alpha^\pm_{out} = A^\pm_{out}e^{ika}$. Then we have

$$Az + Bz^{-1} = \alpha^+_{out} + \alpha^+_{in}, \tag{G.25}$$
$$Az^{-1} + Bz = \alpha^-_{out} + \alpha^-_{in}, \tag{G.26}$$
$$r(Az - Bz^{-1}) = \alpha^+_{out} - \alpha^+_{in}, \tag{G.27}$$
$$r(Az^{-1} - Bz) = -\alpha^-_{out} + \alpha^+_{in}. \tag{G.28}$$

We can solve these equations by adding and subtracting the first and third and also the second and fourth, obtaining:

$$2\alpha^+_{out} = (1+r)zA + (1-r)z^{-1}B, \tag{G.29}$$
$$2\alpha^-_{out} = (1-r)z^{-1}A + (1+r)zB, \tag{G.30}$$
$$2\alpha^+_{in} = (1-r)zA + (1+r)z^{-1}B, \tag{G.31}$$
$$2\alpha^+_{out} = (1+r)z^{-1}A + (1-r)zB. \tag{G.32}$$

These equations reflect the fact that there are only two linearly independent solutions of the Schrödinger equation. The *in* and *out* states represent two independent bases for the Hilbert space. These equations tell us how to transform between them. The matrix relating them is $S = S_+S_-^{-1}$, where $S_\pm$ the matrices exhibited above, relating the in and out bases to the A, B basis. Using the usual formula for the inverse of a 2×2 matrix, we get

$$S = \frac{1}{(1-r)^2z^2 - (1+r)^2z^{-2}} \begin{pmatrix} (1-r^2)(z-z^{-1}) & -4r \\ -4r & (1-r^2)(z-z^{-1}) \end{pmatrix}. \tag{G.33}$$

Answer to Exercise 4.9: The reflection amplitude is of course the same ratio of reflected to incident wave

$$R = (|\frac{A_{out}^-}{A_{in}^-}|)^2.$$

The out amplitudes are related to the in amplitudes by the S-matrix

$$A_{out}^+ = S_{11}A_{in}^+ + S_{12}A_{in}^- = S_{12}A_{in}^-.$$

$$A_{out}^+ = S_{21}A_{in}^+ + S_{22}A_{in}^- = S_{22}A_{in}^-.$$

The transmission and reflection coefficients are just the absolute squares of the S_{12} and S_{22} coefficients of the S-matrix which is the norm of the vector formed by the second column of the matrix. Unitarity of the S-matrix guarantees that this is 1. We calculated the S-matrix in Exercise 4.8

$$S_{12} = \frac{-4r}{(1-r)^2 z^2 - (1+r)^2 z^{-2}}.$$

Recall that

$$r = \frac{k_0}{k}, \qquad z = e^{ik_0 a}.$$

Thus,

$$T = |\frac{4}{2(1/r + r)\sin(k_0 a) - 4\cos(k_0 a)}|^2.$$

In the range of energies where k_0 is real, but $r < 1$, this oscillates as a function of energy. At points where $\sin(k_0 a) = 0$, $T = 1$, exhibiting the Ramsauer–Townsend effect.

Answer to Exercise 4.11: The solution of the Heisenberg equation of motion is $X(t) = X + \frac{Pt}{m}$. The uncertainty in this operator is the square root of

$$\langle |X^2 + \frac{t}{m}(XP + PX) + \frac{t^2 P^2}{m^2}\rangle - \frac{t^2}{m^2}(\langle P\rangle)^2.$$

For large t, this is dominated by the uncertainty of P in the wave function ψ. Heisenberg's uncertainty relation says that $\Delta P \geq \hbar \Delta X$. Assuming minimal uncertainty, this implies a position uncertainty at time t of order

$$\Delta X(t) \sim \frac{t\hbar}{m\Delta X(0)}.$$

So the 1 m uncertainty is achieved at time

$$t = m(1 \text{ m})\Delta X/\hbar,$$

where $\hbar = 10^{-34} \frac{\text{kg m}^2}{s}$. The time in seconds is

$$t = 10^{34} m \Delta X \text{s},$$

where the mass is expressed in kilograms and the initial position uncertainty in meters. For a baseball with $\Delta X = .001$ and $m = .5$, this is 5×10^{30} s or about 5×10^{23} years. For the moon, with the same initial uncertainty, the time becomes about 4×10^{46} years.

Answer to Exercise 4.13: Write $P_n(x) = \sum_{k=0}^{n} C_{nk} x^k$. By definition, P_n is orthogonal to all of the lower P_k, but since x^k for $k < n$ is a linear combination of the lower P_k, we have

$$\int dx \ r(x) \sum_{k=0}^{n} C_{nk} x^k x^l = \sum_{k=0}^{n} C_{nk} r_{k+l} = 0$$

for each l with $0 \le l < n$. This says that the $n+1$ dimensional vector whose k-th component is C_{nk} is orthogonal to all n of the $n + 1$ dimensional vectors R_l, where $0 \le l < n$ and R_l is the vector whose k-th component is r_{l+k}. C_{nk} is thus given by

$$C_{nk} = c_n \epsilon_{ka^1...a^n} r_{1a^1} \ldots r_{na^n},$$

where ϵ is the $n + 1$ dimensional Levi-Civita symbol. This is equivalent to the determinant formula quoted in the exercise. The norm of the polynomial is easily computed in terms of the moment matrices

$$\int dx \ r(x) P_n^2(x) = \sum_{k,l=0}^{n} C_{nk} r_{kl} C_{nl},$$

and this determines c_n.

Answer to Exercise 4.15: The multiplication operator by the function c is obviously Hermitian if c is real, so it is sufficient to study the other two terms. Using integration by parts $(d \equiv d/dx)$

$$\int rf^*(ad^2g + bdg) = \int [d^2(rf^*a)g - d(rbf^*)g].$$

In order for this to be $\int rg(a^* d^2 f^* + b^* f^*)$, which would give Hermiticity if a and b were real, we need

$$d^2(ra) - d(rb) = 0, \quad \text{and} \quad d(ra) = rb.$$

The two constraints are compatible, and evaluate b in terms of a.

Answer to Exercise 4.17:

$$\int dx \ f(x) \frac{d}{dx} \theta(x) = -\int dx \ \frac{df}{dx} \theta(x),$$

as long as $f(x)$ vanishes at the endpoints of the integration. If the integral of integration does not include zero, this expression vanishes: if it is all on the negative axis $\theta(x) = 0$. If it is all on the positive axis, then the integral is just the difference of the values of f at the endpoints, both of which vanish. If the interval includes zero, then

$$\int dx \ f(x) \frac{d}{dx} \theta(x) = -\int_0^a dx \ \frac{df}{dx} = f(0) - f(a) = f(0).$$

Answer to Exercise 4.18: The residue theorem states that the integral of a function with only poles, around a counterclockwise contour surrounding some of the poles is equal to $2\pi i$ times the sum of the residues of the poles inside the contour. In our case, we have a single pole, at $s = i\epsilon$. If $x > 0$ then in the upper half plane the integrand falls exponentially, so we can add a circle at infinity to make a closed contour, which is followed counterclockwise. We thus pick up the residue at the pole, which gives the value 1. If $x < 0$ we can close the contour in the lower half plane, and get zero, because there are no poles.

Answer to Exercise 4.19: Since the potential goes to infinity outside the well, the contribution to the expectation value of the Hamiltonian in *any* state that has $\psi^*\psi \neq 0$ outside the well will be infinite. Therefore, all finite energy eigenstates vanish outside the well, and they are all bound states. Now integrate the Schrödinger equation in a tiny interval straddling the point $x = a$. We get

$$-[\frac{d\psi}{dx}]_{a-\epsilon}^{a+\epsilon} \propto |V_0|\psi(a + \epsilon),$$

since we have set the potential to zero inside the well. The derivative vanishes at $a + \epsilon$ so the derivative at the wall is related to the wave function at the wall, by a relation of the form $d\psi/dx = c\psi$. The constant c is a limit of the product of the infinite potential and the vanishing wave function outside the wall, and so can be any number. Indeed, the Hamiltonian is a Hermitian operator for any choice of c, as one can verify by integration by parts. A similar boundary condition must be imposed at $x = -a$. The solutions inside the well are of course e^{ikx} with $E = \frac{\hbar^2 k^2}{2m}$. The boundary conditions are

$$ik(Ae^{ika} - Be^{-ika}) = c(Ae^{ika} + Be^{-ika}),$$

$$ik(Ae^{-ika} - Be^{ika}) = c(Ae^{-ika} + Be^{ika}).$$

These lead to two evaluations of B/A, which agree only if

$$e^{4ika} = 1,$$

unless $c = 0$. In that case, both conditions are solved by $\sin(ka) = 0$. In all cases, we have only a discrete spectrum of energies.

Answer to Exercise 4.21: The solutions have the form $A_\pm e^{\pm kx}$ with $k > 0$ for $x > a$ and $x < -a$, respectively. We take $A_- = 1, A_+ = A$. In the region $-b \leq x \leq b$, the solution has the form $C_+ e^{kx} + C_- e^{-kx}$ and in the interval $[-a, -b]$, we have $B_+ e^{ik_0 x} + B_- e^{-ik_0 x}$. Finally, in the interval $[b, a]$, we have $D_+ e^{ik_0 x} + D_- e^{-ik_0 x}$. The eigenvalue is $E = -\frac{\hbar^s k^2}{2m}$ and $\hbar k_0 = \sqrt{2m(V_0 - \frac{\hbar^2 k^2}{2m})}$. Define $\alpha = e^{ak}$, $\beta = e^{bk}$, $a_0 = e^{ik_0 a}$, $b_0 = e^{ik_0 b}$. We have continuity conditions at $\pm a$ and $\pm b$. These are

$$-a: \ \alpha^{-1} = B_+ a_0^{-1} + B_- a_0, \qquad k\alpha^{-1} = ik_0(B_+ a_0^{-1} - B_- a_0),$$

$$-b: \ C_+ \beta^{-1} + C_- \beta = B_+ b_0^{-1} + B_- b_0, \qquad k(C_+\beta^{-1} - C_-\beta) = ik_0(B_+ a_0^{-1} - B_- a_0),$$

$$b: \ C_+\beta^+ C_-\beta^{-1} = D_+ b_0 + D_- b_0^{-1}, \qquad k(C_+\beta^{-1} - C_-\beta) = ik_0(D_+ b_0 - D_- b_0^{-1}),$$

$$a: \ \alpha^{-1}A = D_+ a_0 + D_- a_0^{-1}, \qquad -k\alpha^{-1}A = ik_0(D_+ a_0^- D_- a_0^{-1}).$$

These can be viewed as matrix equations of the form

$$M_1 B = v, \quad M_2 B = M_3 C, \quad M_4 C = M_5 D, M_6 D = w.$$

The vectors v and w are $v = \alpha^{-1}(1, 1)$ and $w = A\alpha^{-1}(1, -1)$. The vectors B, C, D are composed of (B_+, B_-), etc. We can solve for B in two different ways

$$B = M_1^{-1} v = M_2^{-1} M_3 M_4^{-1} M_5 M_6^{-1} w.$$

This gives us two conditions on the two free parameters A and k, so there will only be discrete solutions for k. To argue that the ground state energy is lowered when b is decreased with $b - a$ fixed, we take the derivative

$$(\partial_a + \partial_b)[\langle \psi(a, b)|H|\psi(a, b)\rangle].$$

Using the fact that ψ is an eigenstate, this is

$$E_0(\partial_a + \partial_b)[\langle \psi(a, b)|\psi(a, b)]\rangle + \langle \psi(a, b)|(\partial_a + \partial_b)H|\psi(a, b)\rangle.$$

The first term vanishes because ψ is normalized to 1 for all values of the parameters. The second term is the expectation value of

$$- V_0[\theta(x - b)\theta(b + x)(\delta(a - x)\theta(a + x) + \delta(a + x)\theta(a - x))$$
$$- \theta(x + a)\theta(a - x)(\delta(x - b)\theta(b + x) + \delta(x + b)\theta(b - x))].$$

The first term vanishes because $b < a$ while the second is positive. Thus, the ground state energy is monotonically increasing as b is increased with $b - a$ fixed.

Answer to Exercise 4.23: The wave functions at $\pm\infty$ give rise to currents:

$$J_\pm = i(\psi_{pm} * \partial_x \psi_\pm - \psi_\pm \partial_x(\psi_\pm)) = -2k(A_* \pm A_\pm - B_\pm * B_\pm).$$

This current must be the same at plus infinity as at minus infinity because the solution is time independent and the current conserved. Thus,

$$A_+ * A_+ + B_- * B_- = A_- * A_- + B_+ * B_+,$$

which is the equation saying that the S-matrix preserves the norm of vectors. Consequently S is unitary.

Answer to Exercise 5.1:

$$\langle s|H^2|s\rangle = \|H|s\rangle\|^2, \tag{G.34}$$

since Hermiticity of H implies that $\langle s|H$ is the bra corresponding to the ket $H|s\rangle$. Thus, expectation values of H^2 are always nonnegative, and the same is true for sums of squares of Hermitian operators. Applying this result to the case where $|s\rangle$ is an eigenstate of H, we find that the eigenvalues of H^2 are all nonnegative.

Answer to Exercise 5.3: 5.3a is a trivial exercise in rescaling variables. To solve **5.3b**, write the equation for the logarithm S of the wave function

$$-(\partial_y^2 S + (\partial_y S)^2) + y^2 = 2\epsilon.$$

At large y, this is solved by

$$\partial_y S = \pm\sqrt{y^2},$$

because, given this ansatz, both the second derivative term and the ϵ term are smaller by two powers of y.

5.3c: If $\psi = e^{-\frac{y^2}{2}}v$, then

$$-\partial_y^2 v + y\partial_y v = (2\epsilon - 1)v.$$

Writing $v = \sum_{n=0}^\infty v_n y^n$, we obtain the recursion relation

$$(n + 2)(n + 1)v_{n+2} = (n + 1 - 2\epsilon)v_n.$$

This shows that we can study even and odd solutions separately. At large n, we get

$$v_{n+2} \sim \frac{1}{n}v_n,$$

which is solved by $v_{2k} = \frac{1}{k!}$ in the even case, and the same formula for v_{2k+1} in the odd case. Thus, we get back the bad exponential behavior unless the recursion stops. This can only happen if $2\epsilon = n + 1$, so we have derived the quantization of the energy levels.

Answer to Exercise 5.5: The oscillator ground state wave function is

$$\psi_0 = (\frac{m\omega}{\pi\hbar})^{1/4} e^{-\frac{m\omega x^2}{2\hbar}}.$$

The overlap of two such wave functions is a Gaussian integral, which gives

$$\langle \psi_0(\omega)|\psi_0(\omega')\rangle = (2)^{1/2}(\frac{\omega\omega'}{(\omega + \omega')^2})^{1/4} < 1.$$

Answer to Exercise 5.7: Recall that

$$a^\dagger \equiv \frac{1}{\sqrt{2\hbar m\omega}}(m\omega X - iP)$$

and

$$a \equiv \frac{1}{\sqrt{2\hbar m\omega}}(m\omega X + iP).$$

Introduce the dimensionless variable y by

$$x = \sqrt{\frac{\hbar}{m\omega}}y.$$

Then

$$a^\dagger \equiv \frac{1}{\sqrt{2}}(y - \frac{d}{dy}),$$

$$a = \frac{1}{\sqrt{2}}(y + \frac{d}{dy}).$$

Write the wave function of the n-th eigenstate as

$$\psi_n = (\frac{m\omega}{\pi\hbar})^{1/4} H_n(y) e^{-\frac{y^2}{2}}.$$

Note that this is not the standard mathematical normalization of the Hermite polynomials, which takes the coefficient of the highest power of y to be 1. It is also different from what is called the "physicist's" normalization in the Wikipedia article on Hermite polynomials.

With this normalization

$$a^\dagger H_n e^{-\frac{y^2}{2}} = \frac{1}{\sqrt{2}}(2yH_n - H'_n)e^{-\frac{y^2}{2}} = \sqrt{n+1}^{1/2} H_{n+1} e^{-\frac{y^2}{2}},$$

so

$$(2yH_n - H'_n) = \sqrt{n+1}\sqrt{2}H_{n+1}.$$

Similarly,

$$\sqrt{n}\sqrt{2}H_{n-1} = H'_n.$$

Answer to Exercise 5.9: The uncertainty of A is defined as

$$\langle A^2 \rangle - \langle A \rangle^2.$$

The Hilbert space of the K–G field is the tensor product of individual oscillator Hilbert spaces, and a coherent state is a tensor product of coherent states of different oscillators. If we take operators A_i acting in different tensor factors, then

$$\langle A_i A_j \rangle = \langle A_i \rangle \langle A_j \rangle.$$

Thus, in computing the uncertainties of N and H, which are sums over modes, we need only sum the uncertainties for individual modes. The key part of the calculation is

$$\frac{\langle z|(a^\dagger a)^2|z \rangle}{\langle z|z \rangle} = e^{-z\bar{z}}[\sum_{n=0}^{\infty} \frac{1}{n!}n^2(z\bar{z})^n].$$

Using $n^2 = n(n-1) + n$ we write this as

$$\frac{\langle z|(a^\dagger a)^2|z \rangle}{\langle z|z \rangle} = e^{-z\bar{z}}[\sum_{n=0}^{\infty} \frac{1}{n!}(z\bar{z})^2 + z\bar{z})(z\bar{z})^n] = (z\bar{z})^2 + z\bar{z}.$$

We have dropped terms in the sums that are zero and renamed the two summation variables. A similar and simpler calculation gives

$$\frac{\langle z|(a^\dagger a)|z \rangle}{\langle z|z \rangle} = z\bar{z}.$$

Therefore, we have

$$\langle N^2 \rangle - \langle N \rangle^2 = \sum_{\mathbf{k}} z(\mathbf{k})z^*(\mathbf{k}),$$

$$\langle H^2 \rangle - \langle H \rangle^2 = \sum_{\mathbf{k}} \omega(\mathbf{k})z(\mathbf{k})z^*(\mathbf{k}).$$

This calculation is obviously valid for any dispersion relation $\omega(\mathbf{k})$ for the waves, and so is more general than the Klein–Gordon equation. Note that if zz^* is large in some range of $\mathbf{k}$ the uncertainties, which are the square roots of the above combinations of expectation values, are much less than the expectation values of the operators themselves. This is the basis for the argument that the coherent states behave like classical fields.

Answer to Exercise 5.11: Consider a single annihilation operator. Since $a^2 = 0$, there must be a subspace of states satisfying $a|s_0\rangle = 0$. The block decomposition of a in the basis formed by this subspace and its orthogonal complement is

$$a = \begin{pmatrix} A & 0 \\ B & 0 \end{pmatrix},$$

so that

$$a^\dagger = \begin{pmatrix} A^\dagger & B^\dagger \\ 0 & 0 \end{pmatrix}.$$

As a consequence

$$aa^\dagger \pm a^\dagger a = \begin{pmatrix} AA^\dagger \pm (B^\dagger B + A^\dagger A) & AB^\dagger \\ BA^\dagger & BB^\dagger \end{pmatrix}.$$

So, we must have

$$AB^\dagger = BA^\dagger = 0$$

and

$$AA^\dagger \pm (B^\dagger B + A^\dagger A) = 1 = BB^\dagger.$$

The first of these equations is generally for a rectangular matrix and the second is actually for two square matrices of (possibly) different dimensions. Since the products of the two (possibly) rectangular matrices are nonnegative Hermitian operators and $BB^\dagger = 1$, the only way to satisfy these equations is to set $A = 0$ and choose the plus sign. B is then a unitary operator on a space of dimension half the original space and

$$a = B \otimes \begin{pmatrix} 0 & 0 \\ 1 & 0 \end{pmatrix}.$$

Notice that we have also solved Exercise 5.12, by taking the dimension of the space B acts on to be one dimensional. For multiple fermionic creation and annihilation operators, we just write $a_i = \alpha_i \otimes \sigma_3$ for $i > 1$. The α_i anticommute with each other, commute with B, and square to zero.

Answer to Exercise 5.13: Start from an infinite collection of commuting copies of the Pauli matrices

$$[\sigma_a(i), \sigma_b(j)] = 2i\delta_{ij}\epsilon_{abc}\sigma_c(i).$$

Choose an ordering of the indices i as points on an infinite one-dimensional lattice. Define $a(i) = \sigma_+(i) \prod_{j>i} \sigma_3(j)$. Then each $a(i)$ squares to zero and the different $a(i)$ anticommute because (assuming without loss of generality that $j > i$) the $\sigma_3(j)$ in $a(i)$ anticommutes with $\sigma_+(j)$. If the indices are arranged on a higher dimensional lattice, then the Jordan–Wigner construction provides operators that depend on a choice of a line from the point where the operator $a(i)$ sits, out to infinity.

Answer to Exercise 5.14: The Heisenberg equation of motion is

$$\dot{a} = \frac{i}{\hbar}[H, a] = -i\omega a,$$

so that

$$a(t) = e^{-i\omega t} a.$$

Thus,

$$ae^{-i\frac{H}{\hbar}t}|z\rangle = e^{-i\omega t} z e^{-i\frac{H}{\hbar}t}|z\rangle. \tag{G.35}$$

This tells us that the time evolved state is proportional to a coherent state with $z(t) = e^{-i\omega t} z$. Since the time evolved state has the same norm as the original coherent state, the proportionality constant must be a pure phase. Indeed, that phase comes from the constant term in the Hamiltonian (which doesn't change the Heisenberg operator $a(t)$ but just changes the state by a time-dependent phase). Thus, the phas is just $e^{-i\frac{\omega}{2}t}$. Note that the equation for the coherent state with $z = 0$ is the same as the equation for the ground state of the Hamiltonian, and this is consistent with the evolution law above.

Answer to Exercise 6.1: In components, the inequality reads

$$\left|\sum v_i^* w_i\right| \le \sqrt{\sum v_i^* v_i \sum w_i^* w_i}.$$

We can divide through by the norms of the vectors, so that the inequality says that the scalar product of two unit vectors has absolute value less than 1. The absolute value of the sum is obviously maximized if all terms in it have the same phase, so we can restrict attention to the case where all components are positive. Now consider that $0 \le \sum(v_i \pm w_i)^2 = \sum(v_i^2 + w_i^2 \pm 2v_i w_i) = 2 \pm 2 \sum v_i w_i$. This is precisely the required inequality.

Answer to Exercise 6.3: The equations

$$[A, A^\dagger] = 0$$

are N^2 complex equations for N^2 complex unknowns. Writing $A = H_1 + iH_2$, where H_i are Hermitian, normality implies H_1 commutes with H_2. The most general Hermitian matrix has N^2 real parameters. Given such a matrix, the condition that a second Hermitian matrix commutes with it puts $N^2 - N$ real constraints on those parameters. To see this, note that the equation

$$[H_1, H_2] = 0$$

looks like N^2 real conditions, but it is automatically satisfied if $H_2 = H_1^k$. A generic Hermitian matrix will, by the Cayley Hamilton theorem, satisfy its characteristic polynomial $P_{H_1}(H_1) = 0$, but no lower order equation, so the powers $k = 0, \ldots, N - 1$ will be linearly independent matrices. Thus, the commutator condition is, generically, $N^2 - N$ constraints and a normal operator has $N^2 + N$ real parameters.

A unitary matrix has N^2 real parameters, but in the formula

$$A = U^\dagger D U,$$

we can ignore unitaries that differ by $U_1 = U_D U_2$, where U_D is diagonal in the same basis as D. There are N independent unitary matrices of this form, each of which depends on one real phase. The complex diagonal matrix D depends on $2N$ independent real parameters, so diagonalizable matrices have the same number of independent real parameters as normal matrices. They are all normal. The spectral theorem, whose proof is sketched in the text, shows that all normal matrices are diagonal in some orthonormal basis.

Answer to Exercise 6.5: Insert a parameter t so that we are trying to compute

$$E_1 \equiv e^{t(A+B)} = 1 + t(A + B) + \frac{t^2}{2}(A + B)^2 + \frac{t^3}{3!}(A + B)^3 + o(t^4).$$

On the other hand,

$$E_2 \equiv e^{tA}e^{tB} = (1 + t(A) + \frac{t^2}{2}(A)^2 + \frac{t^3}{3!}(A)^3 + o(t^4))(1 + t(B) + \frac{t^2}{2}(B)^2 + \frac{t^3}{3!}(B)^3 + o(t^4)).$$

The two expressions agree up to order t, but at order t^2 we have

$$\frac{1}{2}(A^2 + B^2 + AB + BA),$$

for E_1 and

$$\frac{1}{2}(A^2 + B^2 + AB),$$

for E_2. To this order, we can fix things by writing

$$E_1 = E_2 e^{-\frac{t^2}{2}[A,B]},$$

but at order t^3 this gives

$$\frac{t^3}{3!}(A^3 + B^3 + AB^2 + BAB + B^2A + BA^2 + ABA + A^2B)$$

$$\neq \frac{t^3}{3!}(A^3 + B^3) + \frac{t^2}{2}(A^2B + AB^2 - (AB + BA)).$$

This can be fixed by writing

$$E_1 = E_2 e^{-\frac{t^2}{2}[A,B]} e^{\frac{t^3}{6}(2[B,[A,B]]+[A,[A,B]])}.$$

The term of order t^k involves k fold commutators. The BCH formula can be derived in an analogous manner. The most interesting case of these identities occurs when the commutator algebra closes after a finite number of iterations. Such a structure is called a finite dimensional Lie algebra.

Answer to Exercise 6.7: The matrix of V in the k basis is

$$V = \begin{pmatrix} 0 & 1 & 0 & \cdots & 0 & 0 \\ 0 & 0 & 1 & 0 & \cdots & 0 \\ \vdots & \vdots & \vdots & \vdots & & \vdots \\ 0 & 0 & 0 & \cdots & & 1 \\ 1 & 0 & 0 & 0 & \cdots & 0 \end{pmatrix}.$$

Its Hermitian conjugate simply has all the ones on the diagonal just below the main diagonal, except for one in the upper right-hand corner. This is the matrix that rotates the clock in the opposite direction, so $V^\dagger = V^{-1}$. The equation

$$UV = VU e^{\frac{2\pi i}{N}}$$

is simply the mathematical formula that says that V shifts the clock by one unit. The matrix V^k has ones on the diagonals k units above the main diagonal and $k-1$ units above the lower left corner. The matrix $U^l V^k$ multiplies the l-th row of V^k by $e^{\frac{2il\pi}{N}}$. Thus, these matrices are all linearly independent. Since there are N^2 of them, they form a basis for all matrices.

Answer to Exercise 6.9: The formula

$$\langle f|g \rangle = \int_{-1}^{1} \mu(x) f^*(x) g(x)$$

defines a scalar product on the space of square integrable functions $\int_{-1}^{1} \mu(x) f^*(x) f(x) < \infty$. The monomials x^n are linearly independent, and if the moments are all finite we have

$$\langle x^n|x^m \rangle = M_{n+m},$$

so they are not orthonormal. Define

$$P_k(x) = \sum_{n=0}^{k} p_{kn} x^n.$$

Then

$$\langle P_k | P_l \rangle = p_{kn}^* M_{n+m} p_{lm}.$$

Considered as a matrix in the indices m, n, M_{n+m} is real and symmetric, and so can be diagonalized by an orthogonal transformation. That is

$$M_{n+m} = O_{np} m_p O_{mp},$$

with

$$O_{np} O_{mp} = \delta_{nm}.$$

If we take

$$p_{kn} = p_{kn}^* = O_{nk} m_k^{-1/2},$$

we have constructed orthonormal polynomials.

Answer to Exercise 6.11: We have

$$U^\dagger(x) U(x) = (x_0 - i x_j \sigma_j)(x_0 + i x_k \sigma_k) = x_0^2 + x_j x_k \sigma_j \sigma_k = x_0^2 + x_j x_k (\delta_{jk} + i \epsilon_{jkl} \sigma_l).$$

The last term vanishes because ϵ_{jkl} is antisymmetric and $x_k x_l$ symmetric under interchange of the summation indices. Thus, $U^\dagger U = 1$. Writing $x_0 = \cos(\theta)$, $x_j = \sin(\theta) e_j$, where $e_j e_j = 1$, we can write $U = e^{i\theta e_j \sigma_j}$. Then $\det U = e^{\text{tr } i\theta e_j \sigma_j} = e^0 = 1$. Since the Pauli matrices and the unit matrix are a complete set, we get the most general matrix by letting x_a be four arbitary complex numbers. Then

$$U^\dagger(x) U(x) = (x_0^* - i x_j^* \sigma_j)(x_0 + i x_k \sigma_k) = x_a^* x_a + i(x_0 * x_j - x_j^* x_0) + i \epsilon_{jkl} x_k^* x_l) \sigma_j.$$

The cross product between the three-vectors x_j and x_j^* is perpendicular to both of those vectors, so unitarity requires it to vanish, and also that $x_0^* x_j = x_j^* x_0$. This means that the complex three vector is an overall phase times a real three vector and that the phases of x_0 and x_j are equal. Thus, the most general unitary 2×2 matrix is an overall phase multiplied by a matrix with real x_a. The determinant of the matrix is just the square of the phase and so is equal to 1 only if the phase is real.

Alternate Answer to Exercise 6.11: By cyclicity of the trace,

$$\text{Tr} \left[U^\dagger(x_a) \sigma_i U(x_a) \right] = \text{Tr} \left[\sigma_i \right] = 0.$$

Since any matrix is a linear combination of the unit matrix and the Pauli matrices, and we have just proven that the coefficient of the unit matrix is zero, we indeed have

$$U^\dagger(x_a)\sigma_i U(x_a) = R_{ij}\sigma_j.$$

Now consider

$$\delta_{ij} = \text{Tr}[\sigma_i\sigma_j] = \text{Tr}\,[U^\dagger(x_a)\sigma_i\sigma_j U(x_a)] = \text{Tr}\,[U^\dagger(x_a)\sigma_i U(x_a)U^\dagger(x_a)\sigma_j U(x_a)]$$
$$= R_{ik}R_{jl}\text{Tr}\,[\sigma_k\sigma_l] = (RR^T)_{ij}.$$

Answer to Exercise 6.13: Since the subalgebra is closed under Hermitian conjugation C commutes with both a and $a^\dagger$ for every member a of the algebra, which implies that $C^\dagger$ also commutes with every element of the subalgebra. Thus, the Hermitian operators $(C \pm C^\dagger)i^{1\mp 1}$, commute with every a. The eigenspaces with fixed eigenvalue, of these operators, would be invariant subspaces, unless all their eigenvalues are equal, which implies that $C \pm C^\dagger$ are both proportional to the unit matrix.

Answer to Exercise 7.1:

$$[L_a, L_b] = \epsilon_{aij}\epsilon_{bkl}[R_iP_j, R_kP_l].$$

Using Leibniz' rule, and the fact that the only nonzero commutator is $[R_i, P_j] = i\hbar\delta_{ij}$, we have

$$[R_iP_j, R_kP_l] = R_i[P_j, R_k]P_l + R_k[R_i, P_l]P_j = i\hbar[\delta_{il}R_kP_j - \delta_{jk}R_iP_l].$$

Now use

$$\epsilon_{aij}\epsilon_{bkl}\delta_{il} = \delta_{ak}\delta_{bj} - \delta_{ab}\delta_{ij},$$

and permutations of this identity, to conclude that

$$[L_a, L_b] = i\hbar\epsilon_{abc}L_c.$$

Answer to Exercise 7.3: The trick for doing this computation is to note that

$$\langle ll|K_+^{l-m}K_-^{l-m}|ll\rangle = \langle ll|K_+^{l-m-1}(\mathbf{K}^2 - K_3^2 + K_3)K_-^{l-m-1}|ll\rangle.$$

The state $K_-^{l-m-1}|ll\rangle$ is an eigenstate of $\mathbf{K}^2$ with eigenvalue $l(l+1)$ and an eigenstate of K_3 with eigenvalue $m + 1$. Thus,

$$\langle ll|K_+^{l-m}K_-^{l-m}|ll\rangle = [l(l+1) - (m+1)^2 + m + 1]\langle ll|K_+^{l-m-1}K_-^{l-m-1}|ll\rangle.$$

Iterating this equation, we get

$$\langle ll|K_+^{l-m}K_-^{l-m}|ll\rangle = \prod_{j=0}^{l-m}[l(l+1)-(m+j)^2+m+j],$$

where we have used the fact that

$$\langle ll|ll\rangle = 1.$$

Answer to Exercise 7.5: The Bohr radius is

$$a_g = \frac{4\pi\epsilon_0\hbar^2}{m_e e^2} \to \frac{\hbar^2}{m^2 GM}.$$

The analog of the Rydberg energy is

$$\frac{\hbar^2}{2ma_B^2} = \frac{m(mGM)^2}{2\hbar^2}.$$

Now $GMm = 6.7 \times 10^{-7}\frac{Mm}{kg^2} = 6.7 \times 10^{48}$ in joule-cm. $m = 6 \times 10^{24}$ kg and $\hbar = 10^{-34}$ joule-s. This gives a gravitational Bohr radius of 2.3×10^{-138} m, and a Rydberg energy of 1.35×10^{188} joules! In fact this calculation is not valid in the real world. The nonrelativistic treatment of the "gravitational Bohr atom" is valid only for very highly excited states E_n for which the effective Bohr radius is na_g. Requiring that expectation value of the velocity be less than that of light gives us this constraint. The expectation value of $\mathbf{p}^2/2m$ in the nth state is of order the binding energy

$$\frac{10^{188}}{n^2} \text{ joules,}$$

and this must be $< mc^2 = 5.4 \times 10^{41}$ joules. Thus, $n > 10^{73}$.

Answer to Exercise 7.7: For a nonzero constant solution, the first two terms in the equation vanish, while the last vanishes only if $q = 0$. The constant is undetermined. For a linear solution, $A + Bx$ the first term vanishes so we have

$$(p+1-x)B + q(A+Bx) = 0,$$

which implies $q = 1$ and $A = (p+1)B$. B is undetermined.

Answer to Exercise 7.9: The Laguerre operator can be written

$$L \equiv x\frac{d^2}{dx^2} + (p+1-x)\frac{d}{dx} = [x\frac{d}{dx} + (p+1-x)]\frac{1}{x}[x\frac{d}{dx}].$$

Now note that, using integration by parts

$$\int dx \ [f^* x^p e^{-x} x \frac{d}{dx} g] = -\int dx \ [x \frac{d}{dx} + (p+1-x)](f^*) x^p e^{-x} g,$$

so that, with respect to this scalar product

$$x \frac{d}{dx} + (p+1-x) = -[x \frac{d}{dx}]^\dagger.$$

Define $D = x \frac{d}{dx}$. Then

$$L = -D^\dagger \frac{1}{x} D.$$

This shows that L is a negative definite Hermitian operator. There is a subtlety in this derivation, because $\frac{1}{x}$ is not defined on the whole Hilbert space, since it blows up on constant functions if p is not large enough. However the action of D to the right, kills the divergence. To compute the norm of L_q^p we recall its explicit power series expansion (this is the expansion for the monic polynomial, rather than the orthonormal one we described in the text):

$$L_q^p = \sum_{k=0}^{q} (-1)^k \frac{\Gamma(q+p+1)}{\Gamma(q-k+1)\Gamma(p+k+1)\Gamma(k+1)} x^k.$$

In evaluating the norm of this function, we have to do the integral

$$\int_0^\infty dx \ e^{-x} x^{p+k_1+k_2} = \Gamma(p+k_1+k_2-1).$$

Answer to Exercise 7.11: The fact that the equation of motion is

$$\frac{d\mathbf{v}}{dt} = -\frac{a}{l} \dot{\mathbf{e}}_\phi,$$

follows by simply combining Newton's equation with the kinematic equation for $\mathbf{e}_\phi$. Since both left and right sides of this equation are total time derivatives,

$$\mathbf{h} \equiv \mathbf{v} + \frac{a}{l} \mathbf{e}_\phi$$

is conserved. $\mathbf{h}$ is sometimes called Hamilton's vector, since the fact that it is conserved is his observation.

Answer to Exercise 7.13:

$$\mathbf{h} \times \mathbf{L} = \mathbf{v} \times \mathbf{L} + a\mathbf{e}_\phi \times \hat{\mathbf{L}},$$

where $\hat{\mathbf{L}}$ is the unit vector in the direction of the angular momentum. The cross product of two orthogonal unit vectors (the angular momentum vector is perpendicular to the plane of the orbit) is the third unit vector in an orthonormal basis, which is just $\pm\hat{\mathbf{e}}$, depending on the order of the cross product. Thus,

$$\mathbf{h} \times \mathbf{L} = \mathbf{v} \times \mathbf{L} - a\mathbf{e}.$$

The LRL vector is usually defined as $m\mathbf{h} \times \mathbf{L}$, while the *eccentricity vector is* $(\mathbf{h} \times \mathbf{L})/a$.

Answer to Exercise 8.1: We compute

$$(-\partial_r^2)[r^l\chi_l] = -[l(l-1)r^{l-2}\chi_l + 2lr^{l-1}\partial_r\chi_l + r^l\partial_r^2\chi_l].$$

$$\frac{2}{r}(-\partial_r)[r^l\chi_l] = -2lr^{l-2}\chi_l - 2r^{l-1}\partial_r\chi_l.$$

Thus, the equation for χ_l is

$$-\partial_r^2\chi_l - \frac{2(l+2)}{r}\partial_r\chi_l - \frac{1}{r}\chi_l = k^2\chi_l.$$

If we now write $\chi_l = e^{\pm ikr}\phi_l$, the $k^2\phi_l$ term cancels from the left and right of the equation and we are left with

$$-r\partial_r^2\phi_l - [2(l+2) \pm 2ikr]\partial_r\phi_l - [1 \pm 2ik(l+2)]\phi_l = 0.$$

Because the independent variable appears only linearly in this equation, if we write ϕ_l as a Fourier or Laplace transform the equation becomes a first order ODE and is exactly soluble. Thus, we have exact integral representations of the solution. This equation is called the *confluent hypergeometric equation.*

Answer to Exercise 8.3: Write the energy in terms of the variable $u = 1/r$ and use the equations

$$\dot{r} = -\dot{u}/u,$$

and

$$\frac{du}{d\theta} = \dot{u}\dot{\theta} = \dot{u}\frac{lu}{m}.$$

Then we get

$$E = \frac{l^2}{2m}[u_\theta^2 + u^2] - \kappa u.$$

Take the derivative w.r.t. θ and assume $l \neq 0$ to get

$$u_{\theta\theta} + u = \frac{2m\kappa}{l^2}.$$

The impact parameter b is defined to be the perpendicular distance between the trajectory of the incoming particle and the origin, which is $r\sin(\theta)$ in the limit of infinite r. The constant on the right-hand side of the differential equation has dimensions of inverse length, so call it L^{-1}. The incoming direction of the particle is $\arctan(b/L)$.

The solution of the equation is

$$u = u_0 \cos(\theta - \theta_0) - \frac{2m\kappa}{l^2}.$$

Asymptotically $u \to 0$, so $\theta \to 0, \pi$ at finite b. This equation defines a hyperbola and u_0 is the point of closest approach of this hyperbola, to the origin. The outgoing direction is $\pi/2 - \arctan(b/L)$. Therefore, the angle of deflection is $\delta = 2\arctan(b/L)$. Thus,

$$b/L = \tan(\delta/2).$$

The scattering cross section $\frac{d\sigma}{d\Omega}d\Omega$, is defined as the number of particles scattered into solid angle $d\Omega$ per unit time, divided by the number of incoming particles per unit time per unit area. Since b and the deflection angle δ are functions of each other we have

$$\frac{d\sigma}{d\Omega}d\Omega = \pi d(b^2) = 2\pi b\,db.$$

On the other hand, the infinitesimal solid angle is $d\Omega = 2\pi \sin(\delta)d\delta$, so that

$$\frac{d\sigma}{d\Omega} = \frac{b}{\sin(\delta)}\frac{db}{d\delta}.$$

In order to write an expression in terms of asymptotic energy and impact parameter, we have to rewrite L in terms of b and energy. The result of this algebra is

$$b = \frac{2\kappa}{E}\cot(\delta/2),$$

so that

$$\frac{d\sigma}{d\Omega} = \frac{\kappa^2}{\sin^4(\delta/2)}.$$

Answer to Exercise 8.5: The spherical Bessel equation is

$$z^2\frac{d^2 j_l}{dz^2} + 2z\frac{dj_l}{dz} + [z^2 - l(l+1)]j_l = 0.$$

Write $j_l = \sum_{l=0}^{\infty} c_n z^n$. Then

$$\sum_{n=0}^{\infty} [n(n+1)c_n z^n + c_n[z^{n+2} - l(l+1)z^n]] = 0.$$

Rewrite this as

$$\sum_{n=0}^{\infty} [[n(n+1) - l(l+1)]c_n z^n +] = \sum_{n=2}^{\infty} c_{n-2} z^n.$$

Equating the coefficients of z^n on both sides we get

$$-l(l+1)c_0 = 0,$$

$$[2 - l(l+1)]c_1 = 0,$$

and

$$c_n = \frac{1}{n(n+1) - l(l+1)} c_{n-2}, \quad n \geq 2.$$

Note that this implies $c_0 = 0$ unless $l = 0$ and $c_1 = 0$ unless $l = 1$. It is also clear that we can solve separately for c_n for even and odd n. This is because the equation is invariant under $z \to -z$. The solution for c_n becomes singular for $n = l$ unless $c_{l-2} = 0$. That is consistent with the recursion relation only if all the lower values of c_n also vanish, so the series for j_l must begin with c_l. Thus,

$$c_n(l) = \prod_{k=1}^{\frac{n-l}{2}} \frac{1}{(l+2k)(l+2k+1) - l(l+1)} c_l(l).$$

Answer to Exercise 9.1: The general form of the Euler–Lagrange equations is (we follow the convention of Lagrangian mechanics and use raised indices on coordinates):

$$\partial_t (\partial_{\dot{q}^i} L) - \partial_{q^i} L = 0.$$

For the charged particle Lagrangian

$$\partial_t (\partial_{\dot{x}^i} L) = m\ddot{x}^i + qA^i,$$

$$\partial_{x^i} L = q\dot{x}^j \partial_{x^i} A^j.$$

The Euler–Lagrange equations are

$$m\ddot{x}^i = q\dot{x}^j (\partial_{x^i} A^j - \partial_{x^j} A^i).$$

We recognize the antisymmetric combination of derivatives of the vector potential as $\epsilon^{ijk} B_k$ (**B** is the magnetic field). So,

$$m\ddot{\mathbf{x}} = q\dot{\mathbf{x}} \times \mathbf{B},$$

the Lorentz force equation.

Answer to Exercise 9.3: The guiding center solutions are

$$\psi(u, z) = e^{\bar{z}u + \bar{u}z}e^{-\bar{z}z}.$$

The scalar product is given by integration over the complex plane. The translation invariant superposition is the integral of $\psi(u, z)$ over the complex u plane and the integral does not converge.

Answer to Exercise 10.1: The wave is not a real wave. It is a way of encoding the predictions of quantum mechanics (QM) for the probability of the particle being at various vertical positions, at horizontal positions to the right of the slits. Like all probability predictions it is tested only by doing repeated experiments with identical initial conditions. When one places a detector capable of discriminating which slit the electron went through, and asks what QM predicts for such a situation, one is asking a different question, namely the *conditional probability of finding the electron at vertical position y given that the detector, a macroscopic object, has registered or not*. These probabilities do not have detectable interference terms because the state of the detector that has registered a macroscopic hit has doubly exponentially small overlap with the state of the detector that has not registered a hit.

Answer to Exercise 11.1: Introduce a basis in the Hilbert space consisting of the even and odd combinations of hydrogen ground states around the two well separated protons, plus an infinite set of wave functions orthogonal to both of these. The expectation value of the Hamiltonian in the "infinity minus two" dimensional subspace is bounded from below by something of order ten Rydbergs. On the other hand, we will show in a moment that the splitting between the two states is of order e^{-d}, so it makes sense to determine the ground state by restricting attention to the matrix elements of the Hamiltonian in the two-dimensional subspace. The two localized hydrogen ground state wave functions are proportional to $e^{-|\mathbf{r}\pm d\hat{\mathbf{z}}|}$. The matrix elements of the Hamiltonian between these two states are real and equal, because the wave functions are real, and given by

$$N \int d^3x \, [e^{-|\mathbf{r}+d\hat{\mathbf{z}}|}(-1 - \frac{1}{|\mathbf{r}+d\hat{\mathbf{z}}|})e^{-|\mathbf{r}-d\hat{\mathbf{z}}|}.$$

N is the squared norm of the exponential function and the -1 comes from acting with the kinetic energy plus potential due to the hydrogen atom on the positive z axis. Both terms are negative, so the off diagonal part of the Hamiltonian in the two-dimensional subspace is $-\epsilon\sigma_1$, where $\epsilon > 0$. The ground state is the positive eigenvalue eigenstate of σ_1, which is the symmetrized state. It is clear that ϵ is exponentially small for large d, because each wave function is of order e^{-d} where the other is order 1.

Answer to Exercise 11.2: To prove the Feynman Hellman theorem note that any eigenstate of the Hamiltonian $H(\lambda)$ is normalized to 1 for all λ. Differentiating the normalization condition with respect to λ we find that

$$\langle \psi(\lambda) | \frac{d\psi}{d\lambda} \rangle = 0. \tag{G.36}$$

Now write

$$E(\lambda) = \langle \psi(\lambda) | H(\lambda) | \psi(\lambda) \rangle. \tag{G.37}$$

Differentiating this with respect to λ we get two terms where the state is differentiated, and its scalar product taken with $H|\psi(\lambda)\rangle = E|\psi(\lambda)\rangle$. The scalar product vanishes, so we are left with

$$\frac{dE}{d\lambda} = \langle \psi(\lambda) | \frac{dH}{d\lambda} | \psi(\lambda) \rangle. \tag{G.38}$$

Answer to Exercise 11.3: Let $|\psi(d)\rangle$ be the normalized ground state of

$$H(d) \equiv H = p_r^2 + p_z^2 - \frac{1}{\sqrt{r^2 + (z-d)^2}} - \frac{1}{\sqrt{r^2 + (z+d)^2}}.$$

Then, since the norm is d independent, $\langle \psi(d) | \partial_d | \psi(d) \rangle + \partial_d \langle \psi(d) | \psi(d) \rangle = 0$. The derivative of the ground state energy $E(d)$ is

$$\partial_d (\langle \psi(d) | H(d) | \psi(d) \rangle) = E(d) \langle \psi(d) | \partial_d | \psi(d) \rangle + \partial_d \langle \psi(d) | \psi(d) \rangle + \langle \psi(d) | \partial_d H(d) | \psi(d) \rangle. \tag{G.39}$$

Now

$$\partial_d H(d) = +\frac{d-z}{\sqrt{r^2 + (z-d)^2}^3} + \frac{d+z}{\sqrt{r^2 + (z+d)^2}^3}. \tag{G.40}$$

The Hamiltonian $H(d)$ is invariant under reflection in z so its eigenstates can be chosen either even or odd under this reflection. We have seen that at large d the ground state is even. This is also true at small d where the wave function approaches the spherically symmetric doubly charged hydrogen ground state. The ground state is even for all d, by continuity. This means that in evaluating the expectation value of $\partial_d H$, the terms involving z vanish by symmetry. Thus,

$$\partial_d E(d) = \langle \psi(d) | \frac{d}{\sqrt{r^2 + (z-d)^2}^3} + \frac{d}{\sqrt{r^2 + (z+d)^2}^3} | \psi(d) \rangle, \tag{G.41}$$

which is manifestly positive.

Answer to Exercise 11.5: Since the splitting between the even and odd combinations of hydrogen wave functions centered around a single proton is exponentially small, it is sufficient to consider the expectation value of the Hamiltonian in either one of these wave functions. Let us choose the left proton as the origin. Then we have to calculate

$$-\int d^3r \ e^{-2r} \frac{1}{\sqrt{r_1^2 + r_2^2 + (r_3 - 2d)^2}} = -\int_0^\infty du \int_{-\infty}^\infty dz \ [ue^{-2\sqrt{u^2+z^2}} \frac{1}{\sqrt{u^2 + (z - 2d)^2}}.$$

For large d, we can approximate the square root by $2d$ as long as u and z are both $\ll d$ in absolute value. When they *are* this large, the integrand is exponentially small, so this approximation is valid up to exponential corrections, throughout the region of integration. There are of course power law corrections to the value of the integral from higher orders in the expansion of the square root. When we substitute in the large d expression for the square root, integral just becomes the normalization integral for the hydrogen wave function, times $-\frac{1}{2d}$. Thus, the leading behavior of the of the Born–Oppenheimer potential is a negative constant plus an attractive Coulomb potential, which exactly cancels the Coulomb repulsion of the protons.

Answer to Exercise 11.7: We have seen in the text and previous exercises that the electron ground state energy for fixed d is a monotonically increasing function of d, varying between -4 and -1 in Rydberg units, as d goes from zero to infinity. The positive Coulomb energy between the protons is of order 1 when their separation is about a Bohr radius, but goes to infinity at $d = 0$ and to zero at $d = \infty$. Thus, the full Born–Oppenheimer potential has a minimum at $d^* \sim 1$.

Answer to Exercise 11.9: We use the first law of thermodynamics at fixed entropy $dE = -PdV$. The entropy in a system of Nq noninteracting electrons is $Nq\ln 2$, so we want to vary the volume at fixed Nq. Thus,

$$\frac{dE}{V} = -\frac{2}{3}\frac{E}{V}dV = -PdV$$

and

$$P = \rho^{5/3} \frac{\hbar^2 (3\pi^2)^{2/3}}{5m}.$$

Answer to Exercise 12.1:

$$\rho_{ji}^* = \sum_I c_{jI} c_{Ii} = \rho_{ij},$$

so ρ is Hermitian. Compute the expectation value of ρ in any state as

$$\sum_{ij} v_i^* \rho_{ij} v_j = \sum_{ijI} v_i^* c_{iI}^* c_{Ij} v_j \geq 0,$$

since it is a sum of absolute squares of complex numbers. Applied to a vector v_i, which is an eigenvector of ρ it implies that all the eigenvalues are positive. The trace of ρ is $\sum_{iI} c_{iI}^* c_{Ii} = 1$, because it is the norm of the full vector in the tensor product Hilbert space. The sum of the eigenvalues is thus 1, and ρ has the properties of a density matrix.

Answer to Exercise 12.3: We are looking for the maximum of the entropy

$$-\mathrm{Tr}\,(\rho \ln \rho)$$

subject to the constraint

$$\mathrm{Tr}\rho N = n_0,$$

where N is the number operator for zero momentum bosons. The expectation value is

$$\sum_{ik} k p_i |\langle k|p_i\rangle|^2,$$

where $|k\rangle$ are the number eigenstates, and $|p_i\rangle$ the eigenstates of the density matrix, while the entropy is $-\sum_i p_i \ln p_i$, and is independent of the scalar products $|\langle k|p_i\rangle|^2 \equiv O_{ki}$. Let us assume the two operators commute, so that the $|k\rangle$ and $|p_i\rangle$ bases are the same. Then the constraint is

$$\sum_{k=0}^{\infty} p_k k = n_0.$$

We minimize the energy subject to this constraint by using a Lagrange multiplier, minimizing

$$-\sum_{k} p_k \ln p_k + L(\sum_{k=0}^{\infty} p_k k = n_0),$$

w.r.t. both p_k and L, obtaining

$$-\ln p_k - 1 + Lk = 0,$$

along with the constraint equation. Thus,

$$p_k \propto e^{Lk}.$$

We determine the proportionality constant by requiring that the sum of all probabilities is one

$$p_k = \frac{e^{Lk}}{1 - e^L},$$

and the value of L by

$$n_0 = \frac{\sum k e^{Lk}}{1 - e^L} = \partial_L \ln(1 - e^L) = \frac{e^L}{1 - e^L}.$$

This implies

$$e^L = \frac{n_0}{1 + n_0}.$$

Answer to Exercise 12.4: In classical statistical mechanics, the partition function is the integral over all momenta and coordinates of $e^{-\beta H(p,x)}$, where H is the classical Hamiltonian. A magnetic field is added by shifting $\mathbf{p}_i \to \mathbf{p}_i - e_i \mathbf{A}(\mathbf{x_i})$, where e_i is the electric charge of the i-th particle and $\mathbf{A}$ is the vector potential whose curl is the magnetic field. Since we can eliminate the vector potential by a shift of all the momentum integration variables, the partition function is independent of the magnetic field. The expectation value of the energy is the derivative with respect to temperature of the energy expectation value and so does not depend on the magnetic field. Thus, there's neither diamagnetism or paramagnetism.

Answer to Exercise 12.5: The full Hamiltonian is

$$H = h(\mathbf{p_i} - \mathbf{e}/\mathbf{cA}(\mathbf{x_i}), \mathbf{x}_i) + H_B.$$

In classical mechanics we have to integrate $e^{-\beta H}$ over all $\mathbf{p_i}, \mathbf{x_i}$ and θ_i. As in Exercise 12.4, we can shift the $\mathbf{p}_i$ integration variable to $\mathbf{p_i} - \mathbf{A}(\mathbf{x_i})$, and the Jacobian of this change of variables is 1. So the partition function factorizes:

$$Z = Z_0 \int d\theta_i e^{-\beta H_B},$$

and the first factor has no B dependence. It affects neither the magnetization nor the susceptibility. The θ_i integrals are all decoupled and identical so

$$\ln Z = N\ln \left[2\pi \int \sin(\theta)d\theta \, e^{-\alpha \cos(\theta)}\right].$$

Note that the integration is over solid angles, which accounts for the $2\pi \sin(\theta)$. Since $\sin(\theta)d\theta = d(\cos(\theta))$ this is

$$\ln Z = N\ln \left[4\pi \int_{-1}^{1} dc\, e^{-\alpha c}\right] = N\ln \left[\alpha^{-1} \sinh(\alpha)\right],$$

up to terms independent of B. The magnetization and susceptibility formulae are simple derivatives of this. At small β, α is also small and we can approximate the partition function by

$$\ln Z \approx N\ln \left[1 + \frac{\alpha^2}{3!}\right].$$

The magnetization is then $N\mu^2 \beta B/3$ and the susceptibility is $N\mu^2 \beta/3$. The signs indicate that we have paramagnetism and one can verify that this is true for all β.

Answer to Exercise 13.1: To calculate the expectation values we write $X = \sqrt{\frac{\hbar}{2m\omega}}(a+a^\dagger)$, and realize that in calculating expectation values of powers of X in energy eigenstates, we only have to take into account terms with equal numbers of powers of a and $a^\dagger$. Thus, effectively

$$X^4 \rightarrow (a^2 a^{\dagger\,2} + a^{\dagger\,2} a^2 + a^\dagger a^2 a^\dagger + aa^{\dagger\,2} a + aa^\dagger aa^\dagger + a^\dagger aa^\dagger a)(\frac{\hbar}{2m\omega})^2.$$

Using the commutation relations we can write all of these expressions in terms of the number operator $N = a^\dagger a = aa^\dagger + 1$.

$$X^4 \rightarrow (a(N-1)a^\dagger + a^\dagger N a + N^2 + (N-1)^2 + 2N(N-1))(\frac{\hbar}{2m\omega})^2.$$

Using the commutation relations again, the first two terms give two more factors of $N(N-1)$, so altogether

$$X^4 \rightarrow (6N^2 - 6N + 1)(\frac{\hbar}{2m\omega})^2.$$

The perturbed E_n is thus,

$$E_n = \hbar\omega(n + 1/2) + b(6n^2 - 6n + 1)(\frac{\hbar}{2m\omega})^2.$$

Answer to Exercise 13.3: The coefficients of the polynomial $P(a)$ are themselves polynomial functions of λ, so even though the operator $A_0 + \lambda A_1$ is not Hermitian for complex λ we can define the roots of $P(a)$ as functions of λ:

$$P(\lambda, a(\lambda)) = 0.$$

Taking the derivative of this equation w.r.t. λ we get

$$\frac{da}{d\lambda} = -\frac{\partial_\lambda P}{\partial_a P}.$$

Both partial derivatives of the polynomial are analytic functions of λ. If we are at a value of λ for which the root $a_k(\lambda)$ is nondegenerate, then $\partial_a P$ is nonzero, so the derivative of $a_k(\lambda)$ is well defined in all directions in the complex plane, so that $a_k(\lambda)$ is analytic. The equation $a_1(\lambda) = a_2(\lambda)$ is a single complex equation for a complex unknown, so it will generally have a solution. On the other hand, more than one degenerate level at a fixed value of λ is nongeneric and might occur only in the presence of symmetries. So, near a level crossing, the solution of the polynomial equation reduces to a quadratic equation and the roots have square root branch points at the value of λ where they coincide. For an $N \times N$ Hermitian matrix we therefore expect each root to have $N - 1$ branch points, where it crosses some other root. Remarkably, this means that if we know the analytic continuation of *any single root* to complex λ, then we can find all the other roots as different branches of the same locally analytic function.

Answer to Exercise 13.5: The perturbation is $\frac{\delta k^2}{2} X^2 = \alpha^2 (a + a^\dagger)^2$, where $\alpha^2 = \frac{\delta k^2}{2} \frac{\hbar}{2(m\omega)} = \frac{\delta k^2}{4k^2} \hbar\omega$. As in Exercise 13.1, we only have to consider terms with equal numbers of powers of a and $a^\dagger$. Thus, in first order perturbation theory we can write

$$\langle n|V|n\rangle = \alpha^2 \langle n|aa^\dagger + a^\dagger a|n\rangle = (2n + 1)\alpha^2.$$

In second order, we have to evaluate

$$\alpha^4 \langle n|(a + a^\dagger)^2 \frac{P}{\hbar\omega(N - n)} (a + a^\dagger)^2|n\rangle.$$

Here, the projection operator $P = 1 - |n\rangle\langle n|$. In this case, because of the projection only the terms

$$\alpha^4 \langle n|(a^2 + a^{\dagger\,2}) \frac{P}{\hbar\omega(N - n)} (a^2 + a^{\dagger\,2})|n\rangle,$$

survive, and in addition we get zero unless we pair the operator a^2 on the left with its Hermitian conjugate on the right, or vice versa. We see that in either case only one of the states $|n\pm2\rangle$ is produced, so the Hamiltonian operator in the denominator is just the number $\pm 2\hbar\omega$. We then have to evaluate the diagonal matrix elements

$$\frac{1}{2\hbar\omega}[\langle n|a^2 a^{\dagger\,2}|n\rangle - langlen|a^{\dagger\,2} a^2|n\rangle)] = \frac{3n}{2\hbar\omega}.$$

So the perturbed levels are

$$E_n = \hbar\omega(n + 1/2) + \alpha^2(2n + 1) - \alpha^4 \frac{3n}{2\hbar\omega}.$$

The exact formula is

$$E_n = \hbar\omega(1 + \frac{\delta k^2}{k^2})^{1/4}(n + 1/2).$$

Plugging in the value of α we see that the two expressions coincide, to second order.

Answer to Exercise 15.1: If we change parameter to $s = f(t)$ then $ds = \frac{df}{dt} dt$ and $dx^\mu/ds = dx^\mu/dt \frac{dt}{ds} = dx^\mu/dt \frac{1}{df/dt}$. The two factors of the derivative of f cancel.

Answer to Exercise 15.3:

$$\mathbf{B}_i^{(n)} = (\nabla \times \mathbf{A}^{(n)})_i = \epsilon_{ijk}\partial_j(\langle n|\partial_k|n\rangle) = \epsilon_{ijk}\partial_j(\langle n|)\partial_k|n\rangle.$$

Now we insert a complete set of states $1 = \sum_{m=1}^{2} |m\rangle\langle m|$, and use the result of Exercise 15.2. We also note the fact that in the formula of 15.2 exchanging m and n is equivalent to complex conjugation:

$$\mathbf{B}_i^{(n)} = i\hbar\epsilon_{ijk} \sum_{m\neq n} \frac{\langle n|\partial_j H|m\rangle\langle m|\partial_k H|n\rangle}{(E_m - E_n)^2}.$$

The $m = n$ term doesn't contribute (avoiding a singularity) because of the antisymmetry under interchange of j and k. There is really only one term in the sum and the energy denominator is $2x$. Thus, $\frac{1}{4x^2}$ factors out of the formula and we can restore the vanishing $m = n$ term to the sum to write this as

$$\mathbf{B}_i^{(n)} = i\hbar\epsilon_{ijk}\langle n|\sigma_j\sigma_k|n\rangle.$$

We can evaluate this using the commutation relations of the Pauli matrices. Now the eigenstates of $\mathbf{x}\cdot\sigma$ have expectation values of σ_j which are just $\pm\hat{\mathbf{x}}_j$, so we get

$$\mathbf{B}_i^{(n)} = \mp\hbar\frac{\hat{x}}{2x^2},$$

which satisfies

$$\nabla\cdot\mathbf{B} = \mp\frac{\hbar}{2}\delta^3(\mathbf{x}).$$

If we think of this as a "magnetic field" it is the field of a magnetic monopole sitting at the origin. This is called the Berry monopole. It is a signal of the degeneracy of the two states at the origin of parameter space.

Answer to Exercise 15.5: The parameter space here is the two-dimensional complex plane $z = re^{i\theta}$ and the relevant component of the Berry potential is

$$A_\theta = i\hbar\int F^*_{Laughlin} \prod_{i=1}^{\nu N}(z_i^* - z^*) \prod_{i=1}^{\nu N}(z_i^*)\partial_\theta[\prod_{i=1}^{\nu N}(z_i - z)\prod_{i=1}^{\nu N}(z_i)F_{Laughlin}]e^{-\sum z_i^* z_i}.$$

Since $F_{Laughlin}$ does not depend on θ we have

$$A_\theta = i\hbar\int F^*_{Laughlin}F_{Laughlin}\prod_{i=1}^{\nu N}(z_i^* - z^*)\prod_{i=1}^{\nu N}(z_i^*)\partial_\theta[\prod_{i=1}^{\nu N}(z_i - z)\prod_{i=1}^{\nu N}(z_i)].$$

The derivative gives a sum of νN terms

$$A_\theta = \hbar\int|\psi_{quasiholes}|^2\sum_{i=1}^{\nu N}\frac{z}{z_i - z}.$$

The integral around θ picks up the pole and we get

$$\int A_\theta d\theta = 2\pi\nu,$$

which is an anyonic Berry phase.

Answer to Exercise 16.1: The Lippman–Schwinger equation is

$$\psi_{\mathbf{k}}(\mathbf{x}) = e^{i\mathbf{k}\cdot\mathbf{x}} - \frac{1}{4\pi}\int d^3y \, \frac{e^{ik|\mathbf{x}-\mathbf{y}|}}{|\mathbf{x}-\mathbf{y}|} U(\mathbf{y})\psi_{\mathbf{k}}(\mathbf{y}),$$

where $U \equiv \frac{2m}{\hbar^2}V$. Define

$$G_0(\mathbf{x}-\mathbf{y}) \equiv \int d^3y \, \frac{e^{ik|\mathbf{x}-\mathbf{y}|}}{4\pi|\mathbf{x}-\mathbf{y}|}.$$

We can solve this by iteration, with the n-th term being

$$\int d^3y_1 \ldots d^3y_n \, G_0(\mathbf{x}-\mathbf{y_1})U(\mathbf{y_1})G_0(\mathbf{y_1}-\mathbf{y_2})U(\mathbf{y_2})\ldots G_0(\mathbf{y_{n-1}}-\mathbf{y_n})U(\mathbf{y_n})e^{i\mathbf{k}\cdot\mathbf{y_n}}.$$

The diagram corresponding to this term is shown in Figure 16.2 at the end of Chapter 16.

Answer to Exercise 16.3: The solutions to the Schrödinger equation inside and outside the well, for angular momentum l are

$$\psi_+ = A_+ j_l(kr) + B_+ h_l(kr),$$

$$\psi_+ = A_- j_l(k_0 r) + B_- h_l(k_0 r),$$

where $k_0^2 = k^2 - \frac{2m}{\hbar^2}V_0$. The second derivative of the wave function is a delta function, so both the wave function and its derivative are continuous at $r = r_0$. In the interior, we must choose $B_- = 0$ since $h_l(kr_0)$ is singular at the origin. Thus,

$$A_+ j_l(kr_0) + B_+ h_l(kr_0) = A_- j_l(k_0 r_0).$$

$$k[A_+ j_l'(kr_0) + B_+ h_l'(kr_0)] = k_0 A_- j_l'(k_0 r_0).$$

We can solve these equations for

$$\frac{B_+}{A_+} \equiv e^{2i\delta_l(k)} = \frac{k_0 j_l(kr_0)j_l'(k_0 r_0) - k j_l'(kr_0)j_l(k_0 r_0)}{k h_l'(kr_0)j_l(k_0 r_0) - k_0 h_l(kr_0)j_l'(k_0 r_0)}.$$

Answer to Exercise 16.5: In the low energy limit, the formula of the previous exercise reduces to

$$\tan \delta_0 = -kr_0\left(\frac{\tan k_0 r_0}{k_0 r_0} - 1\right),$$

and $k_0 \approx \sqrt{-\frac{2m}{\hbar^2 V_0}}$. Remember that V_0 is negative for a spherical well. The scattering cross section, which is also the total cross section at these low energies is

$$\sigma_0 \approx \frac{4\pi}{k^2} \sin^2(\delta_0(k)) \approx 4\pi r_0^2 \left(\frac{\tan k_0 r_0}{k_0 r_0} - 1\right).$$

Answer to Exercise 16.7: Turning the formula for the cross section in terms of δ_l into the Breit-Wigner form is a matter of simple algebra. The formula obviously has a maximum at $E = E_R$. When $E = E_R \pm \Gamma/2$, the cross section has fallen to half of its value at the peak. This is the reason for the name width, short for "full width at half maximum". If we take a Fourier transform of this function, we get a function that behaves like $e^{-\Gamma t}$. Roughly speaking, this tells us that the physics of a Breit–Wigner peak is that the scattering particle gets trapped in a meta-stable bound state for a time of order $1/\Gamma$. A more sophisticated analysis confirms this expectation. The inverse of Γ is therefore called the lifetime of the meta-stable resonance.

Answer to Exercise 17.1: Differentiating w.r.t. $q^j(0)$ we get

$$\partial_{q^i(0)} s(t) = \int_0^t ds \left\{ [\dot{q}^i(s) - \partial_{q^i(s)} S] \partial_{q^j(0)} (\partial_{q^i(s)} S) - [\partial_{q^k(s)} V + \frac{d}{ds} \partial_{q^k(s)} S] \frac{\partial q^k(s)}{\partial q^j(0)} \right\},$$

where we have used $p_i(s) = \partial_{q^i(s)} S$, and integrated by parts in the last term. The vanishing of the two independent terms in square brackets is precisely the statement of the classical equations of motion.

Answer to Exercise 17.3: The integrand in the definition of the Airy function is the exponential of

$$i(px - \frac{\hbar^2 p^3}{6ma}),$$

and the stationary phase point is

$$p = \sqrt{2max/\hbar}.$$

Thus, the integral is approximately

$$Ai \sim e^{\frac{2i}{3} \sqrt{\frac{2ma}{\hbar}} x^{3/2}}.$$

It is important to realize that $V - E$ is going from negative to positive as x increases so that a is negative and this function falls like the exponential of $x^{3/2}$ as x goes to positive infinity. The logarithmic derivative of the Airy function thus behaves like $cx^{1/2}$ for large positive x, where $c = -\sqrt{\frac{2m|a|}{\hbar}}$. The second solution is

$$Bi = bAi,$$

with $b' = e^{\frac{4}{3}\sqrt{\frac{2m|a|}{\hbar}}x^{3/2}}$, so that Bi is exponentially growing at infinity and must be discarded.

Answer to Exercise 17.5: The JWKB solutions are

$$(E - V)^{-1/4}e^{\pm\int dx\sqrt{\frac{2m}{\hbar}(E-V)}}.$$

The integral is taken over the real x axis, starting from a point to the left of the turning point $V = E$. If the potential V is an analytic function,[1] these solutions are analytic functions of x away from the turning point. Thus, we can connect the behaviors to the right and left of the turning point, by following a contour that avoids the turning point. Not too far from the right of the turning point, the exponentially falling piece of the wave function dominates, so we have to analytically continue this solution. There are two possible contours to choose to avoid the turning point, either in the upper or lower half x plane. In either case, we pick up a phase factor from the prefactor $(V - E)^{-1/4}$. The phase is $e^{i\pi/4}$ for the upper contour and $e^{-i\pi/4}$ for the lower contour.

The Schrödinger equation is real in the classically forbidden region, so we can take the solution to be real. The analytic continuation that preserves this reality property is to take the average of the results of the upper and lower contours

$$\psi = C[2m(E - V)]^{-1/4}\cos(\frac{1}{\hbar}\int_a^x dy\ \sqrt{2m(E - V)} + \pi/4).$$

The coefficient C is the coefficient of the exponentially falling solution to the right of the turning point. Note that the integral is taken from a to $x < a$ because that is the analytic form of the solution in the classically forbidden region. We can rewrite it in terms of the conventional direction of integration in the classically allowed region

$$\psi = C[2m(E - V)]^{-1/4}\sin(\frac{1}{\hbar}\int_x^a dy\ \sqrt{2m(E - V)} + \pi/4).$$

Answer to Exercise 17.7: The Bohr–Sommerfeld condition is

$$\int_0^{\frac{E-V_0}{k}} dr\ \sqrt{\frac{2m}{\hbar}(E - V_0 - kr)} = n\pi.$$

Defining $r = \frac{E-V_0}{k}y$, this is

$$\int_0^1 dy\ \sqrt{(1 - y)} = n\pi\frac{k}{\sqrt{2m}}(E - V_0)^{-3/2}.$$

---

[1] Actually, for the purposes of this argument, it is sufficient for V to be continuous on a domain in the complex plane including the contour to be described below. Any such continuous function can be uniformly approximated by polynomials on that domain.

We solve for the binding energies as

$$E_n = V_0 + \left(\frac{2n\pi k}{3\sqrt{2m}}\right)^{2/3}.$$

The masses are

$$m_{(n+1)s} = 2m + E_n.$$

Thus,

$$m_{1s} = 2m + V_0,$$

$$m_{2s} = m_{1s} + \left(\frac{2\pi k}{3\sqrt{2m}}\right)^{2/3},$$

$$m_{3s} = m_{1s} + 2^{2/3}(m_{2s} - m_{1s}).$$

For the charmed bound states this gives

$$m_{3s} = 3.1 + 1.58(.6) = 4.05\text{GeV}.$$

The experimental value is 4.05GeV. For bottom the formula gives

$$m_{3s} = 9.46 + 1.58(.56) = 10.34\text{GeV}.$$

The experimental value is 10.36.

Answer to Exercise 17.8: The Bohr–Sommerfeld conditions are

$$\int_L^R dx \ \sqrt{2m(E - V(x))} = n\pi\hbar.$$

We can make the integral as large as we want by taking E large, so that the turning points L, R are at larger values of $|x|$. In this limit the action S_0 is large, equal to $n\pi\hbar$ and $S_0/\hbar$ is the JWKB approximation to the log of the wave function. The exact equation for the logarithm of the wave function $S/\hbar$ is

$$E = \frac{1}{2}(\partial_x S)^2 + i\hbar\partial_x^2 S.$$

The condition for validity of the JWKB approximation is

$$|\hbar\partial_x^2 S| \ll \frac{1}{2}(\partial_x S)^2.$$

Applying this criterion to the approximate solution $S = S_0$, we get

$$\hbar\partial_x V \ll \sqrt{2m(E - V)^3/2}.$$

Since E_n is very large, and V is independent of n, this inequality will be satisfied better and better at large n, except for a region around the turning point, which shrinks in size, as n goes to infinity.

Answer to Exercise 17.9: A bound state would correspond to a normalizable solution of the Schrödinger equation. However, it is clear that the expectation value of the Hamiltonian in any normalizable state is positive, so the bound state would have to be positive energy. At very large r we can solve the Schrödinger equation with plane wave solutions $e^{\pm ikr}$, with $\hbar^2 k^2 = 2mE$. Thus, there are no bound states, since the plane waves are not normalizable.

Answer to Exercise 17.11: The equation for $\hbar/i$ times the logarithm of the wave function is

$$(\partial_x S)^2 - ig^2 \hbar \partial_x^2 S = 2m(E - V) \equiv p^2(x).$$

Write $S = \sum_{n=0}^{\infty} S_n (ig^2 \hbar)^n$. S_n has dimensions of $[action]^{1-n}$, but contains no powers of $\hbar$. $\partial_x S_0$ is the classical momentum $p(x)$ at point x. The equation for S_n is

$$\partial_x^2 S_{n-1} = \sum_{k=0}^{n} \partial_x S_k \partial_x S_{n+1-k}.$$

This gives

$$\partial_x^2 S_0 = 2\partial_x S_0 \partial_x S_1$$

and

$$\partial_x^2 S_1 = 2\partial_x S_0 \partial_x S_2 + (\partial_x S_1)^2.$$

The solution of the first equation is

$$S_1 = \frac{1}{2}\ln p(x),$$

so the second equation becomes

$$\partial_x S_2 = -\partial_x^2 [\frac{1}{4p}].$$

So

$$S_2 = \frac{\partial_x p}{4p^2}.$$

The second order correction to the log of the wave function is real.

Answer to Exercise 18.1: We have

$$[\frac{p^2}{2} + gx^{2q}]\psi = E\psi.$$

Define $x = ay$, with

$$ga^{2q+2} = 1.$$

Then the equation becomes

$$[\frac{p_y^2}{2} + y^{2q}]\psi = a^2 E\psi.$$

Since the eigenvalues of the new operator are independent of g, we must have $E = a^{-2}\mathcal{E} = g^{\frac{1}{q+1}}\mathcal{E}$, where $\mathcal{E}$ is independent of g.

Answer to Exercise 18.3: To use the variational principle to prove the existence of positive energy bound states, we first show that there is a minimum energy for propagating out to infinity. This is easy. Go far out in the region $x \to \infty$, $a > |y|$. In this region the lowest energy we can have is a plane wave e^{ikx} multiplied by the lowest state $\sin(\pi y/a)$ in the infinite square well. This has energy at least $\frac{\pi^2}{a^2}$. Thus, if we can show there is a normalizable wave function with expectation value of the Hamiltonian below this, then we have proven there is a bound state.

Here's a trial wave function:

$$\psi = (a^2 - xy)e^{-b}\theta(a - x)\theta(a + x)\theta(a - y)\theta(a + y)$$
$$+ e^{-b|y|}(a^2 - ax)\theta(a - x)\theta(a + x)\theta(y^2 - a^2) + e^{-b|x|}(a^2 - ay)\theta(a - y)\theta(a + y)\theta(x^2 - a^2).$$

It is not normalized but is symmetric under interchange of x and y, as well as under reflection of each coordinate. The expression for the expectation value of the Hamiltonian in this state is a bit complicated, because the action of the Laplacian on this function has delta function singularities, which must be taken into account. You can find the gory details of the full solution in the solution manual to the Second Edition of Griffiths (Exercise 7.20).

Answer to Exercise 18.5: The harmonic atom Hamiltonian has the form

$$H = \sum \frac{P_i^2}{2} + \frac{1}{2}(\Omega^2)_{ij} X_i X_j.$$

The matrix Ω^2 is given by

$$(\Omega^2)_{ij} = \Omega^2 \delta_{ij} - \omega^2 1_{ij},$$

where the matrix 1_{ij} has a one in every matrix element. its eigenvalues are $-N$ and 0, with the latter having multiplicity $N-1$. We thus get eigenfrequencies of oscillation $\Omega^2 - N\omega^2$ and Ω^2. The former controls the motion of the center of mass of the electrons, while the latter describes relative motions. We obviously want to keep $\Omega^2 - N\omega^2$ positive to stabilize the system and make sure the center of mass is bound to the "nucleus". We will however insist that $0 < \frac{\Omega^2}{\omega^2} - N \ll N$. For distinguishable particles, the eigenstates of the Hamiltonian are arbitrary excited states of all N oscillators, but we must impose Fermi statistics. Also, since $\Omega^2 \gg \Omega^2 - N\omega^2$, the ground state will be a state in which the relative oscillations are excited as little as possible. The relative coordinates $x_i - x_j$ are redundant. They can all be expressed in terms of the coordinates $x_I - x_j$ for some particular value of I. However, if we first form the Slater determinant $\det [\psi_i(x_I - x_j)]$ using the first $N - 1$ eigenstates of the oscillator with frequency Ω, and then explicitly antisymmetrize under permutations that exchange I with one of the other labels, then we get a totally antisymmetric function, which has energy $\sum_{k=0}^{N-1} \hbar\Omega(k + 1/2)$. We then multiply this by the ground state of the center of mass oscillator, which is invariant under permutations. The total energy is

$$E = \frac{\hbar}{2}(\Omega - N\omega + N^2\Omega).$$

Answer to Exercise 18.7: We have to compute the expectation value of $\sum_{i<j} \sigma^a(i)\sigma^a(j)K_{ij}$, in the ground state of the external field Hamiltonia, which is the state where $h^a(i)\sigma^a(i) = -|h(i)|$. The expectation value of H in this state is just

$$< H >= \sum_{i<j} K_{ij} < \sigma^a(i) >< \sigma^a(j) > .$$

This is because the state is a tensor product over sites and the Hamiltonian has no operator products of operators on the same site. The expectation values of spins are $< \sigma^a(i) >= -\hat{h}^a(i) \equiv s^a(i)$. Note that it is independent of the local field strength. The variational problem then becomes one for classical spins, unit vectors. It is easy to solve if K is negative definite, the ground state wants all spins aligned. However, if some of the bonds are positive, they want to antialign, but the Hamiltonian can want to antialign a spin with two others that are preferentially antialigned as well. Think of three spins with links on an equilateral triangle with all positive bond strengths. The technical term for this sort of situation is *frustration* and spin problems with frustrated bonds have interesting and complex behavior.

Answer to Exercise 19.1: We insert a complete set of eigenstates of the wave number operator and write this as

$$\int d^3k \ \langle y|k\rangle\langle k|x\rangle e^{-\tau\frac{\hbar k^2}{2m}} e^{-\tau/\hbar V(x)}.$$

The Gaussian k integral now gives

$$N(\tau)e^{-\frac{\tau}{\hbar}[\frac{m(x-y)^2}{2\tau^2}-V(x)]}.$$

In the limit $\tau \to 0$ the exponent approaches $d\tau L$. The normalization factor diverges but we have dealt with this in the text.

Answer to Exercise 19.3: The functional equation is

$$\frac{d^2}{dt^2}\frac{\delta Z}{i\delta j(t)} + V'[\frac{\delta}{i\delta j(t)}]Z = j(t)Z.$$

Expand

$$Z[j] = \sum_{n=0}^{\infty} \frac{i^n}{n!} \int dt_1 \ldots dt_n \ [G_n(t_1 \ldots t_n)j(t_1) \ldots j(t_n)],$$

and equate the coefficients of $j(t)j(t_1) \ldots j(t_{n-1})$ on both sides. The first term on the left-hand side gives a coefficient

$$\frac{d^2}{dt^2}G_n(t, t_1 \ldots t_{n-1}).$$

If $V'(X) = \sum a_k X^k$, then the second term gives

$$\sum a_k G_{n+k-1}(t, t, \ldots, t, t_1 \ldots t_{n-1}),$$

where the number of equal arguments of G_{n+k-1} is k. The right-hand side gives

$$i \sum \delta(t - t_k)G_{n-1}(t_1, \ldots t_{n-1}).$$

So the SD equations are

$$\frac{d^2}{dt^2}G_n(t, t_1 \ldots t_{n-1}) + \sum a_k G_{n+k-1}(t, t, \ldots, t, t_1 \ldots t_{n-1}) = i \sum \delta(t - t_k)G_{n-1}(t_1, \ldots t_{n-1}).$$

Answer to Exercise 19.7: The real time Green function equation is

$$(d_t^2 + \omega^2)G(t, s) = \delta(t - s),$$

which has the Fourier transform solution

$$G(t, s) = \int \frac{d\nu}{2\pi} \frac{1}{\omega^2 - \nu^2} e^{i\nu(t-s)}.$$

The integrand has poles at $\omega = \pm\nu$ and we must choose the contour of integration to avoid them. There are four independent choices. One can take the contour in the lower half plane, or the upper half plane, or take it above the negative pole and below the positive one, or vice versa. The differences between these contours are contour integrals that encircle one or more of the poles, which, by the residue theorem are proportional to $e^{\pm i\omega t}$, both solutions of the homogeneous equation.

Answer to Exercise 19.9: We will set $\hbar = 1$ for this We will give the path integral for the operator ordering $H\frac{1}{2}(P_iP_jM_{ij}^{-1}(Q) + M_{ij}^{-1}(Q)P_iP_j)$. Other orderings can be treated in a similar manner. The strategy is to generalize the Trotter product formula, writing

$$\langle q'|e^{-iHt}|q\rangle = e^{itA_1}e^{it^2A_2}\dots,$$

and neglecting A_n $n \geq 2$ for small t. The analog of A_1 can be computed by expanding the operator expression to first order in t. We have

$$\langle q'|H|q\rangle = \int d^n p\, e^{ip(q'-q)}(\frac{1}{2}(M_{ij}(q') + M_{ij}(q)))p_ip_j.$$

Introducing complete sets of intermediate q and p states at intermediate times $t_k = tk/N$ we have

$$\langle q'|e^{-iHt}|q\rangle = \int [d^n q(t)d^n p(t)]e^{i\int_0^t[p\dot{q} - \frac{1}{2}p_ip_jM_{ij}^{-1}(q)]}.$$

This is called the Hamiltonian form of the path integral. Going back over the original derivation, one can see that we actually used this when M_{ij} was just a constant. We can do the integral over $p_i(t)$ exactly, because it is a Gaussian. This gives

$$\langle q'|e^{-iHt}|q\rangle = \int [d^n q(t)\mathrm{det}\mathrm{M}(\mathrm{q}(\mathrm{t}))]e^{\frac{i}{2}q_iq_jM_{ij}(q)}.$$

So the "position dependent mass" changes the local measure on the path integral.

Answer to Exercise 19.11: Since different a_i involve different γ matrices, they all anticommute with each other and with each other's adjoint. Then we have

$$[a_1, a_1]_+ = \frac{1}{2}([\gamma_1, \gamma_1]_+ - [\gamma_2, \gamma_2]_+) = 0,$$

since γ_1 and γ_2 anticommute. Similarly

$$[a_1, a_1^\dagger]_+ = \frac{1}{2}([\gamma_1, \gamma_1]_+ + [\gamma_2, \gamma_2]_+) = 1.$$

Answer to Exercise 20.1: $[\sigma_a(i), \sigma_b(j)] = 0$ because the operators act on different factors in a tensor product Hilbert space. By definition, $C_{ij} = \frac{1+\sigma_3^{(i)}}{2}\sigma_1^{(j)} + \frac{1-\sigma_3^{(i)}}{2}$, so

$$C_{ij}C_{ji}C_{ij} = [\frac{1+\sigma_3^{(i)}}{2}\sigma_1^{(j)} + \frac{1-\sigma_3^{(i)}}{2}][\frac{1+\sigma_3^{(j)}}{2}\sigma_1^{(i)} + \frac{1-\sigma_3^{(j)}}{2}][\frac{1+\sigma_3^{(i)}}{2}\sigma_1^{(j)} + \frac{1-\sigma_3^{(i)}}{2}].$$

This can be simplified by using

$$\sigma_1^2(i) = 1, \quad \sigma_1(i)P_\pm(i) = P_\mp(i)\sigma_1(i) \quad P_+(i)P_-(i) = 0, \quad P_\pm^2(i) = P_\pm,$$

where $P_\pm = \frac{1 \pm \sigma_3^{(i)}}{2}$. The result is

$$C_{ij}C_{ji}C_{ij} = P_+(i)P_+(j) + P_-(i)P_-(j) + \sigma_1(i)\sigma_1(j)(P_+(i)P_-(j) + P_-(i)P_+(j)).$$

In words, the operation on the RHS does nothing if both Q-bits have the same value, and changes the value of both Q-bits if they are different. That is, it is precisely the swap operator S_{ij}.

Answer to Exercise 20.3: The pure state has the form $|s\rangle = \sum c_{aB}|a, B\rangle$, where a and B label bases in the first and the second parts of the system. If the dimensions are unequal, we choose B to be the larger system. In terms of the rectangular matrix c, the reduced density matrices are

$$\rho_A = cc^\dagger, \quad \rho_B = c^\dagger c.$$

Then

$$\mathrm{Tr}[\rho_A^n] = \sum c_{a1}^{B1}(c^\dagger)_{B1}^{a2} \ldots c_{a(n-1)}^{B(n-1)}(c^\dagger)_{B(n-1)}^{a1}.$$

This is obviously the same as $\mathrm{Tr}[\rho_B^n]$. It is a simple generalization of the cyclicity of the trace formula, to rectangular matrices. One can compute the von Neumann entropy as the limit of $\mathrm{Tr}\rho^n/n - 1$. Traces of powers of the density matrix are called Renyi entropies.

Answer to Exercise 20.5: The definition of the Hadamard operator is $H^{(i)} = \frac{1}{\sqrt{2}}(\sigma_1^{(i)} + \sigma_3^{(i)})$. This operator is unitary and transforms from the basis where σ_3 is diagonal to the basis where σ_1 is diagonal. The easiest way to see this is to note that the columns of the Hadamard operators in the σ_3 basis are the orthonormal eigenvectors of σ_1. The definition of C_{ij} is

$$C_{ij} = P_+(i) \otimes 1(j) + P_-(i) \otimes \sigma_1(j).$$

Here $P_\pm(i)$ are the projectors on eigenstates of $\sigma_3(i)$. Writing the unit matrix and the σ_1 operators on the j Q-bit as linear combinations of the projectors on σ_1 eigenstates.

$$C_{ij} = 1(i) \otimes P_+^{(1)}(j) + \sigma_3(i) \otimes P_-^{(1)}(j),$$

where the superscripted $P_\pm$ are projectors on eigenstates of $\sigma_1(j)$. This looks like a C-not operation, except that the roles of σ_1 and σ_3 are interchanged in both Q-bits. Thus, if we conjugate by the product of the Hadamard operators on both Q-bits, we will in fact get C_{ji}, as claimed.

Answer to Exercise 20.6: Clearly the number of independent bit strings is the same as the number of basis vectors in a vector space of dimension 2^{2N}. The identification of a given pair of bit strings with eigenstates of the complete set of commuting operators $\sigma_2(i)$ shows us that the vectors form an orthonormal basis. The mapping $U(\mathbf{f})$ clearly maps every basis vector into another unit vector. Thus, all we have to do is prove that

$$\langle \mathbf{x}, \mathbf{y} + \mathbf{f}(\mathbf{x})|\mathbf{z}, \mathbf{w} + \mathbf{f}(\mathbf{z})\rangle = 0$$

unless $\mathbf{x} = \mathbf{z}, \mathbf{y} = \mathbf{w}$. For the first argument, this is obvious from the definition of the scalar product. Then, for the second argument we must have $\mathbf{y} + \mathbf{f}(\mathbf{x}) = \mathbf{w} + \mathbf{f}(\mathbf{x})$, which implies $\mathbf{y} - \mathbf{w} = \mathbf{0}$.

Answer to Exercise 20.7: We have to prove

$$\prod_{i=1}^{N}[\frac{1}{\sqrt{2}}(\sigma_3(i) + \sigma_1(i))]|\mathbf{x}, \mathbf{y}\rangle = 2^{-N/2}\sum_{\mathbf{z}}(-1)^{\mathbf{x}\cdot\dot{\mathbf{z}}}|\mathbf{z}, \mathbf{y}\rangle.$$

To do this note that for $1 \le i \le N$, the action of $\sigma_3(i) + \sigma_1(i)$ on $|\mathbf{x}, \mathbf{y}\rangle$ gives $(-1)^{x^i}|\mathbf{x}, \mathbf{y}\rangle + |\mathbf{x} + \mathbf{1_i}, \mathbf{y}\rangle$. $\mathbf{1}_i$ is the N bit string with a 1 in the i-th place and zeroes elsewhere. When we act with the tensor product operator on all i, we get an N-fold tensor product of sums over two states with a sign $(-1)^{x^i}$. Expanding this out by the distributive law for tensor products of sums of states, we just get a sum over all N bit strings $\mathbf{z}$ with a sign $(-1)^{\mathbf{x}\cdot\mathbf{z}}$. This is the key point of the algorithm, we can find a state which is a specific linear combination of all 2^N states $|\mathbf{x}, \mathbf{y}\rangle$ with fixed $\mathbf{y}$ by doing only N quantum operations. The rest of this exercise is essentially done in the statement of the problem.

Answer to Exercise 20.8: For every $\mathbf{z}$, $\mathbf{z} + \mathbf{a}$ also appears in the sum

$$\sum_{\mathbf{z}}(-1)^{\mathbf{y}\cdot\mathbf{z}}|y, \mathbf{f}(\mathbf{z})\rangle.$$

The value of $\mathbf{f}$ on these two vectors is the same, by periodicity. If $\mathbf{y} \cdot \mathbf{a} = 1$, then the two terms appear with opposite sign, and cancel. So we only get contributions from $\mathbf{y}$ orthogonal to the period vector.

Answer to Exercise 20.9: The product of $\sigma_3(i) + \sigma_1(i)$ can obviously be implemented in N steps. The operation $U(\mathbf{f})$ on $|\mathbf{z}, \mathbf{0}\rangle$ just requires one consultation of the lookup table defining the function. Thus, the isolation of one string $\mathbf{y}$ satisfying $\mathbf{y} \cdot \mathbf{a} = 0$ only takes N steps. Since the space of bit strings in N-dimensional, kN random choices will find a whole basis in the space for modest values of k, with very high probability. Indeed, for the first

$N/2 - 1$ choices, the probability of *not* finding linearly independent strings goes to zero as N goes to infinity. Afterwards we have to get a bit more lucky but even after we have found a subspace of dimension $N - 1$, the probability that $167N$ further random choices will all lie within that subspace is very small and goes to zero with N. Thus, in a time linear in N we will know all bit strings orthogonal to the period.

Bibliography

[1] L. Mlodinow, *The Drunkard's Walk: How Randomness Rules Our Lives*, Vintage Books, Random House, New York, 2009, ISBN: 978-0-3-7-37517-2.

[2] W. Feller, *An Introduction to Probability Theory and Its Applications*, John Wiley and Sons, 1957, 1968, Wiley Series in Probability and Mathematical Statistics.

[3] T. Bayes, R. Price, An Essay towards solving a Problem in the Doctrine of Chance. By the late Rev. Mr. Bayes, communicated by Mr. Price, in a letter to John Canton, A. M. F. R. S, *Philosophical Transactions of the Royal Society of London*, 1763, 53(0):370–418. doi:10.1098/rstl.1763.0053.

[4] T. Bass, *The Eudaemonic Pie*, Houghton Mifflin Harcourt, 1985.

[5] J. von Neumann, *The Mathematical Foundations of Quantum Mechanics*, Princeton University Press, 1955.

[6] N. A. Wheeler (Ed.), Mathematical Foundations of Quantum Mechanics. https://press.princeton.edu/titles/11352.html

[7] S. J. Gould, *The Streak of Streaks*, New York Review of Books, August 18, 1988; *The Median Isn't the Message*, Discover Magazine, 1985.

[8] E. Noether, *Invariante Variationsprobleme*, Nachr. d. Konig. Gesellsch. d. Wiss. zu Gottingen, Math-phys. Klasse, 1918, 235–257; English translation M. A. Travel, Transport Theory and Statistical Physics, 1971, 1(3):183–220.

[9] P. Michel, Mittag-Leffler, Contribution l'tude de la representation d'une fonction arbitraire par les intgrales dfinies, *Rendiconti del Circolo Matematico di Palermo*, 1910, 30(1):289–335. doi:10.1007/BF03014877.

[10] W. E. Boyce, R. C. DiPrima, *Elementary Differential Equations and Boundary Value Problems* (8th ed.). New Jersey: John Wiley & Sons, 2005, Inc. ISBN 0-471-43338-1. Joseph Fourier, translated by Alexander Freeman (published 1822, translated 1878, re-released 2003). *The Analytical Theory of Heat.*, Dover Publications. ISBN 0-486-49531-0. 2003 unabridged republication of the 1878 English translation by Alexander Freeman of Fourier's work *Thorie Analytique de la Chaleur*, originally published in 1822.

[11] J. -P. Antoine, *Quantum Mechanics Beyond Hilbert Space*, appearing in Irreversibility and Causality, Semigroups and Rigged Hilbert Spaces, Arno Bohm, Heinz-Dietrich Doebner, Piotr Kielanowski, eds., Springer-Verlag, 1996, ISBN 3-540-64305-2.

[12] M. Nauenberg, Quantum wave packets on Kepler elliptic orbits, *Physical Review A* 1989, 40(2):1133–1136.

[13] F. Cooper, A. Khare and U. Sukhatme, Supersymmetry and quantum mechanics, *Physics Reports* 1995, 251:267–385, and references therein.

[14] D.J. Griffiths, *Introduction to Quantum Mechanics* (2nd ed.). Cambridge University Press, 2017.

[15] R. Rajaraman, *Solitons and Instantons*, North Holland Personal Library, Elsevier, Amsterdam, 1982, ISBN-13: 978-0444870476, ISBN-10: 0444870474.

[16] W. Pauli, The connection between spin and statistics, *Physical Review*, 1940, 58(8):716–722. Bibcode:1940PhRv...58..716P. doi:10.1103/PhysRev.58.716; G. Lüders, B. Zumino, Bruno, Connection between spin and statistics, *Physical Review* 1958, 110: 1450. doi: 10.1103/PhysRev.110.1450.

[17] T. Banks, *Modern Quantum Field Theory: A Concise Introduction*, Cambridge University press, 2008; A. Zee, *Quantum Field Theory in a Nutshell*, 2nd Edition, Princeton University Press, 2003.

[18] A. C. Newell, *Solitons in Mathematics and Physics*, CBMS-NSF Regional Conference Series in Applied Mathematics, 1985, ISBN: 978-0-89871-196-7 eISBN: 978-1-61197-022-7.

[19] See the article on Legendre Polynomials by I. Stegun (Chapter 8) in M. Abramowitz and I. Stegun, *Handbook of Mathematical Functions*, National Bureau of Standards Applied Mathematics Series, 55, 1964.

[20] D. R. Hartree, *The Calculation of Atomic Structures*, Wiley and Sons, New York, 1957.

[21] P. Hohenberg, W. Kohn, Inhomogeneous electron gas, *Physical Review*, 1964, 136 (3B):B864–B871. Bibcode:1964PhRv..136..864H. doi:10.1103/PhysRev.136.B864; W. Kohn, L. J. Sham, L. Jeu, Self-consistent equations including exchange and correlation effects, *Physical Review*, 1965, 140(4A):A1133–A1138. Bibcode:1965PhRv..140.1133K. doi:10.1103/PhysRev.140.A1133.

[22] B. H. Brandsen, C. J. Jochain, *Physics of Atoms and Molecules*, ISBN-13: 978-0582356924, ISBN-10: 058235692X ; P. W. Atkins, R. S. Friedman, *Molecular Quantum Mechanics*, 5th Edition, Oxford University Press, London; M. Born, *Atomic Physics*, 8th Edition, Dover Books on Physics, ISBN-13: 978-0486659848, ISBN-10: 0486659844.

[23] S. Goudsmit, G. E. Uhlenbeck, *Physica*, 1926, 6, 273.

[24] L. H. Thomas, Motion of the spinning electron, *Nature*, 117: 514. Bibcode: 1926Natur. 117..514T. doi:10.1038/117514a0.

[25] Collected Papers of L D Landau, Ed. D ter Haar, NY, 1965 (Reprint of Landaus papers); Landau, L. D.; and Lifschitz, E. M.; (1977). Quantum Mechanics: Non-relativistic Theory. Course of Theoretical Physics. Vol. 3 (3rd ed. London: Pergamon Press). ISBN 0750635398 ; H. Murayama, *Landau Levels*, hitoshi.berkeley.edu/221a/landau.pdf.

[26] J. von Neumann, *The Mathematical Foundations of Quantum Mechanics*, Princeton University Press, 1955.

[27] M. A. Schlosshauer, *Decoherence and the Quantum-to-Classical Transition*, (The Frontiers Collection) Springer-Verlag, Berlin, Heidelberg, (2007), and references therein. ISBN-13: 978-3540357735 ISBN-10: 3540357734; C. J. Riedel, W. H. Zurek, M. Zwolak, The objective past of a quantum universe: redundant records of consistent histories, *Physical Review A*, 2016, 93:032126; R. B. Griffiths, *Consistent Quantum Theory*, Cambridge University Press, Cambridge, 2002; R. Omn'es, *Interpretation of Quantum Mechanics*, Princeton University Press, Princeton, 1994; M. Gell-Mann, *The Quark and the Jaguar*, W.H. Freeman, New York, 1994; M. Gell-Mann and J.B. Hartle, *Quantum Mechanics in the Light of Quantum Cosmology*, in Complexity, Entropy, and the Physics of Information, ed. by W. Zurek, Addison Wesley, Reading, MA, 1990, P. C. Hohenberg, An introduction to consistent quantum theory, *Reviews of Modern Physics*, 2010, 82:2835–2844; E. Joos and H. D. Zeh, The emergence of classical properties through interaction with the environment, *Zeitschrift fr Physik B*, 1985, 59:223.

[28] W. H. Zurek, Environment-assisted invariance, entanglement, and probabilities in quantum physics, *Physical Review Letters*, 2003, 90:120404; Probabilities from entanglement, Borns rule from envariance, *Physical Review A*, 2005, 71:052105.

[29] J. B. Hartle, Quantum mechanics of individual systems, *American Journal of Physics*, 1968, 36(8):704–712. doi:10.1119/1.1975096.

[30] A. Einstein, B. Podolsky, N. Rosen, Can quantum-mechanical description of physical reality be considered complete?, *Physical Review*, 1935, 47(10):777–780. Bibcode: 1935PhRv...47..777E. doi:10.1103/PhysRev.47.777. M. D. Reid, P. D. Drummond, W. P. Bowen, E. G. Cavalcanti, P. K. Lam, H. A. Bachor, U. L. Andersen, G. Leuchs, Colloquium: the Einstein–Podolsky–Rosen paradox: from concepts to applications, *Reviews of Modern Physics*, 2009, 81:1727, and references therein. doi :10.1103/RevModPhys.81.1727.

[31] A. Aspect, P. Grangier and G. Roger, Experimental tests of realistic local theories via Bell's theorem, *Physical Review Letters*, 1981, 47(7):460–463, Bibcode:1981PhRvL..47..460A, doi:10.1103/PhysRevLett.47.460; A. Aspect; J. Dalibard; G. Roger, Experimental test of Bell's inequalities using time-varying analyzers, *Phys. Rev. Lett.*, 1982, 49(25): 1804–1807, Bibcode:1982PhRvL..49.1804A, doi:10.1103/PhysRevLett.49.1804.

[32] N. D. Mermin, Simple unified form for the major no hidden variables theorems, *Physical Review Letters*, 1990, 65:1838.

[33] T. J. Hollowood, Copenhagen quantum mechanics, *Contemporary Physics*, 2016, 57(3):289. doi:10.1080/00107514.2015.1111978 [arXiv:1511.01069 [quant-ph]].

[34] *Coherence*, 2014 1h 29m Coherence is an American science fiction film directed by James Ward Byrkit in his directorial debut. Release date: June 20, 2014 (New York City).

[35] J. Wisdom, S. J. Peale, F. Mignard, The chaotic rotation of hyperion, *ICARUS*, 1984, 58:137–152.

[36] M. Born, J. R. Oppenheimer, Zur Quantentheorie der Molekeln [On the Quantum Theory of Molecules], *Annalen der Physik (in German)*, 1927, 389(20): 457–484. Bibcode:1927AnP...389..457B. doi:10.1002/andp.19273892002.

[37] D. M. Ceperley, B. J. Alder, Ground state of the electron gas by a stochastic method, *Physical Review Letters*, 1980, 45:566; J. P. Perdew, Y. Wang, Accurate and simple analytic representation of the electron-gas correlation energy, *Physical Review B*, 1992, 45:13244; J. Sun et al., Correlation energy of the uniform electron gas from an interpolation between high- and low-density limits, *Physical Review B*, 2010, 81:085123; R. G. Parr and W. Yang, *Density-Functional Theory of Atoms and Molecules* (Oxford U. Press, 1989); C. Fiolhais, F. Nogueira, M. A. L. Marques (Editors) *A Primer in Density-Functional Theory*, Lecture Notes in Physics Volume 620, 2003, Springer-Verlag, Hedelberg, Berlin, ISBN: 978-3-540-03083-6 (Print) 978-3-540-37072-7 (Online); J. P. Perdew, K. Burke and M. Ernzerhof, Generalized gradient approximation made simple, *Physical Review Letters*, 1996, 77:3865; J. Sun et al. Density functionals that recognize covalent, metallic,and weak bonds, *Physical Review Letters*, 2013, 111:106401.

[38] T. Banks, *Density Functional Theory for Field Theorists I*, cond-mat arXiv:1503.02925.

[39] A. D. Becke, Density functional exchange energy approximation with correct asymptotic behavior, *Physical Review A*, 1988, 38:3098–3100; A. D. Becke, Density functional thermochemistry. III. The role of exact exchange, *Journal of Chemical Physics*, 1993, 98: 5648–5652.

[40] A. Georges, G. Kotliar, W. Krauth and M. Rozenberg, Dynamical mean-field theory of strongly correlated fermion systems and the limit of infinite dimensions, *Reviews of Modern Physics*, 1996, 68(1):13, Bibcode:1996RvMP...68...13G. doi:10.1103/RevModPhys.68.13; A. Georges and G. Kotliar, Hubbard model in infinite dimensions, *Physical Review B*, 1992, 45(12):6479, Bibcode:1992PhRvB..45.6479G. doi:10.1103/PhysRevB.45.6479.

[41] A. K. Soper and C. J. Benmore, Quantum differences between heavy and light water, *Physical Review Letters*, 2008, 101:065502.

[42] W. L. Faissler, *Introduction to Modern Electronics*, John Wiley and Sons, New York, 1991; R. Wolfson, *Understanding Modern Electronics* Course No. 1162, www.thegreatcourses.com/courses/understanding-modern-electronics.html

[43] D. J. Griffiths, *Introduction to Quantum Mechanics*, 2nd Edition, Cambridge University Press, 2017.

[44] R. Kronig and W. G. Penney, Quantum mechanics of electrons in crystal lattices, *Proceedings of the Royal Society of London A*, 1931, A130:499.

[45] S. Chandrasekhar, The density of white dwarf stars, *Philosophical Magazine* (7th series), 1931, 11:592–596; The maximum mass of ideal white dwarfs, *Astrophysical Journal*, 1931, 74:81–82; *The Highly Collapsed Configurations of a Stellar Mass* (second paper), S. Chandrasekhar, *Monthly Notices of the Royal Astronomical Society*, 1935, 95:207–225; L. D. Landau, *On the Theory of Stars*, in Collected Papers of L. D. Landau, ed. and with an introduction by D. ter Haar, New York: Gordon and Breach, 1965; originally published in Phys. Z. Sowjet. 1 (1932), 285.

[46] Collected Papers of L. D. Landau, Ed. D ter Haar, New York, 1965 (Reprint of Landaus papers) ; Cross, Michael. "Fermi Liquid Theory: Principles" (PDF). California Institute of Technology, http://www.pmaweb.caltech.edu/ mcc/Ph127/c/Lecture9.pdf.

[47] R. Shankar, Renormalization group approach to interacting fermions, *Reviews of Modern Physics*, 1994, 66:129; J. Polchinski, *Proceedings of the 1992 TASI Elementary Particle Physics*, Editors J. Polchinski and J. Harvey, World Scientific, 1992.

[48] L. N. Cooper, Bound electron pairs in a degenerate fermi gas, Physical Review, 1956, 104:1189.

[49] J. Bardeen, L. N. Cooper and J. R. Schrieffer, Microscopic theory of superconductivity, *Physical Review*, 1957, 106:162.

[50] S. A. Hartnoll, A. Lucas and S. Sachdev, *Holographic Quantum Matter*, arXiv:1612. 07324 [hep-th].

[51] S. Popescu, A. J. Short, A. Winter, Entanglement and the foundations of statistical mechanics, *Nature Physics*, 2006, 2:754; S. Deffner, W. H. Zurek Foundations of statistical mechanics from symmetries of entanglement arXiv.org ¿ quant-ph ¿ arXiv:1504.02797, *New Journal of Physics*, 2016, 18: 063013. doi: 10.1088/1367-2630/18/6/063013.

[52] D. N. Page, Average entropy of a subsystem, *Physical Review Letters*, 1993, 71:1291. doi:10.1103/PhysRevLett.71.1291 [gr-qc/9305007].

[53] M. H. Anderson, J. R. Ensher, M. R. Matthews, C. E. Wieman, E. A. Cornell, Observation of bose-einstein condensation in a dilute atomic vapor. *Science*, 1995, 269(5221):198; K. B. Davis and M. -O. Mewes, M. R. Andrews, N. J. van Druten, D. S. Durfee, D. M. Kurn and W. Ketterle, Bose–Einstein condensation in a gas of sodium atoms, *Physical Review Letters*, 1995, 75:3969. doi = 10.1103/PhysRevLett.75.3969, url = http://link.aps.org/doi/10.1103/PhysRevLett.75.3969.

[54] E. P. Gross, Structure of a quantized vortex in boson systems, *Nuovo Cimento*, 1961, 20:454; Hydrodynamics of a superfluid condensate, *Journal of Mathematical Physics*, 1963, 4:195; L. P. Pitaevski, Vortex lines in an imperfect Bose gas, *Zh. Eksp.Teor. Fiz.*, 1961, 40: 646 [Sov. Phys. JETP 13, 451(1961)].

[55] Y. B. Zel'dovich and E.V. Levich, Bose condensation and shock waves in photon spectra, *Sov. Phys. JETP*, 1969, 28:1287–1290; R. Y. Chiao, Bogoliubov dispersion relation for a photon fluid: is this a superfluid? *Optics Communications*, 2000, 179:157–166; E. L. Bolda, R. Y. Chiao and W.H. Zurek, Dissipative optical flow in a nonlinear Fabry-Prot cavity, *Physical Review Letters*, 2001, 86:416–419.

[56] J. Klaers, F. Vewinger and M. Weitz, Thermalization of a two-dimensional photonic gas in a white-wall photon box, *Nature Physics*, 2010, 6:512–515; J. Klaers, J. Schmitt, F. Vewinger and M. Weitz, Bose–Einstein condensation of photons in an optical microcavity, *Nature*, 2010, 468:545–548. doi:10.1038/nature09567.

[57] J. W. S. Rayleigh, *The Theory of Sound* Vol. 1, Dover Publications. The book has been digitized by Google and others and can be found online.

[58] E. Schrödinger, An undulatory theory of the mechanics of atoms and molecules, *Physical Review*, 1926, 28 (Schrödinger's original paper on the wave equation, in which he reports without details the perturbative calculation of the Stark Effect); E. Schrödinger, Ann. Phys. 80, 437, (1926).

[59] L. Brillouin, Les problmes de perturbations et les champs self-consistents, *Journal de Physique et Le Radium*, 1932, 3:373–389. doi: 10.1051/jphys-rad:jphysrad0193200309037300; E. P. Wigner, *On a Modification of the Rayleigh-Schrödinger Perturbation Theory*, Volume A/4 of The Collected Works of Eugene Paul Wigner, 131, ed. A. S. Wightman, Springer-Verlag, Berlin Heidelberg, 1997. isbn="978-3-642-59033-7", doi="10.1007/978-3-642-59033-7-12".

[60] A. Dalgarno, J. T. Lewis, The exact calculation of long-range forces between atoms by perturbation theory, *Proceedings of the Royal Society of London A*, 1995, A233:70.

[61] R. P. Feynman, *Physical Review*, 1939, 56:340; H. Hellmann, A combined approximation procedure for calculation of energies in the many electron problem, *Acta Physicochimica URSS*, 1934.

[62] Google the Wikipedia article on the Wigner-Eckart theorem for proofs, and a host of references.

[63] W. Heisenberg, Zur Theorie des Ferromagnetismus, *Zeitschrift fr Physik*, 1928, 49(9–10):619–636. Bibcode:1928ZPhy...49..619H. doi:10.1007/BF01328601.

[64] N. F. Mott, *Metal-Insulator Transitions*, 2nd edition, Taylor & Francis, London, 1990. ISBN 0-85066-783-6, ISBN 978-0-85066-783-7, and references therein.

[65] F. J. Dyson, The radiation theories of Feynman, Schwinger and Tomonaga, *Physical Review*, 1949, 75:486.

[66] Einstein's route to the statistical properties of photons is described on pages 351–52 of *Introduction to Quantum Mechanics*, 2nd Edition, by D.J. Griffiths, Cambridge University Press, Cambridge, 2017.

[67] D. J. Griffiths, *Introduction to Quantum Mechanics*, 2nd Edition, Cambridge University Press, 2017, Section 9.3.

[68] D. J. Griffiths *Introduction to Electrodynamics*, 4th Edition, Pearson, 2013.

[69] A. Shapere and F. Wilczek, eds. *Geometric Phases in Physics*, World Scientific, Singapore, 1989.

[70] Y. Aharonov and D. Bohm, Significance of electromagnetic potentials in the quantum theory, *Physical Review*, 1959, 115(3):485–489.

[71] D. Chambers, Shift of an electron interference pattern by enclosed magnetic flux, *Physical Review Letters*, 1960, 5:3.

[72] J. P. Preskill, *Notes on Quantum Computation*, www.theory.caltech.edu/people/preskill/ph229/

[73] http://www.physics.rutgers.edu/gmoore/511Fall2014/Physics511-2014-Ch2-Topology.pdf

[74] A. Stern, *Anyons and the quantum Hall effect A pedagogical review* and references therein. (You can download the PDF from the Wikipedia page on anyons). Annals of Physics. 323: 204. arXiv:0711.4697v1Freely accessible. Bibcode:2008AnPhy.323..204S. doi:10.1016/j.aop.2007.10.008.

[75] D. C. Tsui, H. L. Stormer, A. C. Gossard, Two-dimensional magnetotransport in the extreme quantum limit, *Physical Review Letters*, 1982, 48(22):1559. Bibcode:1982PhRvL..48.1559T. doi:10.1103/PhysRevLett.48.1559; H.L. Stormer (1999). *Nobel Lecture: The fractional quantum Hall effect.* Reviews of Modern Physics. 71 (4): 875. Bibcode:1999RvMP...71..875S. doi:10.1103/RevModPhys.71.875; R. B. Laughlin, Anomalous quantum hall effect: an incompressible quantum fluid with fractionally charged excitations, *Physical Review Letters*, 1983, 50(18):1395. Bibcode:1983PhRvL..50.1395L. doi:10.1103/PhysRevLett.50.1395.

[76] H. Jeffreys, On certain approximate solutions of linear differential equations of the second order, *Proceedings of the London Mathematical Society*, 1924, 23:428–436. doi:10.1112/plms/s2-23.1.428; L. Brillouin, La mcanique ondulatoire de Schrdinger: une mthode gnrale de resolution par approximations successives, *Comptes Rendus de l'Academie des Sciences*, 1926, 183:24–26; H. A. Kramers, Wellenmechanik und halbzhlige Quantisierung, *Zeitschrift fr Physik*, 1926, 39(10–11):828–840. Bibcode:1926ZPhy...39..828K. doi:10.1007/BF01451751; G. Wentzel, Eine Verallgemeinerung der Quantenbedingungen fr die Zwecke der Wellenmechanik, *Zeitschrift fr Physik*, 1926, 38(6–7):518–529. Bibcode:1926ZPhy...38..518W. doi:10.1007/BF01397171.

[77] T. Banks and C. M. Bender, Coupled anharmonic oscillators. II. Unequal-mass case, *Physical Review*, 1973, D8:3366, doi:10.1103/PhysRevD.8.3366.

[78] G. 't Hooft, Symmetry breaking through Bell–Jackiw anomalies, *Physical Review Letters*, 1976, 37:8–11. doi: 10.1103/PhysRevLett.37.8.

[79] The Meson Table, 2018 edition, Volume II. http://www.pdg.lbl.gov/

[80] http://www.cond-mat.de/events/correl11/manuscript/Koch.pdf, and references therein.

[81] P. A. M. Dirac, *Quantum Mechanics*, Section 32, 4th Edition, Oxford University Press, 1958.

[82] F. J. Dyson, The S matrix in quantum electrodynamics, *Physical Review*, 1949, 75:1736. Bibcode:1949PhRv...75.1736D. doi:10.1103/PhysRev.75.1736. J. Schwinger, On Green's functions of quantized fields I + II, *Physical Review*, 1951, 37:452–459. Bibcode:1951PNAS...37..452S. doi:10.1073/pnas.37.7.452. PMC 1063400, Ibid. 1953, 728, *Proceedings of the National Academy of Sciences of the United States of America*, 1951, 37:452.

[83] T. I. Banks and C. M. Bender, Anharmonic oscillator with polynomial self-interaction, *Journal of Mathematical Physics*, 1972, 13:1320. doi:10.1063/1.1666140.

[84] I. M. Gelfand and A. M. Yaglom, Integration in functional spaces and it applications in quantum physics, *Journal of Mathematical Physics*, 1960, 1:48. doi:10.1063/1.1703636; G. V. Dunne, Functional determinants in quantum field theory, *Journal of Physics A*, 2008, 41:304006. doi:10.1088/1751-8113/41/30/304006, [arXiv:0711.1178 [hep-th]].

[85] R. P. Feynman and A. R. Hibbs, *Quantum Mechanics and Path Integrals*, McGraw Hill, 1960, ISBN-13: 978-0070206502, ISBN-10: 0070206503.

[86] M. A. Nielsen and I. L. Chuang, *Quantum Computation and Quantum Information*, Cambridge University Press, 2013; E. Rieffel, W. H. Pack, *Quantum Computing : A Gentle Introduction*, MIT Press, Cambridge, MA, 2011; N. D. Mermin, *Quantum Computer Science: An Introduction*, Cambridge University Press, 2007; M. M. Wilde, *Quantum Information Theory*, 2nd Edition, Cambridge University Press, 2017, ISBN-13: 978-1107176164 ISBN-10: 1107176166.

[87] J. P. Preskill, *Notes on Quantum Computation*, www.theory.caltech.edu/people/ preskill/ph229/; E. Dennis, A. Kitaev, A. Landahl and J. Preskill, Topological quantum memory, *Journal of Mathematical Physics*, 2002, 43:4452.

[88] R. Orus, Advances on tensor network theory: symmetries, fermions, entanglement, and holography, *The European Physical Journal B*, 2014, 87: 280. doi:10.1140/epjb/e2014-50502-9,[arXiv:1407.6552 [cond-mat.str-el]]; J. C. Bridgeman and C. T. Chubb, *Hand-waving and Interpretive Dance: An Introductory Course on Tensor Networks*, ArXiv e-prints (Mar, 2016) 1603.03039.

[89] W. Pfaff, B. Hensen, H. Bernien, S. B. van Dam, M. S. Blok, T. H. Taminiau, M. J. Tiggelman, R. N. Schouten, M. Markham, D. J. Twitchen and R. Hanson, Unconditional quantum teleportation between distant solid-state qubits, *Science*, 2014, 345:532–535. doi:10.1126/science.1253512,arXiv:1404.4369 [quant-ph].

[90] D. Deutsch and R. Jozsa, Rapid solution of problems by quantum computation, *Proceedings of the Royal Society of London A*, 1992. doi: 10.1098/rspa.1992.0167.

[91] P. W. Shor, Polynomial-time algorithms for prime factorization and discrete logarithms on a quantum computer, *SIAM Review*, 2006, 41(2):303, http://epubs.siam.org/doi/abs/10.1137/S0036144598347011.

[92] R. Van Meter and C. Horsman, A blueprint for building a quantum computer, *Communications of the ACM*, 2013, 56:84; B. Lekitsch, S.Weidt, A.G. Fowler, K. Mlmer, S. J. Devitt, C. Wunderlich and W. K. Hensinger, Blueprint for a microwave trapped ion quantum computer, *Science Advances*, 2017. e1601540 Open Access CCBY; G. Popkin, Quest for qubits, *Science*, 2016, 354:1090. doi: 10.1126/science.354.6316.1090.

[93] J. D. Bekenstein, Black holes and entropy, *Physical Review*, 1973, D7:2333. doi :10.1103/PhysRevD.7.2333; S. W. Hawking, Particle creation by black holes, *Communications in Mathematical Physics*, 1975, 43:199. doi:10.1007/BF02345020.

[94] T. Jacobson, Thermodynamics of spacetime: the Einstein equation of state, *Physical Review Letters*, 1995, 75:1260. doi:10.1103/PhysRevLett.75.1260.

[95] J. Polchinski, *String Theory*, Vols. 1–2, Cambridge Monographs in Mathematical Physics, 2005.

[96] O. Aharony, S. S. Gubser, J. M. Maldacena, H. Ooguri and Y. Oz, Large N field theories, string theory and gravity, *Physics Reports*, 2000, 323:183. doi:10.1016/S0370-1573(99)00083-6 [hep-th/9905111].

[97] G. 't Hooft, Determinism beneath quantum mechanics, in *Quo Vadis Quantum Mechanics?* eds. A. C. Elitzur, S. Dolev and N. Kolenda, Springer-Verlag, Berlin-Heidelberg, 2005. isbn : 978-3-540-26669-3, doi: 10.1007/3-540-26669-0-8; *The Cellular Automaton Interpretation of Quantum Mechanics. A View on the Quantum Nature of our Universe, Compulsory or Impossible?* arXiv:1405.1548 [quant-ph]; Relating the quantum echanics of discrete systems to standard canonical quantum mechanics, *Foundations of Physics*, 2014, 44:406. doi:10.1007/s10701-014-9788-y.

[98] G. 't Hooft, Beyond relativistic quantum string theory or discreteness and determinism in superstrings, *Subnuclear Series*, 2014, 50:243. doi:10.1142/9789814603904-0016.

[99] L. de Broglie, The reinterpretation of quantum mechanics, *Foundations of Physics*, 1970, 1:5. doi:10.1007/BF00708650 and references therein.

[100] D. Bohm, A suggested interpretation of the quantum theory in terms of hidden variables, I and II, *Physical Review*, 1952, 85(2):166; *Physical Review*, 85(2):180. doi:10.1103/PhysRev.85.166. doi:10.1103/PhysRev.85.180.

[101] D. Drr and S. Teufel, *Bohmian Mechanics–The Physics and Mathematics of Quantum Theory*, Springer (2009); D. Drr, S. Goldstein and N. Zangh", *Quantum Physics Without Quantum Philosophy*, Springer, 2012.

[102] E. Z. Madelung, Quantentheorie in hydrodynamischer Form, *Zeitschrift fr Physik*, 1927, 40:322. doi:10.1007/BF01400372.

[103] H. Everett III, "Relative State" formulation of quantum mechanics, *Reviews of Modern Physics*, 1957, 29:454. doi:https://doi.org/10.1103/RevModPhys.29.454; B. S. DeWitt and R. N. Graham, eds, *The Many-Worlds Interpretation of Quantum Mechanics*, Princeton Series in Physics, Princeton University Press, 1973, ISBN 0-691-08131-X.

[104] G. Lindblad, On the generators of quantum dynamical semigroups, *Communications in Mathematical Physics*, 1976, 48(2):119. Bibcode:1976CMaPh..48..119L. doi:10.1007/BF01608499; A. Kossakowski, On quantum statistical mechanics of non-Hamiltonian systems, *Reports on Mathematical Physics*, 1972, 3(4):247. Bibcode:1972RpMP....3..247K. doi:10.1016/0034-4877(72)90010-9; V. Gorini, A. Kossakowski and E. C. G. Sudarshan, Completely positive semigroups of N-level systems, *Journal of Mathematical Physics*, 1976, 17(5):821, Bibcode:1976JMP....17..821G. doi:10.1063/1.522979.

[105] T. Banks, M. E. Peskin and L. Susskind, Difficulties for the evolution of pure states into mixed states, *Nuclear Physics B*, 1984, 244:125–134, Bibcode:1984NuPhB.244..125B. doi:10.1016/0550-3213(84)90184-6.

[106] E. Wigner, *The Unreasonable Effectiveness of Mathematics in the Natural Sciences*, in Communications in Pure and Applied Mathematics, vol. 13, No. I, 1960. New York: John Wiley & Sons.

[107] H. Georgi, *Lie Algebras in Particle Physics*, Westview Advanced Book Program, Perseus Books Group, 1999. ISBN 0-7382-0233-9.

[108] W. Pauli, ber das Wasserstoffspektrum vom Standpunkt der neuen Quantenmechanik, *Zeitschrift fr Physik*, 1926, 36:336–363. Bibcode:1926ZPhy...36..336P. doi:10.1007/BF01450175.

[109] W. R. Hamilton, The hodograph or a new method of expressing in symbolic language the Newtonian law of attraction, *Proceedings of the Royal Irish Academy*, 1847, 3:344–353.

Index